ALASKA CODFISH CHRONICLE

ALASKA CODFISH CHRONICLE:
A HISTORY OF THE PACIFIC COD FISHERY IN ALASKA

James Mackovjak

Published by
University of Alaska Press
Fairbanks, Alaska

Published by
University of Alaska Press
P.O. Box 756240
Fairbanks, AK 99775-6240

Cover and interior layout by 590 Design.

Cover image: Pacific cod, photo courtesy of NOAA Fisheries.

Library of Congress Cataloging in Publication Data

Names: Mackovjak, James R., author.
Title: Alaska codfish chronicle : a history of the Pacific cod fishery in Alaska / James Mackovjak.
Description: Fairbanks, AK : University of Alaska Press, [2019] | Includes bibliographical references and index. |
Identifiers: LCCN 2018060367 (print) | LCCN 2019002650 (e-book) | ISBN 9781602233904 (e-book) | ISBN 9781602233898 (paperback : alk. paper)
Subjects: LCSH: Pacific cod fisheries—Alaska—History.
Classification: LCC SH351.P24 (e-book) | LCC SH351.P24 M33 2019 (print) | DDC 333.95/6633—dc23
LC record available at https://lccn.loc.gov/2018060367

DEDICATION

To my wife, Ann, whose patience and support made writing this book a pleasure.

CONTENTS

ACRONYMS AND INITIALISMS

ABC—Allowable biological catch (also, acceptable biological catch).

ADF&G—Alaska Department of Fish and Game.

ASMI—Alaska Seafood Marketing Institute.

BiOp—Biological opinion.

BSAI—Bering Sea/Aleutian Islands.

CDQ—Community development quota.

DAP—Domestic annual processing.

EEZ—Exclusive economic zone.

ENGO—Environmental nongovernment organization.

EPIRB—Emergency position-indicating radio beacon.

EU—European Union.

F/V—Fishing vessel.

GHL—Guideline harvest level.

GOA—Gulf of Alaska.

H&G—Headed and gutted.

IPHC—International Pacific Halibut Commission.

ITQ—Individual transferrable quota.

JVP—Joint-venture processing.

LLP—License limitation program.

OFL—Overfishing level.

MSC—Marine Stewardship Council.

NEFCO—New England Fish Company.

NMFS—National Marine Fisheries Service.

NOAA—National Oceanic and Atmospheric Administration.

NPFMC—North Pacific Fishery Management Council.

TAC—Total allowable catch.

TALFF—Total allowable level of foreign fishing.

WGC—Whole gutted cod.

FOREWORD

Codfish has played a major role in the history of the world, our nation, and Alaska.

The firm, white-fleshed cod attracted Viking fishermen across the Atlantic to Newfoundland a thousand years ago. The Basques of Spain fished off Cape Cod years before Columbus ever found his way over here. Since before the American Revolution, a five-foot wooden codfish has hung in the Massachusetts Assembly (now State House). Named the "Sacred Cod," in honor of its role in the state's history, it remains there today. Back then, if you said "fish" in a contract, the courts ruled it meant cod.

High in protein and low in fat, cod supported early Native populations of Alaska and the fur traders who followed. Cod attracted California fishermen to the "slime banks" of the eastern Aleutians during the waning days of Russian America. It inspired Seattle fishermen to push for the purchase of "Seward's Icebox" in 1867. Cod fishermen established Alaska's first permanent fishing stations in the Shumagin Islands, catching and salting cod.

Canned salmon later dominated Alaska's fishing industry and cod was relegated to use as halibut and crab bait until my friend Senator Ted Stevens recognized the true value of cod and other groundfish. Passage of the Magnuson-Stevens Fishery Conservation and Management Act in 1976 later boosted Alaska fishery production by ten-fold. Cod was a big part of that. Now Alaska produces half a billion pounds of codfish annually and does so sustainably, without the overfishing that has threatened the long-term population of this valuable stock elsewhere.

My personal history and appreciation for cod started on New York's Long Island, where I grew up. Dory fishermen and salt cod were still on people's minds. My earliest memory was the stock market crash of 1929. My father, a wealthy architect, was wiped out. His firm went under, he lost his grandfather's farm in Islip, so he went out fishing. I've never forgotten what that was like. We didn't have much of a selection for dinner—I don't ever want chicken and dumplings again—but the fish was good.

My mother cooked creamed cod, codfish balls, beer-battered cod, and more. It is the fish you can do the most different things with. Some fish has no flavor except whatever sauce you put on it, but codfish tastes like codfish. I came to Alaska after World War II and caught Pacific or "P-cod" while fishing for halibut in lower Cook Inlet. We filleted and salted them and packed them in small barrels, either for personal use or local sale. More recently, I have enjoyed beautiful cod fillets from Adak. Slice them up, and deep-fried or baked or however you want to cook them, they're wonderful fish.

Cod populations are known to go up and down. The Aleut word for cod is, "the fish that isn't always there," but as long as you protect the grazing stock, they will come back again. C. L. "Andy" Anderson was my hero. The first commissioner of the Alaska Department of Fish and Game never got the respect he deserved for the system that saved Alaska's fisheries and made them number one. The federal government had so misused our resources. I will always thank Bill Egan who sent Andy to the state constitutional convention to include the words "sustained yield."

Alaska is number one in fisheries management because of the system Andy set up. The department takes care of the resource; the boards that were appointed by the governor and confirmed by the legislature could only allocate what the department said was surplus. That's why we have the fisheries we now have: salmon, pollock, crab, and cod. It's really not that complicated.

In *Alaska Codfish Chronicle*, Jim Mackovjak tells the story of Alaska cod, the people, places, and problems involved in this historic industry, and the management, marketing, and moon pools used in the cod fishery today. Previously, Jim has told the story of early settlers of Gustavus in *Hope and Hard Work*, chronicled Southeast's logging industry in *Tongass Timber*, tells the story of the people and ships that plied waters in the southwestern archipelago in *Aleutian Freighter*, and chronicled the contentious history of commercial fishing in Glacier Bay National Park in *Navigating Troubled Waters*.

Alaska Codfish Chronicle is a welcome addition to the historic record of the development, loss, and rebirth of an Alaska fishery that remains a major part of the Alaska economy today. Jim Mackovjak's use of personal stories of life aboard the schooner fleet and during the dory boat days, historic photographs, and even recipes about how cod has appeared on our dinner tables makes it a fascinating and enjoyable read.

Clem Tillion
Halibut Cove, Alaska

ACKNOWLEDGMENTS

Compiling the history of Alaska's Pacific cod fishery was a big, complex project, and I couldn't have completed it without the help, support, and encouragement of numerous individuals, businesses, and institutions.

To name them all would make this section read like a telephone book, but first among those I would like to thank by name is Frank Norris, the former regional historian for the National Park Service in Alaska and my editor. Frank has a commanding knowledge of Alaska's history and is smart and inquisitive. A writer himself, Frank previously helped me on a number of writing projects, but I thought this one might push his limit. I was wrong. Frank graciously spent countless hours helping me make the codfish story more comprehensible (and probably made me appear a lot smarter than I actually am). No doubt, Frank now knows more about codfish than he ever intended.

Greg Streveler and Judy Brakel, my neighbors in Gustavus who are both knowledgeable about the fishing industry, read a relatively early draft of the work and helped steer me in the right direction, as did Alaska fisheries historian Bob King.

In Kodiak, retired trawler and fishing industry representative Al Burch helped me understand the development of the Pacific cod fishery in his community. And Clem Tillion, who has been involved in Alaska's fisheries for seven decades in a broad range of capacities and has a long history with Pacific cod, helped me understand how the fishery fits into the tapestry that is Alaska's fishing industry. Sam Cotten, former commissioner of the Alaska Department of Fish and Game, and John Jensen, of the Alaska Board of Fisheries and the North Pacific Fishery Management Council, were very interested in my project and advocated on my behalf. Add to this all the individuals and businesses who shared stories, information, and images.

I often asked staff at the Alaska Department of Fish and Game, the National Marine Fisheries Service (NMFS, also known as NOAA Fisheries), and the North Pacific Fishery Management Council for information or to check material I had written for accuracy. To a person, they were unfailingly helpful, and my gratitude runs deep. I also appreciate the prodigious amount of material these agencies make available on the internet.

Scientists at the University of Alaska and the University of Washington and at other institutions provided invaluable information, perspective, and commentary. And the folks at the Alaska Seafood

Marketing Institute and the Alaska consulting firm McDowell Group were also very helpful, as was *Seafood News*'s Peggy Parker.

The four individuals who peer reviewed my draft material—Terry Johnson (professor of fisheries emeritus, University of Alaska Fairbanks), Franz Mueter (associate professor, College of Fisheries and Ocean Sciences, University of Alaska Fairbanks), Dave Witherell (executive director of the North Pacific Fishery Management Council), and Nathanial Howe (executive director of Northwest Seaport, in Seattle)—provided invaluable, critical commentary. Reviewing a 125,000-word manuscript wasn't a quick or easy chore, and I admire their intellectual stamina.

I have no end of respect and appreciation for the many institutions that work to preserve our history. In particular, the San Francisco Maritime National Historical Park, the University of Washington Libraries, the Alaska State Library, the Poulsbo (Washington) Historical Society, and the Anacortes (Washington) Museum were unfailingly helpful in providing me with historical photographs and documents. I also want to thank Marinke Van Gelder, of the Alaska State Court Law Library (Juneau), for help with legal documents. To Marinke, I think my name and "request" are synonymous.

I would also like to thank the Freezer Longline Coalition, American Seafoods, and Icicle Seafoods for their financial contributions to the University of Alaska Press. Those contributions helped push the project forward.

Finally, I want to thank the University of Alaska Press for taking on this project. Nate Bauer, the press's executive director, provided just the right amounts of encouragement and caution as he shepherded my project through the acquisition process.

In addition to all the help they provided, it was for me a great pleasure and honor to associate with those in Alaska's Pacific cod industry, including those who study it, manage it, and report on it. They are a smart, hardworking lot who, by and large, want to do the right thing for the wonderful fish we call Pacific cod and for the environment that is essential to its survival.

I take full responsibility for any errors or omissions contained herein.

INTRODUCTION[1]

> *The growth of the codfish industry is one of the romantic chapters in the development of the Pacific Coast. Established over half a century ago by American mariners and fishermen, its history is one of dauntless enterprise and arduous endeavor, marked by many periods of discouragement.*
>
> —Pacific Fisherman, 1919[2]

Alaska's Pacific cod fishery, in the words of federal fisheries inspector E. Lester Jones in 1914, "is the oldest fishery proper in Alaska."[3] It began in the mid-1860s—during the waning years of Russian control—when San Francisco–based fishermen began catching cod in the Shumagin Islands, along the southern coast of the Alaska Peninsula. The fish were salted on the fishing grounds and then sold on the San Francisco market. The cod fishery was Alaska's most important fishery until it was eclipsed by the canned salmon industry in the late 1880s. By 1914, the halibut and herring fisheries had grown to the extent that the cod fishery ranked as the territory's fourth-largest fishery.

The history of the Pacific codfish industry in Alaska can be divided into three periods: the Salt Cod Era, the Bait Cod Era, and, for want of a better term, the Modern Era.

The Salt Cod Era began with the San Francisco–based fishermen and continued until 1950, the year the sailing schooner *C. A. Thayer* made its final voyage to the Bering Sea. The cod fishery then had two components: an offshore sailing-schooner fishery in the Bering Sea and a shore-station fishery in the western Gulf of Alaska. Shore stations were located primarily along the southern coast of the Alaska Peninsula. In both fisheries, codfish were caught almost exclusively by dory fishermen using handlines. Typically, the schooner fleet fished from about mid-spring until early fall, while the station-based dories fished—depending on weather and availability of cod—pretty much year-round, mostly in the near-shore waters adjacent to the stations. This era's history was especially colorful: each sailing schooner that went north had its story, as did each codfish station and each fisherman. Mostly, they were stories of hardship and danger.

The nearly three decades that followed—the Bait Cod Era, 1950 until about 1978—were far less eventful, with little domestic interest in Alaska cod other than as bait to catch halibut and crabs. Meanwhile, foreign fleets fishing off Alaska's coast ravaged the Pacific cod and other fish resources.

The passage of the Magnuson-Stevens Fishery Conservation and Management Act in 1976 changed everything. Following the lead of Iceland and several South American nations, the legislation established a two-hundred-mile-wide fishery conservation zone along the entire U.S. coast, thereby creating the most extensive maritime domain of any country in the world. Complementing this vast geographic expansion of U.S. maritime sovereignty, the legislation also incorporated provisions that fostered the replacement of the foreign fishing fleets with a domestic fleet augmented by shore-based fish-processing plants. The undertaking was called *Americanization* and marked the beginning of the Modern Era.

Americanization was a complicated, freewheeling endeavor in which the groundfish fisheries off Alaska's coast became overcapitalized in less than a decade of serious fishing.[4] Too many boats were chasing limited fish resources—a situation that was both inefficient and a conservation issue. This led to bureaucratic and congressional efforts to *rationalize* the fisheries—essentially, to bring domestic fish-catching and fish-processing capabilities into balance with the available fish resources, and to manage the fisheries in a manner that fostered fishing and fish-processing operations that were efficient, ecologically sound, and sustainable.

Today, the Pacific cod fishery is, in terms of species and volume, the second-largest fishery in Alaska and is considered among the best-managed fisheries in the world.[5] Modern, often highly automated vessels and shore plants produce frozen headed-and-gutted fish, frozen gutted fish, frozen fillets, and a host of ancillary products. Most of the fishing effort is in the Bering Sea and the central and western Gulf of Alaska, just as it was in 1914. In 2017, the Alaska Pacific cod harvest of 298 thousand metric tons represented about 18 percent of the worldwide codfish catch.[6]

[H]istorical experience remains our principal source of knowledge.

—Thomas Piketty, economist, 2014[7]

[A] book committed to principles is doomed to early obsolescence, while a book of pure observations is never out of date.

—Robert MacArthur, ecologist, 1972[8]

ENDNOTES

[1] Author's note: This book employs the traditional cod-fishery terms *codfishing*, *codfisherman*, and *codfishermen*. The terms *cod* and *codfish* are used interchangeably, as are *salt cod* and *salted cod*. Likewise, the terms *West Coast* and *Pacific coast* are used interchangeably.

[2] "Protect American Cod Fisheries," *Pacific Fisherman* (December 1919): 36.

[3] E. Lester Jones, *Report of Alaska Investigations*, 1914 (Washington, DC: GPO, 1915), 58.

[4] In the North Pacific Ocean, the term *groundfish* comprises about a dozen species, including walleye pollock, Pacific cod, yellowfin sole, rock sole, rockfish, and Atka mackerel. Overlapping terms are *bottomfish*, *whitefish*, *demersal species*, and *underutilized species*.

[5] The largest fishery in Alaska is for walleye pollock (*Gadus chalcogrammus*).

[6] Ben Fissel et al (National Marine Fisheries Service), Stock Assessment and Fishery Evaluation Report for the Groundfish Fisheries of the Gulf of Alaska and Bering Sea/Aleutian Islands Atea: Economic Status of the Groundfish Fisheries off Alaska, 2019 (Seattle: Alaska Fisheries Science Center, April 17, 2019), 3, 178.

[7] Thomas Piketty, *Capital in the Twenty-First Century* (Cambridge, MA, and London, UK: Belknap Press of Harvard University Press, 2014), 575.

[8] Robert H. MacArthur, *Geographical Ecology: Patterns in the Distribution of Species* (New York: Harper & Row, 1972), 2.

Part One

THE SALT COD ERA (1863–1950)

Chapter 1
CODFISH FUNDAMENTALS

[T]he most valuable fish on the [Alaska] coast is the cod.

—George Davidson, geographer, 1867[9]

The salmon is the commonest of common fish in all the rivers of the North Pacific, and is rated accordingly as food only fit for those who cannot get better.

—Frederick Whymper, British artist and explorer, 1868[10]

The Cod is perhaps the most generally diffused and abundant of all [fish of Alaska], for it swims in all the waters of this coast from the Frozen ocean to the southern limit, and in some places it is in immense numbers.

—Sen. Charles Sumner (Massachusetts), 1867[11]

The cod is one of the most valuable of all food-fishes, and in the United States ranks as the most prominent commercial fish. In the matter of persons engaged, vessels employed, capital invested, and value of catch, the taking of cod in the United States is more extensive than any other fishery for fish proper. . . . The approximate annual value of the cod catch in recent years is about $3,000,000, a sum representing the first value of the fish. The weight of the fish as landed from the vessels (fresh, split, and salted) is about 100,000,000 pounds.

—Report of the U.S. Commissioner of Fish and Fisheries, 1897[12]

[T]he committee predicts that the annual catch of [Alaska] cod can be made to exceed that of Newfoundland or any other part of the world.

—Committee on Territories, U.S. Senate, 1903[13]

Pacific cod has a high growth rate and high natural mortality and can support heavy exploitation.

—Food and Agriculture Organization of the United Nations, 2014[14]

PACIFIC COD

Pacific cod are commonly referred to as gray cod (grey cod) or true cod.

In 1810, German naturalist and explorer Wilhelm Tilesius gave Pacific cod its scientific name, *Gadus macrocephalus*. Tilesius had examined Pacific cod in the Kamchatka region of the Bering Sea in 1804. American ichthyologist Tarleton Bean, however, was of the opinion that Pacific cod and the better-known Atlantic cod (*Gadus morhua*) were the same species.

Bean spent a good part of 1880 investigating Alaska's fisheries, and in a subsequent report on Alaska's cod fishery, he suggested that Tilesius had based the species name, *macrocephalus*—the Latin translation of which is "long head"—on an examination of a "deformed individual" with a long head. Similar deformations, Bean pointed out, were commonly found in Atlantic cod—"It is a matter of daily experience to find long-headed and short-headed cod in the same school off the New England coast or wherever the species occurs, as the length of the head is one of the most variable characters," adding that "There are no differences as far as general appearances go between Alaskan and New England cod. It would be impossible to tell one from the other if they were mixed in a tank without tags or some other means of identification."[15]

There were also questions regarding the size of the air bladders of the two fish: the air bladders of Pacific cod appeared to be smaller than those of Atlantic cod, and some thought this subtle difference was an indication that the two fish were different species.[16] Morphological factors aside, DNA analysis shows Pacific cod and Atlantic cod to reasonably be considered different species.[17]

> *Mating codfish are being considered a risk to national security in Norway. During the winter mating season, male cod make grunting noises about every 80 seconds that sound like foreign submarines. It creates havoc with underwater surveillance systems, and the cod problem is considered a threat to the safety and defense of Norway's Navy.*
>
> Laine Welch, *Alaska Fish Radio*, July 2016[18]

Pacific cod are demersal, dwelling at or near the ocean floor. In northern waters, Pacific cod are usually found from late spring until fall on relatively shallow feeding grounds on the middle-upper continental shelf. The fish move in the fall to spawning areas (see below) located primarily in the deep waters of the outer continental shelf and upper continental slope. Mature fish tend to prefer a sand or mud substrate. Fishermen generally agree that the largest and highest-quality cod are found in deeper waters.[19]

The range of the Pacific cod extends around the rim of the North Pacific Ocean from southern California north along the North American coast to the Bering Strait and then south along the Asian coast to the Yellow Sea. The species is rare or uncommon in the southern part of its range.[20]

> *A fishery biologist once calculated how many cod there would be in the world if all the eggs spawned by all the female cod in one spawning season were to hatch and survive to adulthood: the number was astronomical. He concluded that the oceans of the world, from shore to shore and from bottom to surface, would be one mass of wriggling cod, with no room for anything else.*
>
> —Albert C. Jensen, The Cod, 1972[21]

Pacific cod are Alaska's largest groundfish species. The species reaches maturity at an age of four or five years and can attain a maximum age of eighteen years. At the time they attain maturity, Pacific cod range in length from about twenty to twenty-three inches. Older, exceptional specimens can be six feet long and weigh up to about eighty-five pounds. Females may grow larger than males.

The species is relatively fast growing and highly fecund. A large female can produce more than 6,000,000 eggs in a single spawning event, but the usual range in the Bering Sea is between 1,000,000 and 2,000,000 eggs, while the usual range in the Gulf of Alaska is between 860,000 and 3,000,000 eggs.

The reproductive output of a fish population is not, however, a fixed proportion of the mass of the population. Fish size matters. A study published in *Science* in 2018 determined that a single thirty-kilogram female Atlantic cod produces more eggs than twenty-eight two-kilogram females (weighing a total of fifty-six kilograms). Moreover, the eggs produced by large females are larger and have a higher energy content than the eggs produced by small fish. It is likely that this size/fecundity relationship applies also to Pacific cod and suggests that large female fish contribute disproportionately to fish population replenishment. According to the study's authors, "Management based on [the common assumption that reproductive output is a fixed proportion of size, with respect to mass] risks underestimating the contribution of larger mothers to replenishment, hindering sustainable harvesting."[22]

Pacific cod usually spawn in late winter and early spring near the ocean floor at a depth between about 20 and 160 fathoms (a fathom is six feet). The eggs are adhesive and hatch in about fifteen to twenty days. Immature cod feed mostly on small invertebrates, while adult cod are opportunistic generalist predators. They feed mostly on fish (such as juvenile pollock), but also on crabs and are known to consume diving seabirds, though whether they target the seabirds or encounter them opportunistically while foraging for other prey is uncertain. In turn, Pacific cod are fed upon by marine mammals and a variety of fish, including halibut, lingcod, and other Pacific cod.

A combination of prolific reproduction potential, rapid growth, and high natural mortality in Pacific cod supports a fairly high maximum sustainable yield of the species.[23] However, given that most major Atlantic cod populations appear to experience substantial sporadic shifts in abundance, Pacific cod may also be subject to this phenomenon. Fluctuations in oceanic climate regimes may be the cause. (Reportedly, the ancient Aleut name for Pacific cod, *atxidaq*, translates literally into "the fish that stops," perhaps alluding to periodic disappearances of the species.[24])

The flesh of Pacific cod, which is usually served in portions cut from fillets, is white when cooked. It flakes well and has a mild, slightly sweet flavor that lends itself to a variety of preparations.

> *Codfish, for whiteness of colour, and moderate hardness and friability of substance, is commended. It is easily digested, and yieldeth meetly strong nourishment, and not very excremental. Being salted, dried, and so kept, it becomes of harder concoction and worse nourishment.*
>
> —Tobias Venner, 1650[25]

According to the *Seafood Handbook*, which is published online by *SeafoodSource News*, a commercial fishing industry information website, the flesh of Pacific cod is not as sweet as Atlantic cod, and, because the flesh of Pacific cod has a slightly higher moisture content, it is less firm than Atlantic cod.[26]

> *"There are 'fishy' fish, and then there are non-fishy fish. And with its mild, milky flesh, cod is one of the least fishy of them all."*
>
> —Melissa Clark, *New York Times*, 2018[27]

Nutritionally, Pacific cod flesh is high in protein, very low in fat, and contains less cholesterol than chicken or lean beef. The species does not accumulate dangerous levels of methylmercury, which has been associated with impaired neurological development in fetuses and children. Guidelines established by the State of Alaska's Department of Health and Social Services allow for unrestricted consumption of Pacific cod by even the most at-risk groups—women who are or can become pregnant, nursing mothers, and children.[28]

> *"Palatable, digestible, and nutritious, the Cod, as compared with other fish, is as beef compared with other meats."*
>
> —Sen. Charles Sumner (Massachusetts), 1867[29]

> *In comparison with meat and most other food products, Codfish is still one of the cheapest of the substantive and nutritive foods and, even at the present high prices, probably gives the most real food value for the money.*
>
> —Union Fish Company, 1911[30]

The flesh of Pacific cod may be host to roundworm parasites (*Pseudoterranova spp.*) commonly called sealworms (because seals, sea lions, and walruses are intermediate hosts) or codworms. Infestation varies depending on, among other factors, location and depth. The presence of these parasites in Pacific cod generally requires fillets cut from them to be candled and trimmed to locate and remove the parasites.

In addition to high-quality flesh, codfish are the source of a host of ancillary products. Traditional among these are cod tongues and cod liver oil (see below). Modern efforts to utilize what was previously considered fish waste—viscera, bones, trimmings, and so on—have developed a number of useful products that add considerably to the value of the codfish catch (see part 3, chapter 20).

The Atlantic cod, the basis of New England's first industry, has attained an iconic status—the carved wooden Sacred Cod hangs prominently in the Massachusetts State House, and in 1930s Newfoundland issued a pair of postage stamps graced with images of codfish. The images were labeled "Newfoundland Currency" (fig. 1). And in 1986, the U.S. Post Office issued an Atlantic cod postage stamp (fig. 2).

Figure 1. "Newfoundland Currency" postage stamp, 1937. Image courtesy of the author.

Figure 2. U.S. Post Office, Atlantic cod postage stamp, 1986. Image courtesy of the author.

Figure 3. Aleut codfishermen at Unalaska. (Henry W. Elliott, in George Brown Goode, *The Fisheries and Fishery Industries of the United States, Section V: History and Methods of the Fisheries* [Washington, DC: GPO, 1887], plate 36.)

Along the Pacific coast, pioneer scientist and surveyor George Davidson, who in 1867 was aboard the U.S. Revenue Cutter *Lincoln* during a geographical reconnaissance of coastal Alaska to ascertain the "most available channels of commerce," considered the Pacific cod to be "the most valuable fish on [Alaska's] coast." Davidson predicted that "[a]s the banks of Newfoundland have been to the trade of the Atlantic, so will the greater banks of Alaska be to the Pacific; inexhaustible in supply of fish that are equal if not superior in size and quality to those of the Atlantic." Almost a half century later, U.S. fisheries agent E. Lester Jones characterized Alaska cod as being "of first-class quality, and notwithstanding occasional adverse reports it is equal in every way to the Atlantic cod." Despite those scientific accolades, however, the Pacific cod has a long history of being portrayed by the Atlantic cod industry as a second-rate fish.[31]

ALASKA NATIVE UTILIZATION OF PACIFIC COD

Natives living along Alaska's coast utilized codfish, particularly at locations where salmon were not abundant. A prime example is Sanak Island, in the western Gulf of Alaska, which has been occupied by Aleuts for perhaps 6,000 years. At prehistoric village sites on the island, Pacific cod bones dominate midden deposits, some of which date back 4,500 years.[32]

In late July and early August of 1880, Tarleton Bean, the aforementioned American ichthyologist, observed Native fishermen at Iliuliuk (Unalaska, in the Aleutian Islands) returning to the settlement with *bidarka* (kayak) loads of cod. The fish were dried for winter use.[33]

Somewhat at variance with Bean's observation, the official report of the 1888 voyage of the U.S. Fish Commission steamer *Albatross* to determine the extent of the cod-fishing grounds in western Alaska (see chapter 2) noted that the Natives at Unalaska never kept large quantities of cod on hand because a short fishing trip to nearby waters would generally satisfy their immediate needs. The report noted that when in the harbor these Natives mostly fished from dories, but they employed their traditional bidarkas when long distances were to be traversed. In fishing, the Unalaska Natives used regular cod hooks, but their fishing lines were made of any suitable material at hand, including sail twine and old string. Their sinkers were pieces of lead, old spikes, bolts, and stones, and their bait was generally sculpins, flounders, salmon, or clams, whichever among them was most easily obtainable.[34]

Further north, where the sea was frozen during part of the year, Iñupiat Eskimos chopped holes in the ice and fished for cod with hooks and lines. Their traditional hooks were fashioned from ivory or bone and the lines from the sinew of walrus, seal, and caribou.[35] Riley Moore, an anthropologist associated with the Smithsonian Institution who visited St. Lawrence Island in 1912, wrote of the Eskimos there as catching "a great many fish with the modern hook and line or ivory hooks with bits of the red skin from duck feet sewed over them for bait."[36]

Census taker Ivan Petroff reported in 1882 that the Natives and creoles on Kodiak and Afognak islands subsisted partially on codfish.

Figure 4. Eskimos with codfish at St. Lawrence Island, 1897. (Arctic Environmental Information and Data Center, University of Alaska Fairbanks. Original image unable to be located. Image, with text removed, from Stephen C. Jewett, "Alaska's Latent Fishery—Pacific Cod," *Alaska Seas and Coasts*, 1977, 5[1]: 6–8.)

Catching them seemed to be the job of boys and old men, who fished the near-shore waters year-round. Cod were apparently less desirable to the Natives in Prince William Sound, who, according to Petroff, consumed cod only "when no other fish [could] be obtained in abundance." The cod were caught mostly near shore in well-sheltered bays. Petroff estimated the Natives in Prince William Sound ate about half the catch fresh; the remainder was split and dried.[37] Natives in Southeast Alaska likewise ate cod. Frederick Schwatka, who conducted a military reconnaissance of Alaska in 1883, reported cod—along with salmon, halibut, and herring—to be a dietary staple in Southeast Alaska villages.[38]

TRADITIONAL CODFISH PRODUCTS

Salted cod

> *One of the great salt fish foods of North America was salt cod from the Grand Banks and Georges Banks of the Northwest Atlantic. For much of the history of the colonies and later the United States, creamed cod on boiled potatoes was a standing Saturday night supper or Sunday morning breakfast for a good number of the more affluent people of the country.*
>
> —Robert J. Browning, *Fisheries of the North Pacific*, 1980[39]

Price List—Pacific Codfish.

Prices Subject to Change Without Notice.

San Francisco, SEP 29 1904 **NET CASH**

Bundles, - - - - -	50 lbs. each, per lb.	4½
Cases, Regular, - -	100 lbs. " " "	5
" Extra Large, -	100 lbs. " " "	6
" Eastern Style,	100 lbs. " " "	6 50
Boneless, - - - - -	30 lb. boxes " "	6½
"Norway," - - - -	40 lb. " " "	6¼
"Narrow Gauge," - -	40 lb. " " "	6¼
"Silver King," - -	40 lb. " " "	7½
Blocks, "Oriental," -	40 lb. " " "	6¼
Blocks, "Seabright,"	40 lb. " " "	6¼
Tablets, "Crown Brand,"	36 lb. " " "	7
Middles, "Golden State,"	60 lb. " " "	6¼
" "White Seal,"	60 lb. " " "	9
5 lb. Boxes Fancy, Boneless,	Crates, 12 bxs. each " "	9
2 lb. " " "	" " " " " "	9 9½
Desiccated, "Gilt Edge," (½ lb. tins)	per doz.	
Pickled Cod, Barrels, - - - - - - -	each	7 50
Pickled Cod, Half Barrels, - - - -	"	4 50

Assorted and 1 lb. Blocks ¼c. per lb. extra
" " " Tablets " " " "

UNION FISH COMPANY
24 CALIFORNIA STREET

Figure 5. Union Fish Company wholesale price list, 1904. (San Francisco Maritime National Historical Park.)

Salted cod is cod that has been preserved by salting and drying. Among the fishes, cod is especially suitable for salting because its flesh contains only a small amount of oil. (Oil slows the impregnation of salt into the flesh and becomes rancid when exposed to air.) In addition to being simple to produce, salted cod is nutritious, has a long shelf life, and is easily transportable. Prior to cooking, salted cod is *freshened*—rehydrated and desalted—by soaking it in cold, fresh water.

The practice of salting cod began at least five hundred years ago in the North Atlantic, and salt cod became a staple food in northern Europe as well as in Europe's predominately Catholic countries. Europeans, particularly Norwegians and Portuguese, also exported substantial quantities of salted cod to Africa, the Caribbean, and South America.

Figure 6. Lid for one-pound wooden Union Fish Company Alaska codfish box, circa 1910. (Courtesy of David Witherell.)

Beginning in about 1881, codfish processors sometimes sprinkled boric acid on salted cod as a preservative, generally upon fish that was to be shipped abroad or shipped a considerable distance domestically during the summer months. In about 1907, sodium benzoate replaced boric acid.

Salted cod produced on the U.S. West Coast for sale on the domestic wholesale market was produced in a variety of styles that were based on how the fish was cut and how much skin and bone were removed. This fish was generally sold in 30-, 40-, 50-, and 100-pound-capacity wooden boxes.

On the retail market, one-pound and two-pound paper-wrapped packages of skinless/boneless cod, known in the trade as *bricks*, were standard items. The cuts of fish included in a brick determined the fanciness of the pack.

Fish for local consumption or destined for dry-climate countries were wrapped in waxed or parchment paper and packed in cardboard or wooden containers. Salted codfish for export to damp localities was often packed in large tin cans—air-tight packaging that prevented the fish from absorbing moisture and spoiling in damp conditions. In 1949, the Pacific Coast Codfish Company's plant at Poulsbo, Washington, packed salted codfish for shipment to the Philippine Islands in ten-gallon cans.[40]

Pacific Fisherman in 1936 characterized traditional salted cod as a "tolerated but unloved step-child" in the family of groceries. That same year, in an effort to produce a more attractive, easily handled, and stable salted cod product, the Robinson Fisheries Company, of Anacortes, Washington, began packing high-grade, skinless/boneless salted codfish in cans. In the canning process, the fish was carefully wrapped in parchment paper and then placed in enameled cans, which were in turn run through a vacuum sealing line. No cooking was involved. In the fall of 1937, the Pacific Coast Codfish Company began producing an identical product.[41]

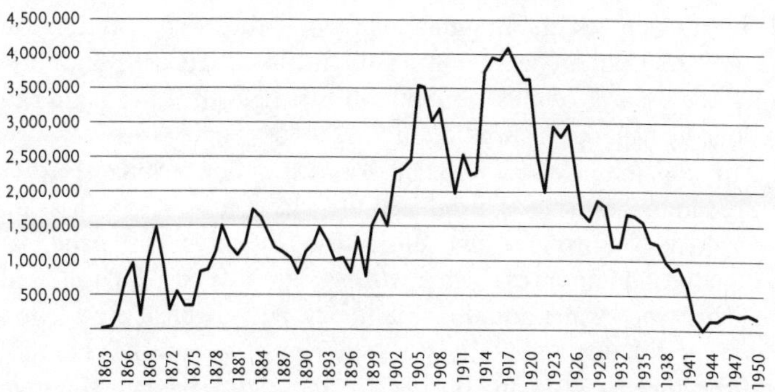

Chart 1. Alaska salted Pacific cod production (numbers of fish), 1863–1950. Data sources: John N. Cobb's 1927 report on Alaska's codfish industry and *Pacific Fisherman*'s 1951 Yearbook. (John N. Cobb, *Pacific Cod Fisheries* (revised edition, 1926), Bureau of Fisheries Doc. No. 1014, appendix 7 to *Report of the U.S. Commissioner of Fisheries* for 1926 [Washington, DC: GPO, 1927], 467; "1950 Pacific Codfish Fleet Catch," *Pacific Fisherman* [January 1951]: 291.)

According to a Robinson Fisheries Company advertisement, the vacuum canning process "eliminates any deterioration of quality in stock and brings salt codfish to the consumer with all its original freshness and flavor."[42] More specifically, the canning process prevented the fish from drying out and prevented the formation of salt crystals on the outside of the package. On the downside, canning codfish was labor intensive and raised the cost of the product, which constricted demand.

According to *Pacific Fisherman*, however, demand for the product in late 1937 was "very active," and in 1939 the trade journal praised the vacuum canning of salted cod as "a major development in the history of one of the oldest fishery products in existence."[43] Despite this auspicious beginning, canned salted cod soon lost favor to the traditional wood- and paper-packaged product and, increasingly, to frozen cod.

A modern advocate for salted cod was world-renowned chef James Beard. In 1967, Beard wrote: "I am so fond of salt cod that I sometimes forget that fresh cod can be prepared in many interesting ways."[44]

Stockfish

Stockfish is unsalted cod that has been gutted and split and then air-dried outdoors, typically on wooden racks. The product has a storage life of several years and is popular in Africa and among Catholics of

Mediterranean descent. In suitable climates—the ideal conditions for making stockfish are temperatures slightly above freezing and little rain—the process of making stockfish is simple and cheap and can be done by fishermen and their families.

In 1909, John Nelson, who had a codfish station at Squaw Harbor, in the Shumagin Islands, produced the first Alaska stockfish as an experiment. Nelson cured his codfish during the colder part of the year by hanging it outdoors over wires with the skin side up, which allowed the fish to shed water and the flesh to dry. His stockfish production that year totaled 13,000 pounds and was shipped in 100-fish bundles held together by wires and burlap. The bundles weighed an average of 100 pounds. Nelson's experiment was a success and he continued to produce stockfish. He was joined in the industry in 1912 by R. H. Johnson, who also had a codfish station at Squaw Harbor.

Stockfish production continued to expand and soon became a staple product in the Shumagin Islands. It was produced by a large number of individual fishermen and, in the words of John Cobb, "meagerly financed" companies during the winter, when fishing for species other than cod was suspended. Annual stockfish production during the years 1916 through 1924 averaged 112,000 pounds, ranging wildly between 12,775 pounds in 1920 to an exceptionally high 678,422 pounds the following year. In 1928, Alaska stations produced 487,000 pounds of stockfish, of which fully 90 percent was dried at Unga.[45]

Figure 7. Stockfish drying at John Nelson codfish station, Baralof Bay, Unga Island, circa 1915. (Courtesy of Thor Lauritzen.)

Pickled cod

Pickled cod is produced in a manner similar to the production of pickled (mild-cured) salmon. Split fish or fish fillets are packed between layers of salt in large wooden barrels known as *tierces*, where they are allowed to absorb salt and expel liquids for several days. The fish are then repacked in fresh salt. When the tierce is full, it is topped off with fresh water, fitted with a lid, and rolled onto its side with a bunged hole facing up. After the tierce has rested in this position for several hours, the bung is removed and brine is poured into the tierce to replace any remaining air inside. The tierce is then ready for shipping.

Beginning in the 1920s, codfish shore stations became increasingly less viable financially and began shutting down. Small, independent salmon salteries recognized an opportunity to expand and diversify their business, and by the mid-1920s had begun salting and pickling codfish. During the years 1930 through 1937—the longest uninterrupted period during which cod was pickled in significant quantities in Alaska—an average of about 99,000 pounds of the product was produced annually along the Central Alaska coast.[46] The production of pickled codfish on a significant scale ended in the early 1940s.

Frozen cod

Industrially freezing seafood (and other food products) became practical after the invention in the late 1920s of the double-belt freezer by Clarence Birdseye, who operated the General Seafood Corporation, in Gloucester, Massachusetts. Birdseye's freezer froze fish quickly, preventing damage to the flesh that was caused by ice crystals that developed during slower freezing processes.

Beginning in late 1937, East Coast firms began shipping large quantities of frozen cod, haddock, and other North Atlantic fish fillets to Los Angeles, where it was well received. The East Coast fish was priced lower than West Coast processors could compete with, due mostly to higher labor costs in the West. At the same time, grocery stores, which were beginning to install frozen food cases, were attracted to the product because it was both convenient and did not require employees who were knowledgeable about fish and able to properly cut it. Despite earlier handicaps, by the early 1940s substantial quantities of frozen fillets—mostly from trawler-caught fish—were being produced on the West Coast. Much to the detriment of salted fish, frozen fish became even more attractive in the years following the end of World War II, as freezers became increasingly commonplace in American homes and ready-to-cook frozen *fish sticks*—mostly made from cod—became available.[47] The market for salted cod, however, has remained substantial in countries such as Portugal, where there is a strong tradition of eating salted cod.

Because Pacific cod is caught far from markets, nearly the entire catch today is frozen at sea or at shore plants close to the fishing grounds as headed-and-gutted (H&G) fish, gutted fish, or fillets. Frozen Pacific cod retains its flavor, moisture, and texture well and has a relatively long shelf life. Much of the frozen Pacific cod sold on the U.S. retail market is, in fishing industry parlance, *twice-frozen* (also, *double-frozen*): fish that has been thawed, reprocessed, and then refrozen, either in lower-48 plants or in Asia.[48]

TRADITIONAL ANCILLARY PRODUCTS

Cod tongues

Cod tongues were commonly eaten in New England, but were not often eaten on the Pacific coast. Fresh tongues were prepared by breading and then frying them, and were likened to fried oysters or scallops. (Sometimes the tongues were boiled briefly to tenderize them before they were fried.) Aboard vessels, cod tongues were cured loosely with salt in barrels and then repacked in barrels to which strong brine was added. The tongues—like salted codfish—were soaked in fresh water prior to preparation.

In the Alaska fisheries, cod tongues were not routinely saved until the Great Depression. The few saved prior to that time were cut at shore stations, mostly by Native boys, or aboard certain schooners, where a boy was paid $3.50 to $5.00 for each barrel of tongues he cut out.

In response to an increased demand for cod tongues, offshore schooners in the early 1930s began saving substantial quantities of tongues. (Shore-station operations by this time were minimal.) During the years 1933 to 1940, the schooner fleet's annual production of tongues ranged from about 12,000 pounds to about 30,000 pounds. In 1940, the fleet produced 25,500 pounds of cod tongues, which were valued at $2,401—a little less than ten cents per pound.[49]

Cod-liver oil

Cod-liver oil is rich in vitamins A and D and was in use as a general medicinal as early as 1840.[50] The oil is also used for tanning hides, as a supplement in animal feed, and as a base for paints. The oil is derived from the livers of codfish and was historically produced in Alaska

Figure 8. Native boy cutting codfish tongues at unidentified shore station. (John N. Cobb Collection, Special Collections, University of Washington Libraries, Seattle, Wash.)

by the simple process of dumping codfish livers into large wooden casks, where they were *rendered*—allowed to rot. The oil released by the rotting livers gradually made its way to the surface, where it was periodically skimmed off, strained, and poured into barrels or tanks for storage and transport. Because both healthy and diseased cod livers were used, the cod liver oil produced in Alaska was generally not of medicinal quality and was mostly sold to tanneries. The operators of a small codfish station near Kodiak from 1923 to 1933 mixed codfish oil with pigment to make paint.

A 1938 Canadian study of Pacific cod determined that livers represented from 2.3 percent to 4.2 percent of the round (live) weight of the fish, and that the oil content of the livers ranged from 24.2 percent to 39.0 percent. Based on the Canadian figures, the liver of a ten-pound Pacific cod yielded between 0.89 and 2.62 ounces of oil.[51]

An 1882 report on the commerce and industries of the Pacific coast stated that some 6,000 gallons of cod liver oil was produced annually aboard the schooner fleet. A portion of this oil was refined for druggists' use and sold for a dollar per gallon; the remainder was sold *crude* to tanners for $0.40 per gallon. (One of the San Francisco firm Lynde & Hough's ancillary products was Okhotsk Sea Cod Liver Oil.) West Coast schooner fishermen did not continue to produce cod liver oil, but they did sometimes save cod livers. The livers were

Figure 9. Worker straining cod liver oil into a holding tank at an unidentified shore station. (John N. Cobb Collection, Special Collections, University of Washington Libraries, Seattle, Wash.)

salted in barrels or metal cans for delivery to processing facilities in San Francisco and Puget Sound (see below).

Small quantities of cod liver oil were, however, regularly produced at shore stations. In 1899, the Alaska Codfish Company made the first attempt to produce medicinal-grade cod liver oil in Alaska. The firm installed a refining plant at its Kelley's Rock (Winchester) codfish station, near Unga, in the Shumagin Islands, and produced about 2,000 gallons of oil that was brought to San Francisco and offered for sale to manufacturers of emulsified cod liver oil. Unfortunately, there was at that time a surplus of the grade of cod liver oil being offered, and the oil was put into storage and the Kelley's Rock refining plant shut down. The oil was sold several years later, but by then the refining plant had fallen into disrepair. It was, however, restarted and operated during at least the 1906 season. A portion of the oil produced that year was of medicinal quality and was sold to hospitals in San Francisco.

The Union Fish Company erected a cod liver oil plant at its Pirate Cove station (Popof Island, also in the Shumagin Islands), and the plant produced a small quantity of oil—none of which was of medicinal quality—during the 1915 season. The venture apparently showed no profit and was permanently shuttered after just one season of operation.[52]

Although there was a tremendously increased demand for vitamin A during World War II, a 1949 U.S. Fish and Wildlife Service report stated that no cod liver oil was produced from Pacific cod in 1941 and that "many tons [of cod livers] were discarded at sea as a waste byproduct of the cod fishery." The price of vitamin D, however, rose substantially just after the end of World War II, spurring interest in Pacific cod livers.[53]

In response, Ed Shields's crew on the schooner *C. A. Thayer* began salting cod livers and storing them in five-gallon cans in the vessel's hold. The oil was rendered at a plant in Seattle.

Going a step further, prior to the 1947 fishing season the Robinson Fisheries Company—anticipating a good market for cod livers—installed a 3,000-cubic-foot cold storage on its schooner, the *Wawona*. The vessel returned that fall with 30,000 pounds of cod livers just, unfortunately, as the market for vitamin oils plummeted. This disappointment might have contributed to the *Wawona*'s 1947 voyage being its last.[54] That same year, two Dutch chemists synthesized vitamin A. (A synthetic version of vitamin D was first produced during the 1920s.)

In modern times, an increase in the demand for natural vitamins and supplements—*nutraceuticals*—has led to a resurgence in the market for cod liver oil (see part 3, chapter 20).

Glue

Gloucester, Massachusetts, resident Benjamin Robinson, father of W. F. Robinson, of the Robinson Fisheries Company, Anacortes, Washington, was credited as the first person on the East Coast to make glue from fish. He first did so in about 1870, and soon after started a business that his sons carried on for at least four decades. The glue was sold in barrels and half-barrels for the factory trade. The LePage Company (now a part of Henkel AG & Company) began processing fish-based glue at Gloucester in 1876.

The Robinson Fisheries Company, in large part a fish by-products operation, carried on the family tradition on the West Coast. In early 1909, the *Anacortes American* newspaper reported that the company had so far produced 8,000 gallons of glue from codfish skins.

At the Pacific Coast Codfish Company plant in Poulsbo, Washington, which processed codfish until 1950, the skin removed from salted codfish being prepared for market was shipped to a glue factory near San Francisco. Likewise, codfish skins from the Union Fish Company's codfish plant near San Francisco were sold to a glue factory.[55]

See part 3, chapter 20, for a discussion of post–Magnuson Act codfish products.

ENDNOTES

[9] George Davidson, *Report of Assistant George Davidson, Relative to the Coast, Features, and Resources of Alaska Territory*, November 30, 1867, in *Russian America*, 40th Cong., 2d sess., H. Ex. Doc. No. 177 (Washington, DC: GPO, 1868), 255.

[10] Frederick Whymper, *Travel and Adventure in the Territory of Alaska* (London: John Murray, Albemarle Street, 1868), 244–45. Whymper, a British artist and explorer, was a member of the Western Union Telegraph Expedition (1865–1867).

[11] "Speech of Hon. Charles Sumner, of Massachusetts, on the cession of Russian America to the United States" (Washington, DC: Congressional Globe Office, 1867), 44.

[12] *A Manual of Fish-culture, appendix to Report of the Commissioner [of Fish and Fisheries] for the Year Ending June 30, 1897* (Washington, DC: GPO, 1898), 195–96.

[13] *Conditions in Alaska*, 58th Cong., 2d sess., 1903, Sen. Rept. No. 282 (Washington, DC: GPO, 1904), 6.

[14] Food and Agriculture Organization of the United Nations, Fisheries and Aquaculture Department, "Species Fact Sheets: *Gadus macrocephalus* (Tilesius, 1810)," accessed February 21, 2014, http://www.fao.org/fishery/species/3011/en.

[15] Tarleton Bean, *The Cod Fishery of Alaska*, in George Brown Goode, *The Fisheries and Fishery Industries of the United States, Section V, The History and Methods of the Fisheries*, vol. I (Washington, DC: GPO, 1887 [1880]), 199, 204.

[16] John N. Cobb, *Pacific Cod Fisheries*, Bureau of Fisheries Doc. No. 830, appendix 4 to *Report of the U.S. Commissioner of Fisheries for 1915* (Washington, DC: GPO, 1916), 5–6.

[17] Steven M. Carr, David S. Kivlichan, Pierre Pepin, Dorothy C. Crutcher, "Molecular systematics of gadid fishes: implications for the biogeographic origins of Pacific species," *Canadian Journal of Zoology*, 1999, 77(1): 19–26.

[18] Laine Welch, *Alaska Fish Radio*, accessed July 6, 2016, http://www.alaskafishradio.com/fish-funnies-from-around-the-world/.

[19] The term *highest quality* is subjective, but usually refers to fish that are healthy, heavily configured (not skinny), and whose flesh is relatively free of parasites.

[20] NOAA Fisheries Service, Alaska Fisheries Science Center, "Pacific Cod Fact Sheet," 2010; Food and Agriculture Organization of the United Nations, Fisheries and Aquaculture Department, "Species Fact Sheets: *Gadus macrocephalus*

(Tilesius, 1810)," accessed February 21, 2014, http://www.fao.org/fishery/species/3011/en; David Witherell and Jim Armstrong (staff, North Pacific Fishery Management Council), *Groundfish Species Profiles, 2015*, 3, accessed November 12, 2018, https://www.npfmc.org/wp-content/PDFdocuments/resources/SpeciesProfiles2015.pdf; John N. Cobb, *Pacific Cod Fisheries*, Bureau of Fisheries Doc. No. 830, appendix 4 to *Report of the U.S. Commissioner of Fisheries for 1915* (Washington, DC: GPO, 1916), 51.

[21] Albert C. Jensen, *The Cod* (New York: Thomas Y. Crowell, 1972), 21.

[22] Diego R. Barneche et al., "Fish reproductive-energy output increases disproportionately with body size," *Science* (May 11, 2018): 642–645.

[23] NOAA Fisheries Service, Alaska Fisheries Science Center, "Pacific Cod Fact Sheet," 2010; Food and Agriculture Organization of the United Nations, "Species Fact Sheets: *Gadus macrocephalus* (Tilesius, 1810)," accessed February 21, 2014, http://www.fao.org/fishery/species/3011/en; NOAA Fisheries Service, Alaska Fisheries Science Center, "Life History, *Gadus macrocephalus*," accessed February 21, 2014, http://access.afsc.noaa.gov/reem/lhweb/LifeHistorySearch.php?SpeciesID=54; Diego R. Barneche et al., "Fish reproductive-energy output increases disproportionately with body size," *Science* (May 11, 2018): 642–645; Tom Ohaus, "Pacific Cod: Chowhound of the Deep," *Pacific Fishing* (September 1991): 20; Sadie E. G. Ulman et al., "Predation on Seabirds by Pacific Cod *Gadus microcephalus* Near the Aleutian Islands, Alaska," *Marine Ornithology* (October 2015): 231–233, accessed November 10, 2015, http://marineornithology.org/PDF/43_2/43_2_231-233.pdf; Sandra K. Neidetcher, Thomas P. Hurst, Lorenzo Ciannelli, and Elizabeth A. Logerwell, "Spawning Phenology and Geography of Aleutian Islands and Eastern Bering Sea Pacific Cod (*Gadus macrocephalus*)," *Deep Sea Research, Part II: Topical Studies in Oceanography* (November 2014): 204–214, accessed November 11, 2017, http://www.sciencedirect.com/science/article/pii/S0967064513004529?via%3Dihub; Ben Fissel et al., *NPFMC Draft Stock Assessment and Fishery Evaluation Report for the Groundfish Fisheries of the Gulf of Alaska and Bering Sea/Aleutian Islands Area: Economic Status of the Groundfish Fisheries off Alaska, 2014* (Seattle: National Marine Fisheries Service, 2015), 213, accessed December 9, 2015, http://www.afsc.noaa.gov/REFM/stocks/plan_team/economic.pdf; Erling Skarr (captain, freezer-longliner *Seattle Star*), personal communication with author, November 25, 2015; Northern Economics, *Commercial Fishing Industry of the Bering Sea*, Tech. Rept. No. 138 (June 1990), 40.

[24] Herbert D. G. Maschner et al., *A 4500-year time series of Pacific cod (*Gadus macrocephalus*) size and abundance: archaeology, oceanic regime shifts, and sustainable fisheries*, NOAA Fishery Bulletin 106, no. 4 (2008): 386–394; Matthew W. Betts, Herbert D. G. Maschner, and Donald S. Clark, "Zooarchaeology of the 'Fish That Stops,'" in *The Archaeology of North Pacific Fisheries*, ed. Madonna L. Moss and Aubrey Cannon (Fairbanks, Alaska: University of Alaska Press, 2011), 188.

[25] Tobias Venner, *Via Recta ad Vitam Longam* (1650), in W. Stephen Mitchell, "The Place of Fish in a Hard-Working Diet," *Fisheries Exhibition Literature*, vol. 1, International Fisheries Exhibition, London, 1883 (London: William Clowes and Sons, Ltd., 1884), 425.

[26] *Seafood Handbook*, accessed September 18, 2015, http://www.seafoodsource.com/seafoodhandbook/finfish/cod.

[27] Melissa Clark, "A Weeknight Dish for Lovers of Non-Fishy Fish," *New York Times*, January 5, 2018.

[28] World Health Organization, "Mercury and Health," accessed July 24, 2014, http://www.who.int/mediacentre/factsheets/fs361/en/; Environmental Protection Agency, "Health Effects of Mercury," accessed July 24, 2014, http://www.epa.gov/hg/effects.htm; Ali K. Hamade, on behalf of the Alaska Scientific Advisory Committee for Fish Consumption, *Fish Consumption Advice for Alaskans: A Risk Management Strategy to Optimize the Public's Health* (Juneau: State of Alaska, Department of Health and Social Services, updated July 21, 2014), 15, 50.

[29] "Speech of Hon. Charles Sumner, of Massachusetts, on the cession of Russian America to the United States" (Washington, DC: Congressional Globe Office, 1867), 44.

[30] Union Fish Company, "Codfish Review," May 15, 1911 (courtesy Union Fish Company).

[31] George Davidson, *Report of Assistant George Davidson Relative to the Coast, Features, and Resources of Alaska Territory*, November 30, 1867, in *Russian America*, 40th Cong., 2d sess., H. Ex. Doc. No. 177, Pt. 1 (Washington, DC: GPO, 1868), 254–257; E. Lester Jones, *Report of Alaska Investigations, 1914* (Washington, DC: GPO, 1915), 58.

[32] Herbert D. G. Maschner et al., *A 4500-year time series of Pacific cod (Gadus macrocephalus) size and abundance: archaeology, oceanic regime shifts, and sustainable fisheries*, NOAA Fishery Bulletin 106, no. 4 (2008): 386–394; Matthew W. Betts, Herbert D. G. Maschner, and Donald S. Clark, "Zooarchaeology of the 'Fish That Stops,'" in *The Archaeology of North Pacific Fisheries*, ed. Madonna L. Moss and Aubrey Cannon (Fairbanks: University of Alaska Press, 2011), 174.

[33] Tarleton Bean, *The Cod Fishery of Alaska* [1880], in George Brown Goode, *The Fisheries and Fishery Industries of the United States, Section V: History and Methods of the Fisheries*, vol. 1 (Washington, DC: GPO, 1887), 201, 203.

[34] Z. L. Tanner et al., "Explorations of the Fishing Grounds of Alaska, Washington Territory, and Oregon, during 1888, by the U.S. Fish Commission Steamer Albatross, Lieut. Comdr. Z. L. Tanner, U.S. Navy, Commanding," *Bulletin of the United States Fish Commission*, vol. 8, 1888 (Washington, DC: GPO, 1890), 21–22.

[35] Alaska Humanities Forum, *Alaska's Heritage*, chapters 2–4, accessed December 29, 2016, http://www.akhistorycourse.org/alaskas-cultures/alaskas-heritage/chapter-2-4-eskimos.

[36] Riley D. Moore, "Social Life of the Eskimo of St. Lawrence Island," *American Anthropologist* (July–September 1923): 339–375.

[37] Ivan Petroff, *Report on the Population, Industries, and Resources of Alaska* [1882], in *Seal and Salmon Fisheries and General Resources of Alaska*, vol. 4 (Washington, DC: GPO, 1898), 268.

[38] Frederick Schwatka, *Report of a Military Reconnaissance in Alaska Made in 1883*, 48th Cong. 2d sess., Sen. Ex. Doc. No. 2 (Washington, DC: GPO, 1885), 63, 66, 71, 74, 113.

[39] Robert J. Browning, *Fisheries of the North Pacific: History, Species, Gear & Processes*, 2nd ed. (Anchorage: Alaska Northwest Publishing Company, 1980), 294.

[40] Mark Kurlansky, *Salt: A World History* (New York: Penguin Books, 2002), 113–114; J. W. Collins, *Report on the Fisheries of the Pacific Coast of the United States*, in *Report of the United States Commissioner of Fish and Fisheries for 1892* (Washington, DC: GPO, 1892), 106; John N. Cobb, *Pacific Cod Fisheries* (revised edition, 1926), Bureau of Fisheries Doc. No. 1014, appendix 7 to *Report of the U.S. Commissioner of Fisheries for 1926* (Washington, DC: GPO, 1927), 453–454; "De Luxe Packaging Puts Codfish On A New Plane As Grocery Item," *Pacific Fisherman* (November 1933): 21; Richard S. Croker, "Alaska Codfish," in *The Commercial Fish Catch of California for the Year 1929*, Division of Fish and Game of California Fish Bulletin No. 30 (Sacramento: California State Printing Office, 1931), 48–55; "Codfish in Big Tins for the Philippines," *Pacific Fisherman* (November 1949): 26.

[41] "Salt Codfish—Now Vacuum-Packed in Tin," *Pacific Fisherman* (November 1936): 25; "'Icicle' Codfish Now Packed in Cans," *Pacific Fisherman* (November 1937): 81.

[42] Robinson Fisheries Co., adv., *Pacific Fisherman* (January 1937): 226.

[43] "Vacuum Pack Proves a Boon to Codfish," *Pacific Fisherman* (January 1939): 242.

[44] James Beard, *James Beard's Fish Cookery* (New York: Warner Paperback Library, 1967), 72.

[45] Ward T. Bower and Harry Clifford Fassett, "The Fishery Industries of Alaska in 1913," *Pacific Fisherman* (January 1914): 51–61, 63, 65–66; "Stockfish," accessed December 14, 2013, http://en.wikipedia.org/wiki/Stockfish; *Fisheries of Alaska in 1909*, Bureau of Fisheries Doc. No. 730 (Washington, DC: GPO, 1910), 41; Barton Warren Evermann, *Fishery and Fur Industries of Alaska in 1912*, Bureau of Fisheries Doc. No. 780 (Washington, DC: GPO, 1913), 65; John N. Cobb, *Pacific Cod*

Fisheries (revised edition, 1926), Bureau of Fisheries Doc. No. 1014, appendix 7 to *Report of the U.S. Commissioner of Fisheries for 1926* (Washington, DC: GPO, 1927), 420; "Reduced Codfish Production Causes Stiffening Market," *Pacific Fisherman* (January 1929): 159, 161.

[46] Bureau of Fisheries, *Alaska Fisheries and Fur-Seal Industries*, annual reports, 1930–1937.

[47] Paul Greenberg, *American Catch: The Fight for Our Local Seafood* (New York: Penguin Press, 2013), 105; "Codfish Marketing Meets Difficulties," *Pacific Fisherman* (January 1939): 241–242; "Codfish in Cans Clicks Cleanly," *Pacific Fisherman* (November 1937): 81; "Atlantic Fillets," *Pacific Fisherman* (May 1938): 57; "Nefco Filleting at Eureka," *Pacific Fisherman* (May 1944): 59.

[48] The provenance of breaded Pacific cod fillets observed by the author in the freezer case at a Walmart in Oregon was a mystery. The label stated only that the fish was "Wild Caught" and was processed in China.

[49] John N. Cobb, "Overcoming Waste in the Fishing Industry," *Pacific Fisherman* (May 1923): 15; Russ Hofvendahl, *Hard on the Wind* (Dobbs Ferry, N.Y.: Sheridan House, 1989), 123; John N. Cobb, *Pacific Cod Fisheries* (revised edition, 1926), Bureau of Fisheries Doc. No. 1014, appendix 7 to *Report of the U.S. Commissioner of Fisheries for 1926* (Washington, DC: GPO, 1927), 442; Ward T. Bower, *Alaska Fishery and Fur-Seal Industries in 1932*, appendix 1 to *Report of Commissioner of Fisheries for the Fiscal Year 1933* (Washington, DC: GPO, 1933), 53; Ward T. Bower, *Alaska Fishery and Fur-Seal Industries in 1933*, appendix 2 to *Report of Commissioner of Fisheries for the Fiscal Year 1934* (Washington, DC: GPO, 1934), 287–288; Ward T. Bower, *Alaska Fishery and Fur-Seal Industries in 1937*, appendix 2 to *Report of Commissioner of Fisheries for the Fiscal Year 1938* (Washington, DC: GPO, 1938), 120–121; Ward T. Bower, *Alaska Fishery and Fur-Seal Industries in 1939*, appendix 2 to *Report of Commissioner of Fisheries for the Fiscal Year 1939* (Washington, DC: GPO, 1941), 154–155; Ward T. Bower, *Alaska Fishery and Fur Seal Industries: 1940* (Washington, DC: GPO, 1942), 49–50.

[50] Charles Butler, *The Fish Liver Oil Industry*, U.S. Fish and Wildlife Service Leaflet No. 233 (March 1948), 5.

[51] Ed Opheim, "Eat Crow," undated report, Baranov Museum, Kodiak, Alaska; "Pacific Cod-Liver Oil is Found High in Vitamin D Potency," *Pacific Fisherman* (January 1939): 56.

[52] John N. Cobb, *Pacific Cod Fisheries* (revised edition, 1926), Bureau of Fisheries Doc. No. 1014, appendix 7 to *Report of the U.S. Commissioner of Fisheries for 1926* (Washington, DC: GPO, 1927), 410, 436, 439, 452–453; John S. Hittell, *Commerce and Industries of the Pacific Coast of North America* (San Francisco: A. L. Bancroft, 1882), 345.

[53] Charles Butler, *The Fish Liver Oil Industry*, U.S. Fish and Wildlife Service Leaflet No. 233 (March 1948), 81; F. B. Sonford and H. W. Nilson, "Vitamin A and D Potencies in the Liver of Pacific Cod (*Gadus Macrocephalus*)," *Commercial Fisheries Review* (U.S. Fish and Wildlife Service, May 1949): 13–15.

[54] "Demand Increasing for Vitamin Oils," *Pacific Fisherman* (November 1939): 69; Ed Shields, *Salt of the Sea: The Pacific Coast Cod Fishery and the Last Days of Sail* (Lopez Island, Wash.: Pacific Heritage Press, 2001), 132; "Codfish," *Pacific Fisherman* (January 1948): 237; "'Wawona' Cod-Liver Return Disappointing," *Pacific Fisherman* (January 1948): 65.

[55] "Just Because the Chowder was a Little Bit Sticky," *Anacortes American*, June 22, 1911; "About LePage," accessed January 16, 2015, http://www.lepageproducts.com/home.aspx#; "Nearly 8,000 Gallons of Codfish Glue," *Anacortes American*, February 18, 1909; Ed Shields, *Salt of the Sea: The Pacific Coast Cod Fishery and the Last Days of Sail* (Lopez Island, Wash.: Pacific Heritage Press, 2001), 187; Richard S. Croker, "Alaska Codfish," in *The Commercial Fish Catch of California for the Year 1929*, Division of Fish and Game of California, Fish Bulletin No. 30 (Sacramento: California State Printing Office, 1931), 48–55.

Chapter 2
EARLY DEVELOPMENT OF THE PACIFIC CODFISH INDUSTRY

In its cod fisheries, Alaska is undoubtedly destined to lead the world, if supply and accessibility are worth anything in computation. The shallow shores of East Behring Sea and the submarine plateaus extend in almost every direction from Alaskan shores and simply swarm with codfish. To compare them with the Atlantic banks would be like comparing the population of China with that of Hudson's Bay Territory.

—Lieutenant Frederick Schwatka, U.S. Army, Alaska explorer, 1886[56]

The known area and productiveness of the various cod banks of Alaska exceed those of Newfoundland and the Atlantic coast generally.

—Lyman E. Knapp, governor of Alaska, 1889[57]

Notwithstanding the fact that the banks are in the North Pacific, the codfisheries are really a San Francisco enterprise, that city being the "home office" of the enterprise.

—Thomas J. Vivian, U.S. Treasury Department, 1891[58]

The codfish banks are extensive and inexhaustible.

—James Sheakley, Governor of Alaska, 1894[59]

Off our Alaska coast are to be found the most extensive cod banks in the world. . . . [I]t has generally been a question of how many [cod] can be sold and not how many can be caught.

—John Cobb, founding director, College of Fisheries, University of Washington, 1918[60]

CAPTAIN MATTHEW TURNER: SALT COD PIONEER

In the years following the discovery of gold in central California in 1848, San Francisco grew quickly from a sleepy hamlet into a thriving commercial center. Many of those who migrated to California during the famed gold rush were of western European, often Catholic, ancestry. For them, salted codfish was a dietary staple. East Coast merchants early on began supplying California with salted Atlantic codfish shipped via the Isthmus of Panama or Cape Horn. The volume grew to be considerable: a total of about 1,000 tons of salted Atlantic cod were imported into San Francisco during the years 1863 and 1864.[61] The cod fishery that developed in Alaska began as an effort to feed the growing California market.

Captain Mathew Turner, the pioneer of the U.S. Pacific cod fishery, was not a professional fisherman but an opportunistic merchant who had some familiarity with the northern seas. In 1857, Turner sailed the 120-ton brig *Timandra* from San Francisco, carrying an assortment of cargo for the Russian port of Nicolaevsk, on the Amur River, which flows into the Okhotsk Sea. It was early in the navigation season, and Turner was detained for three weeks at Castor Bay, waiting for ice to go out of the river. During the wait, the *Timandra*'s crew, as a pastime, began fishing with handlines over the vessel's rail. They were surprised at the abundance of codfish, which averaged about two feet in length. Turner himself had never previously seen codfish, but he was aware of their market value in San Francisco.[62]

In 1859, Turner made another trading voyage to the Amur River, and during that voyage he found cod in great abundance near Sakhalin Island. Having no means to preserve the fish, his crew caught only enough for the ship's use. On a similar trading voyage to the Amur River in 1863, however, the *Timandra* carried fishing gear and twenty-five tons of salt. Turner intended to catch and cure codfish on the return voyage. He found fish initially plentiful in the Strait of Tartary, which connects the Sea of Japan with the Okhotsk Sea, and in a few days of fishing from the vessel's deck, the crew caught ten tons, which they salted in *kenches* (wooden bins). Then the fish suddenly disappeared. Turner found codfish again along the Kamchatka Peninsula coast, but during the second day of fishing both of the ship's anchors were lost, compelling Turner to abandon the fishing venture and to sail for San Francisco. The *Timandra*'s thirty-ton cargo of salted cod—the first-ever cargo of salted cod from the Pacific fishing grounds to be landed on the U.S. West Coast—was dried on Yerba Buena Island, in San Francisco Bay, and then sold locally for fourteen cents per pound.

In 1864, Turner once again sailed the *Timandra* to the northern seas, but this time the vessel carried no trading goods; it was outfitted to catch codfish, which made it the first dedicated fishing vessel from a West Coast port to enter the Pacific codfish industry. The voyage was successful: after fishing the same grounds as the previous year, Turner returned to San Francisco with 100 tons of salted codfish, which he sold at a profit.

Approximately coterminous with Turner's voyage, the year 1863 or 1864—the reports are unclear—marked the first American effort to catch cod in the Bering Sea. During one of those years, the schooner *Alert* journeyed to Bristol Bay, likely primarily to trade with Natives. During the voyage, however, the *Alert* prospected for codfish and it returned to San Francisco with nine tons of salted cod. This might have seemed like a modest amount, but it represented a catch of perhaps six thousand individual fish.[63]

AN INDUSTRY IS BORN

Captain Turner's success inspired others, and in 1865 six fishing vessels sailed from San Francisco to the Okhotsk Sea. All the vessels were small schooners that had been built in New England for the Atlantic fisheries, but had made the journey around Cape Horn to find employment on the Pacific coast. The vessels' names and tonnages were: *Equity* (63 tons), *Flying Dart* (94 tons), *R. L. Ruggles* (75 tons), *J. D. Sanborn* (71 tons), *Mary Cleveland* (91 tons), and *Taccon* (20 tons).[64] As J. W. Collins, who later headed the U.S. Fish Commission's Division of Fisheries, explained, "There was no lack of cod, and even with the method of fishing with hand lines over the vessel's side then in vogue, no difficulty was experienced in filling moderate-sized schooners in a reasonable time."[65]

Turner himself chose to look for codfish in Alaska waters. In late March 1865, he sailed for Alaska on the 45-ton schooner *Porpoise* and arrived at the Shumagin Islands, off the southern shore of the Alaska Peninsula, on May 1. Fishing was good, but, in anticipation of a glutted market when the Okhotsk Sea fleet returned, Turner sailed for San Francisco before filling his vessel to capacity. The *Porpoise* arrived in San Francisco on July 7 with thirty tons of salted cod. It was the first-ever cargo of codfish delivered from the Shumagin Islands, which themselves became an important center of the Alaska codfish industry.[66]

The growth of the San Francisco codfish industry did not go unnoticed in Washington Territory. In January 1866, the territory's legislature passed a memorial that pointed out to President Andrew Johnson the "abundance of cod-fish, halibut and salmon, of excellent quality" along Alaska's coast, and asking that he

> obtain such rights and privileges of the Government of Russia, as will enable our fishing vessels to visit the ports and harbors of its [Alaska] possessions, to the end that fuel, water and power

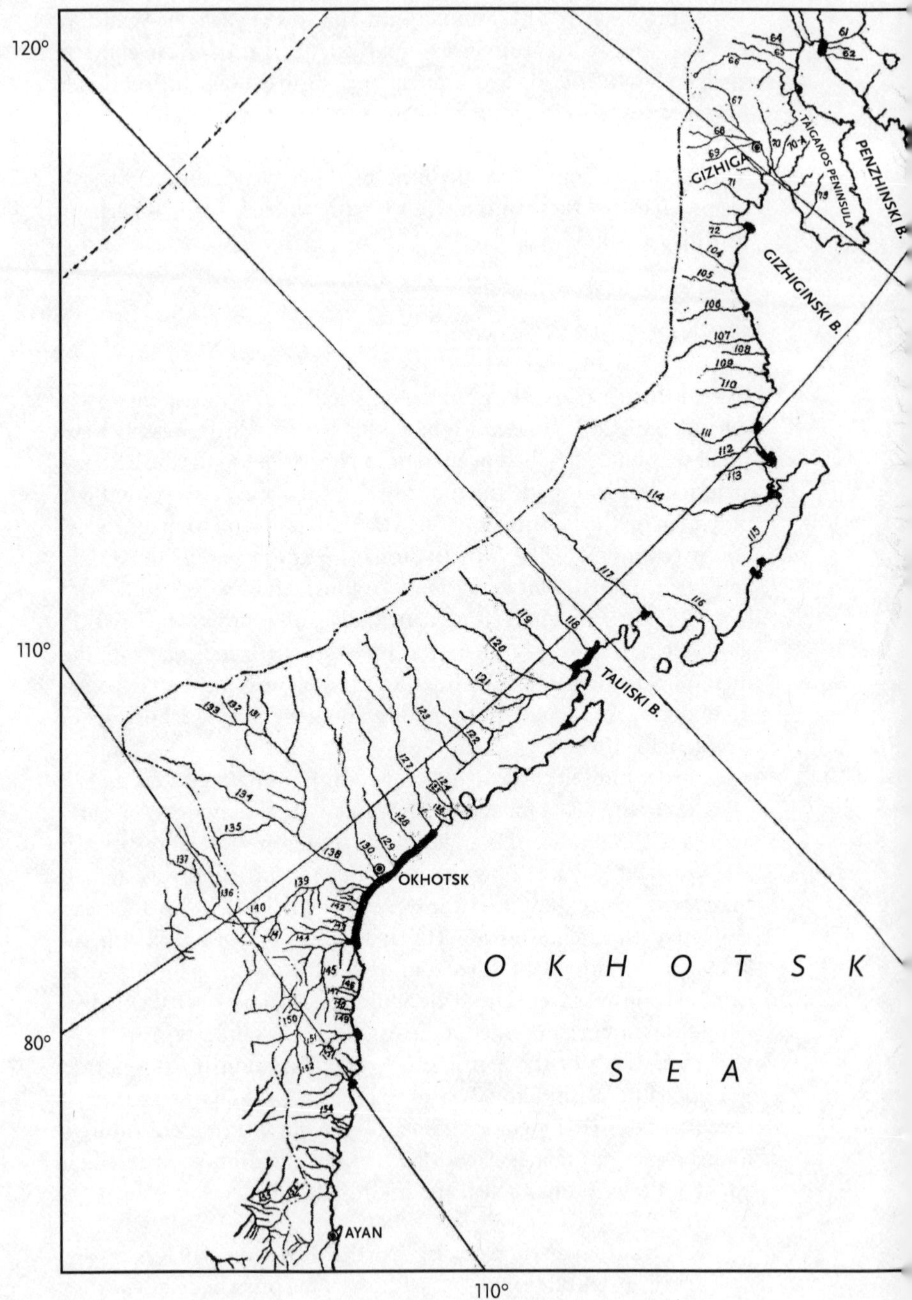
120°
110°
80°
110°
PENZHINSKI B.
TAIGANOS PENINSULA
GIZHIGA
GIZHIGINSKI B.
TAUISKI B.
OKHOTSK
AYAN
O K H O T S K
S E A

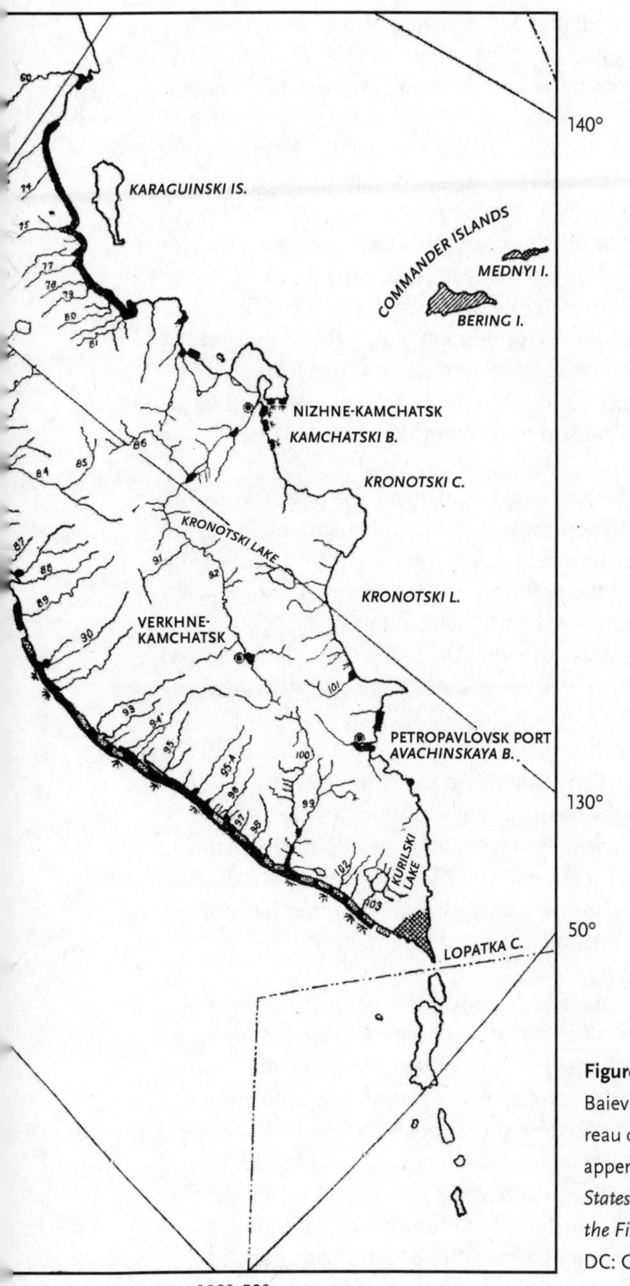

Figure 10. Okhotsk Sea. (Boris Baievsky, Fisheries of Siberia, Bureau of Fisheries Doc. No 1006, appendix 2 to *Report of the United States Commissioner of Fisheries for the Fiscal Year 1926* [Washington, DC: GPO, 1927], fig. 2.)

may be easily obtained; that our sick and disabled fishermen may obtain sanitary assistance, together with the privilege of curing fish and repairing vessels in need of repairs.[67]

The Washington legislature also requested copies of "such fishing licenses, abstract journals and log books as will enable our hardy fishermen to obtain the bounties [income] now provided and paid to the fishermen in the Atlantic states," and that the federal government detail a Navy ship to explore and survey the fishing banks of the Pacific coast from California to the Bering Strait. The *New York Times* considered this memorial to be the foundation of the treaty under which the United States the following year purchased Alaska from Russia.[68]

In March 1866, fully eighteen vessels—a combination of barks, brigs, and schooners, each carrying from three to six dories and a crew of ten to eighteen men—departed San Francisco for the northern fishing grounds. The vessels did not reveal their exact destinations, but planned to remain on the grounds until about September. In all, the vessels were reported to have caught 706,200 cod, with a salted weight of 1,614 tons. (Based on these figures, the average weight of a salted cod was nearly 4.6 pounds—exceptionally high.) Notably, a portion of the catch—255 tons—was caught in Alaska, on grounds discovered by vessels en route to the Okhotsk Sea, with the best catches coming from the vicinity of the Shumagin Islands.[69]

The quality of the delivered product, however, was an issue. The aforementioned George Davidson noted that "[m]any of the persons engaging in the [codfish] business knew nothing of the manner of catching or of curing the fish." Problematic also was the lack of suitable curing facilities at San Francisco or elsewhere on the West Coast.[70] In 1906, C. P. Overton, secretary of San Francisco's Union Fish Company, described the early years of his city's codfish industry:

> [The Californians] were a rough and ready people and when they went into the codfish business they went in on the same plan. Any old vessel would do, any kind of salt would do, any sort of tanks or butts [barrels] in any sort of shed would do. The fish were carelessly cured with stained and dirty salt; they were kept in leaky butts or were even piled up on the floor in dry salt, exposed to the air and not in pickle at all. . . . Under such conditions the finest kind of codfish are necessarily turned out yellow, hard, sour, tough, everything that is bad and such undesirable goods naturally are salable only to camps and among the poorest and least fastidious consumers.[71]

The aforementioned John Cobb wrote that "[i]n certain cases the [early] fishermen received their share of the voyage in fish, which, after being cured in a good, bad or indifferent manner by themselves, were hawked around the city."[72]

During 1867, a record twenty-three San Francisco–based schooners engaged in the Pacific cod fishery. Their combined catch was 947,264 fish, which amounted to 2,164 tons (more than 4 million pounds) of dry-salted product. Based on these figures, the average weight of a salted cod would have been nearly 4.6 pounds, similar to the exceptionally high average of the previous year.

The large catch of 1867 temporarily overstocked the San Francisco market, and the following spring only ten vessels engaged in the fishery. To help relieve the glutted market, a portion—about 50,000 pounds—of the previous year's catch was exported to New York.[73] This was likely the first substantial shipment of salted Pacific cod to the East Coast. Such shipments would continue, but irregularly and with mixed results, as will be discussed later in this work.

The year 1867 also marked what was said to be the establishment of the first permanent business on the West Coast devoted exclusively to the fish trade. Thomas W. McCollam, of San Francisco, saw opportunity in conducting a codfish business "on the most approved methods"—as opposed to the commonly slipshod operations of the time—and bought his first cargo of cod that year. The following year, McCollam journeyed to New England, where he purchased the fishing schooners *Wild Gazelle*, *Flying Mist*, and *Rippling Wave*. The three vessels departed for San Francisco, but the *Rippling Wave* was lost while transiting the Strait of Magellan. The *Wild Gazelle* made its inaugural trip to the Shumagin Islands in 1870, and the *Flying Mist* made its journey the following year. In 1873, McCollam took a partner into his business, which then became Thomas W. McCollam & Company. The following year, the schooner *Alfred Adams* was added to the company's fleet. The fleet's production was augmented with purchases of salted cod from other vessels.[74]

McCollam initially cured cod at Sausalito (just north of San Francisco), but the location was not satisfactory and he soon moved his operation to a site near the mouth of Redwood City Creek, about thirty miles south of San Francisco, where he constructed wharves, storehouses, and *flake* (curing) yards. The flake yards were covered with lattice-like outdoor drying racks—known in the trade as *flakes* or *flakeboards*—that were used to dry fish during sunny weather.[75]

In 1867, more than half of the season's catch was taken in the vicinity of the Shumagin Islands. According to William Healey Dall, an early scientific explorer of Alaska, the quality of the Shumagin Islands cod was higher than those from the Okhotsk Sea.[76] An 1891 U.S. Treasury Department report agreed; it described Okhotsk Sea cod as being "long and thin," while those from the Shumagin Islands were "short and thick." The report went on to praise Shumagin Islands cod as "the best of the Pacific cod, and superior to those of the Labrador coast, while the finer specimens are equal to the best of the Newfoundland fisheries."[77]

By 1870, most of the San Francisco fishermen had ceased traveling to the Okhotsk Sea. Distance was a factor: the distance from San Francisco to the Shumagins—about 1,900 miles—is about half the distance from San Francisco to the Okhotsk Sea. Thus, a cod-fishing expedition to the Shumagins usually lasted about 110 days, while an expedition to the Okhotsk Sea averaged about 170 days. Furthermore, the grounds fished in the Okhotsk Sea were from ten to forty miles offshore. In 1880, ichthyologist Tarleton Bean noted that "the Okhotsk fisherman is cut off from fresh provisions and good harbors; he rides out storms 'hove to' or trusting to his anchors," while in the Shumagins safe harbors were close at hand and there were better opportunities to obtain fresh provisions, wood, and water. Moreover, there were no bothersome Russian officials (see chapter 7).[78]

In part because of the great distance to the grounds, cod-fishing schooners usually made one trip north per year, leaving in the spring and returning in the late summer or early fall. However, if a vessel met with good fortune on the first trip, sometimes a second one would be made. The first vessel to make a second trip was the schooner *Porpoise*, in 1868. That trip, however, yielded only a half cargo. By contrast, because the Grand Banks, near Newfoundland, are closer to fishing ports, East Coast cod-fishing vessels typically made two trips each year, making the Atlantic industry potentially more profitable than the Pacific industry for codfish companies and fishermen alike.[79]

In 1869, overproduction—coupled with chronically inconsistent and often inadequate methods of curing—was once more a problem, a consequence of which was that hundreds of tons of spoiled fish were thrown overboard at San Francisco. Nevertheless, in 1870 fully twenty-two vessels engaged in the fishery and caught a record 1,467,000 codfish.[80] As with the previous year, this production was in excess of what could be absorbed by the West Coast market, and producers looked to selling the surplus product in midwestern and eastern states as well as abroad. (The first transcontinental railroad, which connected San Francisco to the eastern U.S. railroad system, had begun operating in May 1869.)

J. W. Collins explained the pattern of entry into and egress from the Pacific salt cod fishery:

> Paradoxical as it may seem, for some years the season of exceptional success was often the cause of disaster. Large profits generally created a temporary "boom." Firms or individuals hastened to engage in the fishery. Frequently, sufficient care was not exercised in selecting men and vessels. Generally the market was much overstocked at the close of the season. Prices dropped far below the point where they gave remunerative returns to investors. Too often the products could scarcely be sold at any price because of the excess of supply over demand. The result was necessarily disastrous, and those who had hastened to engage in an enterprise because others had been "lucky" usually abandoned it with the utmost precipitation, leaving the field only to those whose "luck" or experience enabled them to succeed under conditions that ruined or discouraged their competitors.[81]

Though the East Coast producers had lost control of the Pacific coast codfish market, San Francisco–based cod interests soon found, to their dismay, that their eastern competitors would continue to control the market throughout the rest of the United States. A certain prejudice worked to the East Coast producers' advantage: as noted in the official report of the 1888 voyage of the U.S. Fish Commission's steamer *Albatross* to determine the extent of the cod-fishing grounds in western Alaska (see below), Pacific cod suffered from "the universal prejudice in favor of supplies coming from old and well-known sources" on the Atlantic coast.[82]

East Coast interests fought hard—and not always honestly— to prevent further intrusion of Pacific cod into their traditional markets. Fish merchants and customers were told that Pacific cod were not cod or were an inferior grade of cod, and that they would not keep. Misstatements such as these made a considerable impression, and the perception that Pacific cod is inferior to Atlantic cod persists to this day.

Unfortunately, the slipshod manner by which some Pacific coast processors prepared their products only reinforced the misstatements. Making matters worse, some of the fish was shipped across the country in ordinary boxcars during the warm season and arrived in very poor condition. The shippers quickly learned their lesson, but the experience did not encourage the few dealers who had been willing to give Pacific cod a trial.[83]

As an "inferior" product, the broad fortunes of the Pacific cod fishery waxed and waned inversely with codfish production in the Atlantic. During lean years on the Atlantic, salt cod from the Pacific sometimes found its way even into the New England market, but during years in which production was good on the Atlantic, the only way to move Pacific cod into the general market was to lower prices, sometimes to levels below the cost of production.

In 1871, the number of San Francisco–based vessels engaged in the cod fishery diminished to eleven, and during the years 1872 to 1878, the number of vessels further diminished, ranging between five and seven. The yearly catch ranged from 300,000 to 550,000 fish, which, if the live weight of the fish averaged 10 pounds, would have amounted to from 3 million to 5.5 million pounds. By contrast, the live weight of Atlantic cod taken by New England fishermen in 1880 was almost 231 million pounds.[84]

Treasury Department agent Henry W. Elliott, in his 1875 report on the conditions and affairs in Alaska, painted a grim picture of Alaska's codfish industry and its potential. According to Elliott, "the quantity and quality [of cod] are insufficient in a business point of view," and "the [cod] fishing grounds are not valuable enough to induce capitalists to engage in taking and curing fish for exportation."[85] Elliott, his report perhaps colored because it was written during a period of decline in the Pacific industry, was wrong; the industry grew considerably over the next four decades, although not without disheartening downturns.

Salt

> *Salt continues to be to a large extent the cornerstone of the fish business.*
>
> —*Pacific Fisherman*, 1916[86]

Almost all the salt used in the U.S. Atlantic coast codfish industry was imported from Europe's Mediterranean Sea region. It was *solar salt*: salt produced by solar evaporation of seawater. During the early years of the twentieth century, about 25,000 tons of salt were used annually at Gloucester alone.[87]

Salt used in the U.S. Pacific coast codfish industry was, likewise, solar salt. It was produced at plants on southern San Francisco Bay, where abundant sunshine and wind during the months of May through October made conditions ideal for the process. About forty tons of seawater was required to produce one ton of salt. A typical salt plant described by *Pacific Fisherman* in 1916 included seven shallow reservoirs defined by fifteen miles of levees. The reservoirs encompassed some 1,600 acres.

In the salt-making process, gates were opened into a nine-hundred-acre reservoir (reservoir No. 1) to allow seawater to flow in. Once the reservoir was filled, the gates were closed and the water was allowed to stand for a couple of weeks to allow its organic contents to precipitate out. Windmills then pumped the water into reservoir No. 2, where solar evaporation of the water concentrated the salt. The remaining water was then pumped sequentially through four increasingly smaller reservoirs, in each of which the evaporation process was repeated. When the water that was left reached reservoir No. 7, the final reservoir, the salt in it was so concentrated that it crystallized and could be collected. During a typical salt-making season, the plant described could produce about 20,000 tons of salt, which was praised by *Pacific Fisherman* as "without doubt the best and most suitable for the curing and salting of fish."[88] Actually, the salt produced at San Francisco Bay was of low quality because of the chemical and organic characteristics of the water from which it was made.

The four prominent San Francisco Bay salt manufacturers were the Morton Salt Company, the Leslie-California Salt Company, the Oliver Salt Company, and the Arden Salt Company. In 1925, the Arden Salt Company constructed a 300-foot *salt dock* on Seattle's waterfront specifically to accommodate the needs of the fishing industry. The company's biggest customer was likely the herring industry, which was followed by the salmon industry and then the codfish industry.

To cure their catch, cod-fishing vessels and shore stations required about 1 ton of coarse salt per one thousand fish. Schooners in 1880 typically carried between 30 and 200 tons of salt, but the larger vessels used a half-century later carried 350 to 500 tons. The cost of salt in San Francisco in 1888 ranged from $7.50 to $8.50 per ton.[89]

In cod fishermen's parlance, a schooner that "came home with all its salt wet" meant the vessel had utilized its entire supply of salt and was carrying a full or nearly full cargo of fish.[90]

Figure 11. Union Fish Company codfish plant. (Courtesy of Union Fish Company.)

UNION FISH COMPANY

In 1876, Thomas W. McCollam & Company moved its operation to Belvedere Island, on Richardson Bay, about five miles north of San Francisco. The facility, called Pescada Landing, included wharves, fish houses, and nearly 14,000 square feet of flake yards. The main building had two stories and measured 100 by 220 feet on the ground floor, which was occupied mostly by rectangular wooden tanks that each held about twelve tons of fish.[91] In 1883, Thomas W. McCollam & Company was renamed the McCollam Fishing & Trading Company.[92]

In about 1898, the McCollam Fishing & Trading Company, with another San Francisco codfish company, Lynde & Hough, and perhaps some additional firms, formed the Union Fish Company as a selling agency for their product.

According to C. P. Overton, general manager of the Union Fish Company in the early 1900s, Lynde & Hough was originally headed by "two enterprising Yankees of the old school" who had started in the codfish industry by selling salted cod on commission, and who would do "anything else that promised a dollar."[93] In 1865, the firm began producing codfish, and, over the years, built up a fleet of vessels

Figure 12. Processing salted cod at Union Fish Company's Belvedere Island codfish plant. (Courtesy of Union Fish Company.)

and established several shore stations in Alaska. In 1872, Lynde & Hough established an outfitting and curing station called California City on a fifty-acre site on the Tiburon Peninsula, about eight miles north of San Francisco.

The facility included a wharf and a large two-story warehouse that was flanked by sheds and other buildings. Each of thirty-six redwood brine tanks on the premises could hold fifteen tons of fish, and the drying racks could accommodate nine tons. During the late 1880s, California City employed from thirty to seventy-five men throughout the year. Among Lynde & Hough's ancillary products were Okhotsk Sea Cod Liver Oil and Dr. Fisherman's Lotion for Man and Beast.

The deaths of Lynde & Hough's senior partners precipitated the merger in 1902 or 1903 of their company with McCollam Fishing & Trading, and the whole business began operating as the Union Fish Company, which remains in the seafood business today. The California City curing station was not incorporated into the merger; it was sold to the U.S. Navy for use as a coaling station.[94] The company's Pescada Landing curing plant on Belvedere Island was renamed Union City.

CODFISH IN ALASKA WATERS

> The only commercial marine fisheries which have been developed in Alaskan waters are those for the cod, and for the seal and other aquatic mammals.
>
> —Richard Rathbun, U.S. Fish Commission, 1888[95]

The Okhotsk Sea, largely because it was familiar grounds and had a record of relatively consistent production, continued to be fished by American vessels, but with decreased regularity and intensity, until 1909, when American fishing in the Okhotsk Sea ceased altogether. (see chapter 7).[96]

The first Alaska-based vessel in the cod fishery was owned by a Captain Haley, of Wrangell, in Southeast Alaska. In 1879, Haley caught cod in Frederick Sound, in central Southeast Alaska, and sold his catch in Wrangell for $100 per ton. In the early 1880s, Haley also purchased for two cents apiece about ten thousand cod from Native fishermen at Killisnoo, near Angoon, in northern Southeast Alaska. The fish were caught in northern Chatham Strait, and Haley dried them and rendered their livers into cod liver oil at the former Northwest Trading Company whaling station at Killisnoo. The dried fish averaged three pounds in weight, and the cod liver oil was said to be of excellent quality. Haley, however, apparently found it difficult to market his products, and he chose not to pursue further his venture into the codfish trade.[97]

Although Pacific cod would continue to be harvested in Southeast Alaska, the quantities would be relatively small.

Until 1882, the American cod-fishing effort in the Pacific had focused on the Okhotsk Sea and on the western Gulf of Alaska, along the southern shore of the Alaska Peninsula. No vessels had gone to the Bering Sea in search of codfish since the voyage of the trading schooner *Alert* in 1864 (see above). The situation changed when the Thomas W. McCollam & Company's schooner *Tropic Bird* inaugurated a regular Bering Sea cod fishery.[98] This vessel was followed several weeks later by the schooner *Isabel*, which was operated by an opportunistic group of San Francisco businessmen. Both vessels made good catches, and, as a result, the next year five vessels fished in the Bering Sea. Unfortunately, the owners of the *Isabel* did not keep it in good repair, and, en route to the fishing grounds in the spring of 1888, the vessel opened up at sea. According to C. P. Overton, of the McCollam Fishing & Trading Company, "As many of the crew as could do so got into the dories, and after suffering many privations about half of them were rescued more nearly dead than alive. [Fourteen men perished.] This ended the venture, and the partners paid up their losses and quit."[99]

Finding commercial quantities of codfish in the Bering Sea presented challenges. As noted in an 1884 letter by C. P. Overton, "The codfish found so far have been in spots, and although well-defined banks undoubtedly exist, they have never been prospected as thoroughly as they should be. The coast, too, is so little known that vessels are obliged to proceed with extreme caution, especially during the foggy weather, prevailing a great part of the summer."[100] When a vessel located good grounds, secrecy became the rule. During the 1885 season, the McCollam vessel *Helen W. Almy* made a good catch on Bering Sea grounds, the location of which the captain kept as a "profound secret."

EARLY FISHERIES EXPLORATIONS IN ALASKA

The United States negotiated the purchase of Alaska from Russia in March 1867. In July of that year, the 165-foot Revenue Cutter *Lincoln*, under Captain W. A. Howard, was dispatched to Alaska to "acquire a knowledge of the country with a view to the due protection of the revenue when it shall have become a part of the United States, and for the information of Congress and the people." In early September, Howard was looking for fishing banks east of Akutan Pass, in the Aleutian Islands. He had directed his crew to take a depth sounding every two hours, and located an "excellent bank" where his crew quickly caught 120 of the "finest and fattest codfish" the captain had ever seen. The fish weighed from eighteen to thirty-five pounds each and were from thirty-two to thirty-seven inches long.[101]

The formal transfer of ownership of Alaska to the United States was completed in October 1867. The acquisition of this vast territory presented challenges for a country burdened with the costs of post–Civil War reconstruction. The government's priorities in Alaska were to establish a presence and to assert authority in its new acquisition to prevent Alaska's resources—to the extent they were known—from being plundered. The government also wanted to learn more about Alaska. To accomplish these goals, the U.S. Coast Survey dispatched the Revenue Cutter *Wayanda* to Alaska in April 1868. Specifically, the *Wayanda*'s mission was to discourage overhunting of fur seals on the Pribilof Islands and to survey the coast.

The 138-foot steamer, under Captain John White, spent some six months cruising Alaska's coast. Among other resources, White found promising quantities of halibut and cod. In his report of the cruise, White wrote of needing a supply of fish for the galley's fare while sounding just south of Kodiak Island, and, in his words

> ordered the sails set back and the lines prepared. What bait? I had a barrel of Puget Sound clams salted for me with this purpose. I took my lead-line as large as my thumb, attached five hooks above the lead, with a clam on each, and fastened it to the davit. Soon the bites, one, two, three, often five were felt. I threw the line over the pulley and put four men to pull, and up would come two, three, and sometimes five cod weighing 30 to 40 pounds apiece. We had out about 20 lines, and caught 250 fish in two hours.[102]

The piecemeal exploration of Alaska's marine waters by cod fishermen and various government vessels was augmented greatly by the fisheries research voyages of the U.S. Fish Commission's steamer *Albatross* in 1888 and 1890.[103] Richard Rathbun, a prominent American scientist associated with the Smithsonian Institution, hailed the endeavor as an "innovation in the support given by government to the development of this particular industry."[104] Rathbun added that "[n]o foreign nation has ever attempted, on more than a very limited scale, to enlighten the fishermen respecting the character, distribution, and abundance along its coasts of the aquatic forms of life which are the objects of their pursuit."[105]

Constructed in 1882 as a deep-sea oceanographic vessel, the 234-foot, iron-hulled, twin-screw *Albatross* was powered by a pair of coal-fired steam engines and was also rigged for sailing. The vessel spent its first five years researching fisheries in the Atlantic Ocean, but in late November 1887—its research value fully established and appreciated—the *Albatross* was dispatched to the Pacific coast to investigate the North Pacific's fishing grounds and to furnish accurate information to the U.S. fishing industry regarding their locations, characteristics, and resources.[106]

The *Albatross* arrived at San Francisco in May 1888. Given that good weather in Alaska could not be expected to continue beyond the summer months, it was decided to send the vessel to Alaska first, after which it would return south to investigate the coasts of Washington Territory, Oregon, and California.[107] On July 4, under the command of U.S. Navy Lieutenant Commander Z. L. Tanner, who had been in charge of all the *Albatross*'s operations on the Atlantic coast, the vessel departed San Francisco for Alaska. This voyage was the first systematic inquiry regarding the distribution and value of Alaska's fish resources.

Figure 13. Pacific cod caught during the U.S. Fish Commission's oceanographic vessel *Albatross*'s 1890 voyage to Alaska. (NOAA Fisheries.)

Specifically, the object of the *Albatross*'s Alaska voyage was to "explore the waters adjacent to the [easternmost] Aleutian Islands and Alaska Peninsula, for the purpose of ascertaining the character and extent of the cod and halibut fishing grounds, and of obtaining all possible information regarding the fishing interests and resources of that region."[108]

The *Albatross* arrived at Unalaska on July 23, 1888, nineteen days after departing San Francisco. The ship's fundamental exploratory process involved running lines of sounding over as much of the region as possible to determine the boundaries and contours of potential fishing banks. The richness of the banks' bottoms—as measured by the character and abundance of animal life and the presence, abundance, and size of edible fish, especially cod—was ascertained by frequent dredging to obtain bottom samples, and by the use of handlines and longlines to catch fish.[109] During the approximately five weeks spent in Alaska waters in 1888, the crewmen aboard the *Albatross* caught, measured, weighed, and sexed nearly three hundred cod.[110] Additionally, harbors and anchorages were described, hazards to navigation were noted, and the operations of the few cod-fishing and salmon-salting stations in the region were described.

In 1890, the *Albatross* made its second voyage to Alaska. The value of Alaska's salmon resources was not yet fully appreciated, and the vessel again focused mainly on cod. The *Albatross*'s 1898 and subsequent voyages to Alaska, however, would focus on salmon, by then Alaska's most prominent and important fishery.[111]

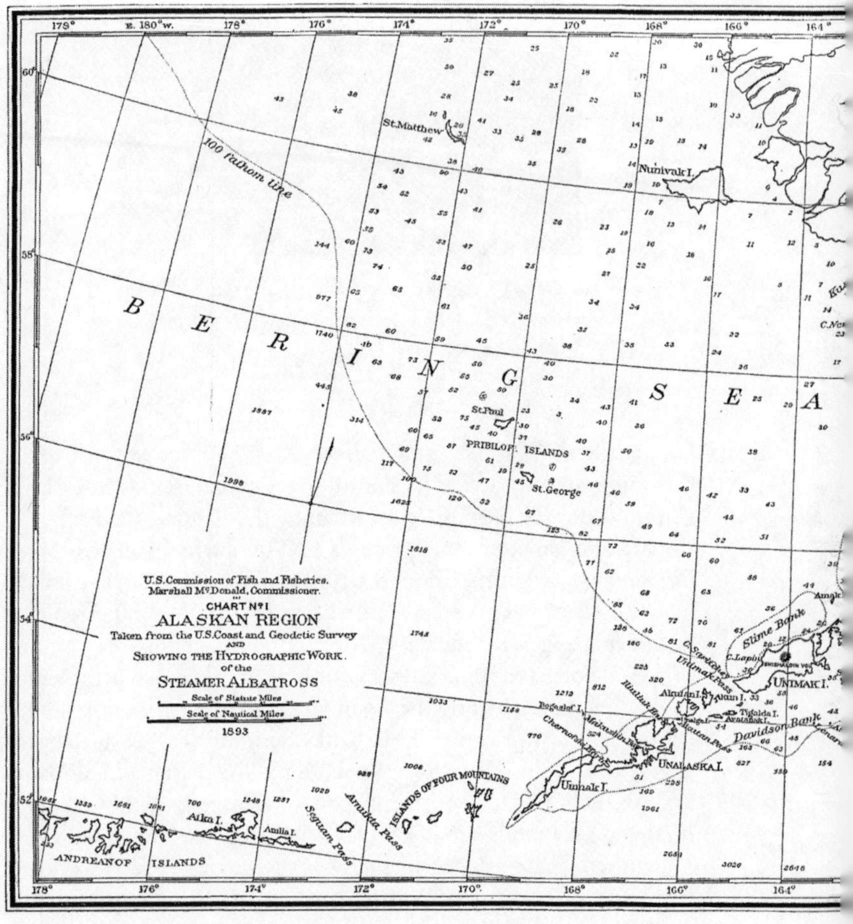
U.S. Commission of Fish and Fisheries.
Marshall McDonald, Commissioner.
CHART No 1
ALASKAN REGION
Taken from the U.S. Coast and Geodetic Survey
AND
Showing the Hydrographic Work.
of the
Steamer Albatross
Scale of Statute Miles
Scale of Nautical Miles
1893
B E R I N G S E A
100 fathom line
St. Matthew
Nunivak I.
St. Paul
PRIBILOF ISLANDS
St. George
Slime Bank
Unimak Pass
UNIMAK I.
Akun I.
Tigalda I.
Davidson Bank
UNALASKA I.
Umnak I.
Bogoslof I.
ISLANDS OF FOUR MOUNTAINS
Amukta Pass
Seguam Pass
Atka I.
Amlia I.
ANDREANOF ISLANDS

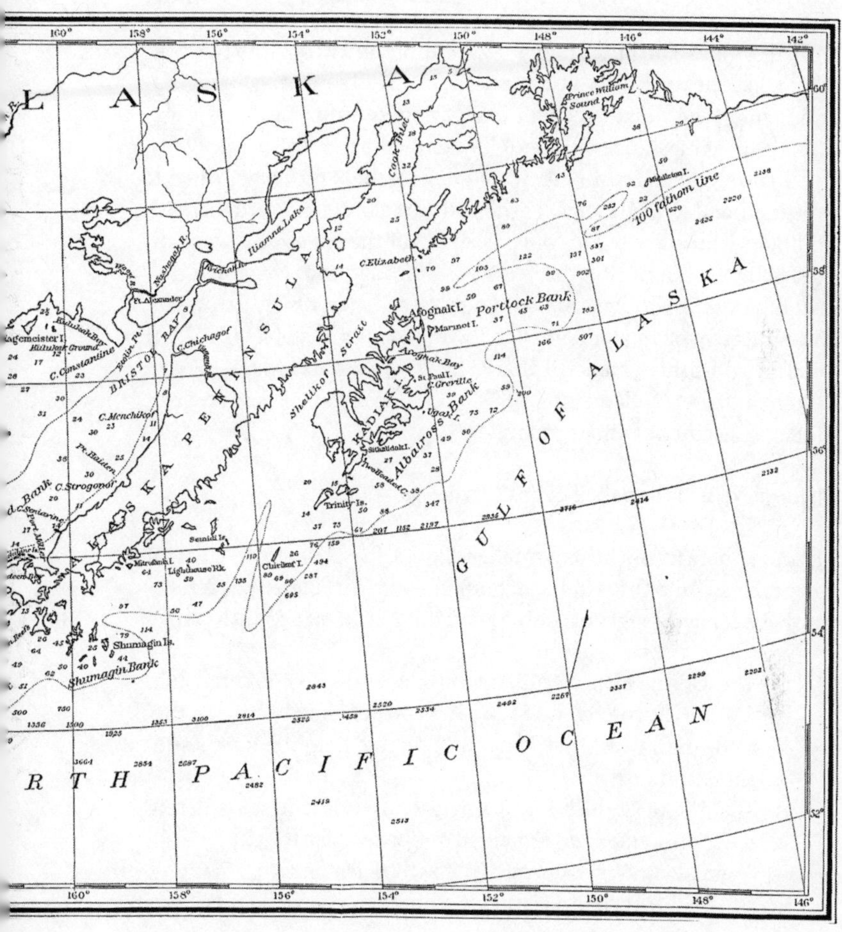

Figure 14. Central and Western Alaska fishing banks, 1893. (U.S. Fish Commission.)

The goal of the *Albatross*'s 1890 voyage, during which nearly three months were spent in the Bering Sea, was to learn more about cod, specifically "to determine the positions and characteristics of the more important cod-fishing grounds"[112] in the Bering Sea and south of the Alaska Peninsula, as well as the waters east of Kodiak Island (Albatross and Portlock banks). In addition to its cod survey work, which, like the work of two years earlier, involved sounding, dredging, and fishing, the *Albatross* spent nearly a month during this voyage doing a reconnaissance of Bristol Bay's shoreline.

In total, the *Albatross*'s crewmen during the 1890 voyage caught, measured, and weighed almost 1,500 cod. The aforementioned Richard Rathbun, who authored an official report of the *Albatross*'s work in the Pacific Ocean during the years 1888 through 1892, regarded the results of the *Albatross*'s 1890 cod survey as "exceedingly favorable and indicative of an abundant supply of good, marketable fish." The primary cod-fishing banks exploited by U.S. fishermen in the Bering Sea circa 1911 extended along the shore from Unimak Pass to Port Moller, a distance of about 250 miles.[113]

Major players in the salted codfish trade

Although there were a host of smaller players, five businesses—two in San Francisco and three in Puget Sound—by virtue of their degree of involvement and longevity, dominated the Alaska salt codfish trade:

1. The Union Fish Company, of San Francisco, entered the Alaska salt codfish trade in 1867 and remained active in the trade until 1938. This company operated both schooners and shore stations.
2. The Alaska Codfish Company, also of San Francisco, entered the trade in 1893 and remained until 1929. Like the Union Fish Company, this firm operated both schooners and shore stations.
3. J. A. Matheson, of Anacortes, Washington, entered the schooner fishery in 1891 and operated until 1930.
4. The Robinson Fisheries Company, also of Anacortes, Washington, entered the schooner fishery in 1904 and operated until 1947.
5. The Pacific Coast Codfish Company, of Poulsbo, Washington, entered the schooner fishery in 1911. This firm saw the Salt Cod Era to its end, in 1950.

ENDNOTES

[56] *Bulletin of the United States Fish Commission, Vol. VI, for 1886* (Washington, DC: GPO, 1887), 462.

[57] *Report of the Governor of Alaska for the Fiscal Year 1889* (Washington, DC: GPO, 1889), 20.

[58] Thomas J. Vivian, *Commercial, Industrial, Agricultural, Transportation, and Other Interests of California* (Washington: Bureau of Statistics, 1891), 486.

[59] *Report of the Governor of the District of Alaska to the Secretary of the Interior, 1894* (Washington, DC: GPO, 1894), 10.

[60] John N. Cobb, "Transportation Gave First Impetus to Pacific Fisheries," *Fishing Gazette* (July 6, 1918): 954, 956, 958.

[61] William H. Dall, *Alaska and Its Resources* (Boston: Lee and Shepard, 1870), 482.

[62] J. W. Collins, *Report on the Fisheries of the Pacific Coast of the United States*, in *Report of the United States Commissioner of Fish and Fisheries for 1892* (Washington, DC: GPO, 1892), 92.

[63] J. W. Collins, *Report on the Fisheries of the Pacific Coast of the United States*, in *Report of the United States Commissioner of Fish and Fisheries for 1892* (Washington, DC: GPO, 1892), 102; Thomas J. Vivian, *Commercial, Industrial, Agricultural, Transportation and Other Interests of California* (Washington: Bureau of Statistics, 1891), 485; John N. Cobb, *Commercial Fisheries of Alaska in 1905*, Bureau of Fisheries Document No. 603 (Washington, DC: GPO, 1906), 9; John N. Cobb, *Pacific Cod Fisheries* (revised edition, 1926), Bureau of Fisheries Doc. No. 1014, appendix 7 to *Report of the U.S. Commissioner of Fisheries for 1926* (Washington, DC: GPO, 1927), 404–405; John N. Cobb, "Motor Vessels in the Alaska Cod Fisheries," *Pacific Motor Boat* (November 1919): 3–8.

[64] John N. Cobb, *Pacific Cod Fisheries* (revised edition, 1926), Bureau of Fisheries Doc. No. 1014, appendix 7 to *Report of the U.S. Commissioner of Fisheries for 1926* (Washington, DC: GPO, 1927), 468.

[65] J. W. Collins, *Report on the Fisheries of the Pacific Coast of the United States*, in *Report of the United States Commissioner of Fish and Fisheries for 1892* (Washington, DC: GPO, 1892), 92–93.

[66] John N. Cobb, *Pacific Cod Fisheries* (revised edition, 1926), Bureau of Fisheries Doc. No. 1014, appendix 7 to *Report of the U.S. Commissioner of Fisheries for 1926* (Washington, DC: GPO, 1927), 404.

[67] Washington Territory memorial, in "Important Annexation: Russian-America Purchased by the United States—Half a Million of Square Miles Acquired," *New York Times*, April 1, 1867.

[68] Ibid.

[69] William H. Dall, *Alaska and Its Resources* (Boston: Lee and Shepard, 1870), 482; George Davidson, *Report of Assistant George Davidson, Relative to the Coast, Features, and Resources of Alaska Territory, November 30, 1867*, in *Russian America*, 40th Cong. 2d sess., H. Ex. Doc. No. 177 (Washington, DC: GPO, 1868), 255–257; *Fishing Grounds of the North Pacific Ocean*, 42d Cong., 2d sess., 1872, Sen. Ex. Doc. No. 34, 7; Joseph Henry, Secretary, Smithsonian Institution, to Mr. Seward, December 23, 1867, *in Russian America*, 40th Cong. 2d sess., H. Ex. Doc. No. 177 (Washington, DC: GPO, 1868), 86–87; "Speech of Hon. Charles Sumner, of Massachusetts, on the cession of Russian America to the United States" (Washington: Congressional Globe Office, 1867), 45.

[70] George Davidson, *Report of Assistant George Davidson, Relative to the Coast, Features, and Resources of Alaska Territory*, November 30, 1867, in *Russian America*, 40th Cong. 2d sess., H. Ex. Doc. No. 177 (Washington, DC: GPO, 1868), 255–257; John N. Cobb, *Pacific Cod Fisheries*, Bureau of Fisheries Doc. No. 830, appendix 4 to *Report of the U.S. Commissioner of Fisheries for 1915* (Washington, DC: GPO, 1916), 28.

[71] C. P. Overton, "Pioneers in the Pacific Coast Codfish Industry," *Pacific Fisherman* (January 1906): 70–71, 75.

[72] John N. Cobb, *Pacific Cod Fisheries*, Bureau of Fisheries Doc. No. 830, appendix 4 to *Report of the U.S. Commissioner of Fisheries for 1915* (Washington, DC: GPO, 1916), 28.

[73] William H. Dall, *Alaska and Its Resources* (Boston: Lee and Shepard, 1870), 482; John N. Cobb, *Pacific Cod Fisheries* (revised edition, 1926), Bureau of Fisheries Doc. No. 1014, appendix 7 to *Report of the U.S. Commissioner of Fisheries for 1926* (Washington, DC: GPO, 1927), 467.

[74] J. W. Collins, *Report on the Fisheries of the Pacific Coast of the United States*, in *Report of the United States Commissioner of Fish and Fisheries for 1892* (Washington, DC: GPO, 1892), 97; John N. Cobb, *Pacific Cod Fisheries* (revised edition, 1926), Bureau of Fisheries Doc. No. 1014, appendix 7 to *Report of the U.S. Commissioner of Fisheries for 1926* (Washington, DC: GPO, 1927), 407, 468.

[75] J. W. Collins, *Report on the Fisheries of the Pacific Coast of the United States, in Report of the United States Commissioner of Fish and Fisheries for 1892* (Washington, DC: GPO, 1892), 97.

[76] William H. Dall, *Alaska and Its Resources* (Boston: Lee and Shepard, 1870), 482.

[77] Thomas J. Vivian, *Commercial, Industrial, Agricultural, Transportation, and Other Interests of California* (Washington: Bureau of Statistics, 1891), 486.

[78] J. W. Collins, *Report on the Fisheries of the Pacific Coast of the United States, in Report of the United States Commissioner of Fish and Fisheries for 1892* (Washington, DC: GPO, 1892), 103; Tarleton Bean, *The Cod Fishery of Alaska [1880]*, in George Brown Goode, *The Fisheries and Fishery Industries of the United States, Section V: History and Methods of the Fisheries*, vol. 1 (Washington, DC: GPO, 1887), 217.

[79] "Pacific Coast Codfish Review," *Pacific Fisherman* (February 1907): 55–58; John N. Cobb, *Pacific Cod Fisheries*, Bureau of Fisheries Doc. No. 830, appendix 4 to *Report of the U.S. Commissioner of Fisheries for 1915* (Washington, DC: GPO, 1916), 27; "Fishermen Like Coast," *Pacific Fisherman* (September 1906): 13.

[80] J. W. Collins, *Report on the Fisheries of the Pacific Coast of the United States*, in *Report of the United States Commissioner of Fish and Fisheries for 1892* (Washington, DC: GPO, 1892), 99; Tarleton Bean, "The Cod Fishery of Alaska" [1880], in George Brown Goode, *The Fisheries and Fishery Industries of the United States, Section V: History and Methods of the Fisheries*, vol. 1 (Washington, DC: GPO, 1887), 210; John N. Cobb, *Pacific Cod Fisheries* (revised edition, 1926), Bureau of Fisheries Doc. No. 1014, appendix 7 to *Report of the U.S. Commissioner of Fisheries for 1926* (Washington, DC: GPO, 1927), 467.

[81] J. W. Collins, *Report on the Fisheries of the Pacific Coast of the United States*, in *Report of the United States Commissioner of Fish and Fisheries for 1892* (Washington, DC: GPO, 1892), 99.

[82] Z. L. Tanner et al., "Explorations of the Fishing Grounds of Alaska, Washington Territory, and Oregon, during 1888, by the U.S. Fish Commission Steamer Albatross, Lieut. Comdr. Z. L. Tanner, U.S. Navy, Commanding," *Bulletin of the United States Fish Commission*, vol. 8, 1888 (Washington, DC: GPO, 1890), 33.

[83] John N. Cobb, *Pacific Cod Fisheries* (revised edition, 1926), Bureau of Fisheries Doc. No. 1014, appendix 7 to *Report of the U.S. Commissioner of Fisheries for 1926* (Washington, DC: GPO, 1927), 454.

[84] Tarleton Bean, *The Cod Fishery of Alaska* [1880], in George Brown Goode, *The Fisheries and Fishery Industries of the United States, Section V: History and Methods of the Fisheries*, vol. 1 (Washington, DC: GPO, 1887), 210; A. Howard Clark (U.S. National Museum), "Fishery Products, and of the Apparatus Used in Their Preparation," catalogue K in *Descriptive Catalogues of the Collections Sent from the United States to the International Fisheries Exhibition, London, 1883* (Washington, DC: GPO, 1884), 1039.

[85] Henry W. Elliott, *A Report on the Conditions of Affairs in the Territory of Alaska* (Washington, DC: GPO, 1875), 166, in *Seal Fisheries in Alaska*, 44th Cong., 1st sess., H. Ex. Doc. No. 83, 1876, 165–166.

[86] "Where Your Salt Comes From," *Pacific Fisherman* (January 1916): 83.

[87] Arvill Wayne Bitting, *Preparation of the Cod and Other Salt Fish for the Market*, U.S. Dept. of Agriculture, Bureau of Chemistry, Bulletin No. 133 (Washington, DC: GPO, 1911), 18.

[88] "Where Your Salt Comes From," *Pacific Fisherman* (January 1916): 83.

[89] J. E. Shields, "How Salt is Made," *Pacific Fisherman* (February 1923): 12–13; Mark Kurlansky, *Salt: A World History* (New York: Penguin Books, 2002), 283, 286–287; Morton Salt Company, adv., *Pacific Fisherman* (January 1928): 193; Leslie-California Salt Co., adv., *Pacific Fisherman* (September 1931): 94; Oliver Salt Co., adv., *Pacific Fisherman* (February 1925): 48; Arden Salt Co., adv., *Pacific Fisherman* (January 1927): 205; "Salt Dock for Seattle," *Pacific Fisherman* (February 1925): 48; Tarleton Bean, *The Cod Fishery of Alaska* [1880], in George Brown Goode, *The Fisheries and Fishery Industries of the United States, Section V: History and Methods of the Fisheries*, vol. 1 (Washington, DC: GPO, 1887), 207; Ed Shields, *Salt of the Sea: The Pacific Coast Cod Fishery and the Last Days of Sail* (Lopez Island, Wash.: Pacific Heritage Press, 2001), 67; J. W. Collins, *Report on the Fisheries of the Pacific Coast of the United States*, in *Report of the United States Commissioner of Fish and Fisheries for 1892* (Washington, DC: GPO, 1892), 71.

[90] "Pacific Coast Codfishing, *Pacific Fisherman* (January 1935): 199.

[91] By 1929, this facility was equipped to dry fish indoors when the weather was not conducive to outdoor drying. In indoor drying, the fish were placed on racks and were dried by furnace-heated air that was drawn over the fish by a large fan.

[92] J. W. Collins, *Report on the Fisheries of the Pacific Coast of the United States*, in *Report of the United States Commissioner of Fish and Fisheries for 1892* (Washington, DC: GPO, 1892), 97–98, 105–106; John N. Cobb, *Pacific Cod Fisheries*, Bureau of Fisheries Doc. No. 830, appendix 4 to *Report of the U.S. Commissioner of Fisheries for 1915* (Washington, DC: GPO, 1916), 28–29; Richard S. Croker, "Alaska Codfish," in *The Commercial Fish Catch of California for the Year 1929*, Division of Fish and Game of California Fish Bulletin No. 30 (Sacramento: California State Printing Office, 1931), 48–55.

[93] C. P. Overton, "Pioneers in the Pacific Coast Codfish Industry," *Pacific Fisherman* (January 1906): 70–71, 75.

[94] C. P. Overton, "Pioneers in the Pacific Coast Codfish Industry," *Pacific Fisherman* (January 1906): 70–71, 75; J. W. Collins, *Report on the Fisheries of the Pacific Coast*

of the United States, in *Report of the United States Commissioner of Fish and Fisheries for 1892* (Washington, DC: GPO, 1892), 106–107; "Salt Fish Review," *Pacific Fisherman* (January 1915): 102–103; John N. Cobb, *Pacific Cod Fisheries*, Bureau of Fisheries Doc. No. 830, appendix 4 to *Report of the U.S. Commissioner of Fisheries for 1915* (Washington, DC: GPO, 1916), 29, 38–39.

95 Richard Rathbun, "Report Upon the Inquiry Respecting Food-fishes and the Fishing-grounds," in U.S. Commission of Fish and Fisheries, part 16, *Report of the Commissioner for 1888* (Washington, DC: GPO, 1892), XLVIII.

96 "Pacific Coast Codfish Review," *Pacific Fisherman* (February 1907): 55–58; John N. Cobb, *Pacific Cod Fisheries*, Bureau of Fisheries Doc. No. 830, appendix 4 to *Report of the U.S. Commissioner of Fisheries for 1915* (Washington, DC: GPO, 1916), 26, 48; C. P. Overton, "Pioneers in the Pacific Coast Codfish Industry," *Pacific Fisherman* (January 1906): 70–71, 75; J. W. Collins, *Report on the Fisheries of the Pacific Coast of the United States*, in *Report of the United States Commissioner of Fish and Fisheries for 1892* (Washington, DC: GPO, 1892), 96; C. P. Overton to unspecified recipient, October 21, 1884, in M. A. Healy, *Cruise of the Revenue Marine Steamer Corwin in the Arctic Ocean in 1884* (Washington, DC: GPO, 1887), 25; "Codfish and Fishers," *San Francisco Call*, October 25, 1885; "Loses Man Overboard," *San Francisco Call*, December 31, 1903.

97 John N. Cobb, *Commercial Fisheries of Alaska in 1905*, Bureau of Fisheries Document No. 603 (Washington, DC: GPO, 1906), 9; Tarleton Bean, *The Cod Fishery of Alaska* [1880], in George Brown Goode, *The Fisheries and Fishery Industries of the United States, Section V: History and Methods of the Fisheries*, vol. 1 (Washington, DC: GPO, 1887), 201–202; Frederick Schwatka, *Report of a Military Reconnaissance in Alaska Made in 1883*, 48th Cong. 2d sess., Sen. Ex. Doc. No. 2 (Washington, DC: GPO, 1885), 11; Eliza Ruhamah Skidmore, *Appletons' Guide-book to Alaska and the Northwest Coast* (New York: D. Appleton and Company, 1893), 88.

98 John N. Cobb, *Pacific Cod Fisheries*, Bureau of Fisheries Doc. No. 830, appendix 4 to *Report of the U.S. Commissioner of Fisheries for 1915* (Washington, DC: GPO, 1916), 26; J. W. Collins, *Report on the Fisheries of the Pacific Coast of the United States*, in *Report of the United States Commissioner of Fish and Fisheries for 1892* (Washington, DC: GPO, 1892), 95.

99 C. P. Overton, "Pioneers in the Pacific Coast Codfish Industry," *Pacific Fisherman* (January 1906): 70–71, 75.

100 C. P. Overton to unspecified recipient, October 21, 1884, in M. A. Healy, *Cruise of the Revenue Marine Steamer Corwin in the Arctic Ocean in 1884* (Washington, DC: GPO, 1887), 25.

101 W. A. Howard [U.S. Revenue Steamer *Lincoln*], in *Russian America*, 40th Cong. 2d sess., H. Ex. Doc. No. 177 (Washington, DC: GPO, 1868), 190, 200.

[102]William Governeur Morris, *Report on the Customs District, Public Service, and Resources of Alaska*, Sen. Doc. No. 59, 45th Cong. 1st sess., in *Seal and Salmon Fisheries and General Resources of Alaska*, vol. 4 (Washington, DC: GPO, 1898), 114.

[103]The U.S. Fish Commission was formally known as the U.S. Commission of Fish and Fisheries. In 1889, the Albatross transported several members of the Senate Committee on Indian Affairs on a three-week inspection tour of the principal Native communities in Southeast Alaska.

[104]Rathbun was also a former U.S. Fish Commission scientist.

[105]Richard Rathbun, "Summary of the Fishery Investigations Conducted in the North Pacific Ocean and Bering Sea from July 1, 1888, to July 1, 1892, by the U.S. Fish Commission Steamer Albatross," *Bulletin of the United States Fish Commission*, vol. 12, 1892 (Washington, DC: GPO, 1894), 127.

[106]Richard Rathbun, in Z. L. Tanner et al., "Explorations of the Fishing Grounds of Alaska, Washington Territory, and Oregon, during 1888, by the U.S. Fish Commission Steamer Albatross, Lieut. Comdr. Z. L. Tanner, U.S. Navy, Commanding," *Bulletin of the United States Fish Commission*, vol. 8, 1888 (Washington, DC: GPO, 1890), 8.

[107]Richard Rathbun, in Z. L. Tanner et al., "Explorations of the Fishing Grounds of Alaska, Washington Territory, and Oregon, during 1888, by the U.S. Fish Commission Steamer Albatross, Lieut. Comdr. Z. L. Tanner, U.S. Navy, Commanding," *Bulletin of the United States Fish Commission*, vol. 8, 1888 (Washington, DC: GPO, 1890), 8.

[108]Z. L. Tanner, "Report Upon the Investigations of the U.S. Fish Commission Steamer Albatross for the Year Ending June 30, 1889," appendix 4, *Report of the United States Commissioner of Fish and Fisheries for the Fiscal Year Ending June 30,1889*, part 16 (Washington, DC: GPO, 1892), 395.

[109]Richard Rathbun, in Z. L. Tanner et al., "Explorations of the Fishing Grounds of Alaska, Washington Territory, and Oregon, during 1888, by the U.S. Fish Commission Steamer *Albatross*, Lieut. Comdr. Z. L. Tanner, U.S. Navy, Commanding," *Bulletin of the United States Fish Commission*, vol. 8, 1888 (Washington, DC: GPO, 1890), 8.

[110]Z. L. Tanner et al., "Explorations of the Fishing Grounds of Alaska, Washington Territory, and Oregon, during 1888, by the U.S. Fish Commission Steamer Albatross, Lieut. Comdr. Z. L. Tanner, U.S. Navy, Commanding," *Bulletin of the United States Fish Commission*, vol. 8, 1888 (Washington, DC: GPO, 1890), 79.

[111] Richard Rathbun, "Summary of the Fishery Investigations Conducted in the North Pacific Ocean and Bering Sea from July 1, 1888, to July 1, 1892, by the U.S. Fish Commission Steamer Albatross," *Bulletin of the United States Fish Commission*, vol. 12, 1892 (Washington, DC: GPO, 1894), 155; Patricia Roppel, "The Steamer *Albatross* and Early Pacific Salmon, *Oncorhynchus* spp., Research in Alaska," *Marine Fisheries Review*, 2004, 66(1).

[112] Richard Rathbun, "Summary of the Fishery Investigations Conducted in the North Pacific Ocean and Bering Sea from July 1, 1888, to July 1, 1892, by the U.S. Fish Commission Steamer Albatross," *Bulletin of the United States Fish Commission*, vol. 12, 1892 (Washington, DC: GPO, 1894), 132.

[113] Ibid, 141, 147, 153; "Anacortes' Codfish Industry Is One of Great Importance," *Anacortes American*, October 12, 1911.

Chapter 3
CODFISH FEVER

With the rapidly growing population of the United States and the increasing demand, the eastern fisheries will not be capable of sustaining the supply, and it is from the cod banks of Alaska that the country must look for its future supply.

—Captain Jarvis, *Pacific Monthly*, June 1906[114]

Fish are what bring the money, and money must come from somewhere to keep things running. Wages go on and all hands in Alaska must eat and it must all come out of the Fish. So get us the Fish. Send us Fish, lots of them. Do not let any chance go by to get Fish.

—C. P. Overton, Union Fish Company, to Captain Walter Tinn, Pirate Cove, Alaska, July 24, 1912[115]

C. P. Overton, who for many years headed the Union Fish Company and whom John Cobb praised as "one of the brightest men engaged in the industry," frequently contributed articles about the codfish industry to *Pacific Fisherman*.[116] In a 1906 article, Overton gave brief descriptions of the experiences of some who had been prominent in the industry. He also provided an epitaph that befitted many of them and helped describe the business environment:

> The bright sun of prosperity shines for a season or two upon the regular stand-bys in the business and it looks very attractive and inviting to some chaps with an old vessel or a little spare money. So they jump in and for a time cut a brilliant dash in the business. So bright are they that the sun of prosperity is all in eclipse and everyone in the trade walks in shadow. When they get tired of this or [go] broke they drop out, and those who are left pick up the scattered ends of the trade, struggle out into the light again, and by and by there is some more prosperity and then a new crop of hopeful investors appears, and so on and on.

To be fair, most of those who failed in the codfish business were legitimate operators attempting to make the best of a perceived opportunity. Some, though, were more colorful than others. Overton described a junk and secondhand goods dealer who had entered the codfish business with the brig *Glencoe*: "Like everything else that old John had, the vessel was poor, the salt was poor, and the fish were, of course, yellow or sour, dried up or slimy, but they went onto the market and helped damn Pacific codfish."[117]

According to Overton, Nick Bichard was "undoubtedly the most picturesque figure in the whole lien." Bichard, a native of the Isle of Jersey (off the English coast), was a pioneer ship owner and merchant of San Francisco, who

> accumulated a fortune during the days of the Civil War and was early in the codfish business with quite a fleet of old vessels, both large and small, and for many years he was a prominent factor in the business. A large, swarthy man, erratic in speech and action, mixing codfish, coal, lumber, and junk, keeping most of his books in his head, he never knew what his cargoes cost him or what they sold for. The codfish business absorbed more and more of his capital; then his real estate, two fine water lots on Stuart Street, the gore lot at California and Market Streets, and other property went the same way; the old vessels wore out and were lost and he finally died peacefully in the night of heart failure, leaving barely enough to bury him.[118]

Bichard's codfish curing and drying operation was located on what was known as Kershaw's Island (now Belvedere), on Richardson Bay, about five miles north of San Francisco.[119]

An 1891 U.S. Treasury Department report mentions A. Anderson & Company as a participant in San Francisco's codfish industry.[120]

Captain Joshua Slocum, whose book *Sailing Alone Around the World* gained him international fame, was an early, though short-term, participant in the Pacific codfish industry. As a boy, Slocum had worked aboard a cod-fishing schooner in the Bay of Fundy. John Cobb recounted Slocum's Okhotsk Sea venture:

> In 1877 Capt. Joshua Slocum, with the schooner Pato (about 45 tons register), was at the Philippine Islands, when he conceived the idea of making a cod-fishing voyage to the Okhotsk Sea and marketing his catch at the islands. Leaving the islands in March, he proceeded to the Okhotsk via Yokohama. Salt

> and fishing gear were obtained from vessels met with in the [Okhotsk] sea, and a cargo of 23,000 fish was soon taken. When the time for sailing arrived the captain decided not to return to the islands, but took his fare to Portland [Oregon] instead, where he sold it at a profitable price.[121]

Cobb noted that this was the only delivery of codfish to Portland, and he thought it odd that no vessels from this city or Astoria (Oregon) participated in the cod fishery.[122]

OPTIMISM IN THE MIDST OF A MARKET DECLINE

"Of the sea fishes [of Alaska] the cod-fish stands foremost in quantity as well as in commercial importance," wrote the U.S. Census Office's Ivan Petroff in his 1882 report on Alaska's population, industries, and resources.[123] The codfish industry, although thriving at the time, would soon begin a decline and be eclipsed by the explosively growing canned salmon industry.

The event that precipitated the growth of the Alaska canned salmon industry and pushed the cod fishery into the background was the success of the Alaska Commercial Company at catching and canning salmon at Kodiak Island in 1888. That year, the company barricaded the Karluk River and caught virtually every red salmon that attempted to enter it. The catch, estimated to number more than 1.2 million fish, allowed the Alaska Commercial Company cannery—which had been constructed the previous year near the river's mouth—to put up 101,000 cases of red salmon (a case is equivalent to forty-eight one-pound cans), the sale of which yielded the company a tremendous profit.[124] Word of the company's success soon reached investors in Seattle and San Francisco, and a boom in cannery construction in Alaska ensued. The salmon industry soon came to dominate Alaska's economy, and remains a major factor to this day.

Codfish production in the nineteenth century peaked in 1883 with a catch of 1,720,000 fish, following which the industry began a period of reduced production that lasted almost two decades. J. W. Collins, head of the U.S. Fish Commission's Division of Fisheries, attributed the decline to market conditions (exacerbated by the inconsistent quality of salted codfish produced on the Pacific coast), keen competition from East Coast producers, and the "tendency for capital to seek investment of the more promising salmon fisheries of Alaska." According to Collins, there was "unquestionably no deficiency of supply" because the fishing grounds were "believed capable of furnishing an unlimited amount of cod."[125] The 1889 catch was 816,000 fish, less than half of what was caught in 1883, and the smallest catch since 1875. Codfish fever, at least for the time being, had run its course.

In his 1892 report on the fisheries of the U.S. Pacific coast, Collins characterized the diminished Pacific codfish industry as being nationally unimportant but still "by no means an unimportant factor in the industries of the far West." Collins added optimistically that the industry—"if not abandoned"— might one day "ultimately attain a status that its present condition gives little reason to hope for."[126]

The industry was not abandoned and even grew a bit, although the remainder of the 1800s was characterized by inconsistent annual catches that ranged from 792,000 to 1,505,000 fish.[127]

A new and subsequently very successful entrant into the industry during the mid-1890s was the Alaska Codfish Company, which was originally organized in San Francisco as the Pacific Marine Supply Company. In 1896, the company, headed by Henry Levi, sent the former whaling schooner *La Ninfa* (*LaNympha*) to the cod-fishing grounds. It returned in late summer, carrying fifty thousand salted codfish. The business prospered and in 1904 changed its name to Alaska Codfish Company.

Among the vessels that made up the company's fleet in 1902 was the 120-ton transporter *Pearl* (one of the smallest codfish industry vessels that regularly traveled between San Francisco and Alaska), which in late October of that year became frozen in ice about two hundred miles from Unga. The crew suffered considerably (several were frostbitten) in the unusually cold spell during which temperatures reached twelve below zero. The *Pearl* reached San Francisco on January 28, 1903, and within two weeks departed again for the northern seas.

In 1905, the Alaska Codfish Company operated five codfish stations and six schooners. In December of that year, the *Pearl* departed San Francisco carrying fishermen and supplies for the company's codfish station at Sanak, in the Shumagin Islands. In all, thirty men were onboard. Unfortunately, the *Pearl* never arrived. The vessel's wreckage was found the following summer on a reef near Sanak Island. The loss of the *Pearl*'s crew was one of the greatest tragedies to befall the Pacific codfish industry during the Salt Cod Era. Unfortunately, the tragedy was matched that same year with the loss of the codfish transporter *Nellie Coleman* (see below).[128]

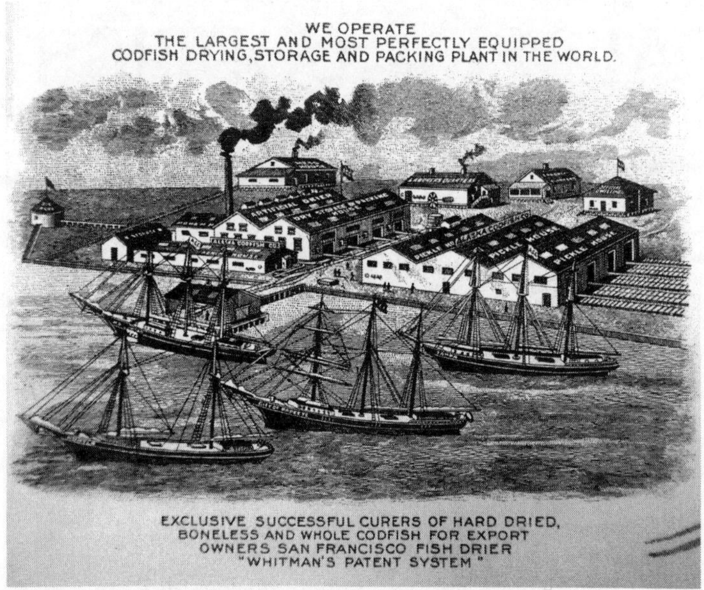

Figure 15. Letterhead, Alaska Codfish Company. (John N. Cobb Collection, Special Collections, University of Washington Libraries, Seattle, Wash.)

A third San Francisco–based firm that entered the codfish industry was the Pacific States Trading Company, which was organized early in 1905. That April, the company sent the schooners *Glen* and *John F. Miller* to the Bering Sea and while they were at sea constructed a curing station on Carquinez Strait, an estuary of San Francisco Bay about thirty miles from San Francisco. The vessels returned that fall carrying a combined 140,000 codfish. Within three years, however, both the *Glen* and *John F. Miller* became casualties (see chapter 4). Pacific States Trading also built and operated several Alaska shore stations (see chapter 6).[129]

PUGET SOUND–BASED CODFISH COMPANIES

It was fully a quarter century after the inception of the Pacific coast codfish industry that Puget Sound interests became engaged in it. Perhaps because by this time San Francisco firms had taken up all the choice locations for shore stations, the Puget Sound firms obtained their supply of codfish almost exclusively from schooners that fished

ALASKA CODFISH COMPANY

PRODUCERS AND PACKERS OF CODFISH

HEAD OFFICE—15 STEUART ST., SAN FRANCISCO, CALIFORNIA, U. S. A.

Cable Address—"MARINE"

We Operate the Largest Export Drying Plant in the World

Leading Export Brands:

"FLAG" BRAND BACALAO
"MARCA FRAGATA" BACALAO "SIN ESPINAS"
"MARCA FRAGATA" BACALAO "CON ESPINAS"
"PACIFIC BELLE" TABLETS
"OCEAN WAVE" BLOCKS
BACALAO IN DRUMS
TIERCES AND BOXES

OUR FLEET:

MOTOR VESSELS—ALASCO
—ALASCO 2
—ALASCO 3
—ALASCO 4
—CHAMPION

SAILING VESSELS—CITY OF PAPEETE
—MAWEEMA
—S. N. CASTLE
—GLENDALE
—BANGOR

Leading Domestic Brands:

FRIGATE STRIPS
NARROW PARAGON
IMPERIAL MIDDLES
PACIFIC BELLE 1-LB. TABLETS
SIBERIA 2-LB. BRICKS
FLAKED CODFISH IN TIN 4, 8 or 16 OZ.
WE SPECIALIZE IN "NO BONE" MIDDLES AND STRIPS PACKED IN 1, 2 and 5-LB. WOOD BOXES

Fishing Stations in Alaska:

COMPANY HARBOR
MOFFAT'S COVE
DORA HARBOR
UNGA
BARANOFF
WINCHESTER
EAGLE HARBOR

Figure 16. Alaska Codfish Company advertisement, *Pacific Fisherman*, Yearbook, January 1923.

on the high seas. And though the Puget Sound firms entered the industry later than the San Francisco–based firms, two of the Puget Sound firms—the Robinson Fisheries Company and the Pacific Coast Codfish Company—outlasted their San Francisco competitors.

J. A. Matheson

The first person to establish an Alaska codfish operation based in Puget Sound was Captain John Matheson. Matheson was born in Nova Scotia in 1849 and began fishing for cod along the Atlantic coast before he was a teenager. In 1872, he moved to Massachusetts. There, in 1882, he began construction with at least one partner of the 142-ton cod-fishing schooner *Lizzie Colby*. Matheson was the managing owner and master of the vessel, which departed on its first Atlantic Ocean fishing voyage in the spring of 1883.[130]

Matheson, whom *Pacific Fisherman* later praised as having "probably a more complete knowledge of the [codfish] industry in all its branches than any one else engaged in it on this coast," recognized Puget Sound's potential as a base for a codfish operation and initiated his move there in 1890.[131] In November of that year, Matheson sent the *Lizzie Colby* to New York, where it took on a load of coal before departing for San Francisco. After an uneventful voyage around Cape Horn, the *Lizzie Colby* arrived in the spring of 1891 in San Francisco, where the coal was discharged and a supply of salt loaded. The *Lizzie Colby* then proceeded to the Bering Sea fishing banks, with orders to return with its cargo to Port Townsend, Washington, and wait there for further orders. Meanwhile, Matheson himself went to Puget Sound, where, after examining various ports, he decided to establish a codfish-processing operation at Anacortes. He quickly acquired waterfront property and promptly constructed a dock and buildings for handling and processing salted codfish.

Matheson's was the first fish-processing operation to be established at Anacortes. The location had many advantages for fisheries-based operation, and by 1909 Anacortes hosted six salmon canneries, two fish by-products plants, and two codfish-processing plants.

The *Lizzie Colby* returned that fall, and its cargo was processed in the new facility. This was the first codfish venture on the West Coast that was not based in San Francisco. The *Lizzie Colby* alone continued to supply the Anacortes plant until 1905, when Matheson added the *Fanny Dutard* to his fleet. Registered at 170 tons, the *Fanny Dutard* became the largest vessel in the Puget Sound–based codfish business. In 1906, the *Lizzie Colby* was retired, and two years later the 252-ton

Figure 17. Unloading schooner *Lizzie Colby*, J. A. Matheson plant, Anacortes, Washington. (Wally Funk Collection, Anacortes Museum, Anacortes, Wash.)

Harriet G. took its place and at the same time superseded the *Fanny Dutard* as the largest vessel in the Puget Sound–based codfish business.

In 1907, Anacortes's local newspaper, the *Anacortes American*, wrote that Matheson's company was the first in Puget Sound to ship a boxcar-load of salted cod to Boston, and that the company was currently shipping salted cod to "Boston and all the principal markets of the country."[132]

In 1910, a group of San Francisco interests, none of which had any experience in the codfish business, formed the Western Codfish Company and attempted to consolidate several Puget Sound cod-fishing companies (see below). In about 1912, the group purchased Matheson's codfish-processing plant and incorporated it as the Matheson Fisheries Company, with Matheson in charge of operations. The Western Codfish Company, however, operated for less than two years.

The Matheson Fisheries Company lasted a little longer. Matheson retired from the business in 1913, and the company ceased sending vessels north after the 1914 season and by the summer of 1915 had disposed of its inventory of codfish and had terminated operations. The company's Anacortes codfish plant was then sold to the Apex Fish Company.

John Matheson, however, quickly reentered the codfish business. In December 1914, the failed company he had headed sold him the schooners *Fanny Dutard* and *Azalea*. Matheson sent the vessels north the following year under his own name (J. A. Matheson). Matheson then reacquired a portion of the property that had been sold to the Apex Fish Company, and in the summer of 1916 tore down his old

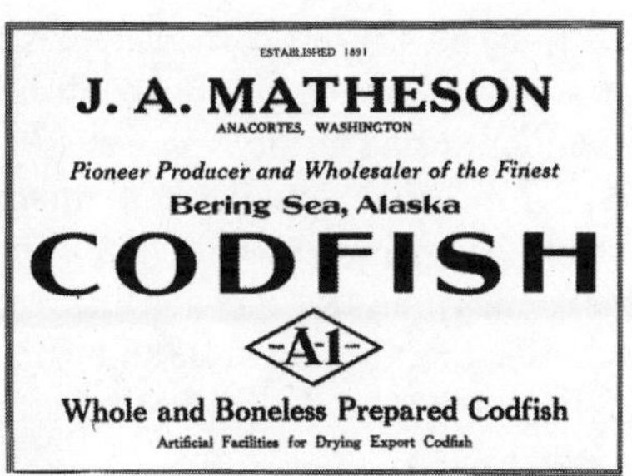

Figure 18. J. A. Matheson advertisement, *Pacific Fisherman*, Yearbook, January 1924.

Figure 19. Schooner *Fanny Dutard* and workers tending drying codfish at Matheson dock, Anacortes, Washington. (Wally Funk Collection, Anacortes Museum, Anacortes, Wash.)

codfish-drying shed—by then an Anacortes landmark—to make room for a modern two-story 54-foot-by-100-foot codfish processing plant that included an indoor drying facility.

J. A. Matheson processed salted cod at Anacortes until 1930, when—after a career that spanned nearly four decades—he quietly withdrew from the codfish industry following that year's fishing season. His *Fanny Dutard*, except for the years 1918 and 1920, had fished each season until it was laid up in Lake Union (Seattle) after the 1930 season. The vessel never sailed again.

The *Azalea*'s career with J. A. Matheson was considerably shorter: it was operated through the 1918 season, after which it was sold to the Robinson Fisheries Company, also of Anacortes. The *Azalea* made its final cod-fishing voyage to the Bering Sea in 1939 (see chapter 8).[133]

Puget Sound & Alaska Commercial Company

The short-lived pioneer of the codfish industry in Seattle was the Puget Sound & Alaska Commercial Company, which began operations in February 1892 and the following month dispatched the 68-ton schooner *Moonlight* to the Bering Sea banks. The vessel returned in August with 175,000 pounds of salt cod, but the company seems to have dissolved shortly afterward.[134]

Oceanic Packing Company

The next codfish company to operate in Seattle, the Oceanic Packing Company, endured a bit longer. The company was organized in 1896, constructed a codfish curing plant in West Seattle, and outfitted the schooner *Emma F. Harriman* and sent it to the Bering Sea. It returned with a full cargo, but, unfortunately, there was little demand at the time in Seattle for salt cod, and the vessel was sent to San Francisco, where its cargo was sold. The following year, the company sent two vessels, the *Blakely* and the *Swan* to the Bering Sea. Both returned with full cargoes that were processed at the company's West Seattle plant. The company codfish venture was interrupted, however, by the Klondike gold rush, which began in 1897. There was a strong demand for vessels to carry gold rush cargo north, and Oceanic Packing chose to engage its vessels in carrying cargo rather than in cod fishing. The *Swan* was wrecked in this endeavor, but the *Blakely* carried cargo for two years before returning to cod fishing in 1899. It successfully fished that season and the next, but the business itself was apparently unprofitable, and the company ceased operations late in 1900.[135]

Lizzie S. Sorrenson

In 1898, John P. Fay, a Seattle lawyer, sent the 84-foot schooner *Lizzie S. Sorrenson* to the Bering Sea to catch and salt cod. The vessel returned with a full cargo that was processed at a plant built at Richmond Beach, on Seattle's waterfront. Apparently the venture was not profitable, as no additional cod-fishing trips were made.[136]

Robinson Fisheries Company

In March 1904, *Pacific Fisherman* announced that William Robinson—whose family had been involved with fisheries in New England and who was the former owner of a fish by-products plant at Anacortes—had, with two local bank officials, incorporated the Robinson Codfish Company, at Anacortes. The new company took over the fish by-products plant of the Robinson & Colt Company, which produced primarily fish-based fertilizer and fish oil, and announced its intention to operate the plant as the Robinson Fisheries Company and to construct on the premises a "first-class" codfish curing plant, which would be operated as the Robinson Codfish Company, a sister firm of the Robinson Fisheries Company. The company's schooner *Alice* was outfitted for cod fishing and dispatched to the Bering Sea that spring. It was joined the following year by the *Joseph Russ*.[137]

During the summer of 1905, the Robinson Codfish Company erected a facility that the local newspaper, the *Anacortes American*, praised as "the largest and most complete codfish plant known today on the Pacific coast."[138] The main building measured 100 feet by 150 feet on the ground floor and was three stories high. Brine tanks on the premises could hold 800 tons of codfish. The facility was equipped outdoor with racks (*flakes*) for drying fish outside on sunny days, and with indoor racks and a hot-air dryer for drying fish inside during wet weather. The dock and wharf—the largest in Anacortes—measured 350 by 350 feet.

According to the *Anacortes American*, "For cleanliness, this plant surpasses all other fish food product plants on the Sound and all of them are clean." The paper added that the room in which all the boning, pressing and wrapping of prepared codfish bricks was done was "as neat as the kitchen of a careful housekeeper."[139] Codfish skins from the plant—traditionally a waste product—were manufactured into glue (see chapter 1) at the Robinson Fisheries Company's by-products plant.

The Robinson Codfish Company's two schooners, each of which carried twenty dories and a crew of about thirty-six men, seemed to do very well at bringing home large cargoes. This was due in part to the

Figure 20. Workers at an unidentified codfish-processing plant. (John N. Cobb Collection, Special Collections, University of Washington Libraries, Seattle, Wash.)

quality of the fishermen employed. In 1906, William Robinson imported forty experienced fishermen from Gloucester, Massachusetts, to crew his company's boats. The *Anacortes American* was effusive in its praise of the men, calling them "as fine a lot of men as ever sailed from any port; refined, courteous, educated, competent and sober."* Robinson soon after began labeling his product "Codfish caught by Gloucester fishermen and packed at Anacortes, the Gloucester of the Pacific." (In 1907, the *Anacortes American*, wrote that "Anacortes is the Gloucester of Puget Sound and will soon be of the Pacific Coast."[140] By 1911, the paper was referring to Anacortes as the "Gloucester of the Pacific."[141])

As an incentive for fishermen to be productive, Robinson gave cash prizes to the top three fishermen on each of the company's vessels ($25 to the first, $15 to the second, and $10 to the third). The top fisherman in 1906 caught 12,650 fish. In 1909, the *Joseph Russ* delivered a record catch: 204,155 fish, all of which were caught in fifty-eight days of fishing.

The *Joseph Russ*'s career came to an abrupt end one dark night three years later, in April 1912, when it wrecked on Chirikof Island,

* Five brothers from Nova Scotia may have been among this group. The men were required by the Robinson Codfish Company to sign two-year fishing contracts.

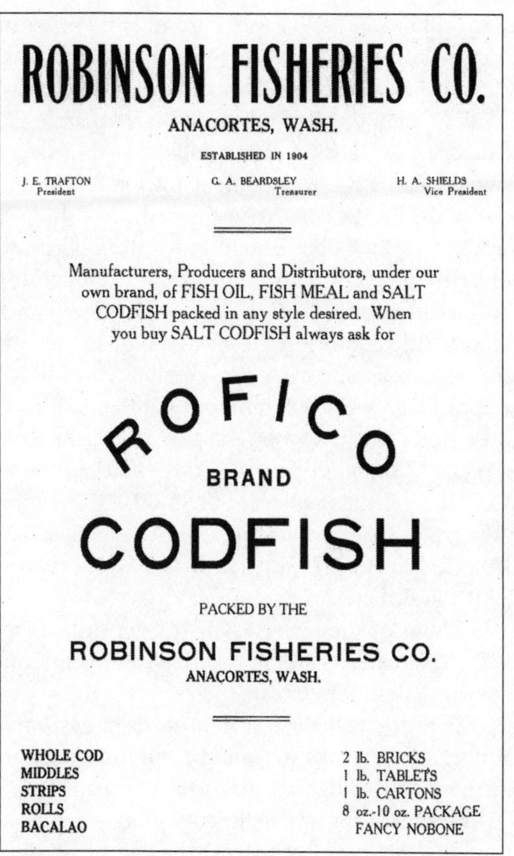

Figure 21. Robinson Fisheries Company advertisement, *Pacific Fisherman*, Yearbook, January 1926.

about eighty miles southwest of Kodiak Island, while en route to the Bering Sea fishing grounds via Unimak Pass. The vessel's first mate was drowned, but its captain, Charles Foss, and the remaining twenty-nine crew members made it safely to shore. Their ordeal, however, was far from over, because now they faced possible death from exposure and starvation on the treeless and windswept island.

On their third day on the island, however, the men found they were not alone; a man and his two sons raised foxes on the island. The following day, the fox farmer visited the shipwrecked men's camp and told the men that the mail steamer *Dora* called at the island, but only a couple of times each year, and that theirs might be a long wait. Three days later, six volunteers set off westward in two dories to seek help, and after eleven days of battling stormy weather on open seas, the men reached Chignik—an as-the-crow-flies distance of about 120 miles from Chirikof Island—only two hours before the *Dora* arrived. The *Dora* promptly steamed to Chirikof Island and rescued the stranded men. Charles Foss would be back at sea the following season and go on to become one of the Pacific coast's most famed cod-fishing captains.[142]

According to John Cobb, who founded the College of Fisheries at the University of Washington and who was an authority on the Pacific codfish industry, the Robinson Codfish Company name was officially changed to Robinson Fisheries Company in 1912. But the change seems to have actually been a combining of the codfish company and the by-products company under one name, and it seems to have come earlier: in its January 1906 review of the Pacific coast codfish industry, *Pacific Fisherman* referred to the fleet of the "Robinson Codfish Co." The journal's review of the industry the following year, however, referred to the "Robinson Fisheries Co." fleet. Similarly, by 1908, William Robinson referred to his codfish-processing operation as the "Robinson Fisheries Co."[143]

William Robinson died in 1916, and his son-in-law, John Trafton, succeeded him as manager of the firm.[144] Trafton continued in that position for more than thirty years.

To replace the *Joseph Russ*, Robinson Fisheries purchased the 156-foot lumber schooner *Wawona* in early 1914 and quickly converted it to cod fishing. Praised by the Anacortes newspaper as "the largest, newest and finest vessel engaged in the codfishing industry out of Puget Sound," the *Wawona* instantly became the flagship of the Robinson Fisheries Company fleet. It departed Anacortes for the Bering Sea on March 31 of that year with Captain Charles Foss at the helm.[145]

Fishing was exceptionally good, and the *Wawona* returned to Anacortes in the fall with 240,000 codfish, weighing about 1,100,000 pounds—the largest codfish cargo ever landed by an American vessel on the Pacific coast. The schooner broke its record the following year, with a catch of 258,323 codfish that weighed about 1,150,000 pounds.[146]

Foss's long cod-fishing career—and his life—ended in August 1935 when he suffered what was apparently a massive stroke while returning to Puget Sound aboard the *Wawona* with his thirty-first consecutive cargo of codfish. Foss was seventy years old. His last radio message to the home office was that the *Wawona* was homeward bound with 307,000 fish onboard.

The day after his death, the *Wawona*'s crew buried Foss at Lost Harbor, on Akun Island, a place Foss had frequently visited to take

Figure 22. Schooner *Wawona* and drying codfish at Robinson Fisheries Co. dock, Anacortes, Washington, 1915. (John N. Cobb Collection, Special Collections, University of Washington Libraries, Seattle, Wash.)

on fresh water and make repairs to his vessel. John Trafton, head of the Robinson Fisheries Company, which still owned the *Wawona*, praised Foss as "an exceptionally fine captain" who "had brought back more codfish than any other man who has ever gone into [the] Bering Sea." Tom Haugen, who had been Foss's first mate for fourteen years, succeeded him as the *Wawona*'s captain.[147]

In 1937, two years after Foss's death, the *Wawona* delivered 363,687 fish after a trip of only four months and five days. The vessel's success that year was attributed to good weather and an abundance of fish. The average catch by each of the *Wawona*'s eighteen dories—all powered by outboard motors—was a little more than 19,000 fish, with the vessel's first mate, Berger Jensen, landing 26,331, the all-time high for the Bering Sea.[148]

The *Wawona*'s cod-fishing career, however, was interrupted by World War II. The U.S. Army Transport Service requisitioned it in August 1942, removed its masts, and used it as a barge. Shortly after the war, the Robinson Fisheries Company bought it back, refitted it for cod fishing, and sent it to the Bering Sea in the spring of 1946. The *Wawona*'s final cod-fishing trip to the Bering Sea was in 1947. The vessel's recorded total catch of 7.3 million codfish during its twenty-nine seasons on Alaska's codfish grounds is the all-time record by a single schooner.[149]

Seattle-Alaska Fish Company

The Seattle-Alaska Fish Company was formed in 1902 by a small group of investors led by Norwegian-born Captain John Grotle. Two years earlier, Grotle, an experienced West Coast halibut fisherman, had purchased the 110-foot schooner *Carrier Dove* in Boston. After taking on a load of blacksmith coal in New Jersey, Grotle sailed to Seattle, where he discharged the coal and outfitted the vessel for cod fishing. The *Carrier Dove* departed in April 1901 for the Shumagin Islands, and returned to Seattle in late August laden with some fifty thousand codfish, which were sold to the Chlopec Fish Company. In late November, the *Carrier Dove*, under Grotle, departed Seattle with a crew of fishermen and supplies to construct a shore station at Falmouth Harbor (Popof Island), a location Grotle had selected during his earlier voyage. Grotle's crew fished that winter from the shore station as well as from his schooner, and he returned to Seattle in early 1902 with some thirty-five thousand codfish. With his associates, Grotle quickly formed the Seattle-Alaska Fish Company and took over the defunct Oceanic Packing Company's West Seattle codfish curing plant. The Seattle-Alaska Fish Company soon after established shore stations at Squaw Harbor and Unga (both on Unga Island), and at Eagle Harbor, on Nagai Island.

In November 1905, the *Seattle Times* reported that the Seattle-Alaska Fish Company had struggled ineffectually for four years to develop a local market for its salted codfish. The local prejudice against Alaska codfish was simply too strong to overcome, and the company resorted to shipping its product by railroad car to Gloucester and other eastern ports. There, after further processing and being labeled as Grand Banks fish, much of it was shipped back to Seattle, where it found ready acceptance in local markets.

In 1903, the company acquired the 97-foot, two-masted schooner *Nellie Coleman* and sent it to the Bering Sea. The vessel fished in the Bering Sea the following two seasons. After its 1905 fishing voyage, the *Nellie Coleman* was sent to the Shumagin Islands to transport the company's Unga shore station's crew and catch to Seattle. Under the command of Captain Andrew Johnson, the vessel departed Unga on November 19. Among the approximately thirty passengers and crew aboard was Johnson's new wife. For the captain and his bride, the voyage was to be something of a honeymoon. That, it definitely was not. There is no record of the vessel's fate, but fifteen bodies—unidentifiable but believed to be from the *Nellie Coleman*—washed ashore near Cape Yakataga, near Yakutat, in early December.

The *Nellie Coleman* was replaced in 1906 by the *Maid of Orleans*, and the *Carrier Dove* was sold two years later. John Grotle made two trips to Alaska (one summer, one winter) until at least 1910, and continued as a codfish schooner captain until his retirement in 1936 (see below). In 1910, the King & Winge Codfish Company (see below) absorbed the Seattle-Alaska Fish Company.[150]

Blom Codfish Company and the Northern Codfish Company

The Blom Codfish Company was organized in Tacoma, Washington, in early 1905 and sent the three-masted, 195-ton schooner *Falcon* to the Bering Sea that May. While the *Falcon* was at sea, the company constructed a curing plant at Quartermaster Harbor, between the south ends of Vashon and Maury islands, in Puget Sound, near Tacoma. The *Falcon* returned in September with a modest-for-her-size cargo of 60,000 pounds. During the fall of 1905, Blom Codfish built a shore station on the north shore of Eagle Harbor, on Nagai Island, one of the Shumagin Islands. The station, however, was abandoned after only two years' operation. The following year, the *Falcon* was replaced by the 138-ton, two-masted schooner *Fortuna*, which was promptly sent to the Bering Sea.

The Blom Codfish Company, according to John Cobb, had a "very checkered career."[151] The company's problems probably peaked in 1910, when Captain Torvald Blom, the company's president, was found dead in a Tacoma park. His throat had been cut. Police were unsure whether he had been murdered or had committed suicide, but there were allegations that Blom had been mixing his personal finances with those of the company and owed the company money.

The *Fortuna* made its last trip for the Blom Codfish Company in 1912. In 1914, the company was dissolved, and its assets, including the *Fortuna*, became the property of Seattle parties who had organized the Northern Codfish Company to carry on the business. The new company sent the *Fortuna* to the Bering Sea in 1915, and the vessel returned with a satisfactory catch. The Northern Codfish Company, however, left the industry early in 1916, and it chartered the *Fortuna* to the Pacific Coast Codfish Company for that year's season, after which the vessel's career in the codfish industry, and the existence of the Northern Codfish Company, seems to have ended.[152]

King & Winge Codfish Company and the Western Codfish Company

The King & Winge Codfish Company, a subsidiary of Seattle's King & Winge Shipbuilding Company, entered the codfish industry in 1905. The 185-ton schooner *Harold Blekum* was readied in the King & Winge shipyard and sent to the Bering Sea in April of that year. Meanwhile, the King & Winge Codfish Company constructed a curing plant in West Seattle, and it later added the schooner *Vega* to its fleet.

In 1910, the San Francisco interests who later purchased J. A. Matheson's Ancacortes operations (see above) attempted to consolidate several Puget Sound cod-fishing companies. The group secured a controlling interest in the King & Winge Codfish Company, which then absorbed the Seattle-Alaska Fish Company. It then began operating as the Western Codfish Company, which advertised itself in October 1911 as "Representing the consolidation of the King & Winge Codfish Co. and the Seattle & Alaska Fish Co." The Western Codfish Company, however, withdrew from active fishing operations early in 1912, marking the end of the cod-fishing industry in Seattle.[153]

Pacific Coast Codfish Company

John Grotle and his fellow stockholders of the Seattle-Alaska Fish Company must have had a plan when they sold out to the King & Winge Codfish Company in 1910 (see above), for the following year, with J. E. Shields, a longtime associate of Grotle, they incorporated the Pacific Coast Codfish Company. The new firm purchased the schooner *John A.*, and Grotle promptly took the vessel north. The firm also began construction of a curing station at Poulsbo, on the western shore of Puget Sound. In addition to space for processing and packaging codfish, the company's Poulsbo plant had storage for 1,000 tons of codfish.[154]

J. E. Shields managed the company, but he gradually bought out his partners and in 1930 became the sole owner.[155] Under Shields's leadership (either as manager or owner), Pacific Coast Codfish Company was active in the salt cod fishery for four decades. In 1913, the company added the schooner *Charles R. Wilson*, followed by the *Maid of Orleans* in 1914. In 1915, the schooner *Fortuna* was added to the fleet. The *C. A. Thayer*—destined to become a historic ship (see chapter 8)—was added ten years later, and on its maiden cod-fishing voyage to the Bering Sea caught 256,160 codfish, one of the largest catches ever made by a schooner. John Grotle was the vessel's captain.[156]

The Pacific Coast Codfish Company also for a time supplied San Francisco curing plants with codfish. The loss of the schooner *Maweema*

Figure 23. Pacific Coast Codfish Company, canned salted codfish label. In 1968, Petersburg Fisheries, an Alaskan-owned seafood company established in 1965, purchased the Icicle brand from J. E. Shields's family. In 1976, the company changed its name to Icicle Seafoods, which became one of the largest seafood processors in Alaska. (Courtesy of Karen Hofstad.)

near St. George Island in the fall of 1928 left the Alaska Codfish Company without a source of codfish. Alfred Greenbaum, who managed the Alaska Codfish Company, contracted with J. E. Shields to supply the company's Redwood City processing plant with two cargoes of salted codfish during the 1929 season. To help fulfill the contract, Shields converted the four-masted lumber carrier *Sophie Christensen*, which he had purchased several years prior, into a codfisher. At 181 feet and 570 tons, the *Sophie Christensen* was the largest of the Pacific coast codfish schooners. It also became, in the words of J. E. Shields's son, Ed, "the pride and joy of old Capt. Shields, the flag ship of his fleet and the jewel of his eyes."[157] On its maiden cod-fishing voyage to the Bering Sea, the *Sophie Christensen* made a good catch of fish, as did the *Charles R. Wilson*, the other vessel contracted to supply the Alaska Codfish Company. The following year, both vessels were contracted to supply codfish to the Union Fish Company, which had absorbed the Alaska Codfish Company (see chapter 8). The *Sophie Christensen* was laid up in Seattle's Lake Union for the 1931 season.

During the 1931 season, however, the crew of the *C. A. Thayer*, again under John Grotle, caught what was at the time a Pacific coast fleet record 302,000 codfish. The following year, the *C. A. Thayer* was replaced by the *Sophie Christensen*.[158] On its first cod-fishing voyage to the Bering Sea for the Pacific Coast Codfish Company, in 1932, its captain was John Grotle, whom *Pacific Fisherman* praised that year as the "dean of the Pacific Coast codfishermen."[159] J. E. Shields appointed himself captain and took the *Sophie Christensen* north in 1933 and each year thereafter until 1942, when the U.S. Army requisitioned the vessel and converted it for use as a supply barge.

John Grotle continued to operate codfish schooners until he retired in 1936, after his thirty-fifth consecutive cod-fishing voyage. He died the following year, and was eulogized by *Pacific Fisherman* as "the last of a school of veteran codfish masters."[160]

Sophie Christensen, 1939: *Then when at anchor on the cod banks off Nelson Lagoon in the early part of the season, the vessel sprang leaks. Father [J. E. Shields] had the Coast Guard tow the vessel into Port Moller where he placed her on the beach and made the necessary underbody repairs at low tide. On floating off the last day at high tide, a severe S.E. squall hit the vessel and carried her very high on the beach. The next several days were severe storms and the vessel became stranded as the storm built a sand bar outside the vessel. Father had to hire a Foss tug to come and propeller dredge the sand so the vessel could be floated off.*

—Ed Shields, 1981[161]

The *C. A. Thayer* remained laid up in Seattle's Lake Union until 1942, when she, too, was requisitioned by the U.S. Army and converted into a supply barge. At war's end, the Pacific Coast Codfish Company purchased the *C. A. Thayer* back from the Army and returned it to the cod fishery in 1946. The vessel made its final cod-fishing voyage to the Bering Sea in 1950, marking the end of the Salt Cod Era (see chapter 8).[162]

ENDNOTES

114 Captain Jarvis, "The Fisheries of Alaska," *Pacific Monthly* (June 1906): 705–708.

115 C. P. Overton, Union Fish Company, to Captain Walter Tinn, Pirate Cove, Alaska, July 24, 1912, John N. Cobb Collection, Special Collections, University of Washington Libraries, Seattle, Washington.

116 John N. Cobb, *Pacific Cod Fisheries*, Bureau of Fisheries Doc. No. 830, appendix 4 to *Report of the U.S. Commissioner of Fisheries for 1915* (Washington, DC: GPO, 1916), 29.

117 C. P. Overton, "Pioneers in the Pacific Coast Codfish Industry," *Pacific Fisherman* (January 1906): 70–71, 75.

[118]Ibid.

[119]Ibid.

[120]Thomas J. Vivian, *Commercial, Industrial, Agricultural, Transportation and Other Interests of California* (Washington: Bureau of Statistics, 1891), 487.

[121]Slocum said none of his crew had ever before seen a codfish. When fishing was at its best, however, some crew members easily caught an average of five hundred fish per day.

[122]John N. Cobb, *Pacific Cod Fisheries*, Bureau of Fisheries Doc. No. 830, appendix 4 to *Report of the U.S. Commissioner of Fisheries for 1915* (Washington, DC: GPO, 1916), 28; J. W. Collins, *Report on the Fisheries of the Pacific Coast of the United States*, in *Report of the United States Commissioner of Fish and Fisheries for 1892* (Washington, DC: GPO, 1892), 97, 103.

[123]Ivan Petroff, *Report on the Population, Industries, and Resources of Alaska* [1882], in *Seal and Salmon Fisheries and General Resources of Alaska*, vol. 4 (Washington, DC: GPO, 1898), 267.

[124]Marshall McDonald, *Report of the Commissioner of Fish and Fisheries Relative to the Salmon Fisheries of Alaska*, 52nd Cong., 1st Sess., Sen. Misc. Doc. No. 192 (Washington, DC: GPO, 1892), 4–5.

[125]J. W. Collins, *Report on the Fisheries of the Pacific Coast of the United States*, in *Report of the United States Commissioner of Fish and Fisheries for 1892* (Washington, DC: GPO, 1892), 91, 100, 108.

[126]J. W. Collins, *Report on the Fisheries of the Pacific Coast of the United States*, in *Report of the United States Commissioner of Fish and Fisheries for 1892* (Washington, DC: GPO, 1892), 100.

[127]John N. Cobb, *Pacific Cod Fisheries* (revised edition, 1926), Bureau of Fisheries Doc. No. 1014, appendix 7 to *Report of the U.S. Commissioner of Fisheries for 1926* (Washington, DC: GPO, 1927), 468–485.

[128]John N. Cobb, *Pacific Cod Fisheries* (revised edition, 1926), Bureau of Fisheries Doc. No. 1014, appendix 7 to *Report of the U.S. Commissioner of Fisheries for 1926* (Washington, DC: GPO, 1927), 407, 473, 495; "Experienced Cold Weather," *San Francisco Call*, January 29, 1903; John N. Cobb, *Commercial Fisheries of Alaska in 1905*, Bureau of Fisheries Doc. No. 603 (Washington, DC: GPO, 1906), 12; "Pacific Codfish Catch Improves During 1929 Season," *Pacific Fisherman* (January 1930): 207, 209; "Codfish," *Pacific Fisherman* (October 1905): 18; John N. Cobb, *Commercial Fisheries of Alaska in 1905*, Bureau of Fisheries Document No. 603

(Washington, DC: GPO, 1906), 14; "Storms in Alaska Will Interfere with the Season's Codfish Catch," *San Francisco Call*, May 29, 1903; "Schooner and Thirty Men Lost," *San Francisco Call*, March 18, 1905; "Discover Wreckage of Schooner Pearl," *Los Angeles Herald*, August 27, 1905; Ed Shields, *Salt of the Sea: The Pacific Coast Cod Fishery and the Last Days of Sail* (Lopez Island, Wash.: Pacific Heritage Press, 2001), 35–36; "Many Corpses Washed Ashore," *[San Jose, California] Evening News*, February 13, 1916; "Codfish Market," *Pacific Fisherman* (September 1908): 33–34.

[129]John N. Cobb, *Pacific Cod Fisheries* (revised edition, 1926), Bureau of Fisheries Doc. No. 1014, appendix 7 to *Report of the U.S. Commissioner of Fisheries for 1926* (Washington, DC: GPO, 1927), 411, 476.

[130]"Fifty Years of Cod Fishing," *Pacific Fisherman* (January 1923): 99.

[131]"Pacific Coast Codfish Review," *Pacific Fisherman* (February 1907): 55–58.

[132]"Captain John E. Matheson," *Pacific Fisherman* (December 1908): 15; "Fifty Years of Cod Fishing," *Pacific Fisherman* (January 1923): 99; "Expansion of Anacortes Codfishing Industry," *Anacortes American*, January 28, 1909; John N. Cobb, *Pacific Cod Fisheries*, Bureau of Fisheries Doc. No. 830, appendix 4 to *Report of the U.S. Commissioner of Fisheries for 1915* (Washington, DC: GPO, 1916), 34; "Our Big Codfishing Industries," *Anacortes American*, Annual Edition, 1907.

[133]"Will Go Into Codfish Business," *Pacific Fisherman* (April 1905): 15; "Codfish Schooners Leave," *Pacific Fisherman* (April 1905): 15; John N. Cobb, *Pacific Cod Fisheries* (revised edition, 1926), Bureau of Fisheries Doc. No. 1014, appendix 7 to *Report of the U.S. Commissioner of Fisheries for 1926* (Washington, DC: GPO, 1927), 414, 479–481; Western Codfish Co., adv., *Pacific Fisherman* (October 1911): 26; "Salt Fish Review," *Pacific Fisherman* (January 1915): 102–103; "Matheson Will Get Back Into Codfish Industry," *Anacortes American*, December 10, 1914; "Matheson Codfish Company Starts Building New Plant," *Anacortes American*, August 10, 1916; J. A. Matheson, adv., *Pacific Fisherman* (January 1918): 32; Gordon Newell, ed., *H. W. McCurdy Marine History of the Pacific Northwest* (Seattle: Superior Publishing Company, 1966), 399.

[134]John N. Cobb, *Pacific Cod Fisheries* (revised edition, 1926), Bureau of Fisheries Doc. No. 1014, appendix 7 to *Report of the U.S. Commissioner of Fisheries for 1926* (Washington, DC: GPO, 1927), 412.

[135]John N. Cobb, *Pacific Cod Fisheries* (revised edition, 1926), Bureau of Fisheries Doc. No. 1014, appendix 7 to *Report of the U.S. Commissioner of Fisheries for 1926* (Washington, DC: GPO, 1927), 413.

[136]John N. Cobb, *Pacific Cod Fisheries* (revised edition, 1926), Bureau of Fisheries Doc. No. 1014, appendix 7 to *Report of the U.S. Commissioner of Fisheries for 1926*

(Washington, DC: GPO, 1927), 413; Alaska Shipwrecks, accessed May 5, 2016, http://alaskashipwreck.com/shipwrecks-a-z/alaska-shipwrecks-l/.

[137] Untitled, *Pacific Fisherman* (March 1904): 7; "Cod Fisheries," *Pacific Fisherman* (April 1904): 12; John N. Cobb, *Pacific Cod Fisheries* (revised edition, 1926), Bureau of Fisheries Doc. No. 1014, appendix 7 to *Report of the U.S. Commissioner of Fisheries for 1926* (Washington, DC: GPO, 1927), 413–414.

[138] "Our Big Codfishing Industries," *Anacortes American*, Annual Edition, 1907.

[139] "Our Big Codfishing Industries," *Anacortes American*, Annual Edition, 1907; "Beautiful Plant Now Complete," *Anacortes American*, July 20, 1905.

[140] "Our Big Codfishing Industries," *Anacortes American*, Annual Edition, 1907.

[141] "Anacortes is 'The Gloucester of the Pacific,'" *Anacortes American*, October 12, 1911.

[142] "Codfish," *Pacific Fisherman* (October 1905): 18; "Codfishermen from Gloucester," *Pacific Fisherman* (March 1906): 28; "Schooner Alice Sailed Tuesday," *Anacortes American*, March 29, 1906; Francis Murphy (grandson of one of the five brothers), personal communication with author, Ketchikan, Alaska, February 28, 2015; "Codfish News," *Pacific Fisherman* (September 1906): 19–20; "Anacortes is 'The Gloucester of the Pacific,'" *Anacortes American*, October 12, 1911; "Codfish News," *Pacific Fisherman* (October 1908): 31; *Fisheries of Alaska in 1909*, Bureau of Fisheries Doc. No. 730 (Washington, DC: GPO, 1910), 43; Barton Warren Evermann, *Fishery and Fur Industries of Alaska in 1912*, Bureau of Fisheries Doc. No. 780 (Washington, DC: GPO, 1913), 65; John Pappenheimer, "Men Against the Sea, A Tale Spun From 1912," *Pacific Fisherman's Journal* (November 1986): 33–35; "Prominent Factors in the Pacific Cod Fishery," *Pacific Fisherman* (January 1927): 130; Gordon Newell, ed., *H. W. McCurdy Marine History of the Pacific Northwest* (Seattle: Superior Publishing Company, 1966), 211.

[143] John N. Cobb, *Pacific Cod Fisheries* (revised edition, 1926), Bureau of Fisheries Doc. No. 1014, appendix 7 to *Report of the U.S. Commissioner of Fisheries for 1926* (Washington, DC: GPO, 1927), 413–414; "Pacific Coast Codfish Review," *Pacific Fisherman* (January 1906): 67, 69; "Pacific Coast Codfish Review," *Pacific Fisherman* (February 1907): 55–58; W. F. Robinson, "The Codfish Industry," *The Coast* (December 1908): 365–367.

[144] Joe Follansbee, *Shipbuilders, Sea Captains, and Fishermen: The Story of the Schooner Wawona* (New York: iUniverse, 2006), 35.

[145] "Prominent Factors in the Pacific Cod Fishery," *Pacific Fisherman* (January 1927): 130; "Schooner Wawona Leaves on Maiden Fishing Trip," *Anacortes American*, April 2, 1914.

[146]John N. Cobb, *Pacific Cod Fisheries*, Bureau of Fisheries Doc. No. 830, appendix 4 to *Report of the U.S. Commissioner of Fisheries for 1915* (Washington, DC: GPO, 1916), 98–99; Joe Follansbee, *Shipbuilders, Sea Captains, and Fishermen: The Story of the Schooner Wawona* (New York: iUniverse, 2006), 37.

[147]"Pacific Codfishing," *Pacific Fisherman* (January 1936): 224–225; Joe Follansbee, *Shipbuilders, Sea Captains, and Fishermen: The Story of the Schooner Wawona* (New York: iUniverse, 2006), 35, 40; "Codfishing—Freakish Season Yields Fair Catches," *Pacific Fisherman* (October 1935): 48.

[148]"Bering Sea Codfish 1937," *Pacific Fisherman* (January 1938): 127–128.

[149]"Codfish Schooners Taken By Transport Service," *Pacific Fisherman* (September 1942): 59; Ward T. Bower, *Alaska Fishery and Fur Seal Industries: 1941*, Statistical Digest No. 5 (Washington, DC: GPO, 1943), 43–44. The *Wawona* fished for codfish each year from 1914 through 1947, except 1921 and 1942–1945. The vessel's career total codfish catch is calculated from data in John N. Cobb's 1926 *Pacific Cod Fisheries* and in reports in *Pacific Fisherman.*

[150]Ed Shields, *Salt of the Sea: The Pacific Coast Cod Fishery and the Last Days of Sail* (Lopez Island, Wash.: Pacific Heritage Press, 2001), 33–36; "31 Years a Codfish Skipper," *Pacific Fisherman* (March 1932): 41–42; "Alaska Cod Sold Only In East. Coast Buyers Will Not Take Fish Until They Are Returned Here Bearing Label Showing Atlantic Origin," *Seattle Times*, November 7, 1905; Gordon Newell, ed., *H. W. McCurdy Marine History of the Pacific Northwest* (Seattle: Superior Publishing Company, 1966), 125; "Many Corpses Washed Ashore," *Evening News* (San Jose, Calif.), February 13, 1916; John N. Cobb, *Pacific Cod Fisheries* (revised edition, 1926), Bureau of Fisheries Doc. No. 1014, appendix 7 to *Report of the U.S. Commissioner of Fisheries for 1926* (Washington, DC: GPO, 1927), 413, 475–476, 495.

[151]John N. Cobb, *Pacific Cod Fisheries* (revised edition, 1926), Bureau of Fisheries Doc. No. 1014, appendix 7 to *Report of the U.S. Commissioner of Fisheries for 1926* (Washington, DC: GPO, 1927), 414.

[152]John N. Cobb, *Pacific Cod Fisheries* (revised edition, 1926), Bureau of Fisheries Doc. No. 1014, appendix 7 to *Report of the U.S. Commissioner of Fisheries for 1926* (Washington, DC: GPO, 1927), 419, 476, 480; Gordon Newell, ed., *H. W. McCurdy Marine History of the Pacific Northwest* (Seattle: Superior Publishing Company, 1966), 112; "Pacific Codfish Department," *Pacific Fisherman* (July 1911), 26; "Think Captain Ended Life," *Los Angeles Herald*, September 11, 1910.

[153]John N. Cobb, *Pacific Cod Fisheries* (revised edition, 1926), Bureau of Fisheries Doc. No. 1014, appendix 7 to *Report of the U.S. Commissioner of Fisheries for 1926* (Washington, DC: GPO, 1927), 414, 479–481; Western Codfish Co., adv., *Pacific Fisherman* (October 1911): 26.

[154] Ed Shields, *Salt of the Sea: The Pacific Coast Cod Fishery and the Last Days of Sail* (Lopez Island, Wash.: Pacific Heritage Press, 2001), 174; "31 Years a Codfish Skipper," *Pacific Fisherman* (March 1932): 41–42.

[155] Richard Walker, "Saltcod and sail: Poulsbo Historical Society celebrates centennial of the Pacific Coast Codfish Co.," *North Kitsap Herald* (Poulsbo, Wash.), May 12, 2011; "Poulsbo's Captain Ed Shields crosses the bar," *North Kitsap Herald* (Poulsbo, Wash.), June 10, 2008; "North Pacific Cod Fishery Yield Above Average of Recent Years," *Pacific Fisherman* (January 1931): 215–216.

[156] John N. Cobb, Pacific Cod Fisheries, Bureau of Fisheries Doc. No. 830, appendix 4 to Report of the U.S. Commissioner of Fisheries for 1915 (Washington, DC: GPO, 1916), 36; "Cod Schooners Bring Large Catches," *Pacific Fisherman* (October 1925): 46.

[157] Ed Shields, "Sailing on the *Sophie Christensen*," *Alaska Fisherman's Journal* (November 1981): 90–92, 94–96.

[158] "Shields Buys C. A. Thayer," *Pacific Fisherman* (April 1925): 46.

[159] "31 Years a Codfish Skipper, *Pacific Fisherman* (March 1932): 41–42.

[160] "Capt. John Grotle Dies," *Pacific Fisherman* (September 1937): 63.

[161] Ed Shields, "Sailing on the Sophie Christensen," *Alaska Fisherman's Journal* (November 1981): 90–92, 94–96.

[162] Ed Shields, *Salt of the Sea: The Pacific Coast Cod Fishery and the Last Days of Sail* (Lopez Island, Wash.: Pacific Heritage Press, 2001), 203; Ward T. Bower, *Alaska Fishery and Fur Seal Industries: 1946*, Statistical Digest No. 17 (Washington, DC: GPO, 1948), 46–47; M. J. Harris, "The History of the *C. A. Thayer*," *Sea Letter*, No. 68 (2007): 4–5.

Chapter 4
SCHOONERS, DORIES, AND CODFISHERMEN

Off our Alaska coast are to be found the most extensive cod banks in the world. Those known as the offshore banks have an area of approximately 29,645 square miles, while the inshore banks have probably from one-third to one-half the area of the former.

—John Cobb, founding director, College of Fisheries, University of Washington, 1918[163]

For many years the oldest fishery in Alaska, and a very important one, that for cod, lagged behind in the utilization of the internal combustion engine. This conservation seems to be inherent in the industry, as to-day seven-tenths of the cod taken on both the Atlantic and Pacific banks are caught and landed in the same manner as our Atlantic ancestors caught and landed them in the Sixteenth century.

—John Cobb, founding director, College of Fisheries, University of Washington, 1919[164]

Figure 24. Union Fish Company schooner *Louise* transiting Golden Gate under full sail. (Courtesy of Union Fish Company.)

SCHOONERS

Many of the vessels that entered the early twentieth-century codfish industry had been built to carry lumber from Washington, Oregon, and Northern California to San Francisco and Los Angeles. They were relatively slow, beamy (broad) sailing vessels designed to carry large cargoes south and to return north without ballast. Several hundred of these vessels had been built. They were operated with small crews and had capacities that ranged from 200 to 800 tons, but they became obsolete after the turn of the century, when steam-powered vessels replaced them. Surplus sailing schooners, some of them very fine ships, were purchased by cod-fishing interests at minimal cost.

John Cobb observed in 1915 that all the fishing vessels in the Pacific cod fishery were schooners, and that not a single one had been built for that purpose. All had entered the fishery "after they had attained varying ages." The vessels ranged in length from 102 feet to 156 feet, with net tonnages ranging from 138 to 413—more than four times that of the vessels that had pioneered the fishery in the 1860s.[165]

One characteristic of the codfish schooner fleet was that many of the vessels were *bald-headed*—they carried no top masts. As *Pacific Fisherman* noted, "You could always tell a codfisherman by her stubby masts, for that fleet fished where the weather made hazards out of topmasts."[166] Another characteristic—which became traditional at least by about 1930—was that their hulls were painted green, and their bulwarks and houses were painted white.

Some schooners were equipped with steam winches to haul the anchor(s). By 1915, gasoline engines had replaced steam power. Gasoline engines were also used to power generators that charged batteries that provided power for radios and lighting.

Codfish schooners generally carried twelve to twenty-four dories and had crews of twenty-two to forty-three men. Some codfish schooners carried cargo during the off-season.[167]

Overall, sailing ships were a good match for the needs of the salted codfish industry. The industry was not especially lucrative, and the fact that the vessels were wind powered kept costs down, although operating a sailing ship did require specialized crew skills that were of diminishing availability as the era of sail receded into the past. Moreover, the wooden construction of the vessels was an important asset in the salted codfish industry. As J. E. Shields, an industry veteran, explained, "You can't put [salted] codfish into an iron ship," because "the salt would destroy the ship and the rust from the iron would destroy the fish."[168]

Figure 25. Doryman with codfish schooner *John A.*, circa 1919–1920. (Glenn Burch manuscript, San Francisco Maritime National Historical Park.)

Unlike fish that must be delivered fresh to the market or a processing plant, salted codfish was not fundamentally a time-sensitive product. A week or a month's delay in returning to port because of a lack of wind or contrary winds meant little other than that the portion of the crew that was salaried—the dress gang (the men who dressed, cleaned, and salted the catch), the cook and his helpers, and the watchman—would have to be paid for the additional time at sea. Although they were sometimes given a small stipend—*run money*—for their part in sailing the vessel to and from the fishing grounds, fishermen, which included a vessel's officers (except for the captain), were paid according to their production, not the duration of the voyage.

During the winter of 1904, the schooner *Czarina* sailed from Pirate Cove, in the Shumagin Islands, to San Francisco—a distance of approximately 1,900 nautical miles—in eleven days. By contrast, in the fall of 1947, the *Wawona* unsuccessfully spent eleven days attempting to beat through Unimak Pass, in the eastern Aleutians, against gale-force headwinds before a U.S. Navy tugboat towed it through.[169]

In about 1920, powerboats that were approximately 25 to 30 feet long came into use in conjunction with some schooner operations. The powerboats distributed the dories over the banks early in the day and, at frequent intervals during the day, tendered the dories' catches back to the schooner for processing. At the end of the fishing day, the powerboats brought the dories back to the schooner.

This method enabled dorymen to fish in locations that were farther from the schooner than were practical to row or sail to and—importantly—to spend more time fishing. By 1923, power vessels in approximately the 40- to 50-foot range were used in conjunction with traditional schooner fishing. In the words of *Pacific Fisherman* in 1923, this was the "modern system of codfishing."[170] Several years later, however, outboard motor–powered dories came into common use (see below), which made tender vessels unnecessary.

SCHOONER LIFE

> **Schooner *Charles R. Wilson*, 1916:** *There we were, 24 men, no ventilation, kerosene lamps burning 24 hours a day, bilge water pushing its stench up the forecastle deck drains, and still everybody happy.*
>
> —G. E. Hermanson, 1972[171]

> **Schooner *Fanny Dutard*, April 22, 1924:** *Luckily, the Cape Flattery weather was nice with a light breeze. The biggest part of the crew were either dead drunk in their bunks or still celebrating the departure. The only sober people aboard were the skipper and the splitter (dead sober), the mate (half sober) and a couple of green kids like myself, who had not yet been initiated into the finer aspects of a cod fisherman's life.*
>
> —Gus Dagg, 1975[172]

> **Schooner *Fanny Dutard*, 1926:** *Everything was done by hand on this vessel, there was no auxiliary power to assist in launching and taking dories on board, or handling the vessels sails etc., in fact conditions were no more modern than those depicted in the Bible.*
>
> —Jimmy Crooks, undated[173]

In the spring of 1937, Russ Hofvendahl, a fifteen-year-old orphan in San Francisco, lied about his age to get a berth on the Union Fish Company's codfish schooner *William H. Smith*, which was bound for the Bering Sea. Hofvendahl recounted his experience in his 1983 book *Hard on the Wind*. The 496-ton, four-masted schooner had been built in 1899 for the lumber trade. Its first cod-fishing voyage was in 1925. The *William H. Smith* carried twenty power dories and a crew of forty-three men.

Hofvendahl first met a number of his crewmates as they were being unceremoniously loaded aboard, drunk. Their drunkenness didn't seem to bode well for the voyage, but the men knew their jobs, and, once the alcohol worked its way out of their systems, they worked very hard. As was the general rule in the entire cod-fishing fleet, no alcohol was permitted aboard the *William H. Smith.*

Many of the *William H. Smith*'s crew—especially the fishermen—were of Scandinavian descent, and all were certified as able-bodied seaman.

Three dozen crewmen slept in bunks—likely in three tiers—in the vessel's cramped, dingy, and marginally ventilated forecastle (below deck in the ship's bow), where the only amenities were one or two kerosene lamps that burned continuously, and a coal stove.

Jimmy Crooks, who crewed aboard the schooner *Fanny Dutard* in 1926, sarcastically acknowledged that his vessel's owner, "out of the goodness of his heart," had provided each bunk with a mattress: a long burlap sack stuffed with straw and referred to as a "donkey breakfast." According to Crooks, the mattresses (which were leftovers from previous voyages) were infested with bed bugs, fleas, and lice. Of the forecastle, Crooks noted that "[tines] crisscrossed the stinking glory hole, and upon these lines wearing apparel both wet and dry was draped," and he complained that the *Fanny Dutard* suffered from an infestation of rats.[174]

Neither the *William H. Smith*'s nor the *Fanny Dutard*'s forecastle had sanitary facilities. Those living there did their call-of-nature business squatted over a bucket or "over the bow," the latter of which would have been a challenge in even modestly foul weather conditions. Men were advised to "Close your mouth and eyes, let it fly, and trust to luck."[175] Aboard the *Fanny Dutard*—and perhaps aboard the *William H. Smith* as well—one of the captain's amenities was a private latrine. It was of the sentry box type: a one holer that was flushed with seawater hauled up in a bucket.

Aboard the *William H. Smith*, the captain, three mates, cook, radioman, and the head splitter (the most skilled member of the codfish processing crew) lived in stern quarters, all above the waterline and with ports that could be opened. Aboard the *Fanny Dutard*, the captain, navigator, and two fishermen who served also as first and second mates lived in cubbyhole-like rooms in the stern. They, too, slept on straw mattresses.

The *William H. Smith* carried fresh water, but only enough for drinking. Two barrels collected rainwater off the forepeak deck and the roof of the deckhouse, but this water was usually brackish from salt spray. Except for the water in the barrels, the crew had no alternative but to bathe and shave with seawater. On the *Fanny Dutard*, one bucket of fresh water was allocated each day for the entire crew's hygienic needs. Galley utensils were cleaned by boiling them in seawater.

Clothing, too, was washed in seawater. Fishermen usually packed only one or two changes of clothing and laundered them as time permitted. Laundered clothing was sometimes hung from a vessel's rigging to dry in the wind. This was called "blowing out the stink."

Schooners sometimes replenished their freshwater supply from rivers and streams. Going ashore in dories to fill barrels with fresh water might be the only time during the voyage that crew members had an opportunity to walk on dry land. Depending on the situation, being ashore could provide other opportunities. Jimmy Crooks recalled that "[f]ires were built with drift wood and like hoboes in the jungle we boiled out our cloths in buckets and tin cans and bathed and scrubbed our bodies with sand and beat the lice in our clothing to death between rocks on the creek banks."[176]

Food on the *William H. Smith* was not fancy, but it was adequate. The vessel had no freezer, but left San Francisco with a supply of frozen meat that was eaten during the first ten days of the voyage north. Once that meat was gone, the men got by on salt pork, bacon, and corned beef. Potatoes and onions were carried too, as were fresh eggs, the latter of which lasted the full five-month duration of the voyage. Salt cod—unsold Union Fish Company product from the previous year—was served at least once each day, boiled. Cod-fishing crews apparently relished their *mug-ups*. Included in the schooner *Wawona*'s provisions for its 1936 cod-fishing voyage was 6,000 pounds of coffee. The coffee was brewed on the coal-fired galley stove in a five-gallon pot. Jean Bagger, the cook on the *Wawona* during the years 1926 through 1936, recalled baking thirty-eight loaves of bread each morning before awakening the crew at about 3:30 a.m.

Aboard the *C. A. Thayer* in the waning years of the Salt Cod Era, dinner—the main hot meal of the day—was served at around 9:30 a.m., just after the fishermen returned to the schooner with their first loads of fish. Supper was served in late afternoon, after the fishermen returned to the schooner with their second loads of fish.

Medical care was extremely limited. According to G. E. Hermanson, who as a teenager was a member of the dress gang on the schooner *Charles R. Wilson* in 1916, the U.S. Coast Guard was supposed to call on each schooner once each month to check whether any men aboard were ill and to bring mail, if there was any. On the *William H. Smith*, a crew member who departed San Francisco suffering from venereal disease was transferred to a Coast Guard cutter at the first opportunity.

News of the outside world was received via mail carried north on Coast Guard cutters and delivered when the opportunity presented itself. Additionally, the *William H. Smith*'s radioman periodically tacked a hand-printed news bulletin to the galley bulkhead.

One of the chores fishermen performed during the voyage north was fabricating lead fishing weights. This was done in the galley, where scrap lead was melted in a steel pot over an open section of the galley stove. For a mold, the men fashioned an elongated cone of newspaper, which was then placed in a bucket and carefully surrounded by rock salt. Molten lead was then poured into the mold, soon after which a gloved crewman shoved the shank of an eye hook into the lead and held it steady until the lead began to harden. The spreader (see chapter 5) would later be attached to this eye hook. The fishing line was attached through a hole drilled through the small end of the weight. It took the crew of the *William H. Smith* three mornings to pour about eighty weights.

The crewmen of *William H. Smith* that season caught 249,509 codfish, with a salted weight of 360.75 tons. (To ensure the owners of cod-fishing vessels provided an accurate weight, the crew of each vessel designated a "delegate" to monitor the weighing of the catch.) The vessel's highliner, Paul Ebbelein, caught 22,732 codfish.[177]

A 1929 state report on California's fisheries noted that many cod fishermen, once ashore, would squander their entire earnings in one big party or would be bilked of it by confidence men. An option for some of these unfortunate men was to work at the codfish curing plant until the schooners left the following spring.[178]

It is doubtful anyone knew at the time, but the 1937 voyages of the Union Fish Company's *William H. Smith* and *Louise* were the last cod-fishing voyages to Alaska by San Francisco–based vessels.

DORIES

> *The dory was a man's home when he was fishing. Each one was an independent boat tailored to suit the man. . . . He spent much of his spare time on the way from Seattle to the fishing grounds redoing things in order to make them just exactly the way he wanted it.*
>
> —Ed Shields, Pacific Coast Codfish Company[179]

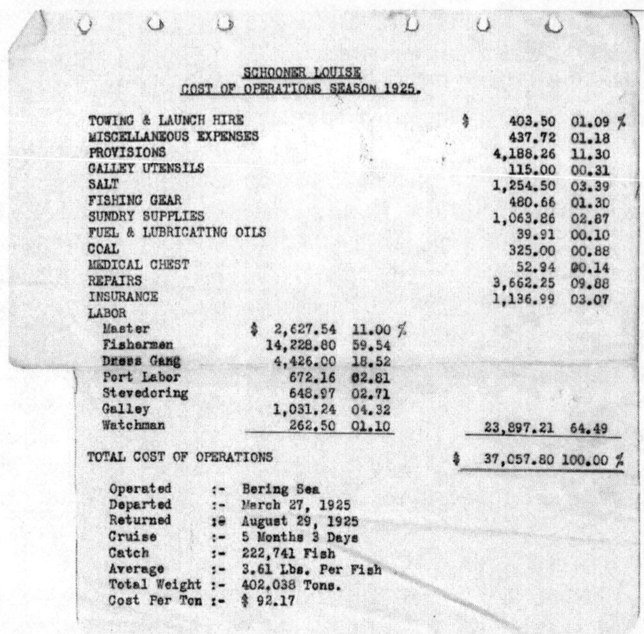

SCHOONER LOUISE
COST OF OPERATIONS SEASON 1925.

TOWING & LAUNCH HIRE			$ 403.50	01.09 %
MISCELLANEOUS EXPENSES			437.72	01.18
PROVISIONS			4,188.26	11.30
GALLEY UTENSILS			115.00	00.31
SALT			1,254.50	03.39
FISHING GEAR			480.66	01.30
SUNDRY SUPPLIES			1,063.86	02.87
FUEL & LUBRICATING OILS			39.91	00.10
COAL			325.00	00.88
MEDICAL CHEST			52.94	00.14
REPAIRS			3,662.25	09.88
INSURANCE			1,136.99	03.07
LABOR				
Master	$ 2,627.54	11.00 %		
Fishermen	14,228.80	59.54		
Dress Gang	4,426.00	18.52		
Port Labor	672.16	02.81		
Stevedoring	648.97	02.71		
Galley	1,031.24	04.32		
Watchman	262.50	01.10	23,897.21	64.49
TOTAL COST OF OPERATIONS			$ 37,057.80	100.00 %

Operated :- Bering Sea
Departed :- March 27, 1925
Returned :- August 29, 1925
Cruise :- 5 Months 3 Days
Catch :- 222,741 Fish
Average :- 3.61 Lbs. Per Fish
Total Weight :- 402.038 Tons.
Cost Per Ton :- $ 92.17

Figure 26. Cost of operation, Union Fish Company schooner *Louise*, 1925 season. (Courtesy of Union Fish Company.)

The earliest codfishermen in Alaska did not fish from dories, but from the decks of schooners. By 1866, however, dory fishing had become standard.

Along with fishermen and their gear, dories were the basic production unit of the cod fisheries. The dories originally used in Alaska were *banks dories*, the traditional boats developed for codfish handlining on the Grand Banks and other fishing banks of the North Atlantic. As a craft from which an individual fisherman could efficiently handline for codfish in often-rough waters, the banks dory had no equal.

A banks dory was a narrow, open boat with a flat bottom, flaring sides, a sharp bow and a V-shaped transom. They were designed to be nested inside each other for transport on the decks of larger vessels. Banks dories were inexpensive to build and were manufactured in great numbers in shops along New England's coast. Although they were sometimes derided as "floating coffins," these dories—in the hands of experienced men—were among the safest and most seaworthy of small craft.

Figure 27. Dory bringing load of codfish to the schooner *John A.*, circa 1919–1920. (Glenn Burch manuscript, San Francisco Maritime National Historical Park.)

Figure 28. "Whiskey Bill" Helgeson handlining cod, Unga, Alaska. (John N. Cobb Collection, Special Collections, University of Washington Libraries, Seattle, Wash.)

Banks dories came in several sizes. Traditionally, a single fisherman fishing with handlines in the Atlantic Ocean used a dory that was 13 feet long on the bottom and a little over 16 feet overall. A single fisherman fishing with handlines in Alaska waters, however, usually used a dory that was 14 feet long on the bottom and a little over 16 feet overall. Such a dory was known as a *14-foot dory* and could reasonably hold about a half ton of codfish, depending upon the weather. The dories, however, were routinely overloaded. For example, in 1916, a fisherman on the *Charles R. Wilson* returned to the schooner after a very successful morning's fishing with only about an inch of freeboard amidships. In his dory were 165 large codfish that, if they averaged 13 pounds, would have had a total weight of 2,145 pounds. Fortunately, the sea was calm.

The inside of the dory was divided into several compartments by *kit boards*—removable dividers that enabled a fisherman to store his fish in a manner that kept them from sliding around as well as allowing him to trim (balance) his dory fore and aft, which were both major safety considerations. Rope straps fastened to both the bow and the stern of the dory facilitated hoisting the dory onto the deck of a schooner, which until about 1920 was done by hand with a block and tackle.

Banks dories were traditionally propelled by the "ash breeze"—rowing. In addition to oars, most fishermen in Alaska also carried in their dory a small sail that could be used when the wind was favorable. (Because the dories had no keels, they could be sailed only downwind.) When weather prevented fishermen from getting back to the schooner, they sometimes fashioned sea anchors from their sails to slow their dories' downwind drift.

In addition to the necessary fishing gear, each dory also carried a small keg (or sometimes a whiskey bottle with a wooden plug) of water, a bailer, a box compass, a horn to signal the dory's location or to call for help, a knife for cutting bait and bleeding fish, a gaff for handling large fish, a ten- to fourteen-pound anchor, and a wooden windlass (*gurdy*) for hauling the anchor.[180]

By the late 1880s, at least one San Francisco codfish company, Lynde & Hough, was manufacturing dories for its Alaska cod-fishing operations. The company's dories corresponded in size and shape to banks dories manufactured in New England, but had heavier bow stems, planking, and timbers. Also, Lynde & Hough's dories were fitted with galvanized iron oarlocks instead of wooden thole pins.

By the early 1890s, the McCollam Fishing & Trading Company, also of San Francisco, was likewise manufacturing the dories.[181] By

1916, several West Coast boat builders were manufacturing banks dories, but a considerable portion of the dories used in the Pacific cod fishery were still manufactured in New England and transported overland to the West Coast. The use of New England–built dories was especially prevalent in Puget Sound: all the dories used by the Pacific Coast Codfish Company, at Poulsbo, Washington, were built to the company's specifications—basically extra-strong construction—on the East Coast.

Dories carried on schooners were assigned to individual fishermen through a drawing held early in each voyage. The dories were numbered, and each dory's number was written on a slip of paper. The papers were then placed in a hat and mixed about. Each fisherman then drew a dory number from the hat and fished from (and maintained) that dory for the entire voyage.

At the end of each day's fishing, all of a schooner's dories were brought aboard, emptied of their removable gear, nested on deck, and fastened securely. When the day's work was done, crewmen secured everything on deck to minimize the potential for damage if a storm suddenly arose.[182]

Motorized dories

Shore-station fishermen (see chapter 6) usually fished in sheltered areas within about five miles of the station. They generally avoided open waters, fearful that a sudden storm might blow them out to sea, from where the likelihood of a safe return to shore was almost nonexistent. Codfish in the early years (the first shore station was established in 1876) were abundant in the immediate vicinity of the stations, but by about 1900, there were signs of depletion, and the situation was worsening year by year. Station fishermen were forced to travel much greater distances to maintain their catches, although the fish they caught averaged smaller than formerly.

To enable reliable travel over these distances, in about 1911 Harry Hyland, a fisherman at Unga, installed a 7.5-horsepower gasoline engine in a dory. With another fisherman aboard to help him, Hyland then began fishing the outer edges of the inshore banks. His innovation was initially not well received among his fellow fishermen, who refused to dress Hyland's fish along with their own. (As will be described in chapter 6, at the end of a day's fishing, the station fishermen formed into "gangs" to jointly dress and salt their combined catch.) But the advantages of the motor could not be denied: Hyland's motorized dory enabled him to fish previously unfished banks, where codfish

Figure 29. Power dory *Satan*, Alaska Codfish Company, Unga, Alaska, 1915. (Silva Collection, Alaska State Library, Historical Section.)

were still abundant. Moreover, he could do so independent of tide and (within limitations) wind, a big factor among fishermen who were traditionally reliant upon oars and sails. Consequently, it was not long before several of the more progressive and industrious fishermen at the station followed Hyland's lead and fitted their dories with gasoline engines. By the late 1910s, there were enough power dories at some stations for the groups of fishermen who operated them to form their own dress gangs, and eventually some stations were forced to provide two separate dress houses, one for traditional dory fishermen and the other for powerboat fishermen. [183]

A problem with the power dories was that the additional weight made them difficult to launch or haul out on beaches, as was commonly done at shore stations. This was especially true during the winter, when inclement weather was more the rule than the exception. *Common* (nonmotorized) dories, on the other hand, were easily handled on the beaches. In his 1919 report on Alaska's fisheries, Ward Bower, of the Bureau of Fisheries, characterized winter fishing at the shore stations as being "carried on rather indifferently since the introduction of power dories."[184]

The schooner fleet—in which dory fishermen frequently fished five or six miles from their schooner and usually made at least two and often three round-trips daily, necessitating perhaps twenty to thirty miles of rowing or sailing—began experimenting with motorized dories in 1914. The schooner *Fortuna*, owned by the Northern Codfish Company, of Seattle, that year carried north a dozen portable gasoline engines suitable for use on banks dories. The engines were provided to the schooner's fishermen with the understanding that the fishermen would pay for them at the end of the season, the cost justified because the motors would theoretically enable the fishermen to make quicker, safer, and more numerous trips between the anchored schooner and nearby fishing spots. This experiment was a failure. According to *Pacific Fisherman*,

> [t]he men did not understand their operation and most of them would not take care of them. Also, when the engines were attached to the dories their weight detracted considerably from the carrying capacity of it, while in slightly rough weather the dories would ship seas which speedily put the motors out of commission.[185]

Despite this rough beginning, as the quality and suitability of small marine motors improved, they were increasingly fitted to dories. In 1919, the Evinrude Motor Company advertised in *Pacific Fisherman* its detachable, heavy-duty 3- and 4-horsepower motors that were "especially adapted to the fishing trade." Station fishermen and San Francisco–based schooner fishermen were the first to employ these motors.

Fitted with an outboard motor, a dory could travel at a speed of about five knots. The speed, however, increased the hazard—especially in a dory heavily laden with codfish—of dipping the bow into a wave and swamping the boat. To prevent this, the power dories were fitted with makeshift canvas spray hoods that covered the forward part of the boat.

A landmark in the usage of outboard motors occurred in 1928, when the *Charles R. Wilson*, owned by the Puget Sound–based Pacific Coast Codfish Company, equipped seventeen of its dories with outboard motors. The investment was successful: overall, the average catch of fishermen in power dories was far greater than that of fishermen in nonmotorized dories. The following year, the Pacific Coast Codfish Company and the Robinson Fisheries Company (Anacortes, Washington) purchased sixty-six 2.5-horsepower outboard motors to outfit the four schooners the companies were sending north that year. To ensure the outboard motors were properly serviced and cared for, each schooner carried an outboard mechanic.

Although some individual fishermen chose to forego the use of motors, during the 1929 season all but one vessel of the offshore codfish fleet (J. A. Matheson's *Fanny Dutard*) were equipped with power

Figure 30. Motorized dory, circa 1926. (Glenn Burch manuscript, San Francisco Maritime National Historical Park.)

dories. Additionally, there were that year fifty-five power dories in the shore-station fleet.[186]

In 1931, Dan McEachern was the highliner aboard the Pacific Coast Codfish Company's *C. A. Thayer* when the vessel delivered what was at that time a Pacific coast record catch of 302,000 codfish. McEachern landed 20,035 fish, and the average catch for the sixteen fishermen aboard the *C. A. Thayer* that season was 15,341 fish. (The dress gang, fishing "over the rail," [see chapter 5] caught almost 26,000 fish.) John Grotle, the *C. A. Thayer*'s captain and himself a legend, credited his vessel's success to having more good fishermen aboard than he had ever seen on a single vessel in his forty-two years in the codfish industry, combined with the fact that all the dories used were equipped with outboard motors. "Good fishermen and motors—that tells the story," said Grotle.[187]

To better accommodate the use of outboard motors, some dories were modified by adding a well, which is basically a boxed-in hole in the bottom of a boat, near its stern. The outboard motor was mounted on the top of the forward side of the well. A drawback of the engine wells was that the dories could no longer be nested on the deck of a schooner. A few could be stored individually on deck, but most were hung on steel davits over a vessel's side. In an apparently unsuccessful attempt to enable nesting of motorized dories, in 1931 the Union Fish Company schooners *William H. Smith* and *Louise* were outfitted with thirty-five new dories that had been constructed (by a San Francisco boatyard) with removable engine wells and were powered by 4-horsepower outboard motors. The wells were bolted to the dories' bottoms through brass flanges and rubber gaskets, and, with the outboard motors, they could be removed at the end of the day's fishing.

The next and final step in the evolution of the Pacific version of the banks dory was to make it larger. Most of the dories built after about 1929 measured 15 feet along the bottom and perhaps 19 feet on top. They were fitted with wells, permanent canvas spray hoods and engine covers, and they could carry about half again as much fish as the traditional banks dories. By 1937, 12-horsepower, air-cooled outboard motors had become somewhat standard on cod-fishing dories. However, during the 1940s, the dories on the Pacific Coast Codfish Company's sailing schooners *Charles R. Wilson* and *Sophie Christensen*, which carried sixteen and twenty-two dories, respectively, were equipped with 9.8-horsepower Johnson outboard motors. Under an agreement with the fishermen, each man owned the motor on his dory.

An ancillary advantage of power dories was that they could provide a modest element of maneuverability to a becalmed schooner or a schooner in tight quarters. In 1935, Charles Foss, captain of the *Wawona*, used power dories to tow his becalmed schooner toward the fishing grounds for more than a day before sufficient wind arose to propel the vessel.

In 1946, half of the eighteen dories carried by the *Wawona* were powered by outboard motors, while the other half were powered by four-cycle, air-cooled inboard engines mounted in housings just aft of amidships on the dories. The effectiveness of this innovation was moot because the *Wawona* made its final cod-fishing voyage the following year.[188]

Figure 31. Fisherman in bow of motorized dory with load of codfish. (Glenn Burch manuscript, San Francisco Maritime National Historical Park.)

CODFISHERMEN

> *Cod fishing is a rather rough life.*
>
> —*Anacortes American*, 1911[189]

> *For any disobedience of lawful commands from said Master or the Officers of the Vessel, whoever they may be, it is agreed to forfeit to the Owner One Thousand Codfish for each offense and for a false report of catch of codfish to the Master, the Fisherman agrees to forfeit the catch so falsely reported.*
>
> —Alaska Codfish Company, 1912[190]

Fishing for codfish in the Bering Sea and Gulf of Alaska was hard, uncomfortable, and dangerous work done far from friends and family and with little certainty of remuneration. The journeys to and from Alaska usually took three to five weeks, depending on the destination and weather, and once a vessel arrived on the fishing grounds—where it would spend several months—severe weather could preclude fishing for weeks on end.

In its early days, however, the fishery was lucrative. Many of the fishermen employed in it were veterans of the Atlantic cod fishery and had sailed to California from New England on fishing schooners purchased for use in the Pacific cod fishery. These trained fishermen, in the words of the U.S. Fish Commission's J. W. Collins, "carried to the Pacific a skill gained by years of experience in the Atlantic fisheries, and hardihood and daring unexcelled."[191]

The Pacific codfish industry declined during the 1880s, and in later years became an occupation that didn't attract much first-class talent. The Bureau of Fisheries 1906 report on Alaska fisheries characterized the crew of a typical codfish schooner as having "a few good fishermen" but mostly comprising "riff-raff picked up along the water fronts of the Pacific Coast cities." (Some of the codfish schooner *Wawona*'s crew were delivered to the vessel on sailing day in 1936 in a police paddy wagon, the men having been bailed out of jail by the vessel's captain.)[192]

The 1905–1906 voyage to the Shumagin Islands of the San Francisco–based Pacific States Trading Company's 107-foot, two-masted schooner *Glen* is an example of how difficult, dangerous, and unprofitable working on a cod-fishing schooner could be. The converted lumber carrier departed San Francisco in September 1905 and, once on the fishing grounds, experienced almost continuously stormy weather that precluded launching its dories. To make matters worse, during a gale on December 20 the *Glen* dragged its anchors and went ashore. Fortunately, no one was injured and the vessel was refloated the following day. Less than two weeks later, on New Year's Day, however, the *Glen* again dragged its anchors and this time went hard ashore. Once again, no one was injured, but it was a week before the *Glen*, with part of its keel missing as well as other damage, was refloated. The captain decided to return to San Francisco for repairs. But the hardships were not yet over: during the stormy, thirty-three-day return voyage, the *Glen*'s main gaff and mainsail were blown away. It arrived in San Francisco on March 8, 1906, leaking badly and without enough fish to even pay the cost of outfitting the voyage, let alone the crew. Nevertheless, perhaps adding insult to injury, as a token payment for a half year of hardship and danger, each crew member received one dollar.[193]

The *Glen* was repaired and put back into service, but the following year would prove to be fatal. On September 30, 1907, the *Glen* was wrecked in East Anchor Cove (known also as Bear Harbor), on Unimak Island. One crew member drowned. The vessel's cargo of thirty-eight thousand codfish, however, was reported to have been intact. The 107-foot, three-masted *John F. Miller* (another company schooner) was sent to salvage as much of the cargo as possible.

On January 8, 1908, while the *John F. Miller* was at anchor in the cove, a storm came up suddenly, and before the vessel could get underway, it was coated with ice. Unmanageable, the *John F. Miller* was driven onto a rocky beach, where it broke in half. Of the twenty-three-man crew, all eight who were on the vessel's forward section drowned, as did two of the fifteen men on the aft section. (Of the ten men who perished, only one was an American. Three were Finns, three were Norwegians, one was a Dane, one was a German, and one was a Swede.)

Stranded on the beach in the winter cold, most of the survivors soon suffered from frostbite. For sustenance they had only raw fish (from the *Glen*) and a paste made from flour and water. (A few bags of flour were the only provisions the survivors had been able to salvage from the *John F. Miller's* wreckage.) In this dire situation, two courageous crew members, both fishermen, put to sea in a dory and with great hardship managed over the course of six days to row to Sand Point, about one hundred miles distant, to summon help. A cod-fishing schooner, likely the *Martha*, was soon dispatched to rescue the eleven remaining men, all of whom survived. News of the tragedy was not received in San Francisco until early March, when the cod-fishing schooner *Czarina* arrived from the north.

The loss of the *Glen* and *John F. Miller* caused the Pacific States Trading Company to suspend operations through 1910. The company sold its inventory of codfish to the Union Fish Company, which also leased the Pacific States Trading Company's cod-fishing schooner *Ottillie Fjord*, two shore stations, and California drying facilities. In 1911, the Pacific States Trading Company reactivated its cod operations, sending the *Ottillie Fjord* to the Bering Sea and operating its shore stations. The company's codfish operations continued without interruption until 1916, when the company permanently withdrew from the business.[194]

John Cobb wrote in 1915 that, except for the company owners, some company officials, and several captains, few native-born Americans were engaged in the industry. Most of those engaged were of Scandinavian birth. Cobb noted also that almost all the captains and mates of the offshore cod-fishing vessels had worked up from the ranks of the fishermen, and he added that

> [o]perating on the codfish banks of Alaska requires considerable local knowledge of the banks, of the prevailing winds, and also of the most convenient spots for shelter and for water. While the majority of [the captains] are good navigators, a few are sadly deficient in this respect, yet their knowledge of Alaska conditions enable them to make about as many successful trips as their fellows who are better grounded in the science.[195]

In 1923, *Pacific Fisherman* noted the death of veteran codfish schooner captain Sven Wallstedt. Wallstedt apparently went to work for the McCollam Fishing & Trading Company (predecessor to the Union Fish Company) in 1887 as mate on a codfish schooner. After six years in that position, he was promoted to captain and for the following thirty years skippered Union Fish Company vessels.[196]

To make their operations more professional and more likely to succeed, cod-fishing companies sometimes crewed their schooners with fishermen imported from the Atlantic coast (see chapter 3). Apparently one codfish company during the 1880s brought in Portuguese fishermen. As a group, Portuguese fishermen had a reputation as being among the most skilled dorymen in the world. Unfortunately, most of the Portuguese fishermen who were brought to Alaska at that time perished in their dories about ten miles off the northwest shore of Unimak Island during a big storm. The grounds on which the fishermen perished became informally known as the "Portuguese graveyard."[197]

Desertions were a frequent problem on cod-fishing schooners, and were thought by government officials to be brought about by the poor class of fishermen employed.[198] Deserters in the early 1900s generally tried to get back to San Francisco on another vessel or tried to row to Nome (about seven hundred miles along the coast from the most-visited Bering Sea banks) and get work there mining gold. *Pacific Fisherman* noted that being in a tiny boat miles from land was a risk "none but daredevils would take."[199] Since deserters necessarily took the dory and fishing gear with them, the monetary loss was also considerable. Usually men deserted alone, but during a 1904 cod-fishing voyage of the Union Fish Company's schooner *Stanley*, fully six fishermen and three members of the dress gang deserted.[200] In part to stem desertions, cod-fishing companies asked the government to station a revenue cutter in the Bering Sea during the cod-fishing season.[201] The request was to no avail.

Sadly, there was sometimes uncertainty about desertions. During the summer of 1910, a fisherman on the schooner *Joseph Russ* disappeared while fishing in dense fog some ten miles from shore. Whenever fog developed while fishermen were out, it was standard practice for idlers (least skilled of the crew members) to take approximately twenty-minute turns pumping a hand-powered foghorn. In this case, the fisherman did not return, and the schooner's captain was uncertain whether he had deserted or was lost at sea. After a frantic but unsuccessful search, the *Joseph Russ* sailed away, leaving the fisherman to his fate.[202]

Sometimes, too, men simply refused to fish. In a letter to *Pacific Fisherman* in November 1904, the Robinson Fisheries Company, of Anacortes, Washington (which had just that year entered the codfish business) spoke of schooner fishermen's "refusal without cause many times to fish, causing the vessel to come home with broken [less than full] cargo, and various annoyances, which jeopardize capital invested in the enterprise." The letter also stated that on many occasions "very sore hands" kept a crew from doing its duty. A hand and wrist affliction apparently common among the dress gang was *gurry boils*, which were thought to be caused by long exposure to a mixture of seawater, fish blood, jellyfish slime, and fish entrails.[203]

Seattle has a well-earned reputation as a center of labor activism, and the Halibut Fishermen's Union of the Pacific, which was founded in that city in 1912, worked primarily to obtain higher wages for halibut fishermen. In January of 1916—the beginning of a potentially very prosperous year for the Pacific codfish industry—fishermen in the offshore codfish fleet organized the Pacific Cod Fishermen's Union, with headquarters in Seattle, and demanded a minimum price of $40 per thousand codfish, plus run money in the amount of $50 per man for crewing a vessel to and from the fishing grounds.

Because of the small size of the codfish industry, however, the cod fishermen's union had limited membership and power, and in February the union merged with the halibut fishermen's union to form the Deep Sea Fishermen's Union of the Pacific, which represented fishermen along the entire West Coast and remains active to this day, though it does not represent Pacific cod fishermen.

Despite the fact that the fishing season was quickly approaching and vessels were being readied to go north, the San Francisco codfish companies refused to negotiate with fishermen's union, and at a meeting on February 26 in San Francisco, the members of the local chapter of the Deep Sea Fishermen's Union voted to strike and urged Bay Area members of a sister union, the Water-Front Workers' Federation of the Pacific Coast, to keep off all cod-fishing vessels.

Recognizing that their businesses were in jeopardy and time was of the essence, the cod-fishing companies quickly entered into negotiations with the fishermen. In what the *Coast Seamen's Journal* called the "Codfishermen's Victory," an agreement was reached in early March.[204] The agreement established the *lay* (remuneration scale for fish caught) for cod fishermen and the wage scales for the crewmen who processed fish in the Pacific codfish schooner fleet.

Under this agreement, only codfish measuring greater than twenty-eight inches in length were given full value. Codfish measuring less than twenty-eight inches but more than twenty-two inches in length were counted two fish for one, while codfish measuring less than twenty-two inches ("snappers") were counted five fish for one. The base rate paid to fishermen was $40.00 per one thousand fish, but as an incentive for production, fishermen who caught twelve thousand or more fish were paid $42.50 per thousand for each fish caught, and fishermen who caught fourteen thousand or more fish were paid $45.00 per thousand. A schooner's officers, who were also fishermen but had additional responsibilities aboard the ship, were paid on a higher scale than regular fishermen: first mate, $55.00 per thousand; second mate, $50.00 per thousand; third mate, $45.00 per thousand. In addition to payments for fish caught, each fisherman received $50 run money.

Salaried crewmen were paid for the entire duration of the voyage. The first splitter was the highest paid, receiving $100 per month. The second splitter's wages were $85 per month. The first salter received $90 per month; the second salter $75 per month. Headers and throaters were paid $35 per month, while members of the dress gang were paid $30 per month. In addition to their wages, salaried crewmen, who often fished over the rail as time permitted, were paid $32.50 per thousand fish. Remuneration for galley help was not addressed in the agreement.[205]

The captains of Puget Sound cod vessels were paid a certain sum per ton for their vessel's delivered catch. On San Francisco cod vessels, the captains generally were engaged by the year and paid a salary. At both locations, the arrangements between the captains and the vessel owners were confidential and varied considerably and depended largely upon the reputation of the captain.[206]

There was a shortage of fishermen to crew the Bering Sea schooner fleet in 1920, which caused keen competition among the codfish companies. (One San Francisco company solved its manpower problem by importing twenty-seven fishermen from the Atlantic coast.) This shortage put the fishermen in an advantageous position, and, despite the fishermen having agreed earlier to fish on the same wage schedule as the previous year, the Deep Sea Fishermen's Union of the Pacific demanded substantially higher wages for the 1920 season.

The Puget Sound operators' response was swift: they declared their vessels would not sail at the new scale and suspended preparations for the quickly approaching season. After negotiations during which the distressed situation of the codfish companies was explained, the union came to an agreement with the codfish companies, and the vessels were provisioned, crewed, and sent to the fishing grounds.

The Bering Sea cod fishermen's lay agreed upon for the 1920 season was $90.00 per thousand fish for the San Francisco fleet. The Seattle fleet settled for two cents per pound, which equaled about $90.00 per thousand fish. Only codfish over twenty-six inches long were "counters." Shorter fish were counted (or weighed) two for one.[207]

By early 1921, the codfish market had eroded, in part because of Japanese competition (see chapter 7). As a result, codfish companies were burdened with a large inventory of the previous year's catch, and the companies refused to pay the same lay as they had agreed to in 1920. In San Francisco, both the Union Fish Company and the Alaska Codfish Company stated that the lay the fishermen demanded did not warrant sending any vessels to the Bering Sea. Although the differences between the codfish companies and the fishermen seemed irreconcilable, in April the fishermen agreed to a lay of 25 percent less than it had been during the previous year. Crew members drawing a monthly pay accepted a similar reduction. Shortly thereafter, three schooners—all from Puget Sound—set sail for the Bering Sea. These vessels were followed in early May by the Union Fish Company's schooner *Louise*, which fished in the vicinity of the company's shore stations in the Shumagin Islands area. The Alaska Codfish Company did not send any fishing schooners north that year.[208]

In 1922, Puget Sound cod fishermen broke away from the union and began negotiating their working arrangements independently. Their lay in 1923 was 1¼ cents per pound for fishermen who caught less than 8,000 pounds during a trip, and 1½ cents per pound for fishermen who caught more than 8,000 pounds. First splitters and first salters were paid $125 per month; second splitters, $90; second salters, $80; and the balance of the dress gang, $40.

There was a report in the spring of 1923 that fishermen at the Shumagin Islands codfish stations owned by San Francisco concerns were on strike, demanding to receive the same lay as was paid on the Puget Sound–based schooners. It is not clear whether the report was accurate, but it was clear the station fishermen did well that year, because production exceeded the production of the previous year. The high production, however, came at an inopportune time because of what the editors of *Pacific Fisherman* termed an "unusual supply" of Norwegian codfish that was being sold in markets traditionally supplied with Pacific coast codfish. This, in turn, resulted in a curtailment of most station operations.[209]

Prospects seem to have been better the following year. The lay for Puget Sound fishermen actually increased in 1924, to 1½ cents per pound for fishermen who caught less than 7,000 pounds during a trip, 1⅝ cents per pound for fishermen who caught between 7,000 pounds and 9,000 pounds, and 1¾ cents for those catching more than 9,000 pounds. First splitters and first salters were paid $135 per month; second splitters, $95; second salters, $85; and the balance of the dress gang, $40. *Pacific Fisherman* reported Puget Sound vessel owners as having little difficulty recruiting crews in 1924.[210]

By 1927, the lay had changed again. The rate was 1½ cents per pound for all fish caught, regardless of size, with the provisions that if the average weight of the fish in a fisherman's catch was 4 pounds or more, the rate was 1⅝ cents per pound, and that for those fishermen who caught more than 9,000 pounds, the rate was 1¾ cents per pound. The pay for the dress gang—except for idlers—increased slightly.[211]

Perhaps reflecting the economic situation during the Great Depression, in 1931 the base rate was cut to 1⅜ cents per pound. The following year it was cut again, to one cent per pound. By 1934, however, the base rate was back up to 1⅜ cents per pound, and the following year it was increased to 1⅝ cents.[212]

By 1939, cod fishermen were represented by Codfish Workers' Union (later the Crab and Codfish Workers' Union), which had been organized at Poulsbo, Washington, the location of the plant of the Pacific Coast Codfish Company. That year, agreement between vessel owners and the union was not reached until March 19, immediately after which the three schooners that fished that season—the *Azalea*, *Sophie Christensen*, and *Wawona*—were outfitted for voyages to the Bering Sea. The agreement that allowed the schooners to sail was praised by a Washington newspaper as providing wage increases "for everybody in the industry"—which, by that time, wasn't overly many.[213]

A few months later, the United States would be at war, and wages in the fishing industry would come under purview of the federal government's National War Labor Board. The Crab and Codfish Workers' Union persisted until at least the mid-1950s.[214]

ENDNOTES

[163] John N. Cobb, "Transportation Gave First Impetus to Pacific Fisheries," *Fishing Gazette* (July 6, 1918): 954, 956, 958.

[164] John N. Cobb, "Motor Vessels in the Alaska Cod Fisheries," *Pacific Motor Boat* (November 1919): 3–8.

[165] John N. Cobb, *Pacific Cod Fisheries*, Bureau of Fisheries Doc. No. 830, appendix 4 to *Report of the U.S. Commissioner of Fisheries for 1915* (Washington, DC: GPO, 1916), 43.

[166] "'Cap' Shields Sails—," *Pacific Fisherman* (August 1962): 35.

[167] Ed Shields, *Salt of the Sea: The Pacific Coast Cod Fishery and the Last Days of Sail* (Lopez Island, Wash.: Pacific Heritage Press, 2001), 178; Russ Hofvendahl, *Hard on the Wind* (Dobbs Ferry, N.Y.: Sheridan House, 1989), 51–52; "Pacific Coast Codfish Review," *Pacific Fisherman* (February 1907): 55–58; "Codfish" Kelly, oral history, October 8, 1958, San Francisco Maritime National Historical Park archive; "Matheson Codfish Ships Sail for the Bering Sea," *Anacortes American*, April 16, 1914; G. E. Hermanson, *Telling It Like It Was* (Bremerton, Wash.: G. E. Hermanson, 1972), 13.

[168] Harlan Trott, *The Schooner That Came Home: The Final Voyage of the C. A. Thayer* (Centreville, Md.: Cornell Maritime Press, 1958), 31–32.

[169] Ed Shields, *Salt of the Sea: The Pacific Coast Cod Fishery and the Last Days of Sail* (Lopez Island, Wash.: Pacific Heritage Press, 2001), 27–28; Joe Follansbee, *Shipbuilders, Sea Captains, and Fishermen: The Story of the Schooner Wawona* (New York: iUniverse, 2006), 36; Harlan Trott, *The Schooner That Came Home: The Final Voyage of the C. A. Thayer* (Centreville, Md.: Cornell Maritime Press, 1958), 31–32; "Bring a Tale of Shipwreck," *San Francisco Call*, March 25, 1904; John N. Cobb, *Pacific Cod Fisheries* (revised edition, 1926), Bureau of Fisheries Doc. No. 1014, appendix 7 to *Report of the U.S. Commissioner of Fisheries for 1926* (Washington, DC: GPO, 1927), 426; "Codfish Vessels Have Trouble Coming Home," *Pacific Fisherman* (October 1947): 89.

[170] "Pacific Codfish Schooners Sail," *Pacific Fisherman* (May 1921): 38; "Pacific American Crews Leaving," *Pacific Fisherman* (March 1920): 58; "Codfish Men Visit Seattle," *Pacific Fisherman* (October 1922): 32–33; "Review of the [1922] Pacific Codfish Season," *Pacific Fisherman* (January 1923): 96.

[171] G. E. Hermanson, *Telling It Like It Was* (Bremerton, Wash.: G. E. Hermanson, 1972), 7.

[172] Gus Dagg, "Codfishing in the Behring Sea—1924," *Fishermen's News* (May 1975): 3, 6–7.

[173] Jimmy Crooks, "Schooner Fanny Dutard," undated, typed recollection of 1926 codfishing voyage to the Bering Sea aboard the schooner *Fanny Dutard*, San Francisco Maritime National Historical Park archive, HDC379, 19 pp.

[174] Jimmy Crooks, "Schooner Fanny Dutard," undated, typed recollection of 1926 codfishing voyage to the Bering Sea aboard the schooner *Fanny Dutard*, San Francisco Maritime National Historical Park archive, HDC379, 19 pp.

[175] Ibid.

[176] Ibid.

[177] Russ Hofvendahl, *Hard on the Wind* (Dobbs Ferry, N.Y.: Sheridan House, 1989), 21, 42, 47, 51–52, 54, 67, 80–83, 91–92; Jimmy Crooks, "Schooner Fanny Dutard," undated, typed recollection of 1926 codfishing voyage to the Bering Sea aboard the schooner *Fanny Dutard*, San Francisco Maritime National Historical Park archive, HDC379, 19 pp.; Joe Follansbee, *Shipbuilders, Sea Captains, and Fishermen: The Story of the Schooner Wawona* (New York: iUniverse, 2006), 51, 55, 90, 92; Gus Dagg, "Codfishing in the Bering Sea—1924," *Fishermen's News* (May 1975): 3, 6–7; Fred R. Bechdolt, "Northward with the Cod Fishermen," *San Francisco Call*, April 16, 1911; Joe Follansbee, "Fishermen's Meals a la 1936," *48° North* (July 2004): 54–55; Ed Shields, Pacific Coast Codfish Company, to Glenn E. Burch, undated transcription on file at San Francisco Maritime National Historical Park archive; G. E. Hermanson, *Telling It Like It Was* (Bremerton, Wash.: G. E. Hermanson, 1972), 6, 13–14; "Berger Jensen Breaks High-Line Record," *Pacific Fisherman* (January 1938): 198.

[178] Richard S. Croker, "Alaska Codfish," in *The Commercial Fish Catch of California for the Year 1929*, Division of Fish and Game of California Fish Bulletin No. 30 (Sacramento: California State Printing Office, 1931), 48–55.

[179] Ed Shields, Pacific Coast Codfish Company, to Glenn E. Burch, undated transcription on file at San Francisco Maritime National Historical Park archive.

[180] J. W. Collins, *Report on the Fisheries of the Pacific Coast of the United States*, in *Report of the United States Commissioner of Fish and Fisheries for 1892* (Washington, DC: GPO, 1892), 102; Russ Hofvendahl, *Hard on the Wind* (Dobbs Ferry, N.Y.: Sheridan House, 1989), 111; Jimmy Crooks, "Schooner Fanny Dutard," undated, typed recollection of 1926 codfishing voyage to the Bering Sea aboard the schooner *Fanny Dutard*, San Francisco Maritime National Historical Park archive, HDC379, 19 pp.; John N. Cobb, *Pacific Cod Fisheries* (revised edition, 1926), Bureau of Fisheries Doc. No. 1014, appendix 7 to *Report of the U.S. Commissioner of Fisheries for 1926* (Washington, DC: GPO, 1927), 423, 427, 437; Ed Shields, *Salt of the Sea: The Pacific Coast Cod Fishery and the Last Days of Sail* (Lopez Island, Wash.: Pacific Heritage Press, 2001), 23–26; G. E. Hermanson, *Telling It Like It Was* (Bremerton, Wash.: G. E. Hermanson, 1972), 12; J. W. Collins, *Catalogue of the Collection Illustrating the Fishing Vessels and Boats, and Their Equipment; the Economic Condition of Fishermen; Anglers' Outfits, Etc.* (Wash-

ington, DC: GPO, 1884), 54–55; Ed Shields, Pacific Coast Codfish Company, to Glenn E. Burch, undated transcription on file at San Francisco Maritime National Historical Park archive; John N. Cobb, "Motor Vessels in the Alaska Cod Fisheries," *Pacific Motor Boat* (November 1919): 3–8; "Our Fisheries," *Monthly Gleason's Companion*, April 1886; Gordon Newell, ed., *H. W. McCurdy Marine History of the Pacific Northwest* (Seattle: Superior Publishing Company, 1966), 237; H. E. Jamison, "Along the Waterfront," *Seattle Star*, April 28, 1937; Fred R. Bechdolt, "Northward with the Cod Fishermen," *San Francisco Call*, April 16, 1911.

[181] Z. L. Tanner et al., "Explorations of the Fishing Grounds of Alaska, Washington Territory, and Oregon, during 1888, by the U.S. Fish Commission Steamer Albatross, Lieut. Comdr. Z. L. Tanner, U.S. Navy, Commanding," *Bulletin of the United States Fish Commission*, vol. 8, 1888 (Washington, DC: GPO, 1890), 23; J. W. Collins, *Report on the Fisheries of the Pacific Coast of the United States*, in *Report of the United States Commissioner of Fish and Fisheries for 1892* (Washington, DC: GPO, 1892), 103, 106.

[182] John N. Cobb, *Pacific Cod Fisheries*, Bureau of Fisheries Doc. No. 830, appendix 4 to *Report of the U.S. Commissioner of Fisheries for 1915* (Washington, DC: GPO, 1916), 44; Ed Shields, Pacific Coast Codfish Company, to Glenn E. Burch, undated transcription on file at San Francisco Maritime National Historical Park archive; Gus Dagg, "Codfishing in the Bering Sea—1924," *Fishermen's News* (May 1975): 3, 6–7; Ed Shields, Pacific Coast Codfish Company, to Glenn E. Burch, undated letter on file at San Francisco Maritime National Historical Park archive.

[183] John N. Cobb, "Motor Vessels in the Alaska Cod Fisheries," *Pacific Motor Boat* (November 1919): 3–8; George M. Bowers, *Report of the Bureau of Fisheries*, 1904 (Washington, DC: GPO, 1905), 102; "Motor Craft of the Pacific Fisheries," *Pacific Fisherman* (January 1919): 55–64; John N. Cobb, *Pacific Cod Fisheries*, Bureau of Fisheries Doc. No. 830, appendix 4 to *Report of the U.S. Commissioner of Fisheries for 1915* (Washington, DC: GPO, 1916), 45.

[184] Ward T. Bower, *Alaska Fisheries and Fur Industries in 1919* (Washington, DC: GPO, 1920), 56.

[185] "Outboard Possibilities Great," *Pacific Fisherman* (June 1929): 52; Ward T. Bower and Henry D. Aller, *Alaska Fisheries and Fur Industries in 1916*, Bureau of Fisheries Doc. No. 838 (Washington, DC: GPO, 1917), 69; "Salt Fish Review," *Pacific Fisherman* (January 1915): 102–103.

[186] Evinrude Motor Co., adv., *Pacific Fisherman* (February 1919), 51; "Reduced Codfish Production Causes Stiffening Market," *Pacific Fisherman* (January 1929): 159, 161; "Pacific Codfish Catch Improves During 1929 Season," *Pacific Fisherman* (January 1930): 207, 209; "Outboard Motors for Puget Sound Fleet," *Pacific Fisherman* (March 1929): 46; Ward T. Bower, *Alaska Fishery and Fur-Seal Indus-*

tries in 1929, Bureau of Fisheries Doc. No. 1086 (Washington, DC: GPO, 1930), 302–303.

[187] "Codfish Fleet Sails North for 5-Month Cruise," *Seattle Daily Times*, April 11, 1930; "Four Boats Take 1,045,000 Codfish; 'C. A. Thayer' Sets All-Time Record," *Pacific Fisherman* (October 1931): 77.

[188] Ed Shields, *Salt of the Sea: The Pacific Coast Cod Fishery and the Last Days of Sail* (Lopez Island, Wash.: Pacific Heritage Press, 2001), 59–60, 68–70; "New Outboard Power Rig for Dories," *Pacific Fisherman* (March 1931): 65; "Veteran Codfish Skipper Commands 'W. H. Smith,'" *Pacific Fisherman* (April 1931): 42; Ralph W. Andrews and A. K. Larssen, *Fish and Ships* (New York: Bonanza Books, 1959), 145 (quotes April 28, 1937 *Seattle Star* article); Joe Follansbee, *Shipbuilders, Sea Captains, and Fishermen: The Story of the Schooner Wawona* (New York: iUniverse, 2006), 39, 75; Ward T. Bower, *Alaska Fishery and Fur Seal Industries: 1944*, Statistical Digest No. 13 (Washington, DC: GPO, 1946), 53; "3 Schooners for Codfish," *Pacific Fisherman* (May 1946): 73.

[189] "Anacortes' Codfish Industry Is One of Great Importance," *Anacortes American*, October 12, 1911.

[190] Fishing Shipping Papers, Alaska Codfish Company schooners *Mary and Ida*, 1900, and *W. H. Dimond*, 1912–1913, San Francisco Maritime National Historical Park archive.

[191] J. W. Collins, *Report on the Fisheries of the Pacific Coast of the United States*, in *Report of the United States Commissioner of Fish and Fisheries for 1892* (Washington, DC: GPO, 1892), 101.

[192] John N. Cobb, *Fisheries of Alaska in 1906*, Bureau of Fisheries Doc. No. 618 (Washington, DC: GPO, 1907), 47; Joe Follansbee, *Shipbuilders, Sea Captains, and Fishermen: The Story of the Schooner Wawona* (New York: iUniverse, 2006), 51.

[193] "Schooner Glen in Hard Luck," *San Francisco Call*, March 9, 1906.

[194] *Fisheries of Alaska in 1907*, Bureau of Fisheries Doc. No. 632 (Washington, DC: GPO, 1908), 48; *Fisheries of Alaska in 1908*, Bureau of Fisheries Doc. No. 645 (Washington, DC: GPO, 1909), 60–61; "Codfish Market," *Pacific Fisherman* (December 1907): 21–22; "Codfish Market," *Pacific Fisherman* (April 1908): 23–24; "Schooner Brings News of Contrary Gus Whom the Dead Saved from Freezing," *San Francisco Call*, May 16, 1908; "Survivors Are in Dire Distress," *San Francisco Call*, March 9, 1908; "Brings Report of Rescue," *San Francisco Call*, March 9, 1908; "Ten Perish, Twenty-Five Stranded in Far North," *Los Angeles Herald*, March 8, 1908; Captain Warren Good and Michael Burwell, *Alaska Shipwrecks: A*

Comprehensive Accounting of Alaska Shipwrecks and Losses of Life in Alaskan Waters (Warren Good, 2018), digital, unpaged; John N. Cobb, *Pacific Cod Fisheries* (revised edition, 1926), Bureau of Fisheries Doc. No. 1014, appendix 7 to *Report of the U.S. Commissioner of Fisheries for 1926* (Washington, DC: GPO, 1927), 411.

[195] John N. Cobb, *Pacific Cod Fisheries*, Bureau of Fisheries Doc. No. 830, appendix 4 to *Report of the U.S. Commissioner of Fisheries for 1915* (Washington, DC: GPO, 1916), 41.

[196] "Veteran Codfish Skipper Dies," *Pacific Fisherman* (November 1923): 17.

[197] Gus Dagg, "Codfishing in the Bering Sea," *Alaska* (September 1975): 13.

[198] John N. Cobb, *Fisheries of Alaska in 1906*, Bureau of Fisheries Doc. No. 618 (Washington, DC: GPO, 1907), 48.

[199] "Codfish Catch," *Pacific Fisherman* (August 1904): 11.

[200] "San Francisco," *Pacific Fisherman* (August 1904): 9.

[201] "Union Fish Co.," *Pacific Fisherman* (November 1904): 8.

[202] "Pacific Codfish Department," *Pacific Fisherman* (September 1910): 22–23; Russ Hofvendahl, *Hard on the Wind* (Dobbs Ferry, N.Y.: Sheridan House, 1989), 184.

[203] "Cod Vessels Need Revenue Boat," *Pacific Fisherman* (November 1904): 18; Joe Follansbee, *Shipbuilders, Sea Captains, and Fishermen: The Story of the Schooner Wawona* (New York: iUniverse, 2006), 80.

[204] Jeff Kahrs, *One Hook at a Time: A History of the Deep Sea Fishermen's Union of the Pacific* (Seattle: Documentary Media, 2015), 23; untitled, *Coast Seamen's Journal* (January 19, 1916): 6; "Important Announcement," *Coast Seamen's Journal* (March 1, 1916): 6; "The Codfishermen's Victory," *Coast Seamen's Journal* (March 8, 1916): 6.

[205] "The Codfishermen's Agreement," *Coast Seamen's Journal* (April 5, 1916): 7, 10.

[206] John N. Cobb, *Pacific Cod Fisheries* (revised edition, 1926), Bureau of Fisheries Doc. No. 1014, appendix 7 to *Report of the U.S. Commissioner of Fisheries for 1926* (Washington, DC: GPO, 1927), 425.

[207] "Cod Fishermen Scarce," *Pacific Fisherman* (May 1920): 55; "San Francisco Fleet Sails," *Pacific Fisherman* (April 1920): 55; "Puget Sound Fleet Held Up," *Pacific Fisherman* (April 1920): 55; "Review of [1920] Pacific Codfish Season," *Pacific Fisherman* (January 1921): 97–98; E. Lester Jones, *Report of Alaska Investigations, 1914* (Washington, DC: GPO, 1915), 58.

[208] "Cod Fishing Operations Suspended," *Pacific Fisherman* (April 1921): 45; "Puget Sound Fleet Uncertain," *Pacific Fisherman* (April 1921): 45; "Pacific Codfish Schooners Sail," *Pacific Fisherman* (May 1921): 38; "Codfish," *Pacific Fisherman* (July 1921): 40; "Alaska Codfish Stations Closed," *Pacific Fisherman* (August 1921): 15; "Review of Pacific Codfish Season, 1921," *Pacific Fisherman* (January 1922): 93; Ward T. Bower, *Alaska Fisheries and Fur-Seal Industries in 1921*, Bureau of Fisheries Doc. No. 933 (Washington, DC: GPO, 1922): 44–45.

[209] "Preparing for Codfish Season," *Pacific Fisherman* (March 1923): 28; "Puget Sound Fleet Sails, *Pacific Fisherman* (April 1923): 33; "Strike at Cod Stations," *Pacific Fisherman* (June 1923): 33; "Codfish Figures Misleading," *Pacific Fisherman* (November 1923): 38; "S.F. Codfish Vessels Leaving," *Pacific Fisherman* (April 1924): 48.

[210] "Sound Schooners Get Away," *Pacific Fisherman* (May 1924): 46.

[211] "Puget Sound Schooners Leave," *Pacific Fisherman* (May 1927): 33.

[212] "Veteran Codfish Skipper Commands '*W. H. Smith*,'" *Pacific Fisherman* (April 1931): 42; "1932 Codfish Lay Cut," *Pacific Fisherman* (May 1932): 57; "1935 Codfish Lay Carries Increases for Men," *Pacific Fisherman* (April 1935): 60.

[213] "Three Codfish Schooners Will Go to Bering Sea," *Pacific Fisherman* (April 1939): 72; "American Codfish Schooners Return From Bering Sea," *Pacific Fisherman* (September 1939): 72; Melvyn Dubofsky, ed., *American Federation of Labor Records, Part 1: Strikes and Agreements File, 1898–1953* (Fredrick, Md.: University Publications of America, 1986), 28; "Everybody Gets a Boost," *Spokane Daily Chronicle*, April 4, 1941.

[214] National Labor Relations Board, *Wakefield's Deep Sea Trawlers, Inc., and Wakefield Fisheries, Inc., and International Longshoremen's and Warehousemen's Union (Fishermen's and Allied Workers' Division) Local No. 3 and Crab and Codfish Workers Union, Seafarers International Union of North America, AFL, Party to a Contract*. Case No. 19-CA-974, June 27, 1955.

Chapter 5
CATCHING AND PROCESSING CODFISH

[T]he great want is not fish, but demand for fish.

—Tarleton Bean, ichthyologist, 1887[215]

The hand-line fishing is quite exciting, and the men take to it like sport.

—*Daily Alta California* (San Francisco), 1879[216]

The competition between fishermen concerning who would catch the most fish was the greatest I have ever seen in any industry.

—Jimmy Crooks, schooner *Fanny Dutard*, 1926[217]

The Fishermen and other members of the crew agree to bleed each and every Codfish caught by them by cutting its throat before taking it from the hook. Also to gaff and pew all Codfish in the head only, and not in the body and to use every means to keep them white and free from blood and other stains.

—Agreement between codfishermen and San Francisco codfish companies, 1918[218]

The orders a codfish schooner's owners gave to its captain were straightforward: fish until the vessel was fully loaded or until the weather was so bad that, in the captain's opinion, the vessel, crew, and cargo would be in danger. By gentlemen's agreement, codfish schooners on the Bering Sea fishing grounds anchored about ten miles apart, which provided a radius of five miles, more or less, around each schooner within which the vessel's dories could fish.[219]

The following is an account by Axel William Smith, captain of the McCollam Fishing & Trading Company schooner *Helen W. Almy* during an 1885 cod-fishing voyage to the Bering Sea:

> Having arrived at the fishing station, twenty dories, with a man in each, are immediately sent out. . . . These dories are small, flat-bottomed boats, being more shells, and if wind and tide permit are ranged in a circle around the ship, but should the wind be blowing in any set quarter, of course, the dories would have to be placed to windward, otherwise they might never be able to regain the ship. Work is commenced

> bright and early in those regions. It is daylight day and night, and by 4 o'clock in the morning the fishermen are all out in their dories actively engaged in fishing, nor is their work relinquished until 8 or 9 o'clock in the evening.[220]

Aboard the *C. A. Thayer*, breakfast was served at 4:00 a.m. Once they finished eating, the fishermen would return to the forecastle to don their *oilskins* (raingear), and wait for the mate to give the command "Throw 'em out"—to put the dories in the water, a process that was efficiently completed in a matter of minutes.[221]

A doryman's fishing gear usually consisted of two cotton handlines, each of which was about 50 fathoms (300 feet) long. Tied to the end of the line was a lead weight about ten inches long and weighing about five or six pounds. Attached to the bottom of the weight was a metal spreader, to each end of which was tied a leader (*snood*) that was about a foot long and made of cotton or hemp. At the end of each leader was a baited hook. Early in the trip, the hooks would be baited with salted herring that had been brought north. The herring was quickly replaced with *shack fish*—fish that were incidentally caught while fishing for codfish. Pacific codfishermen favored halibut, but almost any fish species would do (see below).

To protect their hands, fishermen wore *nippers*—thin, circular bands of woolen cloth covered with white knitted yarn or, alternatively, narrow tubes that were knit from woolen yarn and stuffed with pieces of woolen cloth. Fishermen wore the nippers like bracelets that, when

Figure 32. Codfish hook and gangion (snood), *C. A. Thayer.* (Poulsbo Historical Society Museum, Poulsbo, Wash.)

the time to handle their fishing lines came, they slid down over their palms and around the back of their hands. By the 1930s, nippers were made of medium-hard rubber that was grooved in the center to accommodate the fishing line.[222]

> *[F]or the fish out here in deep water were heavy and fat, while the fish caught in shoal water were skinny and light and usually covered with lice and full of worms.*
>
> —Jimmy Crooks, schooner *Fanny Dutard*, 1926[223]

The largest and best cod in Alaska were usually found in fairly deep water, such as sixty to eighty fathoms. The depth of water fished by handline fishermen, however, usually ranged from fifteen to forty fathoms because of the additional work involved in fishing deeper. Once a doryman had chosen a spot to fish and cast his anchor, he paid out a fishing line between the thole pins (a pair of wooden pins inserted vertically about amidships in the dory's gunwales to serve as fulcrums for oars) on one side of his dory until the lead weight hit the seafloor. The doryman then hauled in a small amount of line to position the baited hooks just above the seafloor. Once this line was secured, the doryman repeated the process with a second line on the dory's opposite side. To bring in a fish (or, if he was lucky, two fish), the doryman stood amidships in his dory and, facing aft, hauled in the line hand over hand, the slack falling at his feet. When the fish reached the surface, the doryman quickly pulled it into the dory's stern section and removed the hook. He'd then quickly rebait the hook and get his gear back in the water. To help prevent unsightly bruises in the fish's flesh, the doryman bled it by cutting its throat. (Traditionally, pursuant to the agreement under which they were hired, fishermen were required to bleed each codfish before it was taken off the hook.)

As fishing progressed, the doryman would transfer a portion of his catch to the bow section to keep the dory on an even keel fore and aft. The amount of fish the dory could carry depended on the weather conditions. When fishing was good, a doryman over the course of a single day might take several loads of fish to the schooner.[224]

Upon returning to the schooner with his catch, a doryman would use a *pew*—a sharp, single-tined fork secured to the end of a broom-length wooden pole—to transfer his fish to the schooner's deck as a crew member kept tally. Regular fishermen were paid on a per-fish basis, based on the length of each fish caught (the fishermen's *lay*, see chapter 4). The schooner fishermen's lay in the late 1880s ranged from $20 to $25 for each one thousand codfish that measured at least twenty-eight inches long (tip of snout to end of tail). Codfish measuring between twenty-six inches and twenty-eight inches counted two for one, while codfish measuring less than twenty-six inches were not counted. The lay varied over the years and among codfish companies.

To encourage productivity, in the 1908 season, the Anacortes, Washington–based Robinson Fisheries Company devised a sliding scale that paid $30.00 per thousand fish to fishermen who caught more than ten thousand fish. Fishermen who caught more than eight thousand fish but fewer than ten thousand received $27.50 per thousand, while fishermen who caught fewer than eight thousand fish received $25.00 per thousand. No fish less than twenty-eight inches in length was counted. That year, the 194,000 codfish (840,000 pounds of salted product) delivered by the company's schooner *Joseph Russ* after a five-month voyage to the Bering Sea were worth about $10,000 to the thirty-five-man crew, with individual earnings (including salaries) ranging from $165 to $600. The 837,035½-pound cargo of codfish delivered by the *Joseph Russ* in 1909 was inventoried upon arrival at $25,111.05—three cents per pound. According to Anacortes's local newspaper, the *Anacortes American*, in 1911 the annual costs of operating a codfish schooner—wages, food, supplies, and so on—totaled about $15,000.[225]

The larger schooners usually carried three officers (mates) who fished in dories the same as the regular fishermen. Because of their additional responsibilities, however, the officers received a higher price for their fish than was received by regular fishermen, or were paid the same prices received by regular fishermen plus a monthly salary. The captain's pay was usually either a salary or a sum based upon the tonnage of salted codfish delivered.

Additionally, the schooners carried two dress gangs—one for the vessel's port side, one for the starboard—to process the catch. A dress gang generally consisted of a throater, a header, a splitter, a man to do final touch-ups of dressed fish, a salter, and from one to three others, the idlers, who mostly did unskilled tasks. All members of the dress gang received a monthly salary that depended upon their skills and responsibilities. In the late 1880s, these salaries ranged from $15 to $40 per month. In 1916, the Deep Sea Fishermen's Union of the Pacific (see chapter 4), which represented cod fishermen, established a codfish dress gang salary scale that ranged from $30 per month for idlers to $100 per month for the head splitter. The vessel's cook, cook's helper, and watchman were also salaried.

Ernie Swanson, idler, codfish schooner *Fanny Dutard*, 1906: I asked about a job and [Captain Jacobsen] said they had an opening for an idler. I didn't know what an idler was and didn't ask. It sounded fine to me. I figured that I was cut out to be an idler and could handle it without any difficulty. . . . [T]he idler was the poorest paid hand aboard and did all the jobs nobody else wanted or had time for. There were plenty of them![226]

All members of the dress gang, as well as the cook, cook's helper, and watchman, were encouraged to fish "over the rail" (see below) of the vessel when they were not busy with other duties. These men were generally paid the same rates as fishermen for the codfish they caught.[227]

An exceptional catch was made during the 1925 season by doryman Dan McEachern, of the schooner *Charles R. Wilson*. McEachern—whom *Pacific Fisherman* later referred to as a "habitual high-liner"—landed 20,072 codfish, of which more than 4,000 were caught during the last week of the season. His high day was 801 fish.[228]

In 1930, power dory (see below) fisherman Axel Hackstad on the *C. A. Thayer* caught 23,600 codfish, and in 1935 Bill Lund, a power dory fisherman on the schooner *Sophie Christensen*, caught 23,581 codfish, 1,060 of which were caught in a single day. If one considers a working day to be seventeen hours, Lund on that day landed, on average, about one fish per minute.[229]

FISHING "OVER THE RAIL"

As noted above, all members of the dress gang, as well as the cook and watchman, were encouraged to fish over the rail of the vessel when they were not busy with other duties.

Usually each over-the-rail fisherman would tend a single line fastened to a peg in the rail called a *soldier*. The over-the-rail fishermen were usually paid the same amount as was paid to the dory fishermen. As the payment for fish was in addition to their regular wages, over-the-rail fishing could add considerably to a crew member's earnings, so some crew members fished whenever they had a few spare moments. Fishing over the rail was also an option when a schooner was riding at anchor in weather too severe to launch the dories.

The watchman, who was often alone on deck, had exceptional opportunities for fishing, and, in at least a couple of instances during the early years of the cod fishery, watchmen were vessel highliners, catching more fish during a trip than anyone else on board.[230]

AN INHERENTLY DANGEROUS OCCUPATION

> *They return at dark with the day's catch, though occasionally owing to the foggy weather, they cannot find their vessel and are obliged to pass the night in the dories or land on some unknown shore and await the coming of the morning light.*
>
> —Thomas J. Vivian, U.S. Treasury Department, 1891[231]

> *[Codfishing] operations in Alaska this year were not of a particularly satisfactory character. The catch was much curtailed on account of inclement weather when it was impossible to fish. Casualties were unusually high, no less than eight fishermen being lost.*
>
> —Barton Warren Evermann, U.S. Bureau of Fisheries, 1912[232]

> *At times the dory would be many miles away from the schooner before a load of fish was obtained, and suddenly and almost without warning a strong wind would spring up from the south east, on such an occasion the fish would have to be dumped overboard and the fisherman would attempt with all speed to return to the vessel. It was very disappointing on such occasions to be forced to throw the fish overboard, and many men met their death by neglecting to do so.*
>
> —Jimmy Crooks, schooner *Fanny Dutard*, 1926[233]

> **Dory fishermen:** *the hardest workers and the most fearless men I have ever had the opportunity to see in action. A person had to be tough to survive and live to tell the tale even for one season alone, but to return year after year to this particular occupation, he had to be absolutely immune to hardship and danger.*
>
> —Jimmy Crooks, schooner *Fanny Dutard*, 1926[234]

Fishing alone in the northern seas in a small dory that was often heavily laden with fish was inherently a very dangerous occupation. High winds or extreme cold could develop suddenly and be overwhelming. Even in the best of circumstances, the smallest mishap or lapse of judgment could prove fatal. Reports on the cod fishery are replete with accounts of dory fishermen who drowned, went missing, or froze. For example, on the schooner *Freemont* in 1907, one crew member froze to death in his dory, and another was adrift in his dory for eight days before being picked up with badly frozen legs. Two years later, six fishermen from the schooner *Harriet G* were drowned when a sudden, especially severe squall drove their dories onto rocks at Dublin Bay, near Unimak Pass, in the eastern Aleutians.[235]

A small sail flapping in an anchored cod-fishing schooner's rigging was a prearranged signal for all fishermen to immediately return to the vessel.

In his 1920 report on Alaska's fisheries, Bureau of Fisheries agent Ward Bower wrote that the loss of life in the codfish industry that year was "comparatively small": three fishermen drowned, and one shore worker was accidentally killed. Alaska's codfish industry that year employed 803 men, which meant almost one in every 200 men in the industry perished (equal to 498 per 100,000).[236]

In sheer numbers, the loss of life in the Alaska cod fishery, however, paled in comparison to the scale of loss in the Atlantic cod fishery. During a two-day gale on the Georges Bank in 1879, thirteen schooners sank and 143 fishermen from Gloucester, Massachusetts, drowned. It was all a matter of scale; far more vessels and fishermen were involved in the Atlantic cod fishery. The Georges Bank cod fishery alone in the fall of 1883 employed seventy-five Gloucester vessels manned by a total of 825 men.[237]

BAIT

> *The cod fishing boats would use halibut for bait, and the halibut boats would use codfish for bait.*
>
> —Ed Opheim, Kodiak, Alaska, undated[238]

The Pacific cod fisherman, wrote John Cobb in 1916, "worries but little about bait. Before sailing enough [salted] herring are taken along for a couple of days' baiting, but the fisherman usually gets enough shack fish the first day to furnish him with plenty of bait for the next day, and so on throughout the season." The "shack fish" Cobb referred to were mostly halibut, sculpins, and flounders. Of these, halibut was preferred and was used almost exclusively by cod fishermen in the Bering Sea, where the large flatfish were abundant. The halibut's tough skin helped keep the bait on the hook. Among some fishermen, the choicest pieces were approximately two-inch squares cut from the bottom (white) side of the fish, while other fishermen preferred elongated pieces cut from the upper (dark) side of the fish.

Fishermen in the Okhotsk Sea preferred to use salmon for bait, but they also used halibut, herring, sculpins, flounders, and clams. The salmon were caught at the mouths of streams and rivers by men using a beach seine carried aboard the schooner for this purpose. John Cobb noted that although clams were abundant in Alaska, fishermen rarely bothered to dig them for bait.[239]

PROCESSING CODFISH

Upon arriving at the fishing grounds, one of the first things the crew did was build a series of *checkers* on the deck. Checkers were open square or rectangular bins used to keep fish organized and from sliding around. They were constructed of wooden planks about two or three inches thick and about twelve inches wide that were placed on edge. The planks had been carried north in the vessel's hold.

The following description of processing codfish aboard a schooner borrows heavily from John Cobb's 1916 report.

A cry of "Dory!" or "Dory comin'!" alerted the dress gang that work would soon begin. Often, the dress gang would be divided into two gangs, one that worked on the port side of the vessel and one that worked on the starboard side. Once his dory was secured alongside

Figure 33. Members of a dress gang, schooner *Wawona*. (Anacortes Museum, Anacortes, Wash.)

the schooner, the doryman pewed his fish aboard through ports about ten feet long and three feet high and located a few inches above deck. There were usually two ports on each side of schooner, so as many as four dories could be unloaded simultaneously. The doryman's fish were counted by an idler who, despite his title, was anything but idle when there were fish to be processed. The idler relayed his count to the crewman who was designated to tally all the fish caught, then pewed the fish into a checker to await processing.

The *throater* began the process by taking a fish by the head and cutting each side of the throat, just behind the gills (the front of the throat had been cut by the fisherman to bleed the fish), and then slit the belly to the vent. The fish was then passed to the *header*, who, while grasping its head and body with the belly side up, pushed the fish's neck against the edge of a table or tub hard enough to break off the head at the first vertebra. The header then opened the belly and tore out the viscera. He then passed the fish to the *splitter*—the worker John Cobb said was the most skilled and important member of the gang. With the fish belly side up, the splitter cut along the backbone from the neck toward the tail, being careful not to cut into the thick flesh along the fish's back. At about three-fifths the distance from the neck to the tail, the splitter cut across the backbone and then back toward the fish's neck, again careful not to cut into the thick flesh along the back. He then removed the forward section of the backbone from the fish, which left the two sides of fish joined at the back.

The split fish was then passed to the *black skinner*, who removed pieces of adhering backbone, clots of blood, and portions of black visceral skin that might have been left on the fish. Once his job was complete, the black skinner dropped the fish into a tub of seawater, where it was stirred about and washed by idlers. Once clean, the fish were slid down a chute into the hold, where the salters received them.

When fish cleaning operations were in progress, swarms of sea gulls surrounded the schooner, hungry for the offal that was being cast overboard. The gulls served a useful purpose in guiding dories back to the schooner when visibility was poor because the noise they made often carried a great distance. (Schooners also carried a foghorn, as did each dory.)

The vessel's hold was divided into compartments—*kenches*—that were about four feet wide and extended the full width of the hold. Usually starting in the forward compartment and working aft, salters *kenched* the fish by placing a layer of fish, skin side down, on a bed of salt, with the fishes' napes and tails alternating, and then salting the layer. The amount of salt used was important: too little salt would

Figure 34. Salting split codfish in a schooner hold. (Glenn Burch manuscript, San Francisco Maritime National Historical Park.)

Figure 35. Kenched salted codfish in a schooner hold. (Glenn Burch manuscript, San Francisco Maritime National Historical Park.)

allow the fish to rot, while too much salt would reduce the amount of fish that could be carried in the hold and waste salt. The layer-of-fish/layer-of-salt process was repeated until the compartment was full, with the top layer of fish placed skin side up. Salt was then poured over the top of the kench.

The salt drew the moisture out of the cod, causing it to shrink, and the resulting liquid drained into the bilge and was pumped out. As the pile of kenched cod lowered, additional layers of cod and salt were added to keep the compartment full.

Codfish caught at shore stations were prepared in a similar manner (see chapter 6).[240]

DISCHARGING CODFISH CARGOES

The Robinson Codfish Company's schooner *Joseph Russ* operated in the early 1900s and was homeported at Anacortes, Washington. Workers—between thirty and forty of them—unloaded, cleaned, and stored its cargo at a rate of thirty tons per day. The salted fish were pewed from the hold, then trundled along the dock in wheelbarrows or carts to a scale, where they were weighed in approximately 600-pound batches. Each fish was then scrubbed with a brush to remove any slime. The cleaned fish were then put in wooden brine tanks for storage until the time came to "work them up for market." Each tank held approximately 20,000 pounds of salted cod. According to William Robinson, "Sometimes a portion of the fish remain in this brine for nearly a year in perfect condition."

In his 2001 book, *Salt of the Sea: The Pacific Coast Cod Fishery and the Last Days of Sail*, Ed Shields, captain of the *C. A. Thayer* during its final cod-fishing voyages to the Bering Sea (see chapter 8), wrote that a crew of approximately thirty-five men unloaded his schooner at a rate of fifty to seventy tons per day. A gasoline-powered deck winch accounted for much of the speed with which the vessel was unloaded. In the unloading process, a cargo net was lowered into the hold, and salted cod were pewed into it. Once filled, the net was hoisted out of the hold and the fish dumped on a large table, where three or four men sorted them by size into carts for weighing and transport to the brine tanks. Shields added that one man selected by the crew was paid to monitor the weighing to ensure the weights were accurate.[241]

Figure 36. Unloading salted cod from a schooner hold. (Courtesy of Union Fish Company.)

Figure 37. Unloading crew washing salted codfish in water-filled dory, Pacific Coast Codfish Company, Poulsbo, Washington, 1913. (Shields Collection, Poulsbo Historical Society, Poulsbo, Wash.)

ALTERNATIVE FISHING GEAR

Gear other than handlines was occasionally employed to catch codfish in Alaska during the Salt Cod Era. This gear included gill nets, beam trawls, otter trawls, salmon traps, and longlines.

In 1913, John Cobb, at the time in the employ of the Union Fish Company, experimented briefly with using anchored gill nets to catch codfish at the company's Pirate Cove station. Gill nets are long, curtain-like nets with a mesh size that allows fish to push their heads but not their bodies through. A fish that has pushed into a gill net is prevented from backing out by its gill covers, which flare out when open. Gill net mesh size varies depending upon the species being targeted. A *corkline* fitted with floats keeps the top of the gill net suspended, while a weighted *leadline* along the gill net's bottom keeps the net spread vertically. Gill nets in Alaska are mostly used to catch salmon, but by 1885 they had been used in Norway to catch cod. Though Cobb's gill nets caught a number of codfish, the experiment was abandoned because the fishermen at Pirate Cove, according to Cobb, were uninterested in the new technology.[242]

A beam trawl is a large, tapered net that is held open at its front by a rigid frame that incorporates a long, horizontal wooden beam. The trawls, which are usually about 40 to 60 feet wide, are dragged across the ocean floor and were in common use by East Coast cod fishermen by the late 1910s.[243] One or more beam trawls were used experimentally on Alaska's cod banks in 1919 and 1920. The experiments were unsuccessful for two reasons: first, because the trawls can be fished only on a smooth seafloor, and the seafloor where the trawls were tested contained rocky patches in which a trawl could easily hang up; second, because, as John Cobb wrote, "the local fishermen are either passively or actively against the introduction of any new and improved methods and would not give the nets a fair trial."[244]

Otter trawls, which were developed in Great Britain in the 1890s and first used in the United States in 1905, were also used experimentally to catch cod in Alaska during the several years preceding the U.S. entry into World War II. The development of otter trawls revolutionized commercial fishing, and they are broadly used today throughout the world in a wide variety of fisheries. Depending upon the application, otter trawls range in width (*sweep*) from about 30 feet to several hundred feet. There are two types of otter trawls: bottom (non-pelagic) trawls, which are dragged along the ocean floor, and mid-water (pelagic) trawls, which, as the name implies, do not touch the ocean floor. The otter trawls employed to catch Pacific cod are bottom trawls. The mouth of a net, which is shaped like a flattened cone, is held open vertically by floats fastened to the *headrope* along the net's top front edge, and a *footrope* along the bottom front edge. The footrope is constructed of chain or steel cable and is fitted at intervals with weighted rubber disks designed to roll along the seafloor to limit contact with the seafloor and to protect the net.

The trawl is kept spread horizontally by a pair of heavy *doors* (*otter boards*, in the parlance of early trawlermen) that function something like underwater kites. Each door is fitted with a steel or chain bridle that is connected to a steel towing cable that runs from a winch on the trawling vessel. A second cable bridle on each door, in turn, connects to the top and bottom of one of the net's two wings. The resistance of the water to the forward motion of the trawling vessel causes the doors to pull horizontally in opposite directions, spreading the net.

Early trawl doors were made of heavy wood bound around the edges by a heavy iron frame; modern trawl doors are made of steel. Fish caught in the trawl accumulate in its *cod end*—a cylindrical section of reinforced heavy netting at the trailing end of the trawl. The cod end is fitted on its bottom and sides with sacrificial *chafing gear*—early on, cowhides; now, typically polyethylene fiber—that protects it from abrasion caused by contact with the seafloor and the trawler's stern ramp.[245]

Pioneer Bering Sea trawlers

In 1938, the halibut schooner *Sitka* experimented with an otter trawl on the codfish grounds but had difficulty finding grounds suitable for the use of the trawl. After his experience, the vessel's captain was of the opinion that the trawl wasn't particularly efficient, but at least one other company was convinced of an otter trawl's potential to catch codfish.

During the early winter of 1939–1940, the Robinson Fisheries Company, an established codfish producer based in Anacortes, Washington, purchased the 105-foot power halibut schooner *Dorothy*. Then, under the supervision of an experienced *fishing master* (basically, the captain of a vessel's fishing operation), the company rigged the vessel with an otter trawl to catch codfish in extraterritorial waters of the Bering Sea. The local newspaper wrote that the new gear would "revolutionize codfishing in the Bering Sea banks." A ten-man crew consisting mostly of experienced trawlermen, three of whom were from the East Coast, manned the *Dorothy*. In addition to the usual provisions, the vessel departed Anacortes in May 1940 carrying large

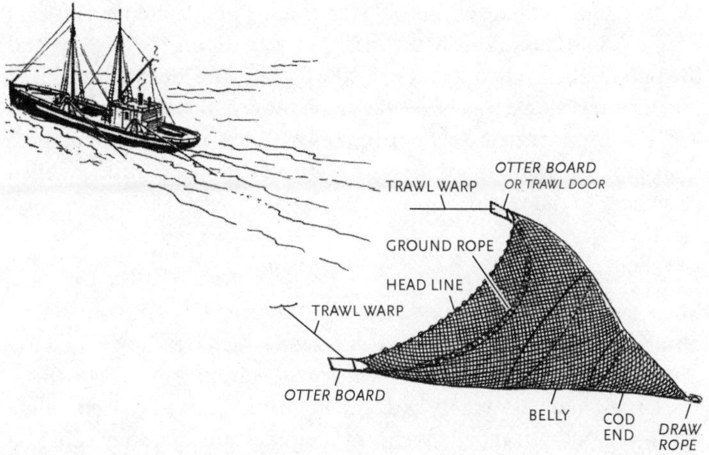

Figure 38. Trawler *Dorothy* with otter trawl. ("'*Dorothy*' Outfitted as Trawler for Bering Sea Codfish Banks," *Pacific Fisherman* [May 1940]: 30.)

bundles of cowhides on deck. The cowhides would be fastened to the bottom of the trawl net to protect it from chafing on the ocean floor.

Once in the Bering Sea, the Robinson Fisheries Company's sailing schooner *Azalea* worked in conjunction with the *Dorothy*. The *Azalea* anchored near the grounds being fished, and its crew dressed, salted, and stored the *Dorothy*'s catch. The experiment, which lasted all summer, was an expensive disappointment. According to the crew, the fish seemed to be above the bottom and could not be taken with the trawl. In all, the *Dorothy* landed only 25 tons of codfish. By contrast, the catch during the same season by the Robinson Fisheries Company's traditional sailing schooner *Wawona*, which carried seventeen dories, was 565 tons of salted product.

The following year, the U.S. Fish and Wildlife Service contracted the *Dorothy* and a smaller vessel, the *Locks*, for a seven-month investigation of the distribution of king crab in Alaska and the potential for catching the crab with various types of gear. In the Bering Sea, two months of trawling yielded significant amounts of incidentally caught finfish, Pacific cod among them. Encouraging was the fact that the cod caught in the deeper waters (up to sixty fathoms) averaged substantially larger than those caught by traditional dory fishermen, whose fishing was generally limited to relatively shallow waters.

Fisheries experimentation, though not all fishing, was suspended in the Bering Sea after the Japanese bombed Pearl Harbor in December

1941. In August 1944, as World War II appeared to be near its end, *Pacific Fisherman* published a series of articles on the growth and the potential of trawling in the Pacific Ocean. The journal noted that the "Greatest single trawlable area along the coastline of Pacific North America lies in the easterly portion of [the] Bering Sea—where there is an extent of water less than 100 fathoms depth perhaps unequalled elsewhere in the world."[246]

Longlines

More successful and dependable than gill nets, beam trawls, and otter trawls were longlines (*lines of trawl*), which were in general use at codfish stations for a number of years beginning in the late 1880s. The largest and best quality cod tend to be found in fairly deep waters, and a very significant advantage of longlines was that they could be set in water deeper than handline fishermen were willing to fish.

Longline gear consisted of two buoy kegs (wooden barrels), two buoy lines, two 12- to 14-pound anchors, and a series of 50-fathom (300-foot) *groundlines* of tarred hemp or cotton that were connected end to end.[247] Fastened to each groundline at approximately 6-foot intervals were 3-foot-long *gangions* (leaders or snoods) of lighter tarred-cotton line, each with a hook at its end. Longline bait was the same as handliners used. Under ordinary conditions, early cod fishermen in Alaska rarely fished longline gear consisting of more than fourteen groundlines—about seven hundred hooks. The groundlines were carried in the dories coiled in wooden half barrels. Because of the amount of gear required and the fact that longlining required two men, dories used for longlining were a bit larger than those traditionally used to fish with handlines. Two fishermen fishing with longline gear usually fished from a dory that was about 15 feet long on the bottom and a little over 19 feet overall ("15-foot dory"), and one codfish company experimented with, but quickly abandoned, using Columbia River–type sailboats for longlining codfish.

Prior to setting longline gear, the grounds were carefully sounded to determine water depths. In setting the gear, the buoy was cast astern, and one doryman rowed ahead while his partner paid out the buoy line over the dory's stern. Once the end of the buoy line was reached, it, as well as the end of the groundline, was tied to an anchor, which was then cast overboard. As the dory was rowed ahead, the aft doryman paid out the groundline, with its baited hooks, over the stern. Once the end of the groundline was reached, a second anchor was attached. To this anchor a second buoy line with its buoy was also attached.

The gear was allowed to *soak* for a few hours before it was hauled, and during good weather the gear would be set and hauled two or even three times daily. Sometimes fishermen set the longline at the end of the day, with the intention—weather permitting—of hauling it first thing in the morning.

Longline gear was usually hauled beginning at the end that, considering tide and/or wind conditions, was easiest. Once the buoy was aboard, the buoy line was hauled in hand over hand over a roller affixed to the dory's bow, and the anchor brought aboard. The groundline was then hauled in, the doryman who hauled it dexterously shaking fish from the hooks while his partner in the stern rebaited the hooks and carefully coiled the groundline in a half barrel so it was ready for the next set.

On a larger scale, the Union Fish Company power schooner *Union Jack* experimentally longlined for codfish in 1913. A mechanical net lifter installed on the vessel for experimental gillnetting for codfish was used to haul a longline. Because of limitations of the vessel and its crew, the experiment was a failure.

One of the problems associated with longlining for cod was the damage done to hooked fish by *sand fleas*—amphipod crustaceans that live on the ocean floor. In areas where they are prevalent, the voracious little creatures can render a codfish worthless in a matter of hours. To help prevent sand flea damage, early cod fishermen sometimes attached floats (hollow glass balls encased in webbing) at frequent intervals to their groundlines to keep hooked fish clear of the seafloor, where the fleas were most numerous.

According to John Cobb, another problem associated with longline fishing was that it was often difficult to find two-man crews who were compatible enough to work long hours together in a very small boat. As well, longlining required more skill than most of the "green" hands in the Alaska codfish industry possessed. The large loss of fish and gear because of stormy weather was also a problem, and by 1926 the Alaska codfish industry had reverted—perhaps completely—to the traditional means of catching codfish: handlines.[248]

CODFISH POETRY

In 1935, Edward Allen, a prominent fishing industry attorney in Seattle and a representative on the International Fisheries Commission, was a guest aboard the small, wooden tramp steamer *Santa Ana* on a voyage that circumnavigated the North Pacific Ocean.[249] The voyage included a stop at Unga, where Allen spent a day fishing from a dory with a local cod fisherman. In his book that chronicled the voyage, Allen wrote that:

> *the [cod-fishing] banks are no place for a weakling. Nor is constant battling with a gale particularly conducive to the cultivation of a mild, refined vocabulary.*
>
> *Whatever may be their vocabularies, these people of Unga were sociable, hospitable, and helpful. They were also entertaining until the subject of cod, which formed the basis of almost every discussion, began to lose its novelty.*[250]

Allen also included in his book a poem about what it would be like to be a codfish:

A CODFISH STEW

How'd you like to be a cod,
In the ocean brine,
Grab a tempting poggy fish,
Yanked up on a line?

Get a knock upon the head,
Thrown into a dory
There to wriggle with your mates,
Slimy and all gory.

Soon your head's cut off ashore,
Ripped right up the belly,
Cleansed and dressed, you then are cast
Into pickle jelly.

There you soak and soak in salt
Shriveling day by day,
Till you're dried and wrapped in sacks,
Shipped to far away—

Where around some festive board,
In the Southland sunny,
Children smack their lips on you,
Aren't codfish funny![251]

ENDNOTES

[215] Tarleton Bean, *The Cod Fishery of Alaska* [1880], in George Brown Goode, *The Fisheries and Fishery Industries of the United States, Section V: History and Methods of the Fisheries*, vol. 1 (Washington, DC: GPO, 1887), 207.

[216] "The California Codfish Trade," *Daily Alta California* (San Francisco, Calif.), June 30, 1879.

[217] Jimmy Crooks, "Schooner Fanny Dutard," undated, typed recollection of 1926 codfishing voyage to the Bering Sea aboard the schooner *Fanny Dutard*, San Francisco Maritime National Historical Park archive, HDC379, 19 pp.

[218] "Codfishermen's Agreement," *Coast Seamen's Journal* (May 8, 1918): 7, 10.

[219] Arnt Oyen, unpublished report on file at Poulsbo Historical Society Museum, Poulsbo, Wash.; Fred R. Bechdolt, "Northward with the Cod Fishermen," *San Francisco Call*, April 16, 1911.

[220] "Codfish and Fishers," *San Francisco Call*, October 25, 1885.

[221] Ed Shields, Pacific Coast Codfish Company, to Glenn E. Burch, undated transcription on file at San Francisco Maritime National Historical Park archive.

[222] J. W. Collins, *Catalogue of the Collection Illustrating the Fishing Vessels and Boats, and Their Equipment; the Economic Condition of Fishermen; Anglers' Outfits, Etc.* (Washington, DC: GPO, 1884), 150; John N. Cobb, *Pacific Cod Fisheries* (revised edition, 1926), Bureau of Fisheries Doc. No. 1014, appendix 7 to *Report of the U.S. Commissioner of Fisheries for 1926* (Washington, DC: GPO, 1927), 426; *Fisheries Exhibition Literature, International Fisheries Exhibition, London, 1883*, vol. 12 (London: William Clowes and Sons, 1884), 13; Arnt Oyen, unpublished report on file at Poulsbo Historical Society Museum, Poulsbo, Wash.; Russ Hofvendahl, *Hard on the Wind* (Dobbs Ferry, N.Y.: Sheridan House, 1989), 101; Joe Follansbee, *Shipbuilders, Sea Captains, and Fishermen: The Story of the Schooner Wawona* (New York: iUniverse, 2006), 70.

[223] Jimmy Crooks, "Schooner Fanny Dutard," undated, typed recollection of 1926 codfishing voyage to the Bering Sea aboard the schooner *Fanny Dutard*, San Francisco Maritime National Historical Park archive, HDC379, 19 pp.

[224] John N. Cobb, *Pacific Cod Fisheries*, Bureau of Fisheries Doc. No. 830, appendix 4 to *Report of the U.S. Commissioner of Fisheries for 1915* (Washington, DC: GPO, 1916), 51; William Governeur Morris, *Report on the Customs District, Public Service, and Resources of Alaska*, Sen. Doc. No. 59, 45th Cong. 1st sess., in *Seal and Salmon Fisheries and General Resources of Alaska*, vol. 4 (Washington, DC: GPO,

1898). 114; George Davidson, *Report of Assistant George Davidson, Relative to the Coast, Features, and Resources of Alaska Territory,* November 30, 1867, in *Russian America*, 40th Cong. 2d sess., H. Ex. Doc. No. 177 (Washington, DC: GPO, 1868), 256; Ed Shields, *Salt of the Sea: The Pacific Coast Cod Fishery and the Last Days of Sail* (Lopez Island, Wash.: Pacific Heritage Press, 2001), 25–26, 96–97; "Codfish and Fishers," *San Francisco Call*, October 25, 1885; John N. Cobb, *Pacific Cod Fisheries* (revised edition, 1926), Bureau of Fisheries Doc. No. 1014, appendix 7 to *Report of the U.S. Commissioner of Fisheries for 1926* (Washington, DC: GPO, 1927), 427–428; Fishing Shipping Papers, Alaska Codfish Company schooners *Mary and Ida*, 1900, and *W. H. Dimond*, 1912–1913, San Francisco Maritime National Historical Park archive; "The Untrawled Banks," *Pacific Fisherman* (August 1944): 63.

225 *The Fisheries of Alaska in 1908*, Bureau of Fisheries Doc. No. 645 (Washington, DC: GPO, 1909), 61; "Review of Pacific Coast Codfisheries for 1908," *Pacific Fisherman, Annual Review* (February 1909): 47; W. F. Robinson, "The Codfish Industry," *The Coast* (December 1908): 365–367; "Sch. *Joseph Russ* Breaks Record," *Anacortes American*, August 26, 1909; "Anacortes' Codfish Industry Is One of Great Importance," *Anacortes American*, October 12, 1911.

226 Ernie Swanson, "From Codfish to Cohoes," *Alaska Sportsman* (October 1964): 8–12, 60–62.

227 J. W. Collins, *Report on the Fisheries of the Pacific Coast of the United States*, in *Report of the United States Commissioner of Fish and Fisheries for 1892* (Washington, DC: GPO, 1892), 101–102; John N. Cobb, *Pacific Cod Fisheries*, Bureau of Fisheries Doc. No. 830, appendix 4 to *Report of the U.S. Commissioner of Fisheries for 1915* (Washington, DC: GPO, 1916), 45–46.

228 "Habitual High-Liner Seeks 12th Title," *Pacific Fisherman* (May 1932): 57; "Increased Catch and Light Demand Mark 1925 Codfish Season," *Pacific Fisherman* (January 1926): 154, 156, 158.

229 Gus Dagg, "Codfishing in the Bering Sea," *Alaska* (September 1975): 13; "Codfish Catch Improves, *Pacific Fisherman* (October 1930): 53; "Pacific Codfishing," *Pacific Fisherman* (January 1936): 224–225.

230 John N. Cobb, *Pacific Cod Fisheries*, Bureau of Fisheries Doc. No. 830, appendix 4 to *Report of the U.S. Commissioner of Fisheries for 1915* (Washington, DC: GPO, 1916), 46; "Our Fisheries," *Monthly Gleason's Companion*, April 1886; J. W. Collins, *Report on the Fisheries of the Pacific Coast of the United States*, in *Report of the United States Commissioner of Fish and Fisheries for 1892* (Washington, DC: GPO, 1892), 101–102; "Our Fisheries," *Monthly Gleason's Companion*, April 1886.

231 Thomas J. Vivian, *Commercial, Industrial, Agricultural, Transportation and Other Interests of California* (Washington: Bureau of Statistics, 1891), 486.

[232] Barton Warren Evermann, *Fishery and Fur Industries of Alaska in 1912*, Bureau of Fisheries Doc. No. 780 (Washington, D.C.: GPO, 1913), 65.

[233] Jimmy Crooks, "Schooner Fanny Dutard," undated, typed recollection of 1926 codfishing voyage to the Bering Sea aboard the schooner *Fanny Dutard*, San Francisco Maritime National Historical Park archive, HDC379, 19 pp.

[234] Jimmy Crooks, "Schooner Fanny Dutard," undated, typed recollection of 1926 codfishing voyage to the Bering Sea aboard the schooner *Fanny Dutard*, San Francisco Maritime National Historical Park archive, HDC379, 19 pp.

[235] *Fisheries of Alaska in 1906*, Bureau of Fisheries Doc. No. 618 (Washington, DC: GPO, 1907), 47; *Fisheries of Alaska in 1909*, Bureau of Fisheries Doc. No. 730 (Washington, DC: GPO, 1910), 43; "Alice Arrived Thursday, 16th," *Anacortes American*, September 23, 1909.

[236] Arnt Oyen, unpublished report on file at Poulsbo Historical Society Museum, Poulsbo, Wash.; Ward T. Bower, *Alaska Fishery and Fur-Seal Industries in 1920*, Bureau of Fisheries Doc. No. 909 (Washington, DC: GPO, 1921), 65–66.

[237] Albert C. Jensen, *The Cod* (New York: Thomas Y. Crowell, 1972), 6; *Bulletin of the United States Fish Commission for 1883*, vol. 3 (Washington, DC: GPO, 1883), 91.

[238] Ed Opheim, "Eat Crow," undated report, Baranov Museum, Kodiak, Alaska.

[239] John N. Cobb, *Pacific Cod Fisheries*, Bureau of Fisheries Doc. No. 830, appendix 4 to *Report of the U.S. Commissioner of Fisheries for 1915* (Washington, DC: GPO, 1916), 47–48; Russ Hofvendahl, *Hard on the Wind* (Dobbs Ferry, N.Y.: Sheridan House, 1989), 173; Joe Follansbee, *Shipbuilders, Sea Captains, and Fishermen: The Story of the Schooner Wawona* (New York: iUniverse, 2006), 69; J. W. Collins, *Report on the Fisheries of the Pacific Coast of the United States*, in *Report of the United States Commissioner of Fish and Fisheries for 1892* (Washington, DC: GPO, 1892), 104, 106.

[240] G. E. Hermanson, *Telling It Like It Was* (Bremerton, Wash.: G. E. Hermanson, 1972), 8; John N. Cobb, *Pacific Cod Fisheries*, Bureau of Fisheries Doc. No. 830, appendix 4 to *Report of the U.S. Commissioner of Fisheries for 1915* (Washington, DC: GPO, 1916), 56–59; Russ Hofvendahl, *Hard on the Wind* (Dobbs Ferry, N.Y.: Sheridan House, 1989), 89, 95, 110; Jimmy Crooks, "Schooner Fanny Dutard," undated, typed recollection of 1926 codfishing voyage to the Bering Sea aboard the schooner *Fanny Dutard*, San Francisco Maritime National Historical Park archive, HDC379, 19 pp.

[241] Terry Slotemaker, *Fidalgo Fishing* (Anacortes, Wash.: Anacortes Museum, 2009), 4; W. F. Robinson, "The Codfish Industry," *The Coast* (December 1908): 365–367; "Sch. Joseph Russ Breaks Record," *Anacortes American*, August 26, 1909; Ed

Shields, *Salt of the Sea: The Pacific Coast Cod Fishery and the Last Days of Sail* (Lopez Island, Wash.: Pacific Heritage Press, 2001), 172.

[242] John N. Cobb, *Pacific Cod Fisheries* (revised edition, 1926), Bureau of Fisheries Doc. No. 1014, appendix 7 to *Report of the U.S. Commissioner of Fisheries for 1926* (Washington, DC: GPO, 1927), 426, 432–436.

[243] Although beam trawls are not currently used for cod fishing, at the time of this writing a small fishery in Southeast Alaska utilizes beam trawls to catch shrimp.

[244] John N. Cobb, Pacific Cod Fisheries (revised edition, 1926), Bureau of Fisheries Doc. No. 1014, appendix 7 to *Report of the U.S. Commissioner of Fisheries for 1926* (Washington, DC: GPO, 1927), 435–436.

[245] Albert C. Jensen, *The Cod* (New York: Thomas Y. Crowell, 1972), 128–129; Wm. H. Dumont and G. T. Sundstrom, *Commercial Fishing Gear of the United States*, U.S. Fish and Wildlife Service Circular 109 (Washington, DC: GPO, 1961), 9–10; Marine Stewardship Council, *Alaska Pacific Cod—Bering Sea and Aleutian Islands*, January 2015, accessed February 11, 2015, http://www.msc.org/track-a-fishery/fisheries-in-the-program/certified/pacific/bering-sea-and-aleutian-islands-pacific-cod/fishery-name; David Witherell, Michael Fey, and Mark Fina (staff, North Pacific Fishery Management Council), *Fishing Fleet Profiles* (April 2012), 17–18.

[246] "Shore Codfishing," *Pacific Fisherman* (January 1939): 242; "Codfish Fleet Prepares for Bering Sea Season," *Pacific Fisherman* (April 1940): 69; "'*Dorothy*' Outfitted as Trawler for Bering Sea Codfish Banks," *Pacific Fisherman* (May 1940): 30; "Power Schooner *Dorothy* Sails for Bering Sea," *Anacortes Daily Mercury*, May 7, 1940; "Codfish," *Pacific Fisherman* (January 1941): 261; Ward T. Bower, *Alaska Fishery and Fur Seal Industries: 1941*, Statistical Digest No. 5 (Washington, DC: GPO, 1943), 43–44; "Supplemental King Crab Industry Foreseen as Survey Ends," *Pacific Fisherman* (October 1941): 65–67; "The Untrawled Banks," *Pacific Fisherman* (August 1944): 63; "The Future Looks WNW," *Pacific Fisherman* (August 1944): 48–49.

[247] For halibut fishermen in Alaska, groundline is simply the line used to make "skates"—300-fathom lengths of groundline to which gangions and hooks are affixed. When fishing, as many as a dozen or more skates are fastened end to end.

[248] John N. Cobb, *Pacific Cod Fisheries* (revised edition, 1926), Bureau of Fisheries Doc. No. 1014, appendix 7 to *Report of the U.S. Commissioner of Fisheries for 1926* (Washington, DC: GPO, 1927), 426–436; Z. L. Tanner et al., "Explorations of the Fishing Grounds of Alaska, Washington Territory, and Oregon, during 1888, by the U.S. Fish Commission Steamer Albatross, Lieut. Comdr. Z. L. Tanner, U.S.

Navy, Commanding," *Bulletin of the United States Fish Commission*, vol. 8, 1888 (Washington, DC: GPO, 1890), 31; Mark Kurlansky, *Cod: A Biography of the Fish That Changed the World* (New York: Penguin Books, 1997), 124, 132; "Review of the Pacific Codfish Season, 1924," *Pacific Fisherman* (January 1925): 146, 148–149.

[249] The International Fisheries Commission was the predecessor to today's International Pacific Halibut Commission, which is charged with developing and maintaining Pacific halibut stocks in U.S. and Canadian waters.

[250] Edward Weber Allen, *North Pacific: Japan, Siberia, Alaska, Canada* (New York: Professional & Technical Press, 1936), 107–108.

[251] Ibid.

Chapter 6
CODFISH STATIONS

> *The late Thomas W. McCollam, of the McCollam Fishing & Trading Co., of San Francisco, was the first to perceive the advantages to be obtained from establishing stations close to the cod banks, where the fishermen could go out daily in dories to the adjacent banks and the catch be stored ashore until a cargo accumulated, when a vessel could be sent north to bring them to San Francisco.*
>
> —John N. Cobb, 1915[252]

Alaska codfish stations were primarily located in an arc that stretched from the Shumagin Islands southwest along the Alaska Peninsula to the southeastern shore of Unimak Island, with most of the stations being in the Shumagin and Sanak islands.

The abundance of cod in the vicinity of the Shumagin Islands was well known, but, as noted above, Thomas McCollam, of the San Francisco–based McCollam Fishing & Trading Company, was the first in the industry to perceive the advantages of establishing a cod-fishing station in the Shumagins. In 1876, McCollam purchased a hunting camp, complete with several buildings and a wharf, at Pirate Cove, a very pretty and well-sheltered harbor at the north end of Popof Island. He converted the camp into Alaska's first codfish shore station. Originally manned by a company agent and about eight fishermen, Pirate Cove would gradually become the largest and most important codfish station in Alaska.

Codfish inhabited the nearby banks year-round, although "school fish" left in August or September and returned in January or February, with the best fishing usually in the latter month. In 1880, codfish caught near Pirate Cove were reported to have typically weighed between eight and twelve pounds, with the largest weighing fifty pounds.[253] Circa 1918, the Union Fish Company, successor to McCollam's firm, praised the Pirate Cove station as "without exception, the best location for all-the-year-round codfishing in Alaska." Among its amenities was a general merchandise store.[254]

Station fishermen furnished their own gear, but McCollam's company furnished dories and provided free lodging and meals.[255] All fishing was done during the daylight hours, with fishermen typically rising between 3:00 a.m. and 4:00 a.m. during the summer season, and between 4:00 a.m. and 5:00 a.m. during the winter season. After

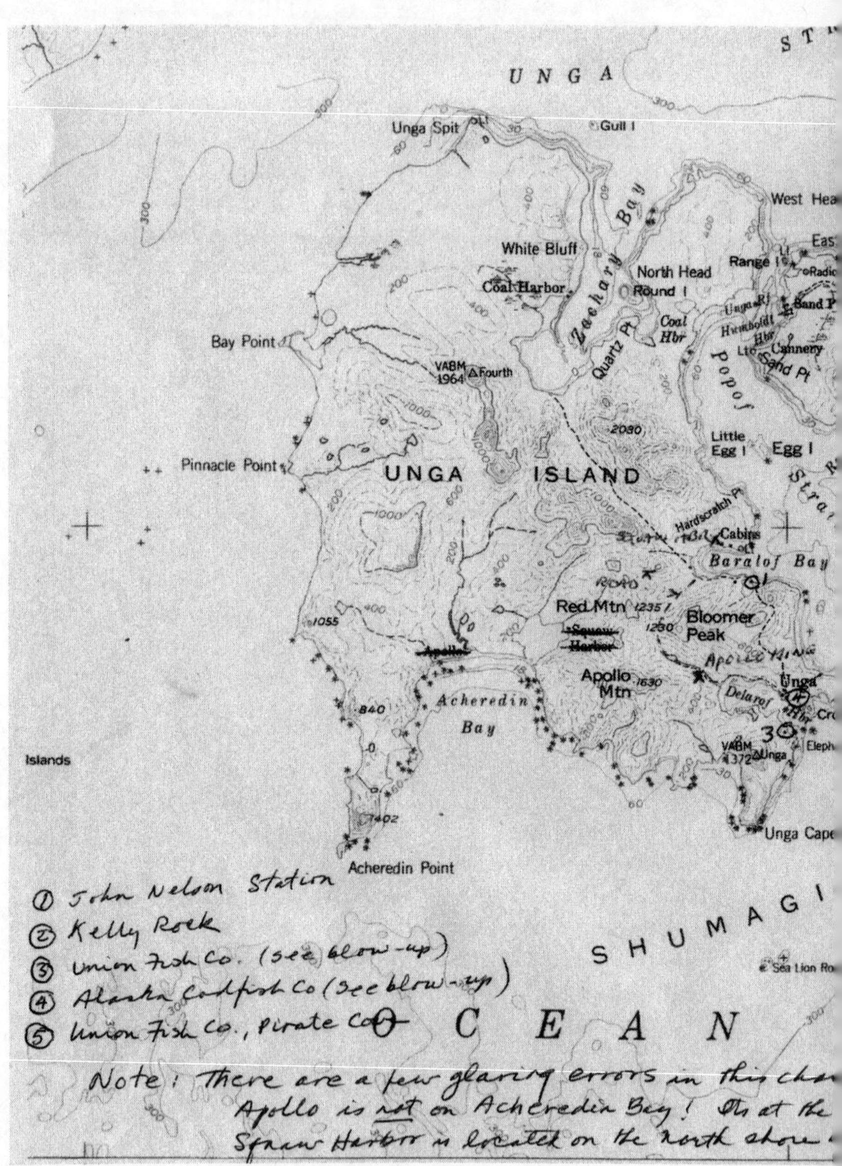
UNGA
Unga Spit
Gull I
Zachary Bay
White Bluff
Coal Harbor
North Head
Round I
Range I
Coal Hbr
Quartz Pt
Bay Point
VABM 1964 Fourth
Popof
Cannery
Sand Pt
Little Egg I
Egg I
Pinnacle Point
UNGA ISLAND
2080
Hardscratch Pt
Cabins
Baralof Bay
Red Mtn
1235
1250
Bloomer Peak
Squaw Harbor
Apollo
1055
Apollo Mtn
Acheredin Bay
840
Unga
VABM 1372 Unga
Islands
Unga Cape
1402
Acheredin Point
① John Nelson Station
② Kelly Rock
③ Union Fish Co. (see blow-up)
④ Alaska Codfish Co (see blow-up)
⑤ Union Fish Co., Pirate Co
OCEAN
Note: There are a few glaring errors in this
Apollo is not on Acheredin Bay! Its at the
Squaw Harbor is located on the north shore

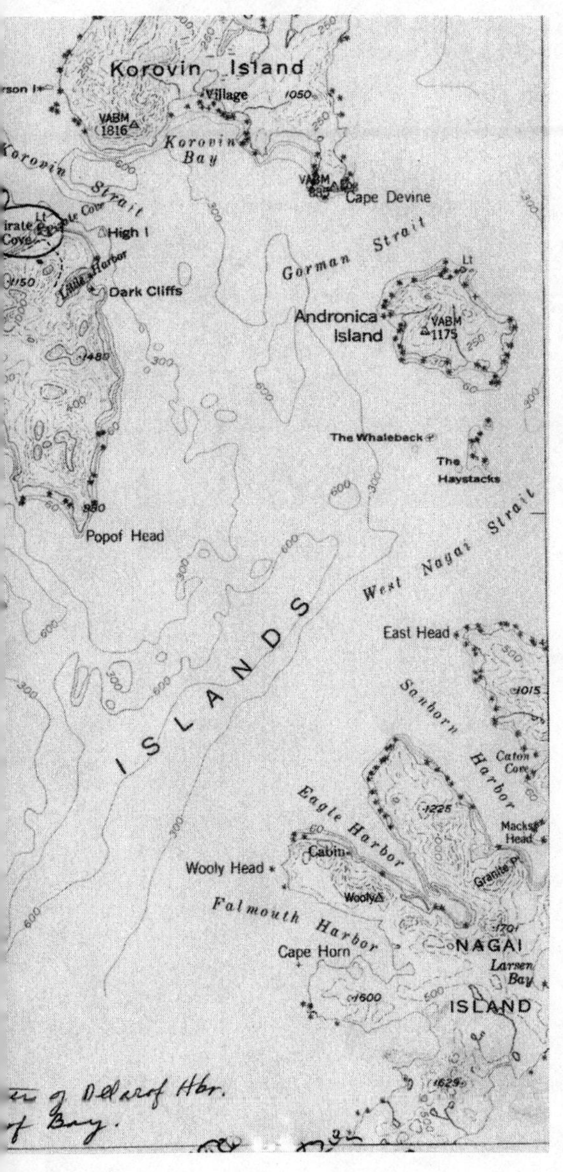

Figure 39. Unga Island and Popof Island, annotated by Thor Lauritzen, whose grandfather operated an independent cod-fishing station at Unga. (Courtesy of Thor Lauritzen.)

Figure 40. Pirate Cove cod-fishing station, 1913. (John N. Cobb Collection, Special Collections, University of Washington Libraries, Seattle, Wash.)

breakfast, the fishermen rowed to nearby fishing grounds, where they fished for several hours before straggling back to the station with their catches between 9:00 a.m. and noon. (The sea tends to be calmer in the early morning hours, which is an important consideration for men working alone in small, often overloaded, boats.)

Once alongside the wharf, a fisherman used a pew to transfer his fish from his dory into a long, slatted (for drainage) low-walled wooden platform located about halfway between the low-tide line and the wharf's deck. The fisherman then pewed his fish from the platform onto the wharf, while the station agent tallied the catch.

> **Fostering competition:** *In the bunkhouse is hung a board ruled so as to show the name of each fisherman and his catch from day to day . . . thus giving the men an opportunity to know just how they stand.*
>
> —John Cobb, 1916[256]

The average tidal range in the Shumagin Islands is about six feet, but the maximum range is about twelve feet. During the highest tides,

Figure 41. Fishermen at Pirate Cove preparing to pew codfish onto the dock. (John N. Cobb Collection, Special Collections, University of Washington Libraries, Seattle, Wash.)

the box would be submerged for a short while, but the higher water would allow fishermen to pew their fish directly from their dories onto the wharf.

Of the fishermen employed at the codfish stations, John Cobb observed in 1916 that "[w]hile a small proportion of the white men are excellent fishermen of the type required for hand-line fishing from dories, the majority of them are ordinary beach combers picked up on the water fronts of San Francisco and Seattle, or men of practically no acquaintance with the sea even, let alone any fishing knowledge." ("Mexican Frank" Rodgers, as he was known as at Unga, had hardly been "picked up." Rather, someone laced cocoa he was drinking in San Francisco with a strong sedative, and Rodgers awoke in the forecastle of a schooner bound for Unga.)[257]

Cobb noted that the station agents, the men in charge of the codfish stations, were generally fishermen who had worked their way up through the ranks. He characterized some as being "excellent workers, with considerable native shrewdness," but added that they had little opportunity for advancement. Cobb lamented the fact that these men had "very little opportunity to keep in touch with the world's progress." He believed this isolation largely explained why "the codfish

Figure 42. Dory fishermen unload codfish at Alaska Codfish Company wharf at Unga, 1910. (Silva Collection, Alaska State Library, Historical Section.)

industry of the Pacific coast is but little further advanced to-day, so far as methods of catching and curing the fish are concerned, than it was 40 years ago."[258]

"Quite a few" Natives, according to Cobb, were employed at the stations. Cobb considered them to be "among the best of the station fishermen, as they are usually well acquainted with the locations of many isolated spots which, while rich in cod, yet cover sometimes but a few feet or yards in extent and are difficult to find without certain landmarks being well fixed in the mind." However, wrote Cobb, the Native fishermen were "very apt to quit when the whim seizes them." He added that, in his experience, both white and Native cod fishermen were "apt to quit on very slight or no provocation at all, the desire for a change of scene at frequent intervals seeming, in their eyes at least, to be one of the essentials of the industry."[259]

Cobb noted that "quite a few" white fishermen had married Native women and the couples resided together in small cottages or shacks on the station grounds.[260] E. Lester Jones, a Bureau of Fisheries agent who visited a codfish station on Unga Island as part of an investigation of Alaska's fisheries in 1914, wrote that the fishermen there "appear to be prosperous and happy, and that they have a comfortable living is evident."[261]

Jones, however, failed to mention one important drawback to codfish station life: liquor. According to John Cobb, liquor was for a

Figure 43. Cod fisherman's home, Sanak Island, circa 1913. (John N. Cobb Collection, Special Collections, University of Washington Libraries, Seattle, Wash.)

number of years "one of the heaviest handicaps under which Alaska [codfish] station owners suffered." In 1913, there was one saloon at Sand Point and two at Unga. Fully seven cod-fishing stations were in close proximity. Drunken carouses by fishermen were common and could last for weeks. Sometimes the result was fatal. Cobb wrote that frequently a fisherman "would meet with an untimely end through the capsizing of his dory while returning in an intoxicated condition from a visit to one of these saloons, or be frozen to death or meet with a fatal fall while traversing intoxicated the rough and slightly marked trails between the stations and the towns." Cobb recalled that one codfish station fisherman was known as "Whiskey Jack" because of his "excessive indulgence in this fluid," while another was known as "Whiskey Bill" because of his "constant preaching of the merits of temperance." Liquor—at least legal liquor—became harder to get in the Shumagin Islands in 1914, when government officials refused to renew the saloons' licenses or to grant new ones.[262] This was four years before Alaska adopted the so-called Bone Dry Law, which prohibited the manufacture or consumption of alcohol in the territory, and six years before the Volstead Act instituted Prohibition throughout the United States. Prohibition ended in 1933.

Homebrew was apparently common in Unga during the mid-1930s. The village school teacher at the time said that holidays, of which there were many, always lasted three days: one day to get drunk; two days to get sober.[263]

Figure 44. Unga, 1914. Both the Alaska Codfish Company and the Union Fish Company operated cod-fishing stations at Unga. (Silva Collection, Alaska State Library, Historical Section.)

Figure 45. Pacific cod in pickle, Unga, 1914. (E. Lester Jones, Report of Alaska Investigations, 1914 [Washington, DC: GPO, 1915], 56.)

Circa 1890, McCollam Fishing & Trading Company (see below) station fishermen were paid $27.50 per one thousand codfish, each of which had to be at least twenty-six inches long. Codfish from twenty-four to twenty-six inches long were counted two for one; all codfish less than twenty-four inches long were discarded.

Dinner—the main hot meal of the day—was served at noon, after which the fishermen divided themselves into gangs to process the catch in a *dress house* built on the wharf. Once dressed, the fish were trundled in wheelbarrows to a generally long, low building known as a *butt house*, where they were *pickled* in brine-filled wooden tanks, each with a capacity of about three thousand to four thousand medium-sized fish. An alternative to storing fish in brine was kenching them—dry salting them in wooden bins. The preserved codfish were accumulated until a transporter (schooner) from San Francisco called to take the catch and resupply the station with salt and other necessities (see below).[264]

In 1882, the average daily catch for a fisherman in the Shumagin Islands was reported to be about 200 fish, with an exceptional day yielding 500 to 600 fish. During 1880, seven shipments of salted codfish, aggregating 432,000 fish and weighing 1,728,000 pounds, were transported from the Shumagin Islands to San Francisco.[265]

In 1884, McCollam's expanded business, the McCollam Fishing & Trading Company, constructed several new buildings and a new dock at Pirate Cove. It might have been at this time that the company replaced its kenches with wooden tanks. McCollam Fishing & Trading also substantially increased its inventory of and expanded the variety of trading goods at the station. Pirate Cove almost immediately became the commercial center for the region, and some fishermen, instead of returning to San Francisco at the close of the main fishing season, began to winter at Pirate Cove, where they fished for cod as the weather permitted and hunted furbearers.[266]

In 1886, McCollam Fishing & Trading established a branch fishing station at Pavlof Harbor, on Sanak Island, about one hundred miles from Pirate Cove. The company later opened two additional stations on Sanak Island, the first in 1890 at Kasatska, on the south side of the island, and the second at Port Stanley. The stations on Sanak Island were operated during only the winter months, the only time of the year sufficient quantities of codfish were close enough to these stations to be fished from dories. In 1892, McCollam Fish & Trading established a station at Sanborn Harbor, Nagai Island, in the Shumagin Islands, where fishing was carried on from the middle of spring until late summer.

The Sanak Island stations at both Kasatska and Port Stanley were abandoned after operating only a few years. The Sanborn Harbor station, however, operated continuously until at least 1926. In 1898, McCollam Fishing & Trading established a station at Dora Harbor, on Unimak Island, in the eastern Aleutians. Cod fishing at Dora Harbor was conducted during only the summer months.

McCollam Fishing & Trading was not the only firm that established cod-fishing stations in Alaska. Lynde & Hough, also of San Francisco, established shore stations at Unga (Delarof Harbor, Unga Island) in 1888 or 1889, and at Squaw Harbor (Baralof Bay, Unga Island), Henderson Island, and Company Harbor (Sanak Island) in 1889. The following year, the firm established stations at Nelson Island (Sanak Islands), Chicago Bay (Alaska Peninsula), and at Ikatan (Unimak Island).

Lynde & Hough also had a station at Sand Point, on Popof Island, in the Shumagin Islands, which it established in 1887. This station, however, was mainly a trading and supply facility where the firm's vessels could land their cargoes and resupply for another trip without having to return to San Francisco.

The Chicago Bay, Nelson Island, and Henderson Island stations were short-lived. In 1902 or 1903, Lynde & Hough and McCollam Fishing & Trading merged to form the Union Fish Company (see chapter 2), and the new firm established a codfish station at Eagle Harbor (Nagai Island) in 1903. About the same year, the firm established a codfish station at Wedge Cape, also on Nagai Island, and during the summer of 1916 the company established a shore station on Tigalda Island, which lies just off the east entrance of Unimak Pass.

In 1896, another San Francisco firm, the Alaska Codfish Company, purchased the Alaska Commercial Company's trading station at Company Harbor and converted it into a codfish station. The Alaska Codfish Company that year also established a station at Kelley's Rock (Winchester), on Unga Island. The Kelley's Rock station, which operated year-round, was by far the most productive of the Alaska Codfish Company's stations. The following year, the Alaska Codfish Company established stations at Moffet Cove, a few miles east of Company Harbor, and at Dora Harbor, on Unimak Island.

Figure 46. Token, obverse and reverse, Alaska Codfish Company, Unga, Alaska. (Courtesy of Kaye Dethridge and Irene Shuler, Alaskatokens.com.)

Figure 47. Two hundred fifty thousand pounds of kenched salt cod at Unga, 1914. (E. Lester Jones, *Report of Alaska Investigations, 1914* [Washington, DC: GPO, 1915], 56.)

In the fall of 1905, the Pacific States Trading Company, a newly established San Francisco firm, established codfish stations on Herendeen Island, Northwest Harbor (Big Koniuji Island), and at Ikatan, on Unimak Island. These stations were operated continuously until 1909, when they were closed due to poor market conditions. The Northwest Harbor station was reopened in the fall of 1911 and was operated until early in 1916, when the Pacific States Trading Company permanently ceased operations (see chapter 4).

In 1906, the Alaska Codfish Company bought the Alaska Commercial Company's trading station at Unga, on Delarof Harbor, on Unga Island, and began cod-fishing operations in the fall. The next year, the Union Fish Company built a codfish station on the opposite side of the harbor. Cod fishing at Unga was carried on year-round.

In June 1909, the Union Fish Company's schooner *Stanley* arrived in San Francisco carrying 952,000 pounds of salted codfish, to date the largest cargo of codfish that had ever been delivered to that port. The fish had been collected from the company's shore stations during the winter and spring.

The first Puget Sound–based company to establish a codfish station in Alaska was the Seattle-Alaska Fish Company, which in the spring of 1903 built a station at Falmouth Harbor, on Nagai Island. This firm opted to use Columbia River boats for fishing instead of the traditional

Figure 48. Alaska Codfish Company's codfish station at Squaw Harbor, Unga Island, circa 1913. Note stockfish drying on dock. (Silva Collection, Alaska State Library, Historical Section.)

banks dories. Columbia River boats were slim, double-ended vessels that were commonly used by salmon fishermen on the lower Columbia River. They were typically about 26 feet long and were powered by oars and sail. (The classic 32-foot Bristol Bay boat is basically a longer, beamier, and deeper version of a Columbia River boat.) When fishing codfish, the Columbia River boats were manned by two to four men. The Falmouth Harbor station, however, proved to be too far from the fishing grounds, so it was quickly relocated to Squaw Harbor, on Unga Island. This station was operated only intermittently until it was closed in 1910, the same year the Seattle-Alaska Fish Company was bought out by the Seattle-based King & Winge Codfish Company (see chapter 3). The Squaw Harbor facility was later purchased and improved by the Alaska Codfish Company.

The shore stations at Sand Point, Squaw Harbor, and Unga suffered a setback in November 1905, when a storm wrecked thirty-five small boats in their harbors. The following year, notwithstanding this setback, the Bureau of Fisheries recognized the shore-station cod fishery as more important and more profitable than the offshore fishery. (Growth of the schooner fleet during the following years, however, rendered this distinction relatively short-lived.) In 1906, there were fully nineteen codfish stations in operation in Alaska. Sixteen were in the Shumagin and Sanak islands, which offered relatively sheltered harbors in close proximity to rich fishing grounds; the remaining three were on

Unimak Island, in the eastern Aleutians. The number of dories used at individual stations ranged from about four to a dozen or more.[267]

John Cobb, who as a former Bureau of Fisheries field agent had studied Alaska's codfish industry extensively, apparently had confidence in its future, and during the years 1912 and 1913 he worked as a manager for the Union Fish Company. While employed by Union Fish, Cobb made two voyages to the company's codfish stations in the Shumagin Islands. Cobb later became editor of *Pacific Fisherman* and founded the College of Fisheries at the University of Washington. In 1916, even though Cobb was no longer employed by the Bureau of Fisheries, his detailed report on the Pacific cod fisheries was published as an appendix to the annual report of the U.S. Commissioner of Fisheries. An updated version of the report was similarly published in 1927.[268]

In 1918, the Seattle-based Bering Sea Fisheries Company constructed a station on Unalaska Island, though this station may not have operated until the following year. To provision this station and a station the company established at Unga in 1919 or 1920, and to carry salted cod to Seattle, the Bering Sea Fisheries Company employed the steamer *Dora*, which for many years had carried mail and freight to central and western Alaska. This arrangement continued until December 1920, when the *Dora*, steaming north, struck a rock

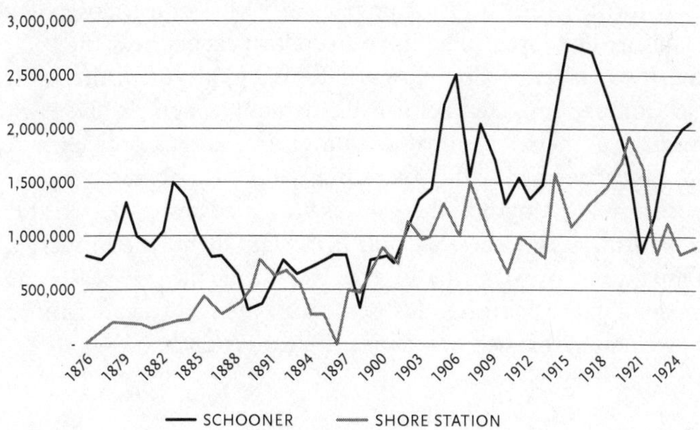

Chart 2. Schooner and shore station salted Pacific cod production, numbers of fish, 1876–1924.

Data source: John N. Cobb, *Pacific Cod Fisheries* (revised edition, 1926), Bureau of Fisheries Doc. No. 1014, appendix 7 to *Report of the U.S. Commissioner of Fisheries for 1926* (Washington, DC: GPO, 1927), 467, 484.

on the northeast coast of Vancouver Island, British Columbia, and was lost. Coastal Alaskans revered the *Dora*, and in his report on Alaska's fisheries for that year, Bureau of Fisheries agent Ward Bower's note of the event bordered on a eulogy: "[A]fter 40 years of almost continuous service, at times under very trying conditions, there was lost the most historic vessel plying Alaskan waters." The Bering Sea Fisheries Company declared bankruptcy in the fall of 1921 and was subsequently liquidated.[269]

By 1920, the Bellingham, Washington–based Pacific American Fisheries, a large firm that operated a number of salmon canneries and other fish-processing facilities in Alaska and Washington, had become engaged in salting codfish in conjunction with its salmon salting operations. The company had stations at Ikatan (on Unimak Island) and Port Moller (on the Bering Sea shore of Alaska Peninsula). Fishermen at these stations used small gasoline-powered boats that each carried four or five dories to get to and from the grounds. Each crew salted its own fish, which were shipped by steamer in tierces (large wooden barrels) to Puget Sound.

Due to poor market conditions, the Pacific American Fisheries Company suspended its codfish operations in 1921. In 1923, however, J. O. W. Brown, who had been superintendent of the company's operations and had in the past worked as a manager at the Matheson Fisheries Company, reconstituted the codfish operations under his newly formed Brown Fish Company. Brown worked closely with his former employer. In addition to codfish caught near the Pacific American Fisheries Company's stations by dorymen, the Brown Fish Company utilized codfish incidentally caught in the Pacific American Fisheries Company's salmon traps. And Brown's salted cod was carried by Pacific American Fisheries Company steamers to Anacortes, Washington, where it was processed and packaged at the Pacific American Fisheries Company's plant there. The Brown Fish Company received 175 tons of salted cod during the 1929 season, and remained in the codfish industry through the 1931 season, although its production that year was a meager 4,200 codfish.[270]

In 1923, the Seattle-based San Juan Fishing & Packing Company, which had been processing fish in Alaska since 1901 and was mostly involved with salmon, halibut, and herring, established a codfish station at Unalaska, on Unalaska Island. The motor vessel *San Jose* (fourteen net tons), which carried eight dories, supplied the station with codfish. The station was closed after the 1925 season.

Also on Unalaska Island, late in 1923 the Aleutian Livestock Company, which raised sheep on the island, established a codfish station at Chernofski, on the west side of the island. The power vessel *Daisy* (thirty net tons) was used in conjunction with the station, which closed after the 1924 season.[271]

OTHER CODFISH STATION LOCATIONS

Kodiak Island

Codfish were also caught in the waters around Kodiak Island, but the industry that developed there was minor in comparison to that of the Shumagin Islands area and the Bering Sea. The earliest report of a cod fishery at Kodiak was in 1879. That year, the *New York Times* quoted the mayor of Sitka, who wrote that two Kodiak fishermen, using hook-and-line gear, had caught twenty-two thousand codfish over a six-month period. Additionally, an 1890 U.S. Treasury Department report noted that some cod fishing was done at Kodiak, and that Natives there were paid seventy-five cents to one dollar per day to head, split, and salt the fish.[272]

In 1883, Ivan Petroff, who while an agent of the U.S. Census Office had during 1880 and 1881 explored much of Alaska and had noted the Pacific codfish industry's commercial importance, built a fishing station on the Sitkalidak Island shore of Sitkalidak Strait, just off the southeast shore of Kodiak Island and close to the Native village at Old Harbor. The station did not operate long, but while operating it shipped what John Cobb termed "considerable quantities" of salted cod to San Francisco.

Another short-lived codfish venture on Kodiak Island was an experiment by the Alaska Commercial Company, which had salmon canning and other interests on the island. In 1910, the company shipped to San Francisco about ninety tons of codfish that had been caught and cured by island Natives. The fish, however, were small and were sold only with great difficulty. This ended the Alaska Commercial Company's interest in codfish, but in the subsequent several years independent Kodiak Island fishermen occasionally caught and cured codfish that they then shipped to dealers in Puget Sound.

The situation changed with the advent of World War I, when high demand for cod engendered by the war caused the price of cod to rise and drew additional Kodiak fishermen into the business. John Cobb wrote of "a small fleet of vessels, mainly powered with gas engines, and a considerable number of small sail and power vessels of less than

5 net tons each" fishing cod on the banks lying off the east shore of the island. The fishermen shipped their cured catch to Puget Sound through local dealers.[273]

The demand for cod—and the cod-fishing effort at Kodiak—fell greatly after the end of the war, but individual operators at Kodiak continued to salt cod on a small scale. Most notable were the W. J. Erskine Company and O. Kraft & Son, both of which operated at least through 1923. The W. J. Erskine Company's saltery was adjacent to its retail goods store in Kodiak; O. Kraft & Son's saltery was at Monashka Bay, a few miles northwest of Kodiak. Among the smaller operators were John Tevik, whose station was also at Monashka Bay, and Chris Opheim, Bill Castell, and John Swenson, whose stations were at Sunny Cove, near Ouzinkie, on Spruce Island (just north of Kodiak).

Chris Opheim, who had begun fishing for cod in the Shumagin Islands during the early 1900s, established his station in 1923. With his three sons, Opheim longlined and handlined for cod from his home-built, 28-foot motor-sailer *Hilda O*, as well as from dories. Their cod were salted in kenches and then shipped in 100-pound-capacity burlap sacks to the Coffin Codfish Company (possibly the W. H. Coffin Fish Company), in Seattle, via scheduled steamers. The shipping cost from Kodiak to Seattle was $5 per ton, and the Seattle fish company paid Opheim $100 per ton of delivered salted cod. Ed Opheim, Chris Opheim's son, recalled that at Sunny Cove, "If there had been no cod there would have been no life. . . . Cod was in everyone's conversation every day." Opheim's station closed in 1933.[274]

Of perhaps a more substantial nature, in 1916 Crescent Hale, whom *Pacific Fisherman* praised as "one of the most competent all-around cannerymen in the Western Alaska district," established the Northern Fisheries Company.[275] The firm was primarily interested in canning salmon, but it established a codfish packing and curing plant at Anacortes, Washington. In January 1917, the Northern Fisheries Company loaded the old cod-fishing schooner *Harold Blekum* with materials to build a shore station on Kodiak Island. Just after midnight on March 4 of that year, the vessel, laden with 100 tons of cargo for Seattle, fetched up on a reef in Ugak Bay, on Kodiak Island's east shore, and became a total loss. The Northern Fisheries Company that year also sent four cod-fishing vessels to Kodiak: the former halibut power schooner *Progress* (115 tons), the schooner *Charles E. Brown* (64 tons), the power schooner *Hunter* (60 tons), and the power schooner *Valdez* (10 tons). The *Progress* caught some 125,000 codfish that year. About 80,000 of the fish were shipped by steamship to Anacortes, and the *Progress* carried the rest there in the fall. The other three vessels at intervals delivered their catches—a total of 42,327 fish—at Kodiak, where Northern Fisheries had expanded a small herring saltery into a general fish salting plant. Unfortunately, the *Hunter* was wrecked in August, with the total loss of the fish onboard. Fortunately, no lives were lost.[276]

The following year, in what the Bureau of Fisheries termed "the most notable development in the cod fisheries of central Alaska" in 1918, the Northern Fisheries Company established a shore station at Kodiak. The company brought in several experienced fishermen, whose success inspired a few local white and Native men to engage in cod fishing. Fishing was done from a small fleet of vessels, mainly powered with gasoline engines, mostly on the near-shore banks lying off the east coast of the island.[277] Among the vessels were the *Progress*, *Charles E. Brown*, and *Valdez*. (The *Progress* again delivered its catch to Anacortes.) Also in 1918, the Northern Fisheries Company purchased the venerable Union Fish Company, of San Francisco, which, together with its progenitors, had a five-decade history in the codfish industry and owned five schooners, eleven Shumagin Islands shore stations, and an extensive processing and curing plant on San Francisco Bay (see chapter 2).[278]

In 1919, in what the Bureau of Fisheries termed "the most conspicuous withdrawal" that year from the Alaska codfish industry, the Northern Fisheries Company shuttered its Kodiak operation, probably because of a combination of the increased availability of Atlantic cod during the years following the end of World War I and the company's desire to focus on the well-situated shore stations it had acquired with the purchase of the Union Fish Company (see chapter 7).[279] The trade journal *Western Canner and Packer* reported in January 1920 that the Robinson Fisheries Company, also of Anacortes, had leased the Northern Fisheries Company's codfish plant at Anacortes.[280]

Kenai Peninsula

A modest amount of salted cod was also produced by the Alaska Ocean Food Company, which began construction of a cold storage facility at Port Chatham, near the southwest tip of the Kenai Peninsula, in 1919. The company's intent was to process halibut, salmon, herring, and cod. Apparently while the company's plant was under construction, its 5-ton power vessel *Trio* was busy fishing for cod in nearby waters. The vessel carried dories and a crew of seven men, including the captain. During the 1919 season—the first of the company's two seasons in the codfish business—Alaska Ocean Food produced approximately

200,000 pounds of salted cod. The product was shipped south in 100-pound-capacity Sitka spruce boxes that were manufactured on site in the company's sawmill. Alaska Ocean Food withdrew from the codfish industry after the 1920 season, likely—as had most small producers of Alaska codfish—because of poor market conditions.[281]

SHORE-STATION VESSELS

Transporters

Shore stations were serviced by transporting vessels of the same type (schooner) and size used in the offshore cod fishery. These vessels brought supplies—including large quantities of salt—to the stations, and returned south laden with salted codfish. Often offshore cod-fishing vessels themselves served as transporters during the winter season. John Cobb noted the peril involved: "As stormy weather with plenty of fog is the rule in the North Pacific Ocean, many of these vessels have met with an untimely end on the inhospitable shores of this region."[282] One such vessel was the Union Fish Company's 143-foot schooner *Stanley*.

In mid-October 1909, the *Stanley*, with a crew of eight, departed San Francisco carrying 150 tons of salt as well as lumber and other supplies for the company's shore stations in the Sanak and Shumagin islands. It would spend the late fall and winter months collecting salted cod from the stations.

The accounts vary slightly, but on the morning of March 28, the *Stanley*, laden with salted cod, was near Sanak Island, fighting a fierce storm that had been raging for four days and included snow and sleet. The *Stanley*'s captain, Bartholomew Koehler, determined his least perilous option was to anchor at the entrance to Pavlof Harbor and, hopefully, ride out the remainder of the storm. With waves washing over the *Stanley*'s decks and the lee shore not far off, Koehler ordered his crew to drop the vessel's two anchors.

Not long after, the anchors failed or were lost, causing the *Stanley*'s bad situation to instantly become dire. Captain Koehler valiantly lashed himself to the helm and, with his two mates, who had lashed themselves to the rigging, made a desperate attempt to prevent the vessel from grounding. Their effort was futile, and the *Stanley* grounded on the rocky shore. Fortunately, the vessel was not immediately dashed to pieces, though frigid breakers continuously washed over it.

The *Stanley*'s grounding was witnessed helplessly by fishermen on the island. When the sea calmed the following morning, several fishermen made their way in dories to the vessel. The two mates had succumbed to the cold and perished during the night, while the captain survived the night but died shortly after being freed from the helm. A fourth casualty was the vessel's cook, who had been washed overboard and drowned. Four of the *Stanley*'s crewmen survived the ordeal. When their rescuers came alongside, the men jumped one by one overboard into the frigid water. Their rescuers described them as being "more dead than alive" when pulled from the water. The survivors were given medical care ashore and were later taken to Unga aboard the steamer *Dora*; the bodies of the four dead seamen were buried on Sanak Island. The *Stanley*, along with its cargo of salted cod, was a total loss.[283]

The first power transport vessel used in the Pacific codfish industry was the steamer *Newport*, which was likely chartered by the Alaska Codfish Company to carry cargo to and from the company's codfish stations to San Francisco during the summer of 1906. The use of the steamer was heralded as innovative but was apparently not repeated in the next few years.[284]

In 1913, the Union Fish Company, of San Francisco, launched the 145-foot power schooner *Golden State*, which it had built specifically to service its codfish stations in Alaska. The *Golden State* was the largest motor-propelled vessel in the fishing industry on the U.S. Pacific coast and could carry more than 500 tons of cargo. The three-masted vessel was manned by a crew of eight and could be propelled by its sails or by its four-cylinder, 150-horsepower Union gasoline (distillate) engine. The engine, however, was inadequate against a headwind but was very useful during calm weather and when entering or leaving harbors. In addition to quarters for the crew, the *Golden State* had cabin accommodations for ten passengers.[285]

The *Golden State*'s maiden voyage was a big success, the vessel returning to San Francisco in mid-October after a thirteen-day passage from Pirate Cove (in the Shumagin Islands) with a cargo of 175,000 codfish as well as five thousand barrels of salted salmon. Until 1927, when it was transferred to the Union Fish Company's tuna operations in California waters, this reliable vessel made as many as three round-trips annually between San Francisco and the company's codfish stations in Alaska.[286]

Figure 49. Union Fish Company codfish transporter *Golden State*. (Courtesy of Union Fish Company.)

Station-to-Station Vessels

As firms established multiple stations, the need developed to transport goods and fish between them. The first station-to-station vessel might have been the 20-ton schooner *Unga*, which, with a crew of five, was brought into service by 1889 by the McCollam Fishing & Trading Company to service its stations in the Shumagin Islands area.[287] Sailing vessels such as the *Unga*, however, were constrained by their inability to make a journey to even the nearest station if the winds were not favorable and by the difficulty of maneuvering them in confined harbors. In 1905, the Union Fish Company (successor to the McCollam Fishing & Trading Company) constructed and sent north the launch *Union Flag* (7 net tons), the first power vessel for the station-to-station service. This vessel, which was equipped with a 40-horsepower Union gasoline engine, was the first power vessel to be utilized in any capacity in the Pacific codfish industry in northern waters. (Small power vessels had been used by the industry for a number of years to move product around San Francisco Bay.) The *Union Flag* was headquartered at the company's Pirate Cove station and remained in service at least until 1919.[288]

In 1912, the Union Fish Company contracted for the construction by the Nilson & Kelez shipyard, in Puget Sound, of its second power vessel for use exclusively in the Pacific codfish industry. The fully rigged schooner *Union Jack* was 73 feet long and had a beam of 18 feet. It was fitted also with an 80-horsepower gasoline engine and was designed to engage in fishing as well as to service the company's codfish stations.

Figure 50. Crew members of the *Union Jack*, 1913. (John N. Cobb Collection, Special Collections, University of Washington Libraries, Seattle, Wash.)

The *Union Jack*'s deep draft, however, apparently limited its ability to service some of the codfish stations, and it was sold during the winter of 1914–1915. The small amount of cod fishing done by the *Union Jack* was a failed experiment in mechanized longlining (see chapter 5).

To replace the *Union Jack*, in 1914 Union Fish built the *Pirate*. At 64 feet long, the *Pirate* was shorter than its predecessor, but it had a 23-foot beam and could carry about 100 tons of cargo. Like the *Union Jack*, the *Pirate* was designed as a combination vessel, one that would fish during the summer months and carry freight between the company's shore stations during the rest of the year. In addition to the 80-horsepower gasoline main engine, the *Pirate* was equipped with a 9-horsepower engine that powered its cargo windlass.[289]

In about 1909 or 1910, the Alaska Codfish Company, of San Francisco, purchased the approximately 72-foot launch *Nonpareil*, which had been used previously for general work in San Francisco Bay. The two-masted schooner, powered by a 40-horsepower Union engine, was promptly sent north and headquartered at the company's Unga station. The *Nonpareil* was the first power vessel to be engaged directly in catching Pacific codfish, fishing during the summer months with a crew of eight men and carrying six dories. During the rest of the year, it transported supplies and fish between the company's Alaska stations. The *Nonpareil* met its end on March 12, 1915, when it went ashore on Unga Island and was wrecked. Fortunately, there were no casualties.[290]

To replace the *Nonpareil*, Alaska Codfish had the *Alasco* built at the J. H. & D. F. Madden shipyard at Sausalito. Although it carried sails, the 60-foot *Alasco* (sometimes referred to as *Alasco I*) was powered as well by an 80-horsepower Union gasoline engine. No problems were encountered during its inaugural voyage to Alaska, which was made during the winter of 1917–1918 and took less than twenty days. The *Alasco* was quickly followed by two additional, although slightly smaller, station vessels: the *Alasco II* (50 feet long, powered by a 40-horsepower Union engine) and the *Alasco III* (55 feet long, powered by a 40-horsepower Enterprise engine) that had also been built at the Madden shipyard. Both vessels entered service in 1918. A fourth station vessel, the *Alasco IV* (similar to the *Alasco III*), entered service during the winter of 1920.

The four Alasco vessels remained in Alaska year-round. To make repairs and to store the vessels when not in use, Alaska Codfish constructed a marine ways at Squaw Harbor in 1919.[291]

Motorized cod-fishing boats and dory tenders

Whereas some fishermen chose to install motors on dories, other fishermen at the stations began in about 1915 to fish from small launches powered by gasoline engines of 2 to 12 horsepower. The type of boats used ranged from modified Columbia River boats (see above) that had been fitted with engines to an assortment of nondescript types. As with motorized dories, the launches were usually manned by two men. According to John Cobb in 1926, the larger codfish companies did not encourage the use of powerboats because they feared that the men who operated them would become too independent and would eventually break away to become station owners themselves.[292]

This sentiment, however, apparently diminished in the face of the reality that motorized boats—at least in certain situations—were considerably more productive and efficient than the traditional banks dories. Given that reality, the Alaska Codfish Company, one of the West Coast's largest codfish firms, contracted in 1919 for the construction of nine 30-foot power fishing boats for use at its stations in the Shumagin and Sanak islands.[293]

Another innovation in shore-station fishing occurred about that same year, when several station operators began using motorized vessels about the size of large salmon purse seine vessels (at that time about 40 to 50 feet long) to transport four or five dories to the fishing grounds each day and to bring the dories, with their catch, back to the station at the end of the day's fishing. This system enabled the fishermen to be more independent of the weather and to reliably cover a wider area than the traditional dory effort.

In 1922, small halibut schooners joined in transporting dories. In contrast to the day-fishing of previous years, by this time the transporters with their dories made fishing trips that lasted about two weeks before they returned to the station to deliver their catch (which was salted on board) and take on supplies. The longer trips enabled the fishing effort to be conducted over a considerably broader area than was available to day fishermen.[294]

ENDNOTES

[252] John N. Cobb, *Pacific Cod Fisheries*, Bureau of Fisheries Doc. No. 830, appendix 4 to *Report of the U.S. Commissioner of Fisheries for 1915* (Washington, DC: GPO, 1916), 37.

[253] Tarleton Bean, *The Cod Fishery of Alaska* [1880], in George Brown Goode, *The Fisheries and Fishery Industries of the United States, Section V: The History and Methods of the Fisheries*, vol. 1 (Washington, DC: GPO, 1887), 205, 216.

[254] Union Fish Co., organization and assets, circa 1918, John N. Cobb Collection, Special Collections, University of Washington Libraries, Seattle, Wash.

[255] A favorite meal at the station was Scotch dumplings: cod stomachs that were filled with a mixture of finely chopped cod livers, a few vegetables, and a little flour, and then tied off and boiled.

[256] John N. Cobb, *Pacific Cod Fisheries*, Bureau of Fisheries Doc. No. 830, appendix 4 to *Report of the U.S. Commissioner of Fisheries for 1915* (Washington, DC: GPO, 1916), 58.

[257] John N. Cobb, *Pacific Cod Fisheries*, Bureau of Fisheries Doc. No. 830, appendix 4 to *Report of the U.S. Commissioner of Fisheries for 1915* (Washington, DC: GPO, 1916), 41–42; Ray Rodgers (Frank Rodgers's son), personal communication with author, Seward, Alaska, circa 1971.

[258] John N. Cobb, *Pacific Cod Fisheries*, Bureau of Fisheries Doc. No. 830, appendix 4 to *Report of the U.S. Commissioner of Fisheries for 1915* (Washington, DC: GPO, 1916), 41.

[259] Ibid., 42.

[260] Ibid.

[261] E. Lester Jones, *Report of Alaska Investigations, 1914* (Washington, DC: GPO, 1915), 58.

[262] John N. Cobb, *Pacific Cod Fisheries*, Bureau of Fisheries Doc. No. 830, appendix 4 to *Report of the U.S. Commissioner of Fisheries for 1915* (Washington, DC: GPO, 1916), 40–42.

[263] Edward Weber Allen, *North Pacific: Japan, Siberia, Alaska, Canada* (New York: Professional & Technical Press, 1936), 109–110.

[264] John N. Cobb, *Pacific Cod Fisheries*, Bureau of Fisheries Doc. No. 830, appendix 4 to *Report of the U.S. Commissioner of Fisheries for 1915* (Washington, DC: GPO, 1916), 58–60; John N. Cobb, "Motor Vessels in the Alaska Cod Fisheries," *Pacific Motor Boat* (November 1919): 3–8; J. W. Collins, *Report on the Fisheries of the Pacific Coast of the United States*, in *Report of the United States Commissioner of Fish and Fisheries for 1892* (Washington, DC: GPO, 1892), 102.

[265] Ivan Petroff, *Report on the Population, Industries, and Resources of Alaska* [1882], in *Seal and Salmon Fisheries and General Resources of Alaska*, vol. 4 (Washington, DC: GPO, 1898), 269.

[266] J. W. Collins, *Report on the Fisheries of the Pacific Coast of the United States*, in *Report of the United States Commissioner of Fish and Fisheries for 1892* (Washington, DC: GPO, 1892), 98.

[267] John N. Cobb, *Pacific Cod Fisheries*, Bureau of Fisheries Doc. No. 830, appendix 4 to *Report of the U.S. Commissioner of Fisheries for 1915* (Washington, DC: GPO, 1916), 38–40; John N. Cobb, *Pacific Cod Fisheries* (revised edition, 1926), Bureau of Fisheries Doc. No. 1014, appendix 7 to *Report of the U.S. Commissioner of Fisheries for 1926* (Washington, DC: GPO, 1927), 416–417; John N. Cobb, *Fisheries of Alaska in 1906*, Bureau of Fisheries Doc. No. 618 (Washington, DC: GPO, 1907), 45–46; "Largest Cargo of Codfish," *San Francisco Call*, June 26, 1909; "Storm Plays Havoc at Unga," *San Francisco Call*, March 13, 1906; C. P. Overton, "The Year 1910 in the Pacific Codfish Industry," *Pacific Fisherman* (February 1911): 23.

[268] John N. Cobb, *Pacific Cod Fisheries*, Bureau of Fisheries Doc. No. 830, appendix 4 to *Report of the U.S. Commissioner of Fisheries for 1915* (Washington, DC: GPO, 1916); John N. Cobb, *Pacific Cod Fisheries* (revised edition, 1926), Bureau of Fisheries Doc. No. 1014, appendix 7 to *Report of the U.S. Commissioner of Fisheries for 1926* (Washington, DC: GPO, 1927).

[269] John N. Cobb, *Pacific Cod Fisheries* (revised edition, 1926), Bureau of Fisheries Doc. No. 1014, appendix 7 to *Report of the U.S. Commissioner of Fisheries for 1926* (Washington, DC: GPO, 1927), 415; Ward T. Bower, *Alaska Fisheries and Fur-Seal Industries in 1920*, Bureau of Fisheries Doc. No. 909 (Washington, DC: GPO, 1921): 64, 66; "Review of the Pacific Codfish Season 1921," *Pacific Fisherman* (January 1922): 93.

[270] "Pacific American Crews Leaving," *Pacific Fisherman* (March 1920): 58; "Review of the Pacific Codfish Season, 1924," *Pacific Fisherman* (January 1925): 146, 148–149; "Matheson Codfish Ships Sail for the Bering Sea," *Anacortes American*, April 16, 1914; "Prominent Factors in the Pacific Cod Fishery," *Pacific Fisherman* (January 1927): 130; August C. Radke, *Pacific American Fisheries, Inc.: History of a Washington State Salmon Packing Company, 1890–1966* (Jefferson, N.C.: McFarland, 2002), 101; "Review of [1923] Pacific Codfish Season," *Pacific Fisherman* (January 1924): 108–110; "Review of the Pacific Codfish Season, 1924," *Pacific Fisherman* (January 1925): 146, 148–149; "Prominent Factors in the Pacific Cod Fishery, *Pacific Fisherman* (January 1927): 130; "1,100 Tons Cod Anacortes 1929," *Anacortes American*, December 12, 1929; "Salt Codfish," *Pacific Fisherman* (January 1932): 227–228; "North Pacific Cod Fishery Yield Above Average of Recent Years," *Pacific Fisherman* (January 1931): 215–216.

[271] "Review of the [1923] Pacific Codfish Season," *Pacific Fisherman* (January 1924): 108–110; Ward T. Bower, *Alaska Fishery and Fur-Seal Industries in 1924*, Bureau of Fisheries Doc. No. 992 (Washington, DC: GPO, 1925), 137–138; "Codfish Production Curtailed Despite Good Fishing Season," *Pacific Fisherman* (January 1927): 204, 206, 208.

[272] "Alaska Fishing-Grounds," *New York Times*, July 15, 1879; Thomas J. Vivian, *Commercial, Industrial, Agricultural, Transportation and Other Interests of California* (Washington, DC: GPO, 1891), 487.

[273] John N. Cobb, *Pacific Cod Fisheries* (revised edition, 1926), Bureau of Fisheries Doc. No. 1014, appendix 7 to *Report of the U.S. Commissioner of Fisheries for 1926* (Washington, DC: GPO, 1927), 416–417, 419–420, 480–481. Ivan Petroff's report is: *Report on the Population, Industries, and Resources of Alaska* [1882], in *Seal and Salmon Fisheries and General Resources of Alaska*, vol. 4 (Washington, DC: GPO, 1898), 167–450.

[274] "Review of the [1923] Pacific Codfish Season," *Pacific Fisherman* (January 1924): 108–110; Dennis A. Johnson, "Alaska's Early Cod Fishery," *Alaska Journal* (Spring 1979): 75–79; Ed Opheim, "Eat Crow," undated report, Baranov Museum, Kodiak, Alaska.

[275] "Further Expansion of Northern Fisheries, Inc.," *Pacific Fisherman* (May 1918): 35, 37.

[276] John N. Cobb, *Pacific Cod Fisheries* (revised edition, 1926), Bureau of Fisheries Doc. No. 1014, appendix 7 to *Report of the U.S. Commissioner of Fisheries for 1926* (Washington, DC: GPO, 1927), 414–415; "Kodiak Island Shipwrecks," part 2, *Kodiak Daily Mirror*, January 30, 2015; Gordon Newell, ed., *H. W. McCurdy Marine History of the Pacific Northwest* (Seattle: Superior Publishing Company, 1966), 292; "Salt Fish Department," *Pacific Fisherman* (January 1919): 124a.

[277] Ward T. Bower, *Alaska Fisheries and Fur-Seal Industries in 1918*, Bureau of Fisheries Document No. 872 (Washington, DC: GPO 1919), 60.

[278] John N. Cobb, *Pacific Cod Fisheries* (revised edition, 1926), Bureau of Fisheries Doc. No. 1014, appendix 7 to *Report of the U.S. Commissioner of Fisheries for 1926* (Washington, DC: GPO, 1927), 414–415.

[279] Ward T. Bower, *Alaska Fisheries and Fur Industries in 1919*, Bureau of Fisheries Document No. 891 (Washington, DC: GPO, 1920), 55.

[280] "Robinson Fisheries Company Leases Anacortes, Wash., Plant of Northern Fisheries," *Western Canner and Packer* (January 1920): 67.

[281] "Alaska Ocean Food Company's Plans," *Pacific Fisherman* (July 1919): 52; "Codfish from Port Chatham," *Pacific Fisherman* (October 1919): 46; Ward T. Bower, *Alaska Fishery and Fur-Seal Industries in 1920*, appendix 6 to *Report of the U.S. Commissioner of Fisheries for 1921* (Washington, DC: GPO, 1921), 64; Ward T. Bower, *Alaska Fishery and Fur-Seal Industries in 1921*, appendix 10 to *Report of the U.S. Commissioner of Fisheries for 1922* (Washington, DC: GPO, 1922), 44.

[282] John N. Cobb, *Pacific Cod Fisheries*, Bureau of Fisheries Doc. No. 830, appendix 4 to *Report of the U.S. Commissioner of Fisheries for 1915* (Washington, DC: GPO, 1916), 43–44.

[283] "Four Lose Lives in Graveyard of Ships," *San Francisco Examiner*, April 29, 1910; "Lost Life in a Blizzard," *Press Democrat* (Santa Rosa, Calif.), June 2, 1910; *Fisheries of Alaska in 1910*, Bureau of Fisheries Doc. No. 746 (Washington, DC: GPO, 1911), 38; "Pacific Codfish Department," *Pacific Fisherman* (June 1910): 21; "Stanley (1910)," Alaska Shipwrecks, accessed October 22, 2017, http://alaskashipwreck.com/shipwrecks-a-z/alaska-shipwrecks-s/.

[284] "Codfish News," *Pacific Fisherman* (July 1906): 22–23; Paul Gooding, "The World's New Cod Banks," *Pacific Monthly* (April 1907): 432–441.

[285] John N. Cobb, *Pacific Cod Fisheries*, Bureau of Fisheries Doc. No. 830, appendix 4 to *Report of the U.S. Commissioner of Fisheries for 1915* (Washington, DC: GPO, 1916), 43–44.

[286] Ward T. Bower and Henry D. Aller, *Alaska Fisheries and Fur Industries in 1914*, Bureau of Fisheries Doc. No. 819 (Washington, DC: GPO, 1915), 54; "Ship Reported Lost Returns With Fish," *San Francisco Call*, October 14, 1913; John N. Cobb, "Motor Vessels in the Alaska Cod Fisheries," *Pacific Motor Boat* (November 1919): 3–8; Ward T. Bower, *Alaska Fishery and Fur-Seal Industries in 1927*, Bureau of Fisheries Doc. No. 1040 (Washington, DC: GPO, 1928), 139; "Review of Pacific Codfish Season," *Pacific Fisherman* (January 1924): 108–110; "Codfish Schooner in Tuna Trade, *Pacific Fisherman* (July 1927): 38.

[287] J. W. Collins, *Report on the Fisheries of the Pacific Coast of the United States*, in *Report of the United States Commissioner of Fish and Fisheries for 1892* (Washington, DC: GPO, 1892), 100.

[288] "Motor Craft in the Pacific Fisheries," *Pacific Fisherman Year Book* (January 1919): 62.

[289] John N. Cobb, "Motor Vessels in the Alaska Cod Fisheries," *Pacific Motor Boat* (November 1919): 3–8.

[290] John N. Cobb, "Motor Vessels in the Alaska Cod Fisheries," *Pacific Motor Boat* (November 1919): 3–8.

[291] J. H. & D. F. Madden Co., adv., *Pacific Fisherman* (January 1920): 166; "Husky Tenders for Alaskan Waters," *Motor Boat* (June 10, 1920): 20; "Madden Building Nine Power Boats," *Sausalito News*, August 23, 1919; "Cod Fishing Suspended," *Pacific Fisherman* (November 1921): 35; Ward T. Bower, *Alaska Fishery and Fur-Seal Industries in 1918*, appendix 7 to *Report of the U.S. Commissioner of Fisheries for 1918* (Washington, DC: GPO, 1920), 59; Ward T. Bower, *Alaska Fishery and Fur-Seal Industries in 1920*, Bureau of Fisheries Doc. No. 909 (Washington, DC: GPO, 1921), 64.

[292] John N. Cobb, *Pacific Cod Fisheries* (revised edition, 1926), Bureau of Fisheries Doc. No. 1014, appendix 7 to *Report of the U.S. Commissioner of Fisheries for 1926* (Washington, DC: GPO, 1927), 424.

[293] "Madden Building Nine Power Boats," *Sausalito News*, August 23, 1919; Ward T. Bower, *Alaska Fisheries and Fur Industries in 1919* (Washington, DC: GPO, 1920), 56.

[294] "Pacific Codfish Schooners Sail," *Pacific Fisherman* (May 1921): 38; "Shields Back from California," *Pacific Fisherman* (April 1922): 46–47; "Review of the [1922] Pacific Codfish Season," *Pacific Fisherman* (January 1923): 96.

Chapter 7
A FLOURISHING INDUSTRY: 1903–1919

No industry on the Pacific Coast has a brighter future than that of codfishing.

—*Anacortes American*, 1907[295]

In 1903, after two decades in which the market for codfish caught in the Pacific was for the most part limited to what could be consumed west of the Rocky Mountains, the market began expanding because of poor catches on both sides of the Atlantic. At Gloucester, Massachusetts, landings of salt cod declined from 30 million pounds in 1902 to 18 million pounds in 1905. Meanwhile, production of salt cod from Alaska waters increased from 9 million pounds in 1902 to 12 million pounds in 1905.[296]

By late 1904, the U.S. Pacific coast salted codfish industry was successfully selling its product—in direct competition with established East Coast firms—in Chicago. Product was also being sold in Hong Kong, South America, Australia, New Zealand, and the Philippines.[297]

The lively demand for Pacific coast codfish notwithstanding, the industry still had a substantial problem: most of the new entrants into the industry had little or no previous experience in the business and were not schooled in properly curing and packaging codfish. The result was that East Coast merchants rejected considerable quantities of salted Pacific codfish.[298]

In 1910, C. P. Overton, head of the Union Fish Company, explained in *Pacific Fisherman* the historical relationship between the Atlantic and Pacific codfish industries:

> The fishing season of 1905–1906 on the Atlantic showed a great shortage in the catch. Dealers were scouring the country for codfish. Word went out that the codfish banks of the Atlantic were exhausted and the demand from that side for our Pacific codfish seemed to confirm it. All kinds of offers were made here, especially for hard cured fish in carload lots. After all the surplus stock was gone there were still loud cries for more, and at good figures. To many people it looked as if those high prices were nailed to the mast, and as there are always those ready to rush in where angels fear to tread, the next year witnessed a great expansion of the industry on the

> Coast. Old concerns increased their fleets and new companies entered the business to share the golden profits. The year 1907 was the height of the boom. Nine companies were in the field operating 19 vessels and 15 shore-fishing stations in Alaska.
>
> But, as has been pointed out before in this journal, the Atlantic shortage and Eastern demand is a matter of one season only and recurs at irregular intervals several years apart. . . .
>
> Before the end of the first year of expansion it was evident that about two-thirds of the catch had been made for a market which no longer existed. Fish on the Atlantic Banks were again as plentiful as usual, and the loud call for Pacific Codfish was heard no more. This meant a carry over into the next season which was again one of over production. Codfish, unlike whiskey, does not improve with age.[299]

BACK TO THE OKHOTSK SEA

One result of the improved market in the early 1900s was the first fishing effort by American vessels in the Okhotsk Sea since 1896. The American absence had its roots in 1884, when fully eleven American vessels fished in the Okhotsk Sea and returned with disappointingly small cargoes. During each of the following two years (1885 and 1886), only four American cod-fishing vessels visited the Okhotsk Sea, and during the following decade (1887–1896), the American effort there was only one or two vessels per year. Complicating matters, in 1892 the Russian government began intermittently enforcing a regulation that required any vessel fishing within thirty miles of its shores to obtain a license. American cod fishermen considered the regulation to be a form of harassment.

An example of this harassment occurred in 1895, when the schooner *Hera*, which had not obtained a license from the Russian government, was the only American vessel fishing for codfish in the Okhotsk Sea, its fishing effort taking place from ten to thirty miles off the Russian shore. While fishing, the *Hera* was boarded by a Russian official who ordered the captain to cease fishing and to report to the district governor to obtain a license. The vessel's captain protested, stating that he considered Russia's seaward boundary to extend only ten miles seaward from shore, and that he had not fished within those waters. The Russian official was not persuaded, and he threatened to seize the *Hera* if his order was not obeyed. Seeing no alternative, the

Hera's captain made way for port, where he lodged a protest with the governor. As had happened at sea, his protest was not accepted, and the governor demanded a payment of $1,000 in gold for a license, without which the *Hera* would not be permitted to leave the port. In a compromise, the governor accepted the captain's personal draft on the vessel's owner's funds in the amount $1,000. Released, the *Hera* was able to complete its fishing season and then returned to San Francisco with nearly 700,000 pounds of salted codfish. The captain's draft was forwarded to the Russian consul in San Francisco for collection, but payment was never made, because the *Hera*'s owners determined the note was written under duress.

One American vessel fished in the Okhotsk Sea in 1896. After that, it was 1903 before American vessels again ventured into those waters.[300] Until 1907, they apparently did so without any significant confrontations with Russian officials. That year, however, the Russians were aggressive in enforcing the license regulation. Four American vessels—*John D. Spreckles*, *Freemont*, *City of Papeete*, and *S. N. Castle*—had just begun their season's fishing when the Russian gunboat *Mandjur* appeared. The *John D. Spreckles* and *S. N. Castle* were boarded by a Russian officer who, claiming the vessels were within the thirty-mile limit, ordered them to cease fishing and took their documentation papers. The officer threatened to seize the vessels if his orders were not obeyed. With very few fish on board, both vessels departed immediately for San Francisco.

Several days later, the *Mandjur*'s captain boarded the *Freemont* and took its papers and ordered it to leave the Russia-claimed waters. The *Freemont*, nevertheless, continued to fish. A week later, the Russian captain boarded the *City of Papeete*, took its papers, and gave it the same orders. The captain of the *City of Papeete* responded by showing the Russian captain a legal opinion written several years before by John Hay, while Hay was U.S. secretary of state. Hay's opinion had been provided to the *City of Papeete*'s captain by Alfred Greenbaum, president of the Alaska Codfish Company, which owned the vessel and anticipated this sort of problem. According to Hay, under international law any vessel had the right to fish at any location three or more miles from shore. Apparently, the Russian captain determined that the opinion might have some validity, and he promptly steamed to the mainland to seek advice from his superiors. The *City of Papeete* continued to fish.

Three weeks later, the *Mandjur* returned to the *City of Papeete* and the Russian captain returned the vessel's papers and informed its captain that the thirty-mile limit would not be enforced. Two days later, the Russian gunboat steamed alongside the *Freemont* and returned its papers as well as those of the *John D. Spreckles* and *S. N. Castle*.

Without incident, three American vessels fished in the Okhotsk Sea in 1908, and the last American vessel to fish there, the *Freemont*, did so—also without incident—in 1909.[301] As explained in chapter 2, the sailing time from Puget Sound ports and San Francisco to the Okhotsk Sea was prohibitively long in comparison to the sailing time to the Bering Sea grounds or the Shumagin and Sanak islands grounds, and it was this extra distance that caused the curtailment of cod fishing by U.S. vessels in the Okhotsk Sea after 1909.

SALTED CODFISH'S PACIFIC HEYDAY

The mid-1910s represented the heyday of Alaska's codfish industry. The domestic and export markets were strong, codfish were abundant, and the weather was generally cooperative.

"The cod industry in Alaska has been in a flourishing condition this year," wrote Ward T. Bower and Harry Clifford Fassett, both of the Bureau of Fisheries, late in 1913. The men reported that all the vessels in the codfish fleet had returned fully laden and the shore stations had done well. They credited this good fortune largely to favorable weather conditions at the shore stations and in the Bering Sea.[302] Emblematic of the industry's prosperity was the construction by the Union Fish Company, of San Francisco, of the 145-foot schooner *Golden State*, which came into service in 1913. The *Golden State* was the largest motor-propelled vessel in the fishing industry on the U.S. Pacific coast (see chapter 6).[303]

In the words of *Pacific Fisherman*, the following year began "under most favorable auspices." The journal noted that "[t]he greater part of the catch of 1913 had been marketed at remunerative prices, the local domestic demand was in excellent condition, while the demand from the East and from foreign markets, except Mexico [because of civil unrest], was good and continually increasing." The codfish schooner fleet, which consisted of ten vessels in 1913, increased to fourteen in 1914.[304]

In a report prepared after his 1914 examination of Alaska's fisheries, Deputy U.S. Commissioner of Fisheries E. Lester Jones wrote that "[t]here was no part of the fishing industry that pleased me more" than the territory's codfish industry. Alaska's codfish, wrote Jones, were "of first-class quality," and he characterized the process of dressing and salting codfish as "clean and wholesome."[305]

The outbreak of World War I in late July of 1914 enhanced the export market considerably. Initially, codfish exporters found it difficult to obtain payment for their shipments due to a war-induced breakdown of established credit arrangements, but this temporary hindrance was overshadowed by the greatly increased demand for salted codfish.

According to John Cobb, the war presented a "golden opportunity" for the Pacific coast codfish industry. European production, a large portion of which had previously been marketed in the West Indies and in Central America and South America, was quickly diverted to the combatants in Europe. As well, high prices attracted to Europe a considerable portion of the codfish caught in Canada, Newfoundland, and the U.S. Atlantic seaboard. Pacific cod was used to fill the void in the Americas. Because of this strong export demand, the value of salted cod from the Pacific was higher than it had been for many years.[306]

Moreover, fishing in 1916 was excellent, with a near-record 3,767,900 salted cod delivered from Alaska to Puget Sound and San Francisco.[307] In their 1916 Bureau of Fisheries report on Alaska's fisheries, Ward Bower and Henry Aller characterized the industry as "very prosperous." At about the same time, *Pacific Fisherman* proclaimed 1916 to be "one of the most successful years" in the industry's history, owing to "the European war opening up new markets to our fishermen which had been closed to them before because of the impossibility of competing with the European and Eastern codfish." The war-induced shortage of cod from the eastern Atlantic pushed the price for salted Pacific cod higher than it had been for many years.[308]

The year 1916 was, however, marred by the drowning of four fishermen in the Bering Sea, along with the murder of two fishermen and the wounding of three others by the cook at the Union Fish Company's Sanborn Harbor (Nagai Island) codfish station. The cook also set afire the station's bunk house and cook house before taking his own life by remaining in one of the burning buildings.[309]

Although prosperous, the scale of Alaska's codfish industry paled in comparison to the territory's canned salmon industry. The codfish industry in 1916 employed 778 persons (including those in the offshore fleet) and produced 14,302,364 pounds of codfish (including 22,488 pounds of canned codfish), which was worth $518,797. By contrast, the canned salmon industry that year—which set a record—employed 19,240 persons and produced 4,900,627 cases of salmon (each case contained 48 one-pound cans) worth $23,269,429.[310]

The comparatively small scale of the codfish industry did not mean the salmon canners did not take notice. In the spring of 1918, the Northern Fisheries Company, which was owned by Crescent

Hale, a well-known Bristol Bay canneryman, purchased the Union Fish Company.[311]

In 1919, Ward Bower, the federal fisheries agent in Alaska, noted that Alaska's cod fishery "maintains a remarkably uniform production year after year," with approximately two-thirds of the catch being made in the Bering Sea, and the remainder taken mainly off the southern shore of the Alaska Peninsula.[312] Bower was clearly referring to the years 1914 through 1918, during which the total number of codfish caught in Alaska ranged from 3.7 million to 4.1 million (see chapter 6).

The Alaska codfish industry continued to enjoy great prosperity while the war raged in Europe, but it declined quickly when European fisheries returned to normal after the war's end. Little could Bower have known that 1914 through 1920 would be the peak years of the Alaska salted codfish era, and that 1921 would be the beginning of what John Cobb called "the great slump."[313]

TAXES

Alaska obtained limited self-government when it became a territory in 1912. The following year, Alaska's first territorial legislature established a system of taxation and licensing that included a tax on canned salmon production ("case tax"), but it did not tax any other fisheries product. In 1915, the legislature expanded the territorial tax law to include, among other items, a tax of 2½ cents per 100 pounds on all salted fish. Perhaps because of the perception that the codfish industry was thriving, a 1919 amendment of the tax law increased the tax on salted codfish to 10 cents per 100 pounds. This rate persisted until 1949, when Alaska's legislature comprehensively overhauled the territory's tax system. Under the new system, the tax on all salted fish was equal to 1 percent of the value of the fish purchased or otherwise obtained for processing by salting. This tax was commonly referred to as the "raw fish tax," and it persisted beyond the end of the salt cod era in 1950.[314]

JAPANESE "INVASION" OF ALASKA'S CODFISH INDUSTRY

The increased demand for cod caused by World War I did not go unnoticed by Japanese fishermen, who were close by the rich fishing banks of the Okhotsk Sea. As *Pacific Fisherman* noted spitefully in late 1919, the American cod-fishing fleet had been "long ago driven from [the Okhotsk Sea] through harassment by the Russians and fouling of the waters by a myriad of Japanese boats." In fact, Americans had fished in the Okhotsk Sea just ten years prior, in 1909 (see above).[315]

The Japanese codfish industry's attempt to establish itself in the United States was partly a result of the Japanese defeat of Russia in the Russo-Japanese War of 1904–1905. The war had formally ended in October 1905 with the signing of the Treaty of Portsmouth, a provision of which required Russia "[to grant] to Japanese subjects rights of fishery along the coasts of the Russian possessions in the Japan, Okhotsk, and Bering Seas." In 1907, Russia followed the treaty's dictates by granting Japan the rights to fish along the Russian coasts—excluding rivers and inlets—of the Sea of Japan, the Okhotsk Sea, and the Bering Sea.[316] Two years later, in 1909, the schooner *Freemont*—the last American vessel to fish for codfish in the Okhotsk Sea—reported the presence there of a considerable Japanese cod-fishing fleet.[317] Japanese exploitation of the cod resources along the Siberian coast steadily increased during the following years, and during the mid-1910s a considerable export trade was developed, particularly with South America and the West Indies.

The Japanese were at the same time seeking to develop an American market for their salted codfish, and sent small shipments to the United States in 1916 and 1917.[318] The Japanese were also studying all aspects of the American codfish industry, and at least one effort was covert. According to Crescent Hale, president of the Union Fish Company, a Japanese cook at the company's Pirate Cove station in 1918 "turned out to be one of the wealthy men of Japan and he is now the managing admiral of all of the Japanese fishing fleets engaging in the cod fishing industry off the Siberian Coast." Hale was convinced that "The Japanese are determined to control the business and are perfectly frank about it. They are not satisfied to control the fishing off the coast of Siberia. They want it all." He added that his firm in 1919 had refused a fix-your-own-price offer to sell out to a "big Japanese concern."[319]

The door that opened U.S. markets to unrestricted Japanese imports was a February 1918 Department of Commerce directive designed "to promote the vigorous prosecution of [World War I] and to make the utmost use jointly of all the resources of the nations now cooperating." The order suspended the law that forbade landings in U.S. ports by foreign fishing vessels coming directly from the fishing banks. Specifically, it allowed "Canadian fishing vessels and those of other nations now acting with the United States to enter from and clear for the high seas and the fisheries, disposing of their catch and taking on supplies, stores, etc."[320] Japan was one of the "other nations."

Given the strong demand for cod, in August 1918 several cod-fishing schooners of the Pacific Trading Company, a Japanese firm, landed a total of 894,000 salted codfish and 1,600 bales of stockfish at San Francisco. To further process the fish, Pacific Trading had acquired the abandoned Pacific States Trading Company (see chapter 3) codfish curing plant on Carquinez Strait, an estuary of San Francisco Bay. The venture did not turn out well and ended by late 1919. According to *Pacific Fisherman*:

> The company, owing partly to imperfect knowledge of the market and partly to inferiority of the goods, found itself unable to dispose of all the importations to advantage, and finally closed out its business, selling the fish to some of the local operators, who were able to find a foreign outlet.[321]

This failure, however, did not deter another Japanese firm, the Japan Fisheries Company, which was working to gain a foothold in Puget Sound. A seemingly unlikely partner in this effort was the Robinson Fisheries Company, of Anacortes, Washington. The Japanese company and Robinson Fisheries had negotiated what was stated to be an "experimental" arrangement under which Robinson Fisheries would receive, process, and market Japanese-caught codfish.[322] In actuality, the Japanese in April 1919 had entered into an agreement to purchase Robinson Fisheries, which at that time was owned chiefly by the Scandinavian-American Bank, for $135,000.[323]

In January 1919, the schooner *Houzan Maru* made the first delivery of Japanese codfish to the Anacortes plant. This delivery was followed less than two weeks later by a delivery from the schooner *Kashimi Maru*. Both vessels had caught their fish in the Okhotsk Sea.[324] According to *Pacific Fisherman*, the fish was poorly cured and, because it did not find a ready market on the Pacific coast, was shipped east and reprocessed for the export trade.[325] Nevertheless, shipments of Japanese codfish continued to arrive: on one day in October 1919, four Japanese schooners arrived in Puget Sound carrying an aggregate 700 tons of codfish.

Pacific Fisherman by this time seemed resigned to a long-term Japanese presence in the industry, and predicted that codfish imports from Japan would increase substantially in the near future, so long as the market remained firm. "A number of Japanese companies," wrote *Pacific Fisherman* in October 1919, "have extensive cod fishing facilities which are being constantly improved. They have made careful study of the most modern American methods of handling the goods, and are rapidly getting into a position to produce in Japan a product closely approximating that which is familiar to American consumers."[326]

Pacific Fisherman's seeming resignation, however, turned within two months into alarm. In a December 1919 article titled "Japanese Invade American Codfish Industry," the journal noted that the Japanese competition had "caused a suspension of new development in Alaska," and that "old established concerns which have survived the vicissitudes of the industry for nearly half a century are profoundly concerned for the future." According to *Pacific Fisherman*, the Alaska codfish industry was "in danger of extinction through the invasion of American markets by Japanese codfish, produced by coolie labor at practically one-tenth the cost of American product."[327]

The solution to this invasion, according to *Pacific Fisherman* and codfish industry representatives, was for Congress to levy a protective tariff. *Pacific Fisherman* in late 1919 itself filed a protest with the federal government against the control by Japanese of American codfish enterprises.[328] Relief was not forthcoming, and Japanese imports increased the following year. The result was that Pacific coast codfish firms generally curtailed their operations, and winter station fishing was suspended. (J. A. Matheson, of Anacortes, for example, did not send his schooner *Fanny Dutard* north in 1920.) Ward Bower, Alaska agent for the Bureau of Fisheries, warned that the Alaska codfish industry was facing a "decline of serious proportions" as a result of the Japanese imports. Despite the reduced fishing effort, however, fishing was good and the American catch in 1920 was 17 percent greater than the previous year.[329]

In the spring of 1921, *Pacific Fisherman* reported a rumor that the Japanese who had been operating in partnership with the Robinson Fisheries Company had "called the deal off and withdrawn from the field."[330] The rumor proved to be true, and, as *Pacific Fisherman* later explained, the Japanese "had evidently become convinced that they had made a bad bargain, and declined to complete the purchase." Moreover—bad bargain or not—the Japanese venture became untenable on July 15, 1921, when, to the great relief of Pacific coast codfish producers, the Department of Commerce withdrew its February 1918 order that allowed unrestricted imports of Japanese codfish. The Japanese left the Puget Sound codfish industry as abruptly as they had left the industry in San Francisco. The Japan Fisheries Company later sued the Robinson Fisheries Company for an unpaid balance on 1,800 tons of codfish, and the two firms eventually settled the issue out of court.[331] In 1923, J. W. Beardsley's Sons, a well-known Newark, New Jersey, manufacturer of shredded codfish—a product made from trimmings and used primarily to make codfish cakes—purchased a half interest in Robinson Fisheries from John Trafton, who continued to manage the firm (see chapter 3).[332]

In all, Japanese fishing vessels delivered 2,598,500 pounds of salted cod to San Francisco and Puget Sound markets during the years 1918 through 1920.[333] Japanese firms would continue to export roughly similar quantities of salted cod to the U.S. Pacific coast market until World War II, but the product came not from direct deliveries by Japanese fishing vessels to West Coast ports; it arrived instead through normal channels of international trade, which required an import duty.

RADIO COMMUNICATIONS

Until 1919, communication between codfish companies' home offices and their shore stations was limited to word of mouth and to letters and reports carried on vessels that serviced the stations. Communications with the offshore schooner fleet was opportunistic. Vessels meeting at sea would exchange information, and vessels heading south would often carry letters from crew members of vessels that had not yet finished fishing.

For the Union Fish Company, communications improved dramatically in early 1919, when the company installed a shortwave radio station at Unga that was able to communicate telegraphically (in code) with the company's home office in San Francisco. The following year, radio stations were installed at Pirate Cove and Sand Point, allowing the shore stations to communicate among themselves as well as with San Francisco.

Shortwave telegraphic equipment came into wide use by salmon canneries in Alaska about this time, but the technology did not seem to have been employed broadly in the codfish industry.

In 1930, the Robinson Fisheries Company installed radios—the first to be used in the codfish fleet—on its schooners *Wawona* and *Azalea*. These radios were powered by dry-cell batteries and were used mainly to make weekly radio telegram reports through a station at Dutch Harbor to the company's home office in Anacortes, Washington; to contact the Coast Guard in the case of an emergency; and as a source of general news, including sports. Radios soon became standard equipment aboard codfish schooners.[334]

ENDNOTES

[295] "Our Big Codfishing Industries," *Anacortes American*, Annual Edition, 1907.

[296] "Pacific Codfisheries Need Protection," *Pacific Fisherman* (October 1904): 21.

[297] "Pacific Codfish Statistics," *Pacific Fisherman* (May 1903): 10; "San Francisco," *Pacific Fisherman* (September 1903): 16; "Cod Fisheries," *Pacific Fisherman* (April 1904): 12.

[298] "Pacific Coast Codfish Review," *Pacific Fisherman* (January 1906), 67.

[299] C. P. Overton, "The Year 1910 in the Pacific Codfish Industry," *Pacific Fisherman, Annual Review* (February 1911): 23.

[300] John N. Cobb, *Pacific Cod Fisheries*, Bureau of Fisheries Doc. No. 830, appendix 4 to *Report of the U.S. Commissioner of Fisheries for 1915* (Washington, DC: GPO, 1916), 33, 85, 90; "Codfish and Fishers," *San Francisco Call*, October 25, 1885.

[301] John N. Cobb, *Pacific Cod Fisheries*, Bureau of Fisheries Doc. No. 830, appendix 4 to *Report of the U.S. Commissioner of Fisheries for 1915* (Washington, DC: GPO, 1916), 33–34, 85, 97; "Russians Seize Papers of Fishers in Okhotsk," *San Francisco Call*, September 30, 1907.

[302] Ward T. Bower and Harry Clifford Fassett, "The Fishery Industries of Alaska in 1913," *Pacific Fisherman* (January 1914): 51–61, 63, 65–66.

[303] "Motor Craft of the Pacific Fisheries," *Pacific Fisherman* (January 1919): 55–64.

[304] "Salt Fish Review," *Pacific Fisherman* (January 1915): 102–103.

[305] E. Lester Jones, *Report of Alaska Investigations, 1914* (Washington, DC: GPO, 1915), 58.

[306] John N. Cobb, *Pacific Cod Fisheries* (revised edition, 1926), Bureau of Fisheries Doc. No. 1014, appendix 7 to *Report of the U.S. Commissioner of Fisheries for 1926* (Washington, DC: GPO, 1927), 460.

[307] "Salt Fish Department," *Pacific Fisherman* (January 1917): 99.

[308] Ward T. Bower and Henry D. Aller, *Alaska Fisheries and Fur Industries in 1916*, Bureau of Fisheries Doc. No. 838 (Washington, DC: GPO, 1917), 44, 68–70; "Salt Fish Department," *Pacific Fisherman* (January 1917): 99.

[309] Ward T. Bower and Henry D. Aller, *Alaska Fisheries and Fur Industries in 1916*, Bureau of Fisheries Doc. No. 838 (Washington, DC: GPO, 1917), 70.

[310] Ibid., 38, 44, 70.

[311] "Further Expansion of Northern Fisheries, Inc.," *Pacific Fisherman* (May 1918): 35, 37; John N. Cobb, *Pacific Cod Fisheries* (revised edition, 1926), Bureau of Fisheries Doc. No. 1014, appendix 7 to *Report of the U.S. Commissioner of Fisheries for 1926* (Washington, DC: GPO, 1927), 415.

[312] Ward T. Bower, *Alaska Fisheries and Fur-Seal Industries in 1918* (Washington, DC: GPO, 1919), 58.

[313] John N. Cobb, *Pacific Cod Fisheries* (revised edition, 1926), Bureau of Fisheries Doc. No. 1014, appendix 7 to *Report of the U.S. Commissioner of Fisheries for 1926* (Washington, DC: GPO, 1927), 460, 468–485.

[314] Act to Establish a System of Taxation, Create Revenue, and Provide for Collection Thereof for the Territory of Alaska (Chap. 52, *Session Laws of Alaska, 1913*); chap. 76, *Session Laws of Alaska, 1915*; chap. 33, *Session Laws of Alaska, 1919*; chap. 97, *Session Laws of Alaska, 1949*.

[315] John N. Cobb, *Pacific Cod Fisheries*, Bureau of Fisheries Doc. No. 830, appendix 4 to *Report of the U.S. Commissioner of Fisheries for 1915* (Washington, DC: GPO, 1916), 34; John N. Cobb, *Pacific Cod Fisheries* (revised edition, 1926), Bureau of Fisheries Doc. No. 1014, appendix 7 to *Report of the U.S. Commissioner of Fisheries for 1926* (Washington, DC: GPO, 1927), 460. "Japanese Invade American Codfish Industry," *Pacific Fisherman* (December 1919): 23–24.

[316] Article 11, Treaty of Portsmouth (October 14, 1905); Russian-Japanese Convention Concerning Fisheries, July 15, 1907.

[317] John N. Cobb, *Pacific Cod Fisheries*, Bureau of Fisheries Doc. No. 830, appendix 4 to *Report of the U.S. Commissioner of Fisheries for 1915* (Washington, DC: GPO, 1916), 34.

[318] "Japanese Invade American Codfish Industry," *Pacific Fisherman* (December 1919): 23–24; "Japanese Schooners Bring Codfish Cargoes," *Pacific Fisherman* (February 1919): 25.

[319] "Japanese Invade American Codfish Industry," *Pacific Fisherman* (December 1919): 23–24.

[320] Order of February 20, 1918, in *American-Canadian Fisheries Conference, 1918* (Washington, DC: GPO, 1918), 5.

[321] Ward T. Bower, *Alaska Fisheries and Fur-Seal Industries in 1918* (Washington, DC: GPO, 1919), 58–61; "Japanese Schooners Bring Codfish Cargoes," *Pacific Fisherman* (February 1919): 25; "Japanese Invade American Codfish Industry," *Pacific Fisherman* (December 1919): 23–24.

[322] "Japanese Invade American Codfish Industry," *Pacific Fisherman* (December 1919): 23–24.

[323] "Japs Sue Robinson Fisheries," *Pacific Fisherman* (July 1921): 40.

[324] "Japanese Schooners Bring Codfish Cargoes," *Pacific Fisherman* (February 1919): 25; "Jap Codfish Schooners Arrive," *Pacific Fisherman* (November 1920): 53.

[325] "Codfish Market Firm," *Pacific Fisherman* (August 1919): 55.

[326] "Japanese Codfish Arriving," *Pacific Fisherman* (October 1919): 47.

[327] "Japanese Invade American Codfish Industry," *Pacific Fisherman* (December 1919): 23–24.

[328] "Protect American Cod Fisheries," *Pacific Fisherman* (December 1919): 36.

[329] "Jap Codfish Schooners Arrive," *Pacific Fisherman* (November 1920): 53; Ward T. Bower, *Alaska Fishery and Fur-Seal Industries in 1920*, Bureau of Fisheries Doc. No. 909 (Washington, DC: GPO, 1921), 63–64.

[330] "Deal Abandoned by Japs?" *Pacific Fisherman* (June 1921): 40.

[331] "Japs Sue Robinson Fisheries," *Pacific Fisherman* (July 1921): 40; "Japanese Suit Settled," *Pacific Fisherman* (July 1922): 52; John N. Cobb, *Pacific Cod Fisheries* (revised edition, 1926), Bureau of Fisheries Doc. No. 1014, appendix 7 to *Report of the U.S. Commissioner of Fisheries for 1926* (Washington, DC: GPO, 1927), 460.

[332] "Beardsley Enters Pacific Field," *Pacific Fisherman* (July 1923): 32.

[333] "Reduced Codfish Production Causes Stiffening Market," *Pacific Fisherman* (January 1929), 159, 161.

[334] "Installs Wireless at Unga," *Pacific Fisherman* (March 1919): 62; "Radio Station for Unga," *Pacific Fisherman* (May 1920): 55; "Radio Takes Important Place in Alaska Fishery Operations," *Pacific Fisherman* (July 1928): 14; "North Pacific Cod Fishery Yield Above Average of Recent Years," *Pacific Fisherman, Statistical Number* (January 1931): 215–216; Eveready Columbia Dry Batteries, adv., *Pacific Fisherman* (January 1934): 5; Joe Follansbee, *Shipbuilders, Sea Captains, and Fishermen: The Story of the Schooner Wawona* (New York: iUniverse, 2006), 41–42, 49, 63, 98.

Chapter 8
ALASKA'S CODFISH INDUSTRY FADES

On this coast, it has not been a question of being able to secure cargoes but has been one of finding a market for the [codfish] catch; a vastly greater catch could be made were the market available for it.

—John Cobb, founding director, College of Fisheries, University of Washington, 1927.[335]

Citing the high prices paid to fishermen, as well as slack demand and Japanese competition (see chapter 7), both the Union Fish Company and the Alaska Codfish Company chose to close their shore stations at the end of the 1920 summer season. The companies reopened the stations early in 1921, and in June of that year, the Alaska Codfish Company's schooner *Maweema* returned from the company's codfish stations carrying a record 336,000 codfish.

Operations at the codfish stations of the Union Fish Company and the Alaska Codfish Company the following year were very different: the two companies pooled their operations. And rather than sending crews to the stations, they relied on resident fishermen, the production of which the companies shared equally.[336]

Except for the Union Fish Company station at Pirate Cove, which operated at a reduced level, all the Shumagin Island stations owned by San Francisco concerns were shut down by November 1923. Fishermen at the Pirate Cove station were employed all winter to fish as weather permitted, but the station's cook house was closed and the men boarded themselves. Three shore stations, however, were reopened in 1924, and fully fourteen operated in 1925. But there was a downside to the increased scale of operations: marketing the additional production. In a January 1926 headline, *Pacific Fisherman* noted that the 1925 season was characterized by an "Increased Catch and Light Demand."[337] The industry was producing more codfish than it could profitably market, and the shore stations owned by the Union Fish Company and the Alaska Codfish Company were on their way out.

In 1926, Bureau of Fisheries agent Ward Bower observed that "the profitable operation of cod shore stations is becoming increasingly difficult each year." The Union Fish Company, a major factor in the codfish industry and whose predecessor business in 1876 established the first codfish station in Alaska, chose in 1926 to permanently close its one remaining station (Pirate Cove) and used its flagship transporter, the *Golden State*, for offshore fishing. (The *Golden State* was transferred to the company's California tuna operations in 1927.)[338]

The other major traditional shore station operator, the Alaska Codfish Company, operated only its Eagle Harbor (on Nagai Island) station in 1926, but it operated both that station and its station at Unga in 1927. In January 1927, the company's Henry Levi attributed the decreased profitability of shore stations to a lack of harmony and cooperation among the codfish companies and to the "excessive cost of maintaining the stations, together with the high lay [price per fish] paid to the fishermen, and the losses incurred from pilferage." The quality of station-produced salted cod must have also been a factor: John Cobb had noted in 1926 that fishermen at the stations resented having to dress their fish, and often did so in a hasty and careless manner.[339] (Aboard schooners, codfish were dressed by a salaried dress gang.)

Some simple math may further explain the waning interest in operating shore stations. Twenty independently owned and two company-owned shore stations operated in 1926. Together they employed 94 men and produced 931,228 pounds of product (about

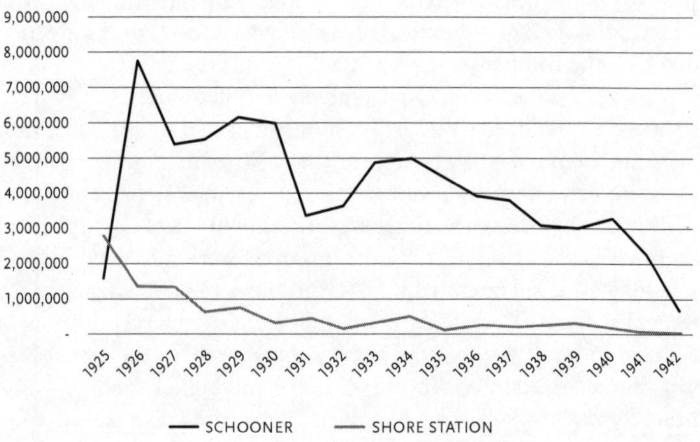

Chart 3. Schooner and shore station salted Pacific cod production, pounds of processed fish, 1925–1942. Shore station production includes salted codfish, stockfish, pickled codfish, and salted codfish tongues. Schooner production includes salted codfish and salted codfish tongues.

Data sources: U.S. Bureau of Fisheries and U.S. Fish and Wildlife Service annual Alaska fisheries reports.

11 percent of total Alaska production). The offshore fleet, meanwhile, consisted of fourteen vessels manned by 348 men, and produced 7,711,085 pounds of product (about 89 percent of total Alaska production).[340] Roughly calculated, the per-employee production at the shore stations that year was about 10,000 pounds, while in the offshore fleet it was more than twice that, about 22,000 pounds.

The codfish catch in 1927 was 1,632,928 fish, some 664,728 fewer than were caught the previous year. The diminished catch was attributable to a reduced effort—eight schooner trips versus eleven the previous year—and to a scarcity of fish in the Bering Sea. The reduced effort was attributed to weak export markets and low prices quoted by Atlantic producers, while the scarcity of fish was likely attributable to natural causes.[341]

Regarding markets, Henry Levi, of the Alaska Codfish Company, was of the opinion that "the modern development of apartment-house living" was causing a falling off in the domestic market for salted codfish.[342]

The permanent closure of the Alaska Codfish Company's stations at Eagle Harbor and Unga after curtailed operations during the 1927 season marked the end of lower-48 companies operating codfish stations in Alaska. For the remainder of the Salt Cod Era, shore-station fishing would be carried on only by independent fishermen or small firms, often in conjunction with salmon and herring pickling operations. In 1928, the twenty shore stations that salted cod employed just fifty-three men and produced 578,173 pounds of salted cod products, a little less than half of the amount produced the previous year.[343]

The Alaska Codfish Company suffered a great loss in August 1928, when its schooner *Maweema*, sailing in dense fog to San Francisco with some 260,000 codfish aboard, struck a reef at St. George Island, in the Bering Sea, and went to pieces. The vessel's thirty-one crew members were able to safely make their way to shore in dories and were later picked up by a U.S. Coast Guard vessel. The entire cargo, however, was lost. Built in 1895 as a lumber carrier, the 398-ton, three-masted *Maweema* (sister ship to the Robinson Fisheries Company's *Wawona*) had entered the service of the Alaska Codfish Company in 1915.[344]

In part due to the loss of the *Maweema*'s cargo, the Pacific codfish catch in 1928—1,580,000 fish—was the smallest since 1901, which caused a shortage of codfish in San Francisco and a consequent rise in the market value of the product.[345]

In 1929, encouraged by favorable market conditions, nine schooners—one more than in 1928—engaged in fishing for codfish, while the shore station effort that year was about the same as in 1928. Together,

Figure 51. Codfishermen, Gus Sjoberg's dock, Unga, Alaska, 1930s. (Courtesy of Peggy Arness.)

the two components of the industry produced 1,882,200 codfish, which were readily absorbed by the market.[346]

In 1932, Ward Bower, the Bureau of Fisheries agent in Alaska, characterized the shore-station fishery as having been "comparatively unimportant for several years." Production at the stations in 1932—dry-salted cod, stockfish, and pickled cod combined—was especially meager: a total of just less than 200,000 pounds. Only 24 men were employed at the few stations that operated that year. By contrast, the offshore fishery employed 135 men and produced 3.6 million pounds of dry-salted cod.[347]

Fortunately, the shore-based operators could get at least some of their product to market via vessels such as the 130-foot steamship *Starr*, which from 1921 until 1938 provided almost year-round monthly mail and freight service between Seward and about two dozen coastal outports from Seldovia to Unalaska. Freight delivered to Seward could be shipped to Seattle on scheduled Alaska Steamship Company vessels.[348]

Adding to the Alaska Codfish Company's woes after the loss of the *Maweema* in 1928, the company's codfish curing plant at Redwood City was destroyed by fire in July 1929. Alaska Codfish, which had operated cod-fishing schooners in Alaska since 1896 (and perhaps earlier), did not send schooners north in 1929, but it contracted to purchase the catch of the schooners *Charles R. Wilson* and *Sophie*

Christensen, which were owned by the Pacific Coast Codfish Company and J. E. Shields, respectively. The vessels were based in Puget Sound, but they delivered their cargoes directly from the Bering Sea to San Francisco and then returned to Puget Sound, where they were laid up until the following spring. The fish delivered by the *Charles R. Wilson* and *Sophie Christensen* in 1929 were processed in the Union Fish Company's plant on Belvedere Island, just north of San Francisco.[349]

Perhaps the loss of the *Maweema* and the fire at Redwood City were the proverbial last straws for Henry Levi. In 1893, Levi had organized what became the Alaska Codfish Company, and he had been active in its management ever since, but he retired in 1929. The firm was said to have been reorganizing, but actually seems to have been in the process of being absorbed by the Union Fish Company. The Alaska Codfish Company did not send any of its schooners north in 1930, but contracted to purchase the 1930 season's catch of the J. E. Shields's schooners *Charles R. Wilson* and *Sophie Christensen*.[350]

Shields had from early on been purchasing stock from the original owners of the Pacific Coast Codfish Company as opportunities arose, and by 1928 he had become the sole owner of the company. Though he had managed the Pacific Coast Codfish Company for a number of years, Shields was not a seafaring man—he had never sailed as a sailor or mate. Nevertheless, he appointed himself master of the *Sophie Christensen* in 1933 and took the vessel north each year through 1941. Shields eventually received a master's license—the only maritime license he ever held—but sailed only as captain of his own vessels. In 1934, Shields's son, Ed, at age eighteen, made his first trip to the Bering Sea as a cod fisherman. Ed Shields earned a master's degree in engineering at Harvard University, but cod fishing was apparently in his blood, and he was captain of the *C. A. Thayer* on the 1950 voyage that marked the end of the traditional salt codfish industry in Alaska, and in 2001 he authored the book *Salt of the Sea: The Pacific Coast Cod Fishery and the Last Days of Sail*, which recounted his experiences as a cod fisherman as well as some of the history of the industry.[351]

COD FISHING DURING THE GREAT DEPRESSION

The Bering Sea codfish fleet in 1933 comprised five schooners. Of these, three—the *Sophie Christensen* (operated by the Pacific Coast Codfish Company), and the *Wawona* and the *Azalea* (both operated by the Robinson Fisheries Company)—were based in Puget Sound, while the other two—the *Louise* and the *William H. Smith* (both operated by the Union Fish Company)—were based at San Francisco. The fleet that year employed 196 men and ninety-three dories and landed 4.86 million pounds of salted cod as well as about 30,000 pounds of salted cod tongues.[352]

The *Sophie Christensen* that year, on its first voyage under its owner and self-appointed captain, J. E. Shields, set the all-time record for the greatest number of fish and the greatest tonnage delivered by a Pacific codfish schooner—453,356 fish that aggregated 672 tons (1,344,000 pounds). The average weight of the salted fish was 2.96 pounds. Based on a recovery rate of 34.8 percent established in the 1980s, the live weight of the average fish was 8.5 pounds, so the total live weight of the cod caught during the trip was about 3.85 million pounds.[353]

The exceptional number of fish was skewed, however, because the Pacific Coast Codfish Company—to the delight of its fishermen—accepted fish down to twenty-three inches long. By contrast, the Robinson Fisheries Company in 1933 had a minimum length of twenty-five inches, and the minimum length at the Union Fish Company was twenty-eight inches.[354]

The weight and conformation of a codfish of a given length can vary considerably for a host of reasons, including health, genetics, location, and season. According to data gathered near Unimak Pass during the early 1980s, a codfish twenty-three inches long would have a live (bled) weight of 5.51 pounds, a twenty-five-inch codfish would weigh 7.27 pounds, and a twenty-eight-inch codfish would weigh 10.59 pounds. Weighing an average of 8.5 pounds, the *Sophie Christensen*'s fish would have averaged a little over twenty-six inches long.[355]

The *Sophie Christensen*'s great success was marred by the loss during a heavy squall of Sven Markstrom, a veteran cod fisherman known in the industry as "The Terrible Swede."[356]

The Bering Sea schooner fleet in 1935 comprised seven schooners, employed 199 men, and produced 4,930,701 pounds of salted cod and 11,675 pounds of tongues. The shore-station production that year—an aggregate of 492,905 pounds of dry-salted cod, stockfish, and pickled cod—was the largest since 1929. Fifty-four men, 25 of whom were Alaska Natives, worked at shore stations.[357]

Despite a long stretch of severe weather that prevented the dories from fishing, the five schooners that fished in the Bering Sea during the 1935 codfish season managed to eke out fair catches.[358]

Five schooners—three Puget Sound–based and two San Francisco–based—fished for codfish in the Bering Sea during the 1936 season and delivered 3.88 million pounds of dry-salted cod. The following spring, however, found the leaders of the Pacific codfish industry less than enthusiastic about its prospects, mainly because of competition from salted codfish imported from Japan. J. E. Shields, of the Pacific Coast Codfish Company, announced that he would not take his company's schooner *Sophie Christensen* north, and none of the industry's other cod-fishing vessels that might have gone north had begun outfitting.

Despite Shields's assertion that the *Sophie Christensen* would not go north in 1937, it did so, as did the *Wawona*, also from Puget Sound, and the *Louise* and *William H. Smith*, from San Francisco. Moreover, the weather cooperated and there were plenty of fish, and the four schooners "wet all their salt" in four months rather than the usual five. They returned home early with a catch of 3.78 million pounds, nearly the same as five schooners had caught the previous year.[359]

The 1938 Bering Sea codfish fleet comprised three schooners, all of which hailed from Puget Sound: the *Charles R. Wilson* and *Sophie Christensen*, of the Pacific Coast Codfish Company, and the *Wawona*, of the Robinson Fisheries Company.[360] Citing "unsettled labor conditions," the Union Fish Company canceled the voyages of its schooners *William H. Smith* and *Louise*. For the first time since 1870, the Union Fish Company or one of its predecessors did not send at least one vessel north.[361]

For the Union Fish Company, of far more immediate concern than Japanese competition was a fire in early November 1938 that destroyed its codfish curing packing plant on Belvedere Island, which had been established in 1876 and was the last remaining codfish operation on San Francisco Bay. At the time of the fire, about 800 tons of codfish were on hand, much of which was lost along with a number of dories and other gear.[362]

After the closure of its shore stations during the previous decade, the death of Crescent Hale, its long-time owner, in 1937, and the fire at its San Francisco curing plant, the year 1938 marked the end of the company's long history in the Pacific codfish industry. Moreover, the Union Fish Company was the last San Francisco–based company involved in the Pacific codfish industry. Its withdrawal meant that the industry that had begun in San Francisco and had been part of the city's economy and culture for some seven decades, no longer had a local presence. The cessation of codfish operations at San Francisco was attributed to competition with salted codfish imported from Japan. The Union Fish Company's two schooners, the *William H. Smith* and the *Louise*, were put up for sale in 1939.[363]

The schooner catch in 1938 was 3.07 million pounds, approximately commensurate with the diminished size of the fleet, and was all landed in Puget Sound ports.[364]

Japanese fishing in the eastern Bering Sea

> *In compensation for the scarcity of natural resources in Japan, God has endowed her with a unique gift of excellent fishing talent to take care of her population. Their destiny is, therefore, to develop that art further and further, and to exploit the fisheries products even from the open sea where the international law gives them absolute freedom and protection for such operation. In other words, they may operate floating fisheries on the open sea, and prepare the products on board the mother ship. This is the only way for them to obtain the chance of more employment by means of which they will take care of their overpopulation.*
>
> — T. Takasaki, Toyo Seikan Kaisha, Ltd., Osaka, Japan, 1937[365]

As noted in chapter 7, the Japanese had caused great alarm among American cod-fishing interests on the Pacific coast during the late 1910s, when they began delivering salted cod to San Francisco and Puget Sound. There was little alarm, however, in 1929, when at least one Japanese floating cannery, accompanied by a fleet of smaller fishing boats equipped with tangle nets, began high-seas crabbing operations in the eastern Bering Sea. The American fishermen had little interest in deep-sea crabs, and the Japanese were operating legally beyond the three-mile U.S. territorial limit.[366]

Japanese vessels returned the following year, and by 1933 the Japanese had expanded their eastern Bering Sea operations to include the manufacture of fish meal and fish oil from bottom-dwelling fish—including cod and halibut—taken by trawling. The fish catch in 1933 was later estimated to be about 3,300 (metric) tons; the 1935 catch was an estimated 28,600 tons. Yet little concern was aroused. In fact, in 1936 the Union Fish Company purchased 360 tons of salted codfish from the Japanese trawl fleet. The fish—a foreign product for which an import duty was paid—was transported from the Bering Sea fishing grounds to San Francisco on the company's schooner *Golden State*, which had been retired from the codfish industry since 1927.[367]

More important than codfish to American fishing interests were Japanese intentions of stationing floating salmon canneries in Bristol Bay. Salmon would be provided by a fleet of catcher boats using high-seas gill nets, some of which would be more than three miles long. Fortunately for the American industry, the Japanese plans never reached fruition.[368]

The pre–World War II Japanese effort and catch in the eastern Bering Sea peaked in 1937. That year, three floating fish-processing plants, accompanied by two dozen catcher vessels—at least a dozen of which were bottom trawlers that ranged from 75 to 150 feet in length—caught an estimated 43,400 tons of fish.[369]

Cod fishermen in the Bering Sea, who often fished close to Japanese vessels, provided firsthand accounts of the Japanese effort. In the spring of 1937, J. E. Shields, captain of the schooner *Sophie Christensen*, penned a letter to *Pacific Fisherman* that detailed several of his experiences over the previous four years. Shields wrote of having his anchored boat surrounded by tangle nets used by the Japanese to catch crab, of Japanese fishing within the three-mile territorial limit, and of the *Tatsuma Maru*, a factory ship accompanied by eight high-powered trawlers "fishing halibut, codfish, soles and what they could get on the bottom."[370]

In May 1938, Shields reported that a portion of the Bering Sea was "covered with Japanese fishing boats and nets." In fact, the Bureau of Fisheries that year noted only one Japanese floating fish-processing plant in the Bering Sea, the *Toten Maru*, which was accompanied by three 50-foot trawlers and ten launches about 30 feet long. Nevertheless, to combat what he termed an "invasion" by Japanese fishermen, Shields asked his home office to send north two dozen high-powered rifles and ammunition for his boat and a like amount of weaponry for the *Charles R. Wilson*, Shields's other vessel on the codfish grounds. Fortunately, this much-publicized request, which could have had very serious consequences, went unfulfilled.[371]

The 1939 Japanese fishing effort in the eastern Bering Sea was directed entirely at crabs, but in 1940 the effort focused on cod and halibut. That year, the floating processing plant *Kosei Maru*, accompanied by ten trawlers, worked the eastern Bering Sea banks from early May until late August. The catch was an estimated 9,600 tons. Part of the catch of cod was salted, while the remainder of the cod, as well as the halibut, was frozen. Needless to say, the Japanese attack of Pearl Harbor in December 1941 ended, at least for a while, the Japanese fishing effort in the eastern Bering Sea.[372]

WORLD WAR II: THE BEGINNING OF THE END OF THE SALT COD ERA

The outbreak of World War II in Europe in early September 1939 interrupted the production and distribution of Scandinavian codfish, which strengthened the market for salted codfish overall and specifically brightened the prospects of Pacific codfish producers. The stiffening of the codfish market did not, however, immediately benefit the Pacific industry, because the three schooners that constituted the Bering Sea fleet that year (*Azalea*, *Sophie Christensen*, and *Wawona*) were just returning to Puget Sound as the war began, and most of their catch had already been committed at prewar prices. The *Sophie Christensen*'s 375,000-fish catch was the largest of the three schooners, and the catch would have no doubt been even larger had the vessel not spent two weeks in June aground near Port Moller.[373]

Fishing was relatively good in the Bering Sea in 1940, and the weather, too, was good. This combination enabled the three schooners that participated in the fishery—*Charles R. Wilson*, *Sophie Christensen*, and *Wawona*—to all wet their salt and return to Puget Sound by mid-August—record time—with a total catch of 3,208,200 pounds of exceptionally large fish. The Puget Sound codfish producers expected to export a considerable portion of their production to South America, which had been supplied before the war by East Coast codfish concerns that were now sending salted codfish to Europe to mitigate the shortage of Scandinavian codfish.[374]

The same schooners that fished for codfish in the Bering Sea returned for the 1941 season, but bad weather and relatively poor fishing that year resulted in a catch of 2,284,853 pounds, almost a million pounds less than the previous year's catch. Sadly, the season was further marred by the death of fisherman John Markie aboard the schooner *Charles R. Wilson*. At the time of his death, Markie was seventy-seven years old and was said to be the oldest deep-sea fisherman on the Pacific coast. He had fished for the Pacific Coast Codfish Company for twenty-eight years and took ill in his dory on the first day after reaching the banks. Markie died in his bunk the following night and was buried on Little Walrus Island (likely one of the Kudobin Islands, which are just west of Port Moller).[375]

Cod fishing at shore stations in 1941 employed only seventeen persons (twelve whites and five Natives), who in aggregate and often in conjunction with salmon and herring pickling operations, produced 99,666 pounds of codfish products—dry-salted cod, stockfish, pickled cod, and quick-frozen fillets and steaks. This was about half of the amount that had been produced the previous year. *Pacific Fisherman* attributed the reduced production to station fishermen moving into more lucrative fisheries or being engaged in Alaska defense work.[376]

The Japanese attacked Pearl Harbor in December 1941, and in early 1942 the U.S. War Shipping administration, an emergency agency established during World War II, requisitioned the Pacific Coast Codfish Company's schooners *John A.* and *Sophie Christensen*.

The company's *C. A. Thayer* and the Robinson Fisheries Company's schooners *Azalea* and *Wawona* were requisitioned in August of that year. All the vessels were converted into barges for use in the war effort; only two, the *C. A. Thayer* and the *Wawona*, returned to the cod fishery after the war.[377]

> *The 1941 season was marked by a series of misfortunes, including death, bad weather, and relatively poor fishing on the distant Bering Sea banks. However, the greatly reduced catch coupled with interruption, by the European War, of Scandinavian production and distribution served to greatly strengthen the North Pacific codfish market.*
>
> —Ward T. Bower, U.S. Fish and Wildlife Service, 1943[378]

In the spring of 1942, the Pacific Coast Codfish Company sent one schooner, the *Charles R. Wilson*, to the Bering Sea. It was the only cod-fishing vessel to go north that year and was on the fishing grounds in early June, when the Japanese attacked Dutch Harbor. Despite the not-too-distant hostilities, the vessel finished its season and returned to Puget Sound carrying 177,447 salted codfish and 1,900 pounds of salted cod tongues. By 1943, with the war raging around the world, short supplies of codfish pushed the price into what *Pacific Fisherman* termed "the fancy goods bracket."[379]

After the Japanese attack on Dutch Harbor, the Aleutian Islands area had been designated a war zone, and the War Shipping Administration paid merchant seamen serving in the area a war-risk bonus equal to 100 percent of their base wages. When the time arrived to gear up for the 1943 cod-fishing season, the crew of the *Charles R. Wilson*—the only cod-fishing schooner scheduled to go north that year—asked for a 30 percent bonus for fishing in the Bering Sea. J. E. Shields, owner of the Pacific Coast Codfish Company, however, could not pay the bonus without permission of the National War Labor Board. Shields applied for the board's permission and purchased salt for the voyage, but he received no response from the board for more than eight months. By then, he had canceled the voyage and sold the salt. The salted cod production in Alaska in 1943 was limited to a small amount produced by independent operators in the Shumagin Islands.[380]

For the 1944 season, the National War Labor Board granted permission to J. E. Shields to pay a war-risk bonus of 30 percent over the lay received by cod fishermen in 1942, and his company's *Charles R. Wilson*, outfitted with a dozen power dories, was once again the lone cod-fishing schooner on the Alaska banks. Its production that season totaled 633,000 pounds. There was no shore-station production in 1944. In its report that year on Alaska's fisheries, the U.S. Fish and Wildlife Service (successor to the Bureau of Fisheries) indicated the decline of the cod fishery by listing cod under the heading Miscellaneous Fishery Products. Until 1944, each of the federal government's annual reports on Alaska's fisheries, which were first published in 1905, contained at least a brief but separate section on Pacific cod that listed production and other data and noted significant developments in the cod fishery.[381]

The 1945 codfish season was largely a repeat of the previous year's, but *Pacific Fisherman*, with the war winding down, anticipated the future of the codfish industry. The journal expected an expansion of fishing in the Bering Sea and an increase in the catch of codfish. It noted that trawling had become the dominant fishing method for codfish in the Atlantic Ocean, and it expected trawling in the not-too-distant future to compete with handlining on the Pacific grounds.[382]

Shortly after the war ended, the Pacific Coast Codfish Company reacquired the *C. A. Thayer* and *Sophie Christensen*, and the Robinson Fisheries Company reacquired the *Wawona*. The vessels, as noted above, had been utilized as barges. The masts and bowsprits of the *C. A. Thayer* and *Wawona* had been removed, but their hulls were in good repair; the *Sophie Christensen* retained its four masts, but its hull was in poor condition. In time for the 1946 fishing season, the companies replaced the masts and bowsprits on the *C. A. Thayer* and *Wawona*, and outfitted both vessels for cod fishing. (The *C. A. Thayer*'s "new" masts, as well as other gear, came off the *Sophie Christensen*, which was then sold to a Canadian firm for use as a log and lumber barge and eventually met its end on a rocky shore of Vancouver Island.)

Ed Shields, of the Pacific Coast Codfish Company, had hoped to send both the *C. A. Thayer* and *Charles R. Wilson* north in 1946, but he could not find enough skilled fishermen to man both vessels, and he instead chose to send the *C. A. Thayer* because it was in better repair than the *Charles R. Wilson*. The *Charles R. Wilson* never sailed again. The vessel was held in layup in Lake Union (Seattle) until the mid-1950s, when it was sold to a Canadian logger who beached it as a breakwater near Powell River, British Columbia.[383]

Codfish were scarce in 1946; the combined catch of 242,907 fish by the *C. A. Thayer* and *Wawona* was low when compared to the 144,317 fish that the *Charles R. Wilson* fishing alone had caught in 1945. The *C. A. Thayer* and *Wawona* returned to the Bering Sea in 1947, and—in part because of adverse weather— their combined catch of 240,424 fish was similar to the previous year's, except that in addition to salted fish, 1947's catch included about 33,000 pounds of cod livers that were salted in five-gallon cans or held in cold storage (see chapter 1). The price of vitamins was generally high during the early postwar years, and the livers were rendered for cod liver oil at a Seattle plant.[384]

Following the return of the *Wawona* from its 1947 voyage, J. W. Beardsley's Sons, which had acquired a half interest in the Robinson Fisheries Company in 1923 (see chapter 7), purchased John Trafton's remaining interest and became the sole owner of the firm. The *Wawona*'s 1947 voyage had been difficult because of adverse weather that caused extended delays, and its catch of 99,700 codfish paled in comparison to the 140,724 codfish landed by the *C. A. Thayer*. Moreover, the 30,000 pounds of cod livers brought south in the *Wawona*'s cold storage were sold on a depressed postwar market. In early 1948, the Robinson Fisheries Company's new owners decided against sending it north, marking the end of the company's forty-three-year history in the codfish industry. As noted in chapter 3, the *Wawona*'s career catch of 7.3 million codfish is the all-time record by a single schooner.[385] (In 1963, a Seattle group, Save our Ships, began a campaign to preserve the *Wawona* as a floating museum. In 1970, the *Wawona* became the first ship to be listed on the National Park Service's National Register of Historic Places, and it was operated as a floating maritime museum on Lake Union until being dismantled in 2009 because of a lack of funding and its deteriorated condition.)[386]

As a result of the Robinson Fisheries Company's decision to leave the industry, J. E. Shields's Pacific Coast Codfish Company was left as the single remaining producer of salted codfish on the U.S. West Coast.

C. A. Thayer and the "last days of commercial sail"[387]

In his 1948 report, Seton Thompson, the U.S. Fish and Wildlife Service agent in charge of Alaska's fisheries, referred to the *C. A. Thayer* as being the only "old-time schooner" to fish that year for codfish in the Bering Sea. The vessel, however, was accompanied by modern counterparts: the trawlers *Deep Sea* and *Lynn Ann*, and the American factory ship *Pacific Explorer*, which was herself accompanied by eight trawlers. (See part 2.)[388]

Fishing twelve dories, the *C. A. Thayer* caught 210,000 codfish in 1948. The following year, it fished fourteen dories, and the catch increased to 243,350.

But time was running out on a sailing schooner that produced salted codfish. First, there was a shortage of experienced fishermen. According to *Pacific Fisherman*, "In recent years very few men have chosen to follow this dory trade, even power dories, and it has been with the greatest difficulty that the codfish companies have manned their fishing craft." Second, salted codfish was losing the competition for the American housewife's grocery shopping dollar to attractively packaged and easy-to-prepare frozen seafood products—the ubiquitous fish sticks.[389] Production of salted Atlantic cod, however, continued unabated: in 1947, Newfoundland alone produced almost 132 million pounds of salted cod.[390]

In the spring of 1950, as the *C. A. Thayer* was preparing to depart Puget Sound for the Bering Sea, there were reports that this was to be its last cod-fishing voyage. According to *Pacific Fisherman*, J. E. Shields maintained that no decision regarding the *C. A. Thayer*'s future had yet been made. As in the prior, very productive year, the vessel fished fourteen dories in 1950, but it landed only 195,486 codfish—about 20 percent fewer than the previous year. Perhaps the poor catch was the proverbial "last straw" for J. E. Shields, and he decided to end the cod-fishing career of the *C. A. Thayer*.

Officially, the Salt Cod Era—and the commercial use of sailing vessels along the U.S. West Coast, with the exception of the Bristol Bay salmon gill-net fleet[391]—ended on September 4, 1950, the date the *C. A. Thayer* was towed to the Poulsbo, Washington, plant of Shields's Pacific Coast Codfish Company to discharge its final cargo of salted Bering Sea codfish. The following year, the U.S. Fish and Wildlife Service's Seton Thompson attributed "poor market conditions" to the cessation of the Alaska cod fishery.[392]

In 1954, J. E. Shields sold the *C. A. Thayer* to a business that renamed it *Black Shield* and moved it to Hood Canal (on Washington's Olympic Peninsula), where it was moored as a pirate ship museum. In 1956, the State of California purchased the vessel, which was then sailed to San Francisco for extensive repairs and refitting before being put on display at the San Francisco Maritime Museum. In 1978, the *C. A. Thayer* was transferred to the National Park Service and is currently on display at the San Francisco Maritime National Historical Park.[393]

ENDNOTES

[335] John N. Cobb, *Pacific Cod Fisheries* (revised edition, 1926), Bureau of Fisheries Doc. No. 1014, appendix 7 to *Report of the U.S. Commissioner of Fisheries for 1926* (Washington, DC: GPO, 1927), 403.

[336] "Alaska Stations Closed," *Pacific Fisherman* (November 1920): 53; "Codfish," *Pacific Fisherman* (July 1921): 40; "Review of the [1922] Pacific Codfish Season," *Pacific Fisherman* (January 1923): 96.

[337] "Codfish Stations Closed," *Pacific Fisherman* (November 1923): 37; Ward T. Bower, *Alaska Fishery and Fur-Seal Industries in 1924*, Bureau of Fisheries Doc. No. 992 (Washington, DC: GPO, 1925), 137–138; Ward T. Bower, *Alaska Fishery and Fur-Seal Industries in 1925*, Bureau of Fisheries Doc. No. 1008 (Washington, DC: GPO, 1926), 134–135; "Increased Catch and Light Demand Mark 1925 Codfish Season," *Pacific Fisherman* (January 1926): 154, 156, 158.

[338] Ward T. Bower, *Alaska Fishery and Fur-Seal Industries in 1926*, Bureau of Fisheries Doc. No. 1023 (Washington, DC: GPO, 1927), 299; Ward T. Bower, *Alaska Fishery and Fur-Seal Industries in 1927*, Bureau of Fisheries Doc. No. 1040 (Washington, DC: GPO, 1928), 139.

[339] "Codfish Production Curtailed Despite Good Fishing Season," *Pacific Fisherman* (January 1927): 204, 206, 208; John N. Cobb, *Pacific Cod Fisheries* (revised edition, 1926), Bureau of Fisheries Doc. No. 1014, appendix 7 to *Report of the U.S. Commissioner of Fisheries for 1926* (Washington, DC: GPO, 1927), 442.

[340] Ward T. Bower, *Alaska Fishery and Fur-Seal Industries in 1926*, Bureau of Fisheries Doc. No. 1023 (Washington, DC: GPO, 1927), 299–300.

[341] "Marked Curtailment in Pacific Codfish Catch of 1927," *Pacific Fisherman* (January 1928): 192, 194, 196.

[342] Ibid.

[343] Ward T. Bower, *Alaska Fishery and Fur-Seal Industries in 1928*, appendix 6 to *Report of U.S. Commissioner of Fisheries for the Fiscal Year 1929* (Washington, DC: GPO, 1929), 291.

[344] "Schr. 'Maweema' Lost," *Pacific Fisherman* (September 1928): 41; Ward T. Bower and Henry D. Aller, *Alaska Fisheries and Fur Industries in 1915*, Bureau of Fisheries Doc. No. 834 (Washington, DC: GPO, 1917), 55.

[345] "Reduced Codfish Production Causes Stiffening Market," *Pacific Fisherman* (January 1929): 159, 161.

346 "Pacific Codfish Catch Improves During 1929 Season," *Pacific Fisherman* (January 1930): 207, 209; Ward T. Bower, *Alaska Fishery and Fur-Seal Industries in 1929*, appendix 10 to *Report of U.S. Commissioner of Fisheries for 1930* (Washington, DC: GPO, 1930), 302.

347 Ward T. Bower, *Alaska Fishery and Fur-Seal Industries in 1932*, appendix 1 to *Report of the Commissioner of Fisheries for Fiscal Year 1933* (Washington, DC: GPO, 1933), 53.

348 James Mackovjak, *Aleutian Freighter: A History of Shipping in the Aleutian Islands* (Seattle: Documentary Media, 2012), 25–30; "Late Codfish Arrives," *Pacific Fisherman* (November 1923): 38.

349 "Codfish Plant Burned," *Pacific Fisherman* (August 1929): 62; "Codfish Catch Only Fair," *Pacific Fisherman* (October 1929): 49; "Cod Schooners Return to Sound," *Pacific Fisherman* (December 1929): 43.

350 "Pacific Codfish Catch Improves During 1929 Season," *Pacific Fisherman* (January 1930): 207, 209; John N. Cobb, *Pacific Cod Fisheries*, Bureau of Fisheries Doc. No. 830, appendix 4 to *Report of the U.S. Commissioner of Fisheries for 1915* (Washington, DC: GPO, 1916), 29; "Codfish Catch Only Fair," *Pacific Fisherman* (October 1929): 49.

351 "Puget Sound Fleet Augmented," *Pacific Fisherman* (March 1929): 45; "Cod Schooners Return to Sound," *Pacific Fisherman* (December 1929): 43; Richard S. Croker, "Alaska Codfish," in *Commercial Fish Catch of California for the year 1929*, Division of Fish and Game of California Fish Bulletin No. 30 (Sacramento: California State Printing Office, 1931), 53; "North Pacific Cod Fishery Yield Above Average of Recent Years," *Pacific Fisherman* (January 1931): 215–216; "31 Years a Codfish Skipper," *Pacific Fisherman* (March 1932): 41–42; "J. E. Shields Tells 'Em," *Pacific Fisherman* (December 1943): 25; Ed Shields, "The Codfish Industry," in *Poulsbo: Its First 100 Years*, ed. Rangveld Kvelstad (Poulsbo, Wash.: Poulsbo Centennial Book Committee, 1986), 136–138; "Historic Past of Capt. Shields Recalled," *Marine Digest* (July 14, 1962): 20–21; "Poulsbo's Captain Ed Shields Crosses the Bar," *North Kitsap Herald*, June 10, 2008; Ed Shields, *Salt of the Sea: The Pacific Coast Cod Fishery and the Last Days of Sail* (Lopez Island, Wash.: Pacific Heritage Press, 2001), 238 pp.

352 Ward T. Bower, *Alaska Fishery and Fur-Seal Industries in 1933*, appendix 2 to *Report of Commissioner of Fisheries for the Fiscal Year 1934* (Washington, DC: GPO, 1934), 287–288.

353 "Salt Codfish," *Pacific Fisherman* (January 1934): 193.

354 Average weight of salted fish delivered, in pounds: *Sophie Christensen*, 2.96; *Wawona*, 3.22; *Azalea*, 3.24; *Louise*, 3.32; *William H. Smith*, 3.33.

[355] Kenneth J. Hilderbrand, *Model White Fish Project: Trident Seafoods, Akutan, Alaska, 1982/1985* (Anchorage: Alaska Fisheries Development Foundation, 1986), 65–66, 68.

[356] Ward T. Bower, *Alaska Fishery and Fur-Seal Industries in 1933*, appendix 2 to *Report of Commissioner of Fisheries for the Fiscal Year 1934* (Washington, DC: GPO, 1934), 287–288; "Salt Codfish," *Pacific Fisherman* (January 1934): 193; "'Sophie Christensen' Loses One of Her Fishermen," *Pacific Fisherman* (September 1933): 49.

[357] Ward T. Bower, *Alaska Fishery and Fur-Seal Industries in 1934*, appendix 1 to *Report of the Commissioner of Fisheries for the Fiscal Year 1935* (Washington, DC: GPO, 1935), 48–49.

[358] "Codfishing—Freakish Season Yields fair Catches," *Pacific Fisherman* (October 1935): 48.

[359] "American Codfishing May Be Suspended," *Pacific Fisherman* (April 1937): 73; "'Sophie' Has Good Season," *Pacific Fisherman* (September 1937): 63; "S.F. Codfish Craft Take Good Catches," *Pacific Fisherman* (September 1937): 63; Ward T. Bower, *Alaska Fishery and Fur-Seal Industries in 1937*, Administrative Rept. No. 31, appendix 2 to *Report of the Commissioner of Fisheries for the Fiscal Year 1938* (Washington, DC: GPO, 1938), 120.

[360] "Three Vessels Working Bering Sea Codfish," *Pacific Fisherman* (May 1938): 61.

[361] "No S.F. Codfish Operations This Year," *Pacific Fisherman* (April 1938): 73.

[362] "Union Fish Co. Plant Destroyed," *Pacific Fisherman* (January 1938): 198.

[363] "Death of C. P. Hale," *Pacific Fisherman* (April 1937): 40; "Will Sell Cod Schooners," *Pacific Fisherman* (September 1939): 72.

[364] Ward T. Bower, *Alaska Fishery and Fur-Seal Industries in 1938*, Administrative Rept. No. 36, appendix 2 to *Report of Commissioner of Fisheries for the Fiscal Year 1939* (Washington, DC: GPO, 1940), 141–142.

[365] "Japanese Intention to Invade Alaska Salmon Fisheries Is Openly Declared," *Pacific Fisherman* (March 1937): 9–11. A footnote to this article stated: "In the interest of clearness and idiomatic expression, some slight liberties have been taken with the original English translation."

[366] Ward T. Bower, *Alaska Fisheries and Fur-Seal Industries in 1932*, appendix 1 to *Report of Commissioner of Fisheries for the Fiscal Year 1933* (Washington, DC: GPO, 1933): 55.

[367] Ward T. Bower, *Alaska Fisheries and Fur-Seal Industries in 1933*, appendix 2 to *Report of Commissioner of Fisheries for the Fiscal Year 1934* (Washington, DC: GPO, 1934): 290; Russell Nelson Jr., Robert French, and Janet Wall, "Sampling by U.S. Observers on Foreign Fishing Vessels in the Eastern Bering Sea and Aleutian Islands Region, 1977–78," *Marine Fisheries Review* (May 1981): 1–19; "Salt Codfish," *Pacific Fisherman* (January 1937): 225.

[368] "Protect Our Alaska Salmon From Alien Exploitation," *Pacific Fisherman* (September 1936): 21–23.

[369] Ward T. Bower, *Alaska Fisheries and Fur-Seal Industries in 1938*, Administrative Rept. No. 36, appendix 2 to *Report of the Commissioner of Fisheries for the Fiscal Year 1939* (Washington, DC: GPO, 1935): 75; Ward T. Bower, *Alaska Fisheries and Fur-Seal Industries in 1937*, Administrative Rept. No. 31, appendix 2 to *Report of the Commissioner of Fisheries for the Fiscal Year 1938* (Washington, DC: GPO, 1938): 75.

[370] "J. E. Shields Tells of Contacts with Japanese in Bering Sea," *Pacific Fisherman* (April 1937): 39–40; "Japanese Salmon Fishing Observed by Capt. J. E. Shields," *Pacific Fisherman* (August 1937): 25.

[371] "Angry Fishermen off Alaska Want to Fire on Japs," *St. Petersburg Times*, May 26, 1938.

[372] Ward T. Bower, *Alaska Fishery and Fur-Seal Industries: 1940*, Statistical Digest No. 2 (Washington, DC: GPO, 1942), 53–54.

[373] "American Codfish Schooners Return From Bering Sea," *Pacific Fisherman* (September 1939): 72.

[374] "Codfish," *Pacific Fisherman* (January 1941): 161; "Puget Sound Codfish Fleet Returns from Bering Sea Banks," *Pacific Fisherman* (September 1940): 78.

[375] "Codfish," *Pacific Fisherman* (January 1942): 263; Ward T. Bower, *Alaska Fishery and Fur Seal Industries: 1941* (Washington, DC: GPO, 1943), 43–44.

[376] Ward T. Bower, *Alaska Fishery and Fur-Seal Industries: 1941* (Washington, DC: GPO, 1943), 43–44; "Codfish," *Pacific Fisherman* (January 1942): 263.

[377] "Codfish Schooners Taken By Transport Service," *Pacific Fisherman* (September 1942): 59; "J. E. Shields Tells 'Em," *Pacific Fisherman* (December 1943): 25.

[378] Ward T. Bower, *Alaska Fish and Fur-Seal Industries: 1941* (Washington, DC: GPO, 1943), 44.

[379] "J. E. Shields Tells 'Em," *Pacific Fisherman* (December 1943): 25; "Codfish Schooners Taken by Transport Service," *Pacific Fisherman* (September 1942): 59; Gordon Newell, ed., *H. W. McCurdy Marine History of the Pacific Northwest* (Seattle: Superior Publishing Company, 1966), 508–536; "Future of Codfishing Will See Sweeping Change," *Pacific Fisherman* (January 1945): 278; Ed Shields, *Salt of the Sea: The Pacific Coast Cod Fishery and the Last Days of Sail* (Lopez Island, Wash.: Pacific Heritage Press, 2001), 203; "War Bonus Permitted," *Pacific Fisherman* (January 1945): 278; Ward T. Bower, *Alaska Fishery and Fur Seal Industries: 1942*, Statistical Digest No. 8 (Washington, DC: GPO, 1944), 38; "Pacific Codfish," *Pacific Fisherman* (January 1943): 367; "Codfish Market Shows Effect of War," *Pacific Fisherman* (January 1941): 261.

[380] "J. E. Shields Tells 'Em," *Pacific Fisherman* (December 1943): 25; Ward T. Bower, *Alaska Fishery and Fur Seal Industries: 1943*, Statistical Digest No. 10 (Washington, DC: GPO, 1944), 37.

[381] "'Charles R. Wilson' Fishes Solo for Cod," *Pacific Fisherman* (May 1944): 59; Ward T. Bower, *Alaska Fishery and Fur-Seal Industries: 1944*, Statistical Digest No. 13 (Washington, DC: GPO, 1946), 53.

[382] "Future of Codfishing Will See Sweeping Change," *Pacific Fisherman* (January 1945): 278.

[383] Seton H. Thompson, *Alaska Fishery and Fur-Seal Industries: 1947*, Statistical Digest No. 20 (Washington, DC: GPO, 1950), 52–53; Ed Shields, *Salt of the Sea: The Pacific Coast Cod Fishery and the Last Days of Sail* (Lopez Island, Wash.: Pacific Heritage Press, 2001), 132, 203–204; Gordon Newell, ed., *H. W. McCurdy Marine History of the Pacific Northwest* (Seattle: Superior Publishing Company, 1966), 604.

[384] "Codfish," *Pacific Fisherman* (January 1948): 237; "'Wawona' Cod-Liver Return Disappointing," *Pacific Fisherman* (January 1948): 65.

[385] "Beardsleys Now Own Robinson Fisheries Co.," *Pacific Fisherman* (October 1947): 83; "Codfish," *Pacific Fisherman* (January 1948): 237; The *Wawona* fished for codfish each year from 1914 through 1947, except 1921 and 1942–1945. The vessel's total catch during those years is calculated from data in John N. Cobb's 1926 *Pacific Cod Fisheries* and reports in *Pacific Fisherman*.

[386] Joe Follansbee, *Shipbuilders, Sea Captains, and Fishermen: The Story of the Schooner Wawona* (New York: iUniverse, 2006), 123–133.

[387] "Historic Past of Capt. Shields Recalled," *Marine Digest* (July 14, 1962): 20.

[388] Seton H. Thompson, *Alaska Fishery and Fur-Seal Industries: 1948*, Statistical Digest No. 23 (Washington, DC: GPO, 1952), 43.

[389] "1950 Pacific Codfish Fleet Catch," *Pacific Fisherman* (January 1951): 291; "Codfish," *Pacific Fisherman* (January 1948): 237.

[390] U.S. Fish and Wildlife Service, "Newfoundland," *Commercial Fisheries Review* (May 1949): 48–49.

[391] The federal regulation that restricted the Bristol Bay, Alaska, salmon gill-net fleet to sail and oar power was lifted in 1951, after which the use of power boats quickly became universal in the bay.

[392] "'C. A. Thayer' Sails for Bering Cod Banks," *Pacific Fisherman* (May 1950): 49; "1950 Pacific Codfish Fleet Catch," *Pacific Fisherman* (January 1951): 291; "'C. A. Thayer' Codfish Fare Under Recent Average," *Pacific Fisherman* (October 1950): 58; Seton H. Thompson, *Alaska Fishery and Fur-Seal Industries: 1951*, Statistical Digest No. 31 (Washington, DC: GPO, 1954), 51.

[393] "C. A. Thayer To Be Museum In San Francisco," *Marine Digest* (June 9, 1956): 10; "C. A. Thayer History," San Francisco Maritime National Historical Park archive, accessed September 7, 2014, http://www.nps.gov/safr/historyculture/ca-thayer-history.htm.

Part Two

THE BAIT COD ERA (1950–1978)

Chapter 9
THE FISHERY AND ITS MANAGEMENT

CODFISH MANAGEMENT PRIOR TO 1976

The U.S. Fish Commission was established by Congress in 1871 as the federal agency responsible for fishery matters. The agency, however, was given no authority regarding Alaska's fisheries, the management of which Congress retained. This situation persisted until 1905, when Congress assigned responsibility for Alaska's fisheries to the U.S. Bureau of Fisheries (within the Department of Commerce and Labor), which had been established two years earlier.[395] Although federal officials monitored the catch, the management of Alaska's codfish resource was not an issue; officials and the industry assumed that such fluctuations in the abundance of codfish as occurred were attributable to natural causes, and that the resource was healthy and could sustain a greatly increased harvest if markets were available.

In 1905, the Bureau of Fisheries began compiling and publishing annual reports on Alaska's commercial fisheries. Each of these reports (until 1944) contained a section that listed production and other statistics and noted significant developments in the cod fishery.

Until 1924, statistics for the offshore and the shore-station components of Alaska's codfish industry were lumped together in the bureau's annual reports on Alaska's commercial fisheries. Beginning in 1924, however, the bureau began separating the shore-station fishery statistics from those of the offshore fishery. Ward Bower, the bureau's agent, explained that "only those vessels landing their catches in the territory are considered as forming the strictly Alaska cod fleet and included in the investments in this industry. Vessels engaged in cod fishing in Bering Sea and the North Pacific Ocean are shown as the offshore cod fleet."[396]

The shore-station fishery by that time was in decline, and eventually that decline would also occur in the offshore cod fishery. Thirty years earlier, by contrast, codfish had been Alaska's fourth-largest fishery.

In 1930, the U.S. Bureau of Fisheries promulgated the only pre–Magunson-Stevens Act regulation that had the potential to limit fishing for codfish. The regulation prohibited the use of trawls in any commercial fishing operation except for catching shrimp. The regulation, which applied only to U.S. territorial waters (at that time three miles seaward of the shoreline), was relaxed in stages, and its

1942 iteration allowed the use of trawls in commercial fishing for all species except salmon, herring, and Dungeness crabs.[397]

Alaska's territorial legislature in early January 1949 created the Alaska Department of Fisheries. This department was placed under the control of the simultaneously created Alaska Fisheries Board, which consisted of five members—three fishermen, one processor, and one member of the public—who were appointed by the governor and confirmed by the legislature. The Alaska Fisheries Board set policy, which was then implemented by the Department of Fisheries. Specifically, the new department's duties were to assist in the conservation and perpetuation of the territory's fisheries resources; to promote more resident ownership, management, and control of the fisheries; and to cooperate with the U.S. Fish and Wildlife Service.[398] In 1957, Alaska's territorial legislature replaced the Alaska Fisheries Board with the Alaska Fish and Game Commission and replaced the Alaska Department of Fisheries with the Alaska Department of Fish and Game (ADF&G). Alaska became a state on January 3, 1959, and that same year Alaska's state legislature changed the name of the Alaska Fish and Game Commission to the Alaska Board of Fish and Game.[399]

In the Alaska Statehood Act, however, Congress stipulated that Alaska would not acquire control over its fisheries until the "first day of the calendar year following the expiration of 90 legislative days after the Secretary of the Interior certifies to the Congress that the Alaska State Legislature has made adequate provision for the administration, management, and conservation of said resources in the broad national interest." *Pacific Fisherman* expected the process would take three to five years, but, to the surprise of many, control of Alaska's fisheries was transferred from the U.S. Fish and Wildlife Service to the Alaska Department of Fish and Game on January 1, 1960, less than a year after Alaska became a state.[400] (Alaska's halibut fishery, managed by the federal government pursuant to a 1923 treaty with Canada, was an exception.)

Unique among the states, Alaska's constitution states that "[f]ish, forests, wildlife, grasslands, and all other replenishable resources belonging to the State shall be utilized, developed, and maintained on the sustained yield principle, subject to preferences among beneficial uses."[401]

The primary focus of the Alaska Fish and Game Commission and the Alaska Department of Fish and Game was on restoring Alaska's salmon runs, which had declined precipitously since peaking in 1936. The new department basically ignored the codfish resource because it had little economic value and presented no conservation issues.

In 1975, citing the complexity of the issues involved in both fishing and hunting/trapping, Alaska's legislature divided the Alaska Board of Fish and Game into two separate and independent entities: the Alaska Board of Fisheries and the Alaska Board of Game. The Board of Fisheries consists of seven members, each of whom is appointed by the governor, confirmed by the legislature, and serves a three-year term. The board's purpose is the conservation and development of the state's fisheries resources, and it has broad authority to adopt fisheries management regulations, which are then implemented by the Department of Fish and Game, an agency that also researches fisheries.[402]

With the passage of the Magnuson-Stevens Act in 1976 (see part 3, chapter 11), Alaska's separation of responsibilities for fisheries management into an entity that sets policy and an entity that enforces policy and researches the fisheries was duplicated at the federal level. The North Pacific Fishery Management Council (NPFMC) is an approximate equivalent of the Alaska Board of Fisheries, while the National Marine Fisheries Service (NMFS, also known as NOAA Fisheries) is an approximate equivalent of the Alaska Department of Fish and Game. Keeping the responsibilities for fisheries management separate provides a balance that helps prevent policy from becoming hidebound.

TRAWLING FOR CODFISH

As the Salt Cod Era was coming to a close in the postwar 1940s, the fishing industry looked to develop a king crab fishery in the Bering Sea and the Gulf of Alaska. The industry's focus was on the Bering Sea, where prior to the war Japanese fleets had fished with abandon. The tasty crustaceans would be caught in trawls and then frozen or canned. Marketable groundfish incidentally caught in the trawls would be filleted and frozen.

Specifically regarding cod, in January 1945—even before World War II had ended—the editors of *Pacific Fisherman* were confident that a dramatic change to Alaska's cod-fishing industry was quickly approaching. The journal noted that trawling had become the dominant method for catching cod in the Atlantic, and it expected the method would soon be extended to the Bering Sea and Alaska Peninsula waters, with a subsequent great increase in the amount of cod caught. The fishing banks of the Bering Sea, according to *Pacific Fisherman*, constituted "the largest area of trawlable waters in the world."[403]

The most ambitious postwar effort to develop the groundfish fisheries of the Bering Sea was the U.S. government–subsidized *Pacific Explorer* venture, which was conceived early in 1945. The venture revolved around the $3.5 million conversion by the U.S. government of a surplus 423-foot World War I freighter into the *Pacific Explorer*—a floating fish processor/mothership that, supplied by a fleet of fishing vessels, would, befitting its name, engage in exploratory fishing off the coast of Alaska. The *Pacific Explorer*, at the time the world's largest and most expensive fishing vessel, was equipped for both canning and freezing, and its reduction plant could produce fish meal and oil from offal and inedible fish.

The vessel was owned by the Reconstruction Finance Corporation, a U.S. government agency, and it was operated under contract by the Pacific Exploration Company. Nick Bez, a politically well-connected Alaska canneryman, was the principal individual behind Pacific Exploration, which received a $2 million government start-up loan.[404]

The *Pacific Explorer*'s maiden voyage, in January 1947, was not to Alaska but to Costa Rica in pursuit of tuna. It returned to Astoria, Oregon, during the summer with a disappointingly small cargo of frozen tuna. The vessel was refitted and in March 1948 was dispatched to the Bering Sea and Shumagin Islands, where a fleet of a dozen privately owned trawlers delivered their catches to it. The *Pacific Explorer* returned to Astoria in midsummer with a cargo of canned crabmeat, frozen crabs, frozen fillets of various groundfish, salted codfish, and fish oil and fish meal. The cargo was worth more than a million dollars, but more than $4 million had been spent in the venture, and there were more than a million dollars in operating losses. Moreover, private king crab businessmen objected to competing with a government-subsidized operation. Because of these factors, the *Pacific Explorer*'s career in the fishing industry was ended, and the Restoration Finance Corporation sold it at auction in 1951 for $181,387.[405]

A parallel but more successful venture than that of the *Pacific Explorer* was led by Lowell Wakefield, the son of a salmon canning industry executive. The company Wakefield headed, Deep Sea Trawlers (later Wakefield Fisheries), in 1946 began construction of the 140-foot *Deep Sea* at Tacoma, Washington.

The *Deep Sea* was as modern a trawler as had ever been built in the United States and was the first major trawler ever constructed in a West Coast shipyard. It was designed to make long voyages to the northern seas, where, using an otter trawl, it would target king crabs and marketable groundfish, both of which would be processed and frozen onboard. In 1947, *Pacific Fisherman* praised the *Deep Sea* as the "Craft of the Year."[406] Lowell Wakefield was later acknowledged as the pioneer of the Alaska king crab fishery.

In the fall of 1946, while the *Deep Sea* was being constructed, Wakefield's company sent the vessel *Bering Sea* (former *Penobscot*, which had been converted into a trawler in a Tacoma, Washington, shipyard) on an experimental trawling voyage to the Bering Sea. The *Bering Sea* returned in December of that year carrying king crab meat and fish fillets, but this seems to have ended its career with Wakefield.[407]

The maiden voyage of the *Deep Sea* in 1947 was considered moderately successful, the vessel returning to Bellingham, Washington, carrying 150,000 pounds of frozen product, of which about 80 percent was crabmeat and the remainder fish fillets. Most of the fillets, however, were pollock. The vessel made two more trips that year, but catches of fish were consistently light and few cod were taken. According to Wakefield's August 1947 report to his business partners, "We have caught fewer crabs and fish, produced less crabmeat and fillets, sold our product more slowly, and spent money more quickly than we ever anticipated." By about 1950, however, Wakefield's operation—based totally on king crab—became viable, and a boom in king crab fishing ensued.[408]

On a smaller scale, the Robinson Fisheries Company, of Anacortes, Washington, a firm long engaged in the salted codfish business and that had in 1940 sent the trawler *Dorothy* to fish for cod in the Bering Sea (see chapter 5), entered into a partnership with Merton Soma to purchase and convert a surplus World War II navy minesweeper into a trawler that was capable of salting, canning, and freezing its catch. The 136-foot *Lynn Ann* was ready to sail in May 1948. It would target cod for salting but would process crabs taken incidental to cod fishing. Although the vessel's captain, Ellsworth Trafton, expressed confidence that trawling was "today's method of catching cod," the venture seems to have lasted only one season.[409]

As gear for catching king crabs, trawls had what the Alaska Department of Fish and Game termed "destructive non-selectivity": they injured or killed large numbers of female crabs, which had never been legal to retain in the domestic fishery. In 1960 (the year Alaska took over management of its fisheries), the department, concerned about the impact of trawling on crab stocks, banned trawling for king crabs in state waters around Kodiak Island. State waters then were coterminous with the U.S. territorial sea (three-mile limit). Trawling for king crabs was banned in other state waters the following year.

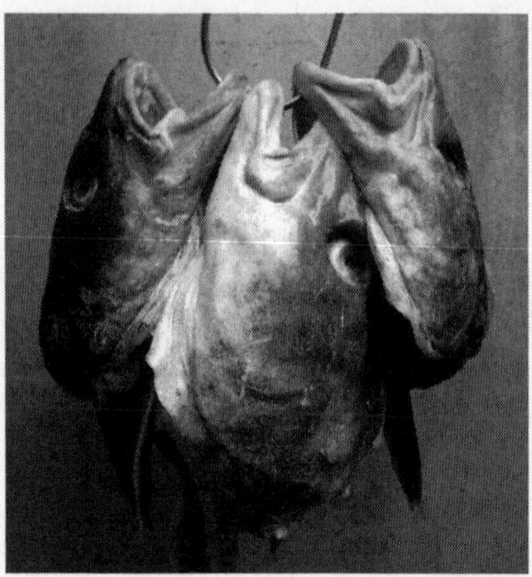

Figure 52. Hanging bait. (Courtesy of Alaska Glacier Seafoods.)

Most of the king crab catch, however, was made beyond the three-mile limit, where neither the state nor the federal government had jurisdiction. To circumvent this management handicap, the state enacted a landing law that made it illegal to land crabs caught by any method except pots. In banning the use of trawls to catch king crabs, the state's exercise of extraterritorial authority effectively precluded for many years any serious domestic effort to catch Pacific cod, even as a secondary target, for food fish.[410]

BAIT COD, MOSTLY

> *Over the next three decades, cod was relegated to the back seat. It was caught either incidentally by the foreign fleet or in isolated pockets by U.S. fishermen, to be sold to crabbers for hanging bait.*
>
> —A. D. Chandler, *National Fisherman*, 1982[411]

Mention of Pacific cod in the 1952 U.S. Fish and Wildlife Service report on Alaska's fisheries was brief: "There was no codfish production in 1952, as in 1951, because of poor market conditions." Domestically, salted cod was continuing to lose favor, while attractive and easy-to-prepare frozen fish was gaining. This trend accelerated as refrigerators and freezers became increasingly commonplace in American homes. Nationwide, Atlantic cod was the whitefish of choice.[412]

There was no mention of cod in the service's 1953 and 1954 reports, but the 1955 report noted as an "unusual feature" the freezing of 102,627 pounds of cod for bait in Southeastern Alaska. The cod was likely used to catch halibut and crabs. Additionally, some 4,700 pounds of cod was frozen for food. This freezing of substantial amounts of cod for bait was apparently a one-off event: the U.S. Fish and Wildlife Service reported that only 5,557 pounds were frozen in 1956.[413]

Cod production listed in the Alaska Department of Fish and Game reports show production declining from 15,443 round pounds (worth $772.00) in 1957 to 3,618 round pounds (worth $181.00) in 1959. In 1960, the department began including cod in the general category of *bottomfish*, a broad category that included lingcod, flounder, rockfish, and about any other finfish species that was caught near the seafloor and did not have its own listing.[414]

Cod was truly off the agency's and the fishing industry's radar. For the next two decades, a few cod would be caught and frozen as bait to catch halibut and crabs. Additionally, cod incidentally caught in the halibut fishery would be welcomed as *gurdy bait* (fish suitable for bait that came up on a halibut-fishing vessel's longline, which was hauled by a winch referred to as a power gurdy) much as the dorymen during the Salt Cod Era used incidentally caught halibut as bait for cod (see chapter 5). Likewise, cod that found their way into crab pots were welcomed and utilized as hanging bait. Similarly, some king crab boats during the 1950s, when the king crab pot fishery was just beginning, carried a small otter trawl that was used to catch cod for bait. This catch cod by individual vessels for use as their own bait was not recorded. During the mid-1960s, however, a few crab boats around Kodiak occasionally targeted cod during the off-season. They sold their catch, which never amounted to much, to a local crab processor, who presumably froze the fish and later sold them to crab fishermen as bait.[415]

AMERICAN DISINTEREST

According to John Wiese, editor of *Alaska Industry* magazine, in 1973 there was an "utter absence of interest" in domestic utilization of Bering Sea and Gulf of Alaska groundfish stocks despite the fact that domestic consumption of groundfish had doubled between 1946 and 1966, and had increased an additional 25 or 30 percent between 1966 and 1973. Wiese attributed this lack of interest to U.S. policy during the Cold War.

According to Wiese, the U.S. policy was to extend economic favors to North Atlantic nations—among them Canada, Iceland, and Norway—in exchange for military concessions, such as port facilities and air bases. The U.S. goal was to maintain a defense posture that would keep Russia and its European allies in check. Because of the U.S. aid, fisheries production in the North Atlantic nations became cheaper than in the United States. Recognizing an opportunity, U.S. seafood distributors made substantial investments and financial commitments in these nations, and they were unwilling to invest in Alaska's groundfish fisheries.[416] And there was more: East Coast interests that produced and sold Atlantic cod continued their century-long campaign to portray Pacific cod as an inferior product.

COD: A LOVE POEM

By Amy Zidulka

I love Charlie because Charlie loves cod
And because, with steady faith,
he loved it
even back when cod was getting 15 cents
a pound.
The crabbing guys would laugh
at Charlie's fried cod dinners;
"Why you eating hanging bait?" they'd
ask.

Cod are hooked without fanfare.
Dentists don't fly up from Wichita
to pose beside a gashed dead cod.
Cameras don't roll
as the first cod of the year
is flown south
to kitchens of Seattle's finest
restaurants.
Cod don't race against the current
toward orgies
of sexual ecstatic deaths.

To know cod is to know mud.
Charlie's eyes can see straight through
the thick ocean skin
of turbulence and waves
straight down
to the soft quiet floor
where cod schools bed.
I love being looked at with those eyes.
Being touched with blunt excited cod
man's hands.[417]

ENDNOTES

[395] Hugh M. Smith (Commissioner of Fisheries), "The Bureau of Fisheries and Some Piscatorial History," *Fishing Gazette* (July 6, 1918): 940; U.S. Fish and Wildlife Service, *History of the U.S. Fish and Wildlife Service*, accessed January 8, 2012, http://www.fws.gov/policy/029fw1.html.

[396] Ward T. Bower, *Alaska Fishery and Fur-Seal Industries in 1924*, Bureau of Fisheries Doc. No. 992 (Washington, DC: GPO, 1925), 137; Ward T. Bower, *Alaska Fishery and Fur-Seal Industries: 1944*, Statistical Digest No. 13 (Washington, DC: GPO, 1946), 53. The first of the Bureau of Fisheries annual reports on Alaska's commercial fisheries was John Cobb's *The Commercial Fisheries of Alaska in 1905*, Bureau of Fisheries Document No. 603.

[397] R. A. Fredin, *History of Regulation of Alaska Groundfish Fisheries*, NOAA NWAFC Processed Report 97-07 (March 1987), 9.

[398] Chap. 68, *Session Laws of Alaska, 1949*.

[399] Alaska Board of Fish and Game and Alaska Department of Fish and Game, Annual Rept., 1959, 3.

[400] *Alaska Statehood Act*, 72 Stat. 339 (P.L. 85-508), §6(e); "Statehood and Alaska's Fisheries," *Pacific Fisherman* (August 1958): 15–16.

[401] *Constitution of the State of Alaska*, 1959, Article 8, §4.

[402] Chap. 206, *Session Laws of Alaska, 1975*; Alaska Department of Fish and Game, "Welcome to the Board of Fisheries," accessed January 19, 2016, http://www.adfg.alaska.gov/index.cfm?adfg=fisheriesboard.main; Karl Ohls, "Alaska Fisheries Board Faces Mounting Pressures," *National Fisherman*, March 1984, 14.

403 "Future of Codfishing Will See Sweeping Change," *Pacific Fisherman* (January 1945): 278; "Deep Sea—A Pioneering Venture," *Pacific Fisherman* (June 1947): 36–38, 40, 43, 45–46.

404 "American Venture Trawling in Bering Sea for Crab and Cod," *Pacific Fisherman* (May 1948): 23–24; "First US King Crab was Packed in 1920, Industry Grew After World War II," *Pacific Fisherman* (June 1965): 23–27; Carmel Finley, "The Forgotten Story of Nick Bez and the Pacific Explorer," accessed September 27, 2014, http://carmelfinley.wordpress.com/2013/09/27/the-forgotten-story-of-nick-bez-and-the-pacific-explorer/.

405 Carmel Finley, "Pacific Explorer—the First Trip—South America," accessed February 13, 2015, https://carmelfinley.wordpress.com/2013/10/05/pacific-exporer-the-first-trip-south-america/; "Pacific Explorer Discharges 20,000 Cases of King Crab," *Pacific Fisherman* (August 1948): 4; Carmel Finley, "The Pacific Explorer returns to Astoria in 1947," accessed September 27, 2014, http://carmelfinley.wordpress.com/2013/10/29/the-pacific-explorer-returns-to-astoria-in-1947/.

406 "Deep Sea—A Pioneering Venture," *Pacific Fisherman* (June 1947): 36–38, 40, 43, 45–46; "Directions of Development in the Fleet, in the Plants [and] in the Products," *Pacific Fisherman* (January 1947): 147.

407 "Bering Sea, Trawler, Trawling Bering Sea," *Pacific Fisherman* (November 1946): 61; "Trawler Bering Sea Completes First Trip," *Pacific Fisherman* (January 1947): 69.

408 "Deep Sea's Maiden King Crab Trip Moderately Successful," *Pacific Fisherman* (September 1947): 50; "Codfish," *Pacific Fisherman* (January 1948): 237; Lowell Wakefield to stockholders, August 11, 1947, in: Mansel G. Blackford, *Modern Small Business: Wakefield Seafoods and the Alaskan Frontier* (Greenwich, Conn.: Jai Press, 1979), 14, 19.

409 "Converted YMS to Trawl Codfish in Bering Sea," *Pacific Fisherman* (April 1948): 1; "American Venture Trawling in Bering Sea for Crab and Cod," *Pacific Fisherman* (May 1948): 23–24; "They'll Trawl and Salt Codfish in Bering Sea," *Pacific Fisherman* (May 1948): 59, 61.

410 Bradley G. Stevens, ed., *King Crabs of the World: Biology and Fisheries Management* (Boca Raton, Fla.: CRC Press, 2014), 111, 113; George W. Gray Jr., Robert S. Roys, Robert J. Simon, and Dexter F. Lall (Alaska Department of Fish and Game), *Development of the King Crab Fishery Off Kodiak Island*, ADF&G, Informational Leaflet No. 52 (April 12, 1965), 3, 13; University of Alaska, Institute of Business, Economic and Government Research, "The Alaskan King Crab Industry," *Review of Business and Economic Conditions* (November 1965): 6.

[411] A. D. Chandler, "Alaska Salt Cod Industry Could Help Offset Crab Bust," *National Fisherman*, February 1982, 19.

[412] Seton H. Thompson, *Alaska Fishery and Fur-Seal Industries: 1952*, Statistical Digest No. 33 (Washington, DC: GPO, 1954), 51.

[413] Seton H. Thompson, *Alaska Fishery and Fur-Seal Industries: 1955*, Statistical Digest No. 40 (Washington, DC: GPO, 1960), 67; Seton H. Thompson and Donald W. Erickson, *Alaska Fishery and Fur-Seal Industries: 1956*, Statistical Digest No. 45 (Washington, DC: GPO, 1960), 64.

[414] Alaska Department of Fish and Game, Annual Rept., 1958, 117; Alaska Department of Fish and Game, Annual Rept., 1959, 112; Alaska Department of Fish and Game, Progress Rept., 1960–1962, 60.

[415] Al Burch, personal communication with author, April 6, 2015; Robert Lochman, personal communication with author, March 13, 2016.

[416] John Wiese, "New Operation May Harvest Alaska's Groundfish," *Alaska Industry* (May 1973): 24, 54, 58–60.

[417] Amy Zidulka, "Cod: A Love Poem," *Pacific Fishing* (December 2001): 26; Amy Zidulka, personal communication with author, September 17, 2018.

Chapter 10
ENJOYING CODFISH: FOUR RECIPES

Alaska cod is a choice seafood with a tender, snow-white flesh, mild flavor and fine, firm texture that flakes readily when cooked. A very lean whitefish, it is low in fat and calories, high in protein and very digestible.

—Alaska Seafood Marketing Institute[418]

Figure 53. Alaska Seafood Marketing Institute cod brochure image, September 1984.

PACIFIC COD PARMESAN

(Courtesy of Gustavus Inn, Gustavus, Alaska.)

1. Cut skinless/boneless fillets into serving-size pieces.
2. Mix enough flour, salt, and pepper to coat fish.
3. Beat enough eggs to coat fish.
4. Mix three parts bread crumbs with one part Parmesan cheese.
5. Roll fish pieces in flour mixture, dip into beaten egg, then coat with bread crumb/cheese mixture.
6. Place on greased cookie sheet and bake at 400° for 10 to 15 minutes. The fish is done when an instant-reading thermometer reads 120° in the thickest part.
7. Serve with lemon wedges and seafood sauce.

CODFISH CAKES

(Courtesy of Koren Bosworth, Juneau, Alaska.)

1. **Binder (continues below).** Chop a small potato (about the size of two eggs) and cook with a little water in microwave oven. Add a little butter, mash, and allow to cool.
2. **Vegetables.** Chop into small pieces and mix together:
   - celery, about 1/2 cup
   - onion, about 1/8 cup
   - bell pepper (the red variety enhances the visual appeal of the cakes), about 1/2 cup
   - 3 to 4 medium chives
   - carrot, about 1/2 cup
   - mung bean sprouts, about 1/2 cup
3. **Codfish.** Cut about one pound of skinless/boneless cod fillets into 1/4-inch cubes and add to vegetable mixture.
4. **Binder.** Mix together:
   - mashed potato
   - 1/4 cup mayonnaise or substitute
   - 1 tablespoon spicy brown mustard
   - 1 beaten egg
5. Add binder to fish/vegetable mixture and mix thoroughly.
6. Add about 1 cup bread crumbs.
7. Shape into patties about 3/4 inches thick, pepper, and fry lightly in canola oil or similar in cast iron pan, flipping patties once. Do not overcook.
8. Season to taste with kosher salt.

PAN-FRIED CODFISH WITH GINGER SALSA

(Courtesy of Nadine Oliver, Seattle, Wash.)

1. **Salsa.** Chop into small pieces and mix together:
   - onion to make about 3/4 cup
   - bell pepper to make about 3/4 cup (the red variety enhances the visual appeal of the salsa)
   - fresh ginger to make about 1/2 cup
   - Sauté lightly in olive oil in a small cast iron pan and set aside.

2. **Codfish.**
    - Cut about one pound of skinless/boneless codfish fillets into serving-size pieces. (Fillets less than about 3/4-inch thick work best.)
    - Pan fry fish in olive oil in cast iron pan, flipping once. Do not overcook.
    - Transfer codfish to serving plate, cover with salsa, then drizzle with lime juice.

CODFISH CADDY GANTY

(Based on a Gustavus Inn, at Gustavus, Alaska, halibut recipe.)

1. **Potatoes.**
    - Slice two pounds of potatoes into 1/4-inch-thick slices.
    - Parboil until potatoes are barely cooked, then set aside.
2. **Codfish.**
    - Lightly salt two pounds of skinless/boneless codfish fillets and place in bowl.
    - Add enough dry white wine to cover fillets, then cover with waxed paper and allow to marinate in a cool place for two hours.
    - Drain the fillets and pat dry with paper towels or a clean cloth.
    - Roll the fillets in dry bread crumbs. (Crumbs made from sourdough French bread enhance the dish's flavor.)
3. **Sauce.**
    - Mix two cups sour cream, one cup real mayonnaise, and one cup chopped onion.
4. **Preparation.**
    - Lightly butter an oven-safe serving dish that can be brought to table.
    - Place a layer of potatoes on bottom of dish.
    - Place codfish atop potatoes.
    - Cover potatoes and fish with sauce.
    - Sprinkle with paprika.
    - Bake at 500 degrees for 15 to 20 minutes or until light brown and bubbly.
    - Serve immediately.[419]

ENDNOTES

[418] Alaska Seafood Marketing Institute, *Alaska Cod: Handling and Preparation of Alaska Cod*, 1984.

[419] Sally Lesh, *A Collection of Recipes from Gustavus Inn* (Gustavus, Alaska: Gustavus Inn, 1973), 1.

Part Three

THE MODERN ERA

SECTION A

The Magnuson-Stevens Act, Its Implementation, and Early Efforts to Process Pacific Cod

INTRODUCTION

The Pacific cod, G. macrocephalus, something of a Johnny-come-lately to the world's attention, has never achieved the high place held by its Atlantic cousin although it has contributed to a substantial fishery through the years.

—Robert J. Browning, 1974[420]

In the cod business, "[y]ou have to calculate your profits in terms of pennies, not dollars.

—Dick Nelson, fish utilization researcher, National Marine Fisheries Service, 1980[421]

With codfish, there is no romance, no big spreads, no big killings, and not much excitement.

—Trans-Pacific Industries, Inc., operator of *Arctic Trawler*, 1981[422]

In entering the cod industry, the Pacific coast is stepping into a huge, highly competitive world trade. All of the established competitors have vast experience and huge investments in their market areas.

—Doug McNair, *Pacific Fishing*, 1982[423]

Pacific coast codfish companies during the Salt Cod Era were small businesses that rarely operated in concert and were not politically active. There seemed to have generally been an abundance of codfish in Alaska's waters, and fishing for them was typically a totally unregulated activity. The only exceptions were the years 1930–1942, during which trawling was banned.[424] Conservation was not an issue, mainly because of the limitations of the type of gear utilized and the relatively modest fishing effort.

For almost three decades following the end of the Salt Cod Era, American fishermen exhibited little interest in Alaska's groundfish resources, Pacific cod among them. By contrast, the Japanese had begun fishing for groundfish off Alaska's coast in the mid-1930s, and they were later joined by other nations. During the early 1970s, the foreign fleets were catching some 2 million metric tons of groundfish annually in the eastern Bering Sea alone. Most of the catch was walleye pollock (*Gadus chalcogrammus*), but it included substantial quantities of Pacific cod as well.[425] In essence, the groundfish fishery in the Bering

Sea had become a high-seas, industrial-scale international fishing derby in which the conservation of fish resources was not a consideration.

The situation changed dramatically in 1976, when the U.S. Congress, concerned about overfishing and sensing an economic opportunity for American fishermen, passed the Fishery Management and Conservation Act of 1976, informally known now as the Magnuson-Stevens Act. The legislation unilaterally extended U.S. jurisdiction over its coastal marine waters to a line two hundred nautical miles from the coastline—an area designated as the Fishery Conservation Zone—"to prevent overfishing, to rebuild overfished stocks, [and] to insure conservation." The Magnuson-Stevens Act also contained an economic mandate: to *Americanize* the fisheries in the newly claimed Fishery Conservation Zone in order to realize the full potential of the U.S. coastal fisheries.[426] The Pacific cod fishery's modern history is a key element within the larger history of the development of the American groundfish fishery off the coast of Alaska.

> **Weather Report:** *Conditions in the Bering Sea were expected to include winds from 35 to 70 knots, with seas between 21 and 50 feet. On Sunday evening, [National Weather Service forecaster Michael] Kutz said a buoy in the western Aleutians had recorded seas of up to 53 feet overnight Saturday near the height of the storm.*
>
> *"Earlier this morning, seas 40 feet and higher stretched all the way across the Bering," Kutz said. "It's still 20-footers out there, which is nothing to sneeze at."*
>
> —*Alaska Dispatch News*, December 13, 2015[427]

Americanizing the groundfish fisheries off the coast of Alaska was no small task. The domestic fishing industry needed to build a fleet of capable vessels and shore-based processing plants; learn how to catch, process, and market pollock, Pacific cod, and other groundfish; and develop an infrastructure to support fishing and fish processing in one of the most remote and harshest environments on Earth—the Bering Sea and the central and western Gulf of Alaska.[428] Because this vast effort was fraught with uncertainty, obtaining the billions of dollars required to finance it was a challenge. And, since the U.S. goal involved the displacement of foreign fishing fleets, geopolitical considerations could not be ignored. Moreover, on short notice, federal regulators were tasked with managing fisheries about which they knew little, and which, in the words of a report prepared by the Fisheries Management Foundation and the Fisheries Research Institute, presented "unusually complex management problems." Rick Lauber, chairman of the North Pacific Fishery Management Council from 1990 to 2000, was more specific: "Never before in the history of the United States has a management plan involved such a volume of fish, value of fish, and so many people," he said.[429]

Alaska's most complex fishery

The present-day Alaska Pacific cod fishery comprises a number of distinct fisheries that extend from Southeast Alaska west across the Gulf of Alaska and into the Aleutian Islands and Bering Sea. It is arguably Alaska's most complex fishery, involving a host of interrelated, sometimes controversial, and often chronologically overlapping issues. At times, all three branches of the federal government—executive, legislative, and judicial—have been actively involved in managing Alaska's federal-waters Pacific cod fishery. For fishery managers, the process of monitoring fisheries in this geographically vast area is a major challenge. The fact that much of the fishing is conducted during the usually stormy winter months adds to the difficulty. Moreover, while Alaska's pollock fishery is significantly larger than its Pacific cod fishery, pollock are targeted by only one gear type: mid-water (pelagic) trawls. By contrast, Pacific cod are targeted by fully eight gear/vessel types that range from outboard motor–powered skiffs using jig gear to massive factory trawlers.[430] Potentially further complicating the fishery's management is recent evidence that the eastern Bering Sea may be home to several distinct spawning populations of Pacific cod.[431]

The fishing gear used to catch Pacific cod in Alaska's waters is classified either as *mobile gear*, which is actively moved through the water, or as *fixed gear*, which fishes while stationary on the seafloor. Trawls are mobile gear, while longlines, pots, and jig gear are classified as fixed gear (even though jig gear is not stationary). Some trawl, longline, and pot vessels are equipped to process their catch onboard.

Vessel size categories established for regulatory purposes and to broaden participation in the fishery further complicate the situation, as does the fact that shore-based fish-processing plants in the Gulf of Alaska receive a collective allocation of Pacific cod, as do groups of Native villages in western Alaska.

Jurisdictional issues add another complicating element. The management of the Pacific cod fishery in the federal waters off Alaska's coast is the responsibility of the North Pacific Fishery Management Council, one of the eight regional councils created under the Magnuson-Stevens Act. The National Marine Fisheries Service (also known as NOAA Fisheries), a federal agency within the Department of Commerce, is responsible for the implementation of the management plans devised by the council. The Alaska state-waters fishery, conducted within three miles of Alaska's coast, is managed by the Alaska Board of Fisheries and the Alaska Department of Fish and Game, often in conjunction with federal officials who manage the Pacific cod fisheries in adjacent federal waters.

The main fishing sector of Alaska's Pacific cod fishery, the freezer-longliner fleet, is limited entry and harvests its allocation of the total allowable catch as a cooperative, with special exemptions from anti-trust laws. In other sectors, such as the trawl catcher-vessel fleets in the Gulf of Alaska and Bering Sea, the fisheries are limited entry and *derby-style*—characterized by short seasons and intense competition for fish. (Derby-style seasons may be divided to help spread the catch throughout the year.) Other cod fisheries, such as the jig and pot fisheries in state waters, remain open-access and derby-style, primarily to provide opportunities for local fishermen.

The controversial issue of bycatch of halibut and Chinook salmon is ever present. Exceeding a bycatch quota can shut down the Pacific cod fishery in a regulatory area.

The Endangered Species Act is at play as well. Fishing is restricted or prohibited in areas designated as important for the endangered population of Steller sea lions. Moreover, the killing during a two-year period of just four short-tailed albatross, an endangered species that sometimes becomes hooked and drowns while attempting to feed on baited hooks as they are being set by longliners, can shut down the Pacific cod longline fishery.

Some of the issues faced by the Pacific cod fishery have been resolved through the regulatory process. Others have been resolved by Congress, or in court, or by the market. Some issues, such as bycatch, are intractable and may never be resolved in a manner satisfactory to all concerned.

Despite its challenges and its complexity, Alaska's Pacific cod fishery is considered among the best-managed fisheries in the world. The fishery is certified as sustainable by the Marine Stewardship Council, a London-based independent fishery certification organization, and by Global Trust, which granted certification in accordance with the United Nations Food and Agriculture Organization's Code of Conduct for Responsible Fisheries. The highly regarded SeafoodWatch program of the Monterey Bay Aquarium, in California, gives Pacific cod caught in Alaska its highest recommendation.[432]

Big money, big politics

Although there were exceptions, engaging in the groundfish fisheries off Alaska's coast wasn't the sort of enterprise an individual or small business with modest resources could accomplish, at least not without substantial help. The capital requirements were intimidating: for example, a large catcher-processor could cost tens of millions of dollars, after which the vessel needed to be fueled, provisioned, and crewed. Moreover, months might pass before the vessel's production reached market. As a 1993 article in *Pacific Fishing* observed, the Alaska groundfish industry "typically favored large, well-heeled entrepreneurs," and the post-Magnuson-Stevens Act era groundfish industry in Alaska is today dominated by medium-sized corporations whose success depends mostly on the efficient employment of sophisticated capital-intensive technology.[433] Some of these firms have prospered impressively. Seattle-based Trident Seafoods, for example, is the largest U.S. seafood processing company, and its Akutan plant in the Aleutian Islands, capable of processing more than 3 million pounds of fish per day, is the largest seafood processing plant in North America.[434]

Alaska's groundfish fishery, of which the Pacific cod fishery is a major component, has made and continues to make large long-term investments in the fisheries, and it understands well that its future depends upon sound management of the fisheries. Beyond this, the industry works constantly and by every conceivable means to ensure the fisheries are conducted in the most economically advantageous manner possible. However, because of competition among gear groups, regions, and so on, the industry does not always speak with a unified voice.

The primary venue for management proposals and action are the North Pacific Fishery Management Council's meetings, which can be contentious and polarized. Also, Congress is lobbied constantly. In addition to amending the Magnuson-Stevens Act a number of times, Congress has passed several stand-alone bills specific to Alaska's groundfish fisheries. Early post-Magnuson-Stevens Act bills were mostly of a general nature—"prevent overfishing," for example—but later legislation was specific and sometimes listed individual vessels for special treatment. Alaska senator Ted Stevens, who served in the U.S. Senate from December 1968 until January 2009, was intimately involved in each piece of legislation that affected Alaska's groundfish fisheries.

Alaska's governor, legislature, and Board of Fisheries are also heavily lobbied by groundfish interests. Jay Hammond, Alaska's governor from December 1974 to December 1982, established the position of Bottomfish Coordinator in his office, and Walter Hickel, in his second term as governor (December 1990 to December 1994) had a "Fish Czar," who focused primarily on groundfish.[435] Though recent governors have not had formal groundfish advisors, all have had people knowledgeable about fisheries on their staffs and could always rely upon the expertise at the Department of Fish and Game and other state agencies, as well as the University of Alaska.

Codfish: The Big Picture

During the 1960s and 1970s, the annual worldwide harvest of Atlantic cod exceeded 2.5 million metric tons, while worldwide annual harvest of Pacific cod was less than 200,000 metric tons. After peaking in 1969, however, Atlantic cod harvests began a dramatic decline. Today, although the Atlantic cod catch on the Georges Banks and other historical New England cod-fishing grounds remains at near-record lows and the Canadian Maritimes cod fishery is in a state of collapse as a result of overfishing, production in the Northeast Atlantic is fairly strong. A bright spot has been the Barents Sea, where mostly Russian and Norwegian vessels in 2013 shared an Atlantic cod harvest of a million metric tons. The cod harvest in the Barents Sea in 2017 was 775,000 metric tons.

In 1980, the U.S. Atlantic cod catch was more than 50,000 metric tons—almost the entire domestic cod catch—but by 2004 the catch had declined to less than 10,000 metric tons. During that same period, however, the annual catch of Pacific cod—virtually all of which was made off the coast of Alaska—had increased from a relatively nominal amount to more than 200,000 metric tons. Today, Pacific cod accounts for more than 95 percent of the U.S. domestic cod harvest.[436]

Internationally, until the 1980s, Japan, whose fleet fished throughout the North Pacific, accounted for most of the world Pacific cod harvest. During the early 1980s, harvests by the Soviet Union in the western North Pacific and the United States in the eastern North Pacific increased rapidly. Later that decade, however, harvests by both Japan and the Soviet Union/Russia began declining, and by 2005 were about half their historical highs. U.S. harvests, in the meantime, remained relatively stable, and by 2005 the U.S. catch accounted for more than two-thirds of the world Pacific cod catch. Worldwide Pacific cod production in 2014 was estimated to be 462,000 metric tons; Alaska production that year was 334,164 metric tons. In 2017, Alaska's Pacific cod production declined to an estimated 284,000 metric tons, about

65 percent of the estimated 435,000–metric ton worldwide production. Meanwhile, Russian production is increasing significantly.[437]

Because the Pacific cod catch in recent years accounts for only about 16 percent of the total amount of cod caught worldwide (an estimated 1,821,000 metric tons in 2014), variations in Alaska's Pacific cod production have relatively little effect on the price and demand for Atlantic cod.[438]

ENDNOTES

[420] Robert J. Browning, *Fisheries of the North Pacific: History, Species, Gear & Processes* (Anchorage: Alaska Northwest Publishing Company, 1974), 95.

[421] "Born Again: A Resurrected Codfishery," *Pacific Fishing* (December 1980): 44–47.

[422] Jon Sabella, "Arctic Trawler Brings Home the Bacon: 1.5 Million Pounds of Cod Fillets and Insight into Bottomfish," *Pacific Fishing* (December 1981): 35–36, 38, 65.

[423] Doug McNair, "Bottomfish Update," *Pacific Fishing* (November 1982): 65–71, 119.

[424] R. A. Fredin, *History of Regulation of Alaska Groundfish Fisheries*, NOAA NWAFC Processed Report 97–07 (March 1987), 9.

[425] Russell Nelson Jr., Robert French, and Janet Wall, "Sampling by U.S. Observers on Foreign Fishing Vessels in the Eastern Bering Sea and Aleutian Islands Region, 1977–78," *Marine Fisheries Review* (May 1981): 1.

[426] Magnuson-Stevens Fishery Management and Conservation Act, April 13, 1976 (90 Stat. 331), §2(a)(6).

[427] Chris Klint, "Monster Bering Sea storm lashes at Aleutians, Western Alaska," *Alaska Dispatch News*, December 13, 2015, accessed December 14, 2015, http://www.adn.com/article/20151213/monster-bering-sea-storm-lashes-aleutians-western-alaska.

[428] In the North Pacific, *groundfish* is a loose term used to collectively describe about a dozen low-valued, mid-water-to-bottom-dwelling species of finfish. Among them are walleye pollock, Pacific cod, yellowfin sole, rock sole, Atka mackerel, and Pacific Ocean perch. The term *bottomfish* is generally synonymous with *groundfish*, as is *whitefish* when used in this context.

429 Daniel D. Huppert, *Managing Alaska Groundfish: Current Problems and Management Alternatives* (Seattle: Fisheries Management Foundation and Fisheries Research Institute, 1988), 1; Laine Welch, "Council 'Freaks' Over ITQs," *Pacific Fishing*, Yearbook (1994): 16, 18.

430 The eight gear/vessel types that target Pacific cod in Alaska are: trawl catcher vessels, trawl catcher-processor vessels, hook-and-line catcher vessels (longliners), hook-and-line catcher-processor vessels (freezer-longliners), pot catcher vessels, pot catcher-processor vessels, jig catcher vessels, and jig catcher-processor vessels.

431 Ingrid Spies, "Landscape Genetics Reveals Population Subdivision in Bering Sea and Aleutian Islands Pacific Cod," *Transactions of the American Fisheries Society*, 2012, 141(6): 1557–1573, accessed November 11, 2017, http://dx.doi.org/10.1080/00028487.2012.711265; Sandra K. Neidetcher, Thomas P. Hurst, Lorenzo Ciannelli, and Elizabeth A. Logerwell, "Spawning phenology and geography of Aleutian Islands and eastern Bering Sea Pacific cod (*Gadus macrocephalus*)," *Deep Sea Research, Part II: Topical Studies in Oceanography* (November 2014): 204–214, accessed November 11, 2017, http://www.sciencedirect.com/science/article/pii/S0967064513004529?via%3Dihub.

432 Marine Stewardship Council, "Pacific Cod, Gulf of Alaska," accessed September 18, 2015, https://www.msc.org/track-a-fishery/fisheries-in-the-program/certified/pacific/gulf-of-alaska-pacific-cod/fishery-name/?searchterm=Pacific%20cod; Marine Stewardship Council, "Pacific cod, Bering Sea, and Aleutian Islands," accessed September 18, 2015, https://www.msc.org/track-a-fishery/fisheries-in-the-program/certified/pacific/bering-sea-and-aleutian-islands-pacific-cod/fishery-name/?searchterm=Pacific%20cod; Global Trust Certification, Ltd., *FAO-Based Responsible Fisheries Management Certification Full Assessment and Certification Report for the Alaska Pacific Cod Commercial Fisheries (200 mile EEZ), April 2013*; Monterey Bay Aquarium, SeafoodWatch, "Pacific Cod," accessed September 18, 2015, http://www.seafoodwatch.org/seafood-recommendations/groups/cod.

433 Joel Gay, "What's This New CDQ?" *Pacific Fishing* (February 1993): 45.

434 Trident Seafoods is involved in all of Alaska's major fisheries.

435 Alaska Sea Grant, *Fishery Education in Alaska, Conference Report*, Alaska Sea Grant Rept. No. 82–4 (Juneau: University of Alaska, December 1980), iii.

436 Gunnar Knapp, Institute of Social and Economic Research, University of Alaska Anchorage, *Selected Market Information for Pacific Cod* (January 12, 2006), 3; John Waldman, "How Norway and Russia Made a Cod Fishery Live and Thrive," *Yale Environment 360* (Yale School of Forestry & Environmental Studies), September 18, 2014, accessed March 13, 2019, http://e360.yale.edu/feature/how_

norway_and_russia_made__a_cod_fishery_live_and_thrive; Thomas Nilsen, "Scientists advise slashing Barents Sea cod, haddock quotas," *Barents Observer*, June 15, 2018, accessed November 5, 2018, https://thebarentsobserver.com/en/ecology/2018/06/scientists-advise-slashing-barents-sea-cod-quotas.

437 James B. Marsh, *Resources & Environment in Asia's Marine Sector* (Washington: Taylor & Francis New York, 1992), 32–33; Gunnar Knapp, Institute of Social and Economic Research, University of Alaska Anchorage, *Selected Market Information for Pacific Cod* (January 12, 2006), 4; Ben Fissel et al., *Stock assessment and fishery evaluation report for the groundfish fisheries of the Gulf of Alaska and Bering Sea/Aleutian Islands area: economic status of the groundfish fisheries off Alaska, 2012* (Seattle: National Marine Fisheries Service, 2014), 248, accessed December 9, 2015, http://www.afsc.noaa.gov/REFM/stocks/plan_team/economic.pdf; NOAA, FishWatch, "Pacific Cod," accessed September 18, 2015, http://www.fishwatch.gov/seafood_profiles/species/cod/species_pages/pacific_cod.htm; Ben Fissel et al., *Stock Assessment and Fishery Evaluation Report for the Groundfish Fisheries of the Gulf of Alaska and Bering Sea/Aleutian Islands Area: Economic Status of the Groundfish Fisheries off Alaska, 2014* (Seattle: National Marine Fisheries Service, 2015), 26, accessed December 9, 2015, http://www.afsc.noaa.gov/REFM/stocks/plan_team/economic.pdf; Tom Seaman, "Groundfish Forum Forecasts 186,000t Drop in Wild Supply in 2019," *Undercurrent News*, October 10, 2018, accessed November 5, 2018, https://www.undercurrentnews.com/2018/10/10/groundfish-forum-forecasts-186000t-drop-in-wild-supply-in-2019/.

438 Alaska Fisheries Science Center, *Wholesale Market Profiles for Alaska Groundfish and Crab Fisheries* (May 2016), 42; Alaska Seafood Marketing Institute, *Seafood Market Bulletin, Spring 2014*, accessed October 13, 2015, http://www.alaskaseafood.org/industry/market/seafood_spring14/2014-alaska-seafood-market-outlook.html; Laine Welch, "Gulf of Alaska cod at record low levels," *National Fisherman*, October 12, 2017, accessed November 5, 2018, https://www.nationalfisherman.com/alaska/survey-shows-gulf-alaska-cod-lowest-levels-ever/.

Chapter 11
THE MAGNUSON-STEVENS ACT

> *The fact that almost seventy percent of the fish caught off the coasts of the United States is taken by foreign fishermen is not in and of itself the most disturbing factor. Rather, it is the fact that foreign fishermen are highly efficient and mobile and can move to other parts of the world if they overfish United States waters. With the use of huge factory vessels and large fleets of smaller fishing boats that deliver their catch to the processing vessels, the foreigners have been virtually vacuuming the seas of precious life and economic value.*
>
> —Senator Warren Magnuson, 1977[439]

The Modern Era of the Alaska cod fishery officially began with the passage of the Fishery Conservation and Management Act of 1976, commonly referred to early on as the Magnuson Act, in recognition of Washington senator Warren Magnuson, the chief architect of the legislation. As part of a 1996 amendment, the legislation was officially retitled the Magnuson-Stevens Fishery Conservation and Management Act in recognition of the efforts and contributions of both Senator Magnuson and Alaska senator Ted Stevens. Today, the legislation is most commonly referred to as the Magnuson-Stevens Act.

The Magnuson-Stevens Act created a 197-nautical-mile-wide *Fishery Conservation Zone* (FCZ) in federal waters along the entire coast of the United States. The FCZ's seaward boundary is 200 nautical miles from the coastline; its shoreward boundary is 3 nautical miles from the coastline—the seaward boundary of state waters.

Despite the slight inaccuracy, the FCZ is almost uniformly referred to as being 200 nautical miles wide—the combination of state and federal waters along the U.S. coast—and is commonly referenced as the "200-mile limit."

With the stroke of President Gerald Ford's pen on April 13, 1976, the United States became a truly maritime nation, with, at 3.4 million square nautical miles, the most extensive maritime domain of any country in the world.[440] The legislation's goal was threefold:

- Eliminate unregulated foreign fishing off America's coast.
- Rebuild and conserve fish stocks.
- Provide opportunities for American fishermen.

The Magnuson-Stevens Fishery Conservation and Management Act is today the primary federal law that governs marine fisheries management in U.S. marine waters. It is required to be reauthorized by Congress every seven years.

HISTORICAL CONTEXT OF THE MAGNUSON-STEVENS ACT: FREEDOM OF THE SEAS, COD WARS

The so-called cod wars between Iceland and Great Britain preceded the Magnuson-Stevens Act and illustrate the context in which the legislation was written.

The unofficial *freedom-of-the-seas doctrine* (*Mare Liberum*), which dates from the early seventeenth century, stated that during times of peace, the sea—except for a narrow belt surrounding a nation's coastline—was free to all and belonged to no one. One of the basic principles of freedom of the seas was *freedom of fishing*: living resources located in the high seas were considered *common property* resources; they were owned by everyone and available for harvest by all.

By tradition, the ocean zone controlled by a nation—its *territorial sea*—extended three nautical miles from its coast. This was approximately the distance a cannonball could travel—the practical reason for accepting that a nation's territory extended only as far as it could be defended. Beyond this boundary, fishermen of every nation had free and open access to all fish stocks, which, as distant-waters fishing fleets developed and fish-catching technology became increasingly sophisticated and effective, were exploited with little consideration for sustainability or the effects upon the marine environment and ecology.

Soon after the end of World War II, the freedom-of-the-seas tradition began to unravel as several nations extended claims over offshore waters, in part because of growing concern over the effects of distant-waters fishing fleets on coastal fish stocks and threats of pollution from cargo ships plying coastal waters.

Iceland was one of the first nations to extend its claim to coastal waters. In 1952, the country unilaterally extended its claim to four miles from its coast, and in 1958 it extended its claim to twelve miles. Iceland's claims were ostensibly to protect cod and other species from overfishing, but were also likely intended to bolster the national fishing industry by excluding competition. The British, whose trawlers had been fishing along Iceland's coast since the 1890s, protested and sent warships to protect their fishing fleet, which was determined to, figuratively, hold its ground, at least a part of it. The British warships kept the fishing fleet in thirty-mile-long rectangular boxes whose perimeters the warships patrolled. While sound militarily, the secure boxes were a disaster for fishermen, who—box or no box—needed to fish where the fish were. Referred to as Cod War I, this confrontation lasted almost three years (until 1961) before the British acceded to Iceland's claim.

Cod War II began in the fall of 1972, when Iceland, citing the serious depletion of several important fish species, extended its coastal waters claim to fifty miles. (At that time, about 20 percent of Iceland's gross national product and more than 80 percent of the country's exports were derived from its fisheries.) This war lasted a year, during which the Icelandic Coast Guard used a secretly developed trawl line cutter to disable dozens of British trawls. The settlement that ended the war granted British fishermen limited permission to fish within the newly claimed area until November 1975. In October 1975, however, Iceland again extended its claim, this time to two hundred miles. Of great concern to Icelanders were cod stocks that both Icelandic and British scientists agreed were being seriously overfished. Cod War III ensued. Lasting seven months, this war was shorter than the previous cod wars. But it was also meaner, characterized by numerous ship rammings and disabling of trawls.[441]

By the time of Cod War III, the British were running against the tide. In November 1975, just a month after the war started, the respected British news magazine *The Economist* wrote that British fishermen had no long-term interest in Icelandic cod, that "within a couple of years 200-mile fishery limits are likely to have international blessing."[442] Nine nations around the world had already claimed 200-mile fishing zones, and the U.S. House of Representatives had passed a bill similar to what would become the Magnuson-Stevens Act. Moreover, in November 1976, the European Economic Community (EEC), of which Great Britain was a member, adopted a resolution that called for member nations to extend their maritime boundaries to 200 miles in concert on January 1, 1977.

Mediation by the Secretary-General of NATO helped bring Cod War III to an end. The final agreement, reached in June 1976, granted the British limited fishing access within Iceland's 200-mile limit for a period of six months.[443]

FOREIGN FISHERMEN RETURN TO THE EASTERN BERING SEA

The Japanese had begun high-seas fishing off Alaska's coast in 1930, but the effort was suspended in 1941, when the Japanese declared war on the United States.[444] In late 1952, however, there were signs the Japanese intended to resume crabbing and trawling in the eastern North Pacific Ocean. The following year, the Japanese processing ship *Tokei Maru* steamed into Bristol Bay with a fleet of tangle net vessels and trawlers and began targeting king crabs.[445] The Japanese were joined by Soviet Union vessels in 1958.

In the early 1960s, Japanese longliners began harvesting Pacific cod in the eastern Bering Sea for the frozen market. About the same time, the Japanese pollock trawl fishery in the North Pacific expanded, and Pacific cod became an important incidental catch in the pollock fishery and were occasionally targeted when high concentrations were located during pollock fishing operations.

The countries' fishing effort grew dramatically in the following years, and in 1964, their combined catch of groundfish in the eastern Bering Sea/Aleutian Islands area was more than 700,000 metric tons.[446] Similarly, the expanded presence at this time of Soviet Union fishing vessels off the coasts of New England, Washington, and Oregon during those Cold War years aroused much public concern.[447]

Further compounding the situation, between July 1, 1964, and July 22, 1966, the U.S. Coast Guard reported thirty-eight illegal incursions by foreign fishing and fishing support vessels into U.S. territorial waters. All but one of the incursions occurred off Alaska's coast. Soviet vessels committed thirty-two of the Alaska incursions; Japanese vessels committed the remaining five.[448]

In October 1966, to curtail the foreign fleets operating off both U.S. coasts, the U.S. Congress, led by senators Edward Bartlett (D-Alaska), Warren Magnuson (D-Wash.), and Edward Kennedy (D-Mass.), passed what is informally known as the Bartlett Act, which extended U.S. fisheries jurisdiction to twelve miles.[449] This was a start, but it wasn't enough: the twelve-mile limit hindered, but did not curtail, foreign fishing operations off the U.S. coasts. In fact, Koreans joined the effort off Alaska in 1967. By 1968, the Japanese, Soviets, and Koreans combined were annually catching more than a million metric tons of fish in the eastern Bering Sea/Aleutian Islands area. By far, most of the catch was pollock. The catch peaked in 1972, when an estimated nearly 2.4 million metric tons were caught. At an estimated 47,000 metric tons (about 104 million pounds), the catch of cod was substantial but represented only 2 percent of that year's total all-species catch.[450]

Globally, the pattern of the foreign fleets was *serial depletion*: the fleets needed large catches to sustain their operations and typically fished an area hard until the local stocks were reduced to unprofitable levels. Then they moved to new grounds to repeat the process.[451]

PASSAGE OF THE MAGNUSON-STEVENS ACT

> *The Magnuson-Stevens Act became law because of Senator Magnuson's and Senator Stevens's personal knowledge and commitment as well as high regard among their colleagues in the Congress.*
>
> —James Walsh, former U.S. Senate staff, former NOAA administrator, 2014[452] [453]

Further extending U.S. jurisdiction to encompass the area being overfished offered a potential solution, but claiming a two-hundred-mile exclusive fishery zone was, in the early 1960s, controversial and had little support from the U.S. government. The official U.S. policy was to assert the right of U.S. citizens to fish on the high seas outside the three-mile limit of all coastal nations. Moreover, the U.S. military was concerned that a U.S. claim of a two-hundred-mile coastal zone might provoke similar claims by other nations and inhibit its ability to operate in coastal areas.[454] But claiming two-hundred-mile-wide coastal zones was not a new idea: as a result of treaty negotiations in the fall of 1954, Peru, Ecuador, and Chile had each claimed exclusive fishing rights for a distance of two hundred miles from their shores.[455]

As with Iceland, the issue of extending U.S. fisheries jurisdiction beyond twelve miles was about conservation as well as economic opportunity. In 1970, Hiroshi Kasahara, a University of Washington College of Fisheries professor, pointed out that "[t]hose [foreign] fleets are simply clobbering our stocks out there. If we don't find the answer for forcing fishing conservation, it isn't only going to hurt our harvesting fish we pursue now, but there won't be anything left when we reach the point of development where we can use those stocks in our industry."[456]

In 1972, *Marine Digest* reported that Alaska governor William Egan thought imposing a two-hundred-mile limit was not the answer but hoped the United Nations Conference on the Law of the Sea scheduled for 1973 might arrive at a solution.[457] It did not. Jay Hammond, a bush pilot and guide with some commercial fishing experience who succeeded William Egan as governor in 1974, was of a different mindset: he considered the only viable alternative to be "immediate unilateral extension through Federal legislation."[458]

The champion of the congressional effort to extend the boundary of U.S. coastal waters to two hundred miles was Washington senator Warren Magnuson, a Democrat. His greatest ally—and the person who helped spur the effort—was Alaska senator Ted Stevens, a Republican. The magnitude of the foreign fishing effort in Alaska's coastal waters in the early 1970s was alarming, and Stevens had seen it for himself. In an address to a fisheries management conference in Washington, DC, in 2003, Senator Stevens recalled:

> When I was a young senator, I went to Kodiak and I commandeered a Navy plane. The Navy had a base there at the time. It was just about Christmas time and we flew up the Pribilofs in a Navy Albatross. As we flew, I counted about ninety Japanese trawlers that were in Alaska waters, and they were (in those days) there year round almost. I got to talking to Senator Magnuson about that and we devised the Magnuson-Stevens Act and went into a series of hearings on it.[459]

In 1973, Senator Magnuson, chairman of the Senate Committee on Commerce, which oversaw the nation's marine fisheries, introduced legislation to extend U.S. fishery management jurisdiction to two hundred nautical miles. The extended jurisdiction, however, would terminate when a law-of-the-sea treaty that recognized the authority of a nation to extend marine jurisdiction came into force. This bill failed to become law, but the opening shot in the fight for extended fishery management jurisdiction had been fired.[460]

In 1974, Senator Stevens reported that government officials had counted 106 Soviet and 72 Japanese vessels fishing off Alaska's coast, a record number.[461] Senator Magnuson reintroduced two-hundred-mile-limit legislation in the fall of 1975—a time by which fully fifteen nations had proclaimed their own two-hundred-mile limits.[462] The bill remained controversial. It was, as historically had been the case, opposed by the State Department, the Navy, and the Air Force. Even Senator Stevens's fellow senator from Alaska, Mike Gravel, a Democrat, did not support the legislation. Senator Gravel argued, according to the *Washington Post*, that "foreign overfishing is being reduced, and can likely be further reduced, by enforcement of international agreements already in place and by the prompt negotiation of further agreements." U.S. president Gerald Ford favored a two-hundred-mile limit, but one achieved through negotiation with other nations, not by unilateral legislation.[463]

Nevertheless, Senator Magnuson and his allies prevailed, and the Fishery Conservation and Management Act of 1976 became law on April 13, 1976. The legislation, which contained no termination provision, went into effect on March 1, 1977, and was a turning point for the beleaguered American fisheries.[464] George Rogers, considered by many to be the dean of Alaska economists, said the legislation "opened up a new frontier for Alaska."[465]

The primary objective of the legislation was to conserve and manage U.S. marine fishery resources for sustained human use.[466] Provisions of the act included:

- Taking "immediate action to conserve and manage the fishery resources found off the coasts of the United States" by declaring management authority over fishery resources in a 197-nautical-mile-wide fishery conservation zone (FCZ) that extended seaward from a line 3 miles off the U.S. coastline ("coterminous with the seaward boundary of each coastal State") to a line 200 miles off the coastline.[467] [468]

- Establishing a national fishery management program to "achieve and maintain, on a continuing basis, the optimum yield from each fishery." (The Magnuson-Stevens Act defines *optimum yield* as the maximum sustainable yield "as modified by any relevant economic, social or ecological factor[s]" that "will provide the greatest overall benefit to the Nation." *Maximum sustainable yield* is defined as the maximum amount of a fish stock that can be harvested year after year without depleting the stock.)[469]

- Encouraging "the development by the United States fishing industry of fisheries which are currently underutilized or not utilized by United States fishermen, including bottom fish off Alaska."[470]

Conservation of fisheries was at the forefront, but the legislation also contained language regarding the challenge of managing the fisheries to maximize benefits to both fishermen and the nation. Congress had determined that

> [m]anagement of fisheries as a common property resource has led to the use of excessive amounts of capital and labor

> in many fisheries. As a result, the profits earned by individual fishermen are low, potential economic benefits to the nation are lost, and the fisheries are depressed industries.[471]

To provide economic opportunities for U.S. fishermen, foreign fishing would over time be eliminated in the Fishery Conservation Zone, based upon the harvesting capability of U.S. fishermen. Foreign fishing would decrease commensurately as U.S. harvesting capabilities increased. The legislation did not mandate a schedule, but Congress hoped that the full *Americanization* of the nation's coastal fisheries would be achieved sooner rather than later.[472]

A shortcoming of the Magnuson-Stevens Act and its early amendments was that there was considerable direction regarding how to control and eventually eliminate the foreign presence, but little regarding how to control domestic fishing. The two-hundred-mile fisheries jurisdiction zone was perceived as needed, in the words of Richard Roe, the National Marine Fisheries Service's Northeast Region director, "to keep the foreigners out, or at least controlled, but not to control the domestic industry."[473]

Consistent with the 1982 United Nations Convention on the Law of the Sea, in 1983 President Ronald Reagan designated the Fishery Conservation Zone claimed by the United States as an *Exclusive Economic Zone* (EEZ), an area in which the United States has exclusive rights for fishing, drilling, and other economic activities.[474] For fisheries purposes, the EEZ is the functional equivalent of the Fishery Conservation Zone. The EEZ along Alaska's coast comprises some 900,000 square miles, an area greater than the combined EEZs of the east and west coasts of the contiguous United States.

ENDNOTES

439 Warren G. Magnuson, "The Fishery Conservation and Management Act of 1976: First Step Toward Improved Management of Marine Fisheries," *Washington Law Review* (July 1977): 427–450.

440 National Oceanographic and Atmospheric Administration, "The United States is an Ocean Nation," accessed September 3, 2018, https://www.gc.noaa.gov/documents/2011/012711_gcil_maritime_eez_map.pdf.

441 Appy Chandler, "In 'Cold War IV,' Iceland is Her Own Worst Enemy," *National Fisherman*, March 30, 1980, 164; "Iceland Pushes Out Her Fishing Limits, Risk-

ing 'Cod War,'" *New York Times*, September 2, 1972; Jon Blair, "The Cod War is Funny Only if Serious Issues are Ignored," *New York Times*, February 1, 1976.

442 "Cod War: Baiting the Wrong Hooks, *Economist*, November 29, 1975, 93–94.

443 *Legislative History of the Fishery Conservation and Management Act of 1976* (Washington, DC: GPO, 1978), 497–498; Mark Kurlansky, *Cod: A Biography of the Fish That Changed the World* (New York: Penguin Books, 1997), 161–169; Robin Rolf Churchill, *EEC Fisheries Law* (Dordrecht, Netherlands: Martinus Nijhoff Publishers, 1987), 14.

444 Ward T. Bower, *Alaska Fisheries and Fur-Seal Industries in 1937*, Administrative Rept. No. 31 (Washington, DC: GPO, 1938), 75.

445 "Crabbing and Codfishing Japanese Plan in 1953," *Pacific Fisherman* (October 1952): 72; "The Japanese Are Back Again," *Pacific Fisherman* (July 1953): 53.

446 Russell Nelson Jr., Robert French, and Janet Wall, "Sampling by U.S. Observers on Foreign Fishing Vessels in the Eastern Bering Sea and Aleutian Islands Region, 1977–78," *Marine Fisheries Review* (May 1981): 1; Grant Thompson, "Simulation Describes Population Dynamics of Pacific Cod in Eastern Bering Sea," *Northwest and Alaska Fisheries Center (National Marine Fisheries Service) Quarterly Report* (April/June 1987): 1–7.

447 James P. Walsh, "The Implementation of the Magnuson-Stevens Fishery Conservation and Management Act of 1976" (Bevan Lecture Series on Sustainable Fisheries, University of Washington, April 24–25, 2014), 7, accessed June 17, 2015, http://www.dwt.com/files/Publication/2bfb5669-3743-4ca2-a5ee-3406cd75ed43/Presentation/PublicationAttachment/1a63c735-e074-4715-afa9-c85cd9703b9f/Bevan%20Presentation.pdf.

448 "Foreign Fishing Intrusions True," *Marine Digest* (July 23, 1966): 39.

449 Act of October 14, 1966, Establishing a Contiguous Fishery Zone Beyond the Territorial Sea of the United States (80 Stat. 908); "No More Deals" (editorial), *Marine Digest* (September 3, 1966): 8. The 1966 legislation was actually an amendment to 1964 legislation (78 Stat. 194) that was informally known as the Bartlett Act. The 1964 legislation prohibited foreign vessels from fishing in the territorial waters of the United States except pursuant to an international agreement to which the United States was a party.

450 Russell Nelson Jr., Robert French, and Janet Wall, "Sampling by U.S. Observers on Foreign Fishing Vessels in the Eastern Bering Sea and Aleutian Island Region, 1977–78," *Marine Fisheries Review* (May 1981): 1; George W. Rogers, *Development of an Alaskan Bottomfish Industry and State Taxes* (Juneau, Alaska: Institute of Social and Economic Research, November 1977), 9.

[451] James Strong and Keith R. Criddle, *Fishing for Pollock in a Sea of Change: A Historical Analysis of the Bering Sea Pollock Fishery* (Fairbanks, Alaska: Alaska Sea Grant, 2013), 6.

[452] James "Bud" Walsh was staff counsel and general counsel for the U.S. Senate Committee on Commerce, Science, and Transportation (1972–1977), where he was responsible for all ocean-related legislation that came before the committee. Mr. Walsh also served as deputy administrator of the National Oceanic and Atmospheric Administration during the Carter administration (1977–1981).

[453] James P. Walsh, "The Implementation of the Magnuson-Stevens Fishery Conservation and Management Act of 1976" (Bevan Lecture Series on Sustainable Fisheries, School of Aquatic and Fishery Sciences/School of Marine and Environmental Affairs, University of Washington, April 24–25, 2014), 17, accessed June 17, 2015, http://www.dwt.com/files/Publication/2bfb5669-3743-4ca2-a5ee-3406cd-75ed43/Presentation/PublicationAttachment/1a63c735-e074-4715-afa9-c85cd97 03b9f/Bevan%20Presentation.pdf.

[454] "Dear Colleague" letter, senators John Tower, Dewey Bartlett, Dick Clark, Gale McGee, Hugh Scott, Robert Griffin, Strom Thurmond, and Barry Goldwater, January 19, 1976, in *Legislative History of the Fishery Conservation and Management Act of 1976*, 94th Cong., 2d sess. (Washington, D.C.: GPO, 1976), 515.

[455] "Ship Seizures Focus World's Attention on 200-Mile Limit," *Pacific Fisherman* (January 1955): 42.

[456] John Wiese, "Fisheries Treaty May Be in Trouble," *Alaska Industry* (June 1970): 45.

[457] "Egan on Coast Limit," *Marine Digest* (September 2, 1972): 25.

[458] Jay S. Hammond, "Alaska Governor Views Fishing Future," *Fisheries* (January–February 1976): 8–10.

[459] David Witherell, ed., *Managing Our Nation's Fisheries: Past, Present and Future, Proceedings of a Conference on Fisheries Management in the United States Held in Washington, D.C., November 2003* (Washington D.C.: NOAA Fisheries, 2004), 13–15.

[460] James P. Walsh, "The Implementation of the Magnuson-Stevens Fishery Conservation and Management Act of 1976" (Bevan Lecture Series on Sustainable Fisheries, School of Aquatic and Fishery Sciences/School of Marine and Environmental Affairs, University of Washington, April 24–25, 2014), 11, accessed June 17, 2015, http://www.dwt.com/files/Publication/2bfb5669-3743-4ca2-a5ee-3406cd-75ed43/Presentation/PublicationAttachment/1a63c735-e074-4715-afa9-c85cd97 03b9f/Bevan%20Presentation.pdf.

461 "Fishing Fleet Hits High," *Marine Digest* (February 23, 1974): 38.

462 "The 200 Mile Limit," *Wall Street Journal*, December 8, 1975.

463 "The Fishing Bill," *Washington Post*, November 4, 1975.

464 Fishery Conservation and Management Act of 1976, April 13, 1976 (90 Stat. 331).

465 George W. Rogers, personal communication with author, Juneau, Alaska, January 29, 2008.

466 Lowell W. Fritz, Richard C. Ferrero, and Ronald J. Berg, "The Threatened Status of Steller Sea Lions, *Eumetopias jubatus*, Under the Endangered Species Act: Effects on Alaska Groundfish Fisheries Management," *Marine Fisheries Review* (March 1995): 14–27.

467 As noted above, waters within three miles of the coastline are state waters, and the fisheries in them are managed by the states. The legislation's definition of "state" includes territories and other U.S. possessions. Coastal waters in U.S. territories and other possessions are managed by the federal government.

468 Fishery Conservation and Management Act of 1976, April 13, 1976 (90 Stat. 331), § 2(b)(1), §§ 10–102, § 3(21), § 404(a).

469 Fishery Conservation and Management Act of 1976, April 13, 1976 (90 Stat. 331), § 2(b)(4), § 3(18)(b).

470 Fishery Conservation and Management Act of 1976, April 13, 1976 (90 Stat. 331), § 2(b)(6).

471 Fishery Conservation and Management Act of 1976, April 13, 1976 (90 Stat. 331), § 2(a)(5).

472 Fishery Conservation and Management Act of 1976, April 13, 1976 (90 Stat. 331), § 201(d).

473 Richard B. Roe, "The Magnuson Fishery Conservation and Management Act—Then and Now," in *Proceedings of the National Industry Bycatch Workshop, February 4–6, 1992, Newport, Oregon* (Seattle: National Resource Consultants, 1992), 47–51.

474 Presidential Proclamation 5030, March 10, 1983 (97 Stat. 1557).

Chapter 12

IMPLEMENTATION OF THE MAGNUSON-STEVENS ACT

Fill the earth and subdue it. Rule over the fish in the sea and the birds in the sky and over every living creature that moves on the ground.

—Genesis 1:27–29

Fishing constitutes a multifaceted business whose participants seldom if ever express a unified view on any single topic.

—Dayton Alverson, fisheries consultant, 1992[475]

Management is particularly complex because of interaction and coordination between respective federal and state fishery management plans and jurisdictions; for example, coordination with the federal government for state-waters Pacific cod fisheries.

—Alaska Department of Fish and Game, 2014[476]

The Pacific Cod fishery is regulated . . . through permits, limited entry, catch quotas (TACs), season, in-season adjustments, gear restrictions, closed waters, bycatch limits and rates, allocations, regulatory areas, record keeping, reporting requirements and observer monitoring.

—David Witherell and Jim Armstrong, staff, North Pacific Fishery Management Council, 2015[477]

The North Pacific is the gold standard of fisheries management.

—Rep. Don Young (Alaska), 2018[478]

The National Marine Fisheries Service (NMFS, also known as NOAA Fisheries), an agency within the U.S. Department of Commerce, has the primary responsibility of conserving and managing marine fisheries in the U.S. EEZ. The Alaska Department of Fish and Game has primary responsibility for fisheries management within Alaska's state waters, which extend seaward three miles from the coastline.

To guide the NMFS's management of fisheries in the EEZ, the Magnuson-Stevens Act created eight regional councils that make decisions based on clearly articulated national standards embodied within the act. Alaska's EEZ groundfish fisheries are within the purview of the North Pacific Fishery Management Council.

The council consists of fifteen members, eleven of whom are eligible to vote. Seven of the voting members are representatives of various fisheries interests and are appointed by the secretary of commerce upon the recommendation of the governors of Alaska and Washington. The governor of Alaska recommends candidates for five seats; the governor of Washington recommends candidates for two seats. The four remaining voting members are the leading fisheries officials from the states of Alaska, Washington, and Oregon, and the NMFS's Alaska region director. Nonvoting members are the executive director of the Pacific States Marine Fisheries Commission, the area director for the U.S. Fish and Wildlife Service, the commander of the Seventeenth Coast Guard District (Alaska maritime region), and a representative from the U.S. State Department.

The council has an Anchorage-based staff of about fifteen, a standing Advisory Panel consisting primarily of fishing industry representatives, and a standing Scientific and Statistical Committee populated by scientists and economists.

Councilspeak: *"As stated in 4.4.3, we recommend basing RPP DMRs for NPT CVs on the DMR for GOA CPs fishing NPT gear."*

—North Pacific Fishery Management Council, DMR Analysis, 2016[479] [480]

The North Pacific Fishery Management Council meets five times each year, usually in February, April, June, October, and December. Three of the meetings are held in Anchorage, one in a fishing community in Alaska, and one either in Seattle, Washington, or Portland, Oregon. Most council meetings last seven days. The late Elmer Rasmuson, then president of the National Bank of Alaska and a champion of Alaska's commercial fishing industry, was the council's first chairman.

The initial fundamental duty of the council was, in consultation with the NMFS and other knowledgeable entities, to determine the optimum yield for each fishery, the U.S. harvesting capability, and the consequent residual amounts available for allocation to foreign nations. As the fisheries became increasingly Americanized, the council began developing fishery management plans. The secretary of commerce has the authority to approve or disapprove the council's plans (or other actions); to, under certain conditions, develop separate plans; and is responsible for the implementation of all plans. The first Gulf of Alaska groundfish fishery management plan was implemented in 1978, and the first Bering Sea/Aleutian Islands area plan was implemented in 1981. Both plans have been amended numerous times.

The North Pacific Fishery Management Council and the National Marine Fisheries Service use three basic reference points for management of groundfish fisheries: the *overfishing level* (OFL), the *allowable biological catch* (ABC), and the *total allowable catch* (TAC). The OFL is based on the fishing mortality rate associated with continuously producing the maximum sustainable yield. This level should never be exceeded. The ABC is the annual catch limit and is set lower than the OFL, providing a buffer that allows for scientific uncertainty in single-species stock assessments and ecosystem considerations and for operational management of the fishery. The TAC, set at or below the ABC, is the target catch level and incorporates management uncertainty as well as economic and social considerations.[481]

Fisheries within the EEZ were to be managed for the benefit of U.S. citizens—Americanized—but off Alaska's coast it was a slow and

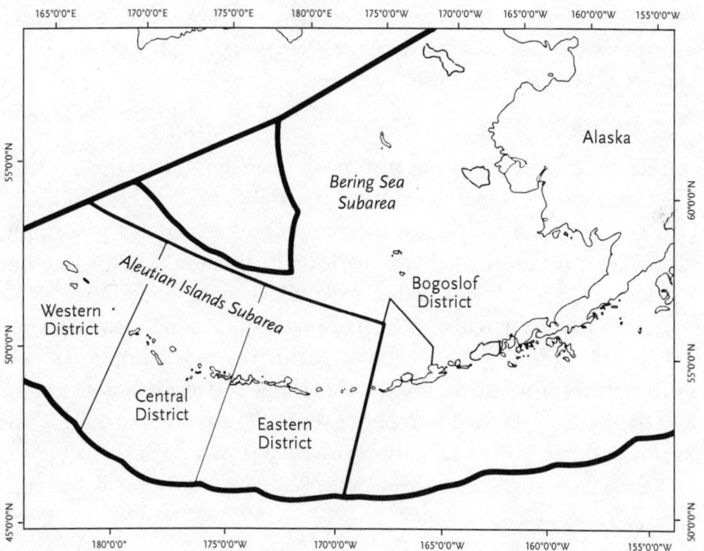

Figure 54. Bering Sea/Aleutian Islands groundfish management area, with subareas and districts. (North Pacific Fishery Management Council.) Note: Within state waters (three nautical miles from the shore), the Pacific cod fishery is managed by the Alaska Department of Fish and Game as a state-waters fishery or as a parallel fishery in which regulations generally mirror NMFS regulations of the adjacent federal Pacific cod fishery (see chapter 23).

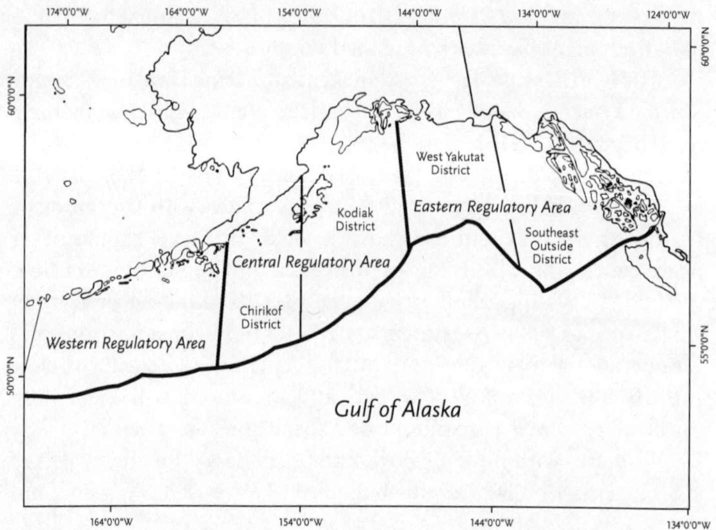

Figure 55. Gulf of Alaska groundfish regulatory area, with subareas and districts. (North Pacific Fishery Management Council.) Note: As in the Bering Sea/Aleutian Islands area, the Pacific cod fishery in state waters in the Gulf of Alaska is managed by the Alaska Department of Fish and Game.

uncertain process. One issue was the sheer magnitude of the fisheries. The foreign fleets were to be phased out, but the U.S. fishing industry had nothing available to replace foreign harvesting and processing capabilities. Adding a further wrinkle to the situation was the fact that markets for groundfish, particularly pollock, were uncertain.

A fortunate coincidence to the passage of the Magnuson-Stevens Act was that the 1977 year class of Pacific cod was exceptional and, once mature, provided an abundance of large cod in the Gulf of Alaska and the Bering Sea for a number of years. (The estimated biomass of Pacific cod peaked in the early 1980s, then slowly declined.)[482]

In 1978, Alaska economist George Rogers summarized the foreign fishing industry's advantages:

> In competition for markets, the foreign fishery has several important advantages. They have had a decade or more experience on the grounds and established specialized and efficient operations which result in economies beyond those of lower labor costs. They have strong United States and international marketing connections and organizations, some of

> which are presently possessed by Alaskan developers. Foreign exploitation of virgin stocks (excepting Pacific halibut) in the Bering Sea and Gulf of Alaska during the 1960s resulted in extremely high catch rates which helped offset much of the development costs. Alaskan fishermen seeking to enter the bottomfish fishery at its present state of depressed catch rates will not enjoy this "natural subsidy," and will have to absorb or pass on to the consumer their full development costs.[483]

Initially, all the available groundfish quotas off Alaska except for a 1,600-metric ton (mt) Pacific cod reserve in the Bering Sea were allocated to foreign fishermen. The reserve was established to provide for the harvest of cod that were used as bait in the crab fisheries.[484]

> **Codfish report, May 1980:** *The Aleutian cod resource is the bright spot in the Pacific Coast bottomfish picture. The fish are large and thus have a high value and specialized markets. The latest fishing and processing ventures for Aleutian cod have targeted on markets as diverse as the United Kingdom fish and chip trade, the Norwegian salt cod market, and U.S. jumbo fillet buyers.*
>
> —*Pacific Fishing*, 1980[485]

JOINT VENTURES: "WILD AND FREE IN THE EEZ."[486]

Americanization of the fisheries, everyone knew, would take time. But the effort was further constrained in the years following passage of the Magnuson-Stevens Act by high inflation, high interest rates, and high fuel costs. Part of the interim solution while foreign operations were phased out and domestic fish-catching and fish-processing capacity developed were *over-the-side* joint ventures in which American trawlers—often retrofitted crab boats—caught and delivered pollock, cod, and other groundfish to foreign fish-processing ships on the fishing grounds, with the final products marketed worldwide.

> *The joint venture argument has an irrefutable logic. There is an excess of modern, high-priced and often heavily mortgaged fishing vessels on the coast, along with a growing constituency of loudly complaining fishermen who can't find room in traditional fisheries. The nation controls millions of tons of fish it can neither process nor consume. The rest of the world, meanwhile, has an*

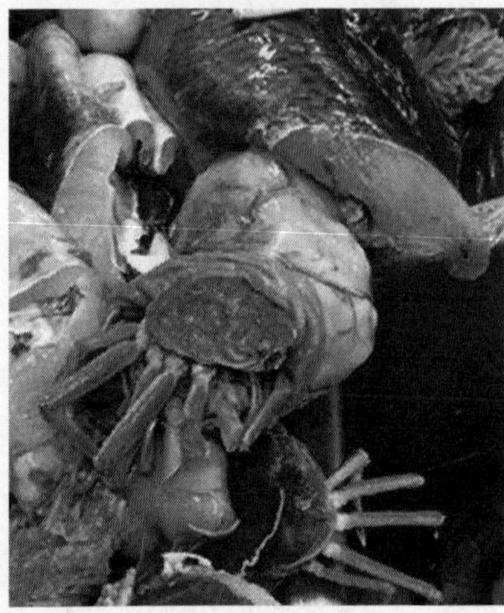

Figure 56. Tanner crab and Pacific cod stomach. (Courtesy of Scott Campbell Jr., F/V *Seabroke*.)

insatiable appetite for fish and a surfeit of floating fish factories that have been idled by the advent of 200-mile limits.

—John Sabella, *Pacific Fishing*, 1982[487]

The crash of crab populations in the Bering Sea and Gulf of Alaska in the early 1980s increased the allure of joint ventures.[488] The shortage of crabs sent crab boat owners, many of them heavily mortgaged, scrambling to find profitable employment for their vessels.[489] Joint-venture trawling for groundfish seemed to offer expedient, though interim, relief from the crabbers' dire financial situation.

The scarcity of crabs coincided with an abundance of cod, and some individuals in the fishing industry speculated that the cod were "gobbling up young king crab," that "crab are down, cod are up, so cod must be the cause." A deckhand on a trawler fishing for cod in Shelikof Strait in the late 1970s recalled that while bleeding cod on deck, he was at times ankle deep in nearly legal-size molting king crab that large cod had eaten but not yet swallowed.[490] More recently, one long-time Kodiak fisherman blamed cod, which he derided as "swimming mouths," for the scarcity of shrimp and Tanner crab around Kodiak Island.[491]

Joint-venture operations began modestly in 1978, when U.S. fishing vessels began delivering fish to Soviet and Korean processing vessels. After that, participation in joint ventures increased rapidly; in 1981, fully thirty-nine U.S. trawlers were involved in joint ventures in the Bering Sea and Gulf of Alaska.[492] The joint-venture effort to catch and process Pacific cod, however, was relatively modest; most of the effort was directed toward pollock.[493] Federal oversight of the joint-venture fisheries was minimal, and some individuals in the industry characterized the situation as "Wild and Free in the EEZ."[494]

Joint ventures worked for U.S. fishermen, but U.S. seafood processors complained that their potential to replace foreign processors—to Americanize the processing sector of the fisheries—was hindered by the joint ventures. Processing preference, however, was not directly related to conservation and management of fish stocks pursuant to the Magnuson-Stevens Act. To foster Americanization of the processing sector, Congress amended the Magnuson-Stevens Act in August 1978 to give first priority for processing any fish harvested in the Fishery Conservation Zone to domestic processing plants and U.S.-flag processing vessels.[495] There was, however, little domestic groundfish-processing capacity in Alaska, either shore-based or floating, so the law had little immediate impact. Pacific cod joint ventures, which had an average annual production of about 1,400 metric tons, ended following the 1988 fishing season.[496]

ENDNOTES

[475] Dayton L. Alverson, "An Industry Perspective on Addressing the Bycatch Problem," in *Proceedings of the National Industry Bycatch Workshop, February 4–6, 1992, Newport, Oregon* (Seattle: National Resource Consultants, 1992), 191–196.

[476] Alaska Department of Fish and Game, Commercial Fishing Information, Westward area, undated, accessed September 26, 2014, http://www.adfg.alaska.gov/index.cfm?adfg=fishingcommercialbyarea.southwest.

[477] David Witherell and Jim Armstrong (staff, North Pacific Fishery Management Council), *Groundfish Species Profiles, 2015*, foreword, accessed December 2, 2015, www.npfmc.org/wp-content/PDFdocuments/resources/SpeciesProfiles2015.pdf.

[478] Jason Huffman, "US House approves Young's Magnuson-Stevens update after sticky provision cut," *Undercurrent News*, July 11, 2018, accessed July 11, 2018, https://www.undercurrentnews.com/2018/07/12/us-house-approves-youngs-magnuson-stevens-update-after-sticky-provision-cut/.

[479] Translation: "As stated in 4.4.3, we recommend basing Rockfish Pilot Program discard mortality rates for non-pelagic trawl catcher vessels on the discard mortality rate for Gulf of Alaska catcher-processors fishing non-pelagic trawl gear."

[480] Linda Behnken, NPFMC member, August 1992–August 2001, in *Celebrating 40 Years of Sustainable Fisheries Management* (Anchorage: North Pacific Fishery Management Council, 2016), 26.

[481] David Witherell and Jim Armstrong (staff, North Pacific Fishery Management Council), *Groundfish Species Profiles, 2015*, foreword, accessed December 2, 2015, www.npfmc.org/wp-content/PDFdocuments/resources/SpeciesProfiles2015.pdf.

[482] Doug McNair, "Cod: The Alaskan Resource," *Pacific Fishing* (January 1984): 37; David Witherell and Jim Armstrong (staff, North Pacific Fishery Management Council), *Groundfish Species Profiles, 2015*, 32, accessed December 2, 2015, www.npfmc.org/wp-content/PDFdocuments/resources/SpeciesProfiles2015.pdf.

[483] George W. Rogers, *Development of an Alaskan Bottomfish Industry and State Taxes* (Juneau, Alaska: Institute of Social and Economic Research, undated, but probably 1977), 11–12.

[484] Ron Berg, in: *North Pacific Fishery Management Council, Celebrating 30 Years of Sustainable Fisheries* (December 2006), 25.

[485] "Bottomfish: The Fishery of the Future?" *Pacific Fishing* (May 1980): 18.

[486] Statement of Guy N. Thornburgh, Pacific Marine Fisheries Commission, reprinted in *Hearings on the Oversight of the Marine Fisheries Management Act*, Committee on Commerce, Science, and Transportation, U.S. Senate, 101st Cong., 1st sess. (Washington, DC: GPO, 1989), 600–601.

[487] John Sabella, "Joint Ventures: Enormous Promise and Broken Promises," *Pacific Fishing* (January 1982): 33–35, 37.

[488] For example, the crab catch in the Bering Sea/Aleutian Islands area peaked in 1980, with the delivery by 377 vessels of 237 million pounds of crab (all species) that was worth $144 million. The catch dropped to 134 million pounds in 1981 and to 67 million pounds in 1982.

[489] "Bottomfish Update," *Pacific Fishing* (November 1982): 65–71; Daniel D. Huppert, *Managing Alaska Groundfish: Current Problems and Management Alternatives* (Seattle: Fisheries Management Foundation and Fisheries Research Institute, 1988), 2.

[490] Doug McNair, "Cod: The Alaskan Resource," *Pacific Fishing* (January 1984): 37; Dan Pratschner, personal communication with author, October 19, 2014.

[491] Anonymous by request, personal communication with author, November 7, 2015.

492 John Sabella, "Joint Ventures: Enormous Promise and Broken Promises," *Pacific Fishing* (January 1982): 33–35, 37.

493 David Witherell and Jim Armstrong (staff, North Pacific Fishery Management Council), *Groundfish Species Profiles, 2015*, 31, accessed December 2, 2015, www.npfmc.org/wp-content/PDFdocuments/resources/SpeciesProfiles2015.pdf.

494 Statement of Guy N. Thornburgh, Pacific Marine Fisheries Commission, reprinted in *Hearings on the Oversight of the Marine Fisheries Management Act*, Committee on Commerce, Science, and Transportation, U.S. Senate, 101st Cong, 1st sess. (Washington, DC: GPO, 1989), 600–601.

495 James P. Walsh, "The Implementation of the Magnuson-Stevens Fishery Conservation and Management Act of 1976" (Bevan Lecture Series on Sustainable Fisheries, School of Aquatic and Fishery Sciences/School of Marine and Environmental Affairs, University of Washington, April 24–25, 2014), 14–15, accessed June 17, 2015, http://www.dwt.com/files/Publication/2bfb5669-3743-4ca2-a5ee-3406cd75ed43/Presentation/PublicationAttachment/1a63c735-e074-4715-afa9-c85cd9703b9f/Bevan%20Presentation.pdf; An Act to authorize appropriations to carry out the Fishery Conservation and Management Act of 1976 during fiscal year 1979, to provide for the regulation of foreign fish processing vessels in the fishery conservation zone, and for other purposes, August 28, 1978 (92 Stat. 519).

496 David Witherell and Jim Armstrong (staff, North Pacific Fishery Management Council), *Groundfish Species Profiles, 2015*, 31, accessed December 2, 2015, www.npfmc.org/wp-content/PDFdocuments/resources/SpeciesProfiles2015.pdf.

Chapter 13
EARLY DOMESTIC CODFISH VENTURES

The Aleutian cod resource is the bright spot in the Pacific Coast bottomfish picture. The fish are large and thus have a high value and specialized markets.

—*Pacific Fishing*, May 1980[497]

Cod represents the most important near-term opportunity. Although the once substantial domestic harvest of Pacific cod virtually disappeared after World War II, so that it has become one of the species considered to be "underutilized" by the domestic industry, its Atlantic counterpart is the most widely consumed fish in this country. Of the 466 million pounds of fillets Americans ate in 1979, for example, cod accounted for more than 38 percent. Of that total, more than 80 percent was imported from nations like Iceland (73 million pounds), Canada (51 million pounds), Norway (8 million pounds) and Denmark (8 million pounds).

—*Pacific Fishing*, March/April 1981[498]

The two major groundfish species in the Bering Sea and Gulf of Alaska are pollock and Pacific cod. American consumers in the 1970s were not generally familiar with pollock, but they were familiar with cod from the Atlantic, which, for culinary purposes, is identical to the Pacific cod. Per pound, Pacific cod was more valuable than pollock, and it presented what seemed to be a greater opportunity for profitable exploitation. Hence, the fledgling groundfish industry off Alaska's coast focused its effort on Pacific cod. The industry's production, which ranged from frozen headed-and-gutted (H&G) fish and fillets to salted fish, was sold on established domestic and foreign cod markets. Kodiak, which had cod-fishing grounds fairly close by and—for better or for worse—was homeport to a large number of underemployed crab and shrimp vessels, was the main locus of early post-Magnuson-Stevens Act shore-based Pacific cod production. In 1978, three companies—two in Kodiak and one in Dutch Harbor—experimented with processing and marketing groundfish. Some of the fish the companies purchased were from crab fishermen who had begun using slightly modified king and Tanner crab pots (see chapter 18) to target Pacific cod. As they did with crabs, these fishermen kept their codfish catch alive in seawater that was continuously pumped into their vessels' live tanks.

Figure 57. Pacific cod filleting operation, Petersburg Fisheries, Petersburg, Alaska, fall 1976. (J. Mackovjak.)

Probably the first significant experiment in producing Pacific cod fillets in Alaska began at the Petersburg Fisheries (later Icicle Seafoods) plant at Petersburg, in Southeast Alaska, in the fall of 1976. The cod processed there were all caught in nearby state waters by seine boats that were outfitted with trawl gear and targeting starry flounder that were, with pollock, the fish the company was focused on. Bob Thorstenson, the plant's manager, was a visionary and likely recognized the potential in processing bottomfish, but at the time he said he just wanted to provide an opportunity for local fishermen during the off season. The cod, as were the pollock, were filleted by German-made Baader machines. (Flounder were hand filleted.) Citing poor markets, Petersburg Fisheries shuttered its bottomfish operation in September 1980. By then, the company had processed some 7 million pounds of locally caught bottomfish, mostly pollock and flounder.[499]

The most ambitious early domestic effort to process codfish was by the New England Fish Company (NEFCO), which in 1978 invested $1.1 million to equip its Gibson Cove plant at Kodiak for codfish processing. (The plant also processed shrimp and crabs.) The company intended to produce whatever cod products the market might demand: fillets, headed-and-gutted, and even split-and-salted. Cod fillets would be individually quick-frozen.

NEFCO, which was founded in Boston in 1868, had a long history in Alaska that began in 1907, when the company built a cold storage plant at Ketchikan. (In 1919, NEFCO froze a few tons of codfish incidentally as part of its halibut freezing operation at its Ketchikan plant.) Eventually NEFCO owned and operated a number of fish-processing facilities along the U.S. West Coast, from Oregon to western Alaska. The company was considered to be one of the most progressive American fish processors.

As an incentive for companies to engage in groundfish processing, the Alaska Department of Commerce and Economic Development not long after the passage of the Magnuson-Stevens Act implemented a loss-guarantee program. Under this program, the department and NEFCO signed a contract that guaranteed NEFCO against loss for up to three cents per pound of groundfish processed, with a $145,000 ceiling on the total potential payment.[500] In exchange, NEFCO agreed to supply the state with information generated by its codfish-processing venture.

Codfish processing at Gibson Cove began in April 1978. The fish were caught by small trawlers—primarily 65- to 90-foot shrimp trawlers that had been converted to trawling for cod—that fished along the west shore of Kodiak Island and in Shelikof Strait. The cod were gutted as soon as they came on deck but received no additional processing on the boats. The dock price at Kodiak during the summer of 1979 ranged from $0.19 to $0.21 per pound. (Some trawlers sold round cod to crab fishermen for bait. These fish brought a higher price than cod sold for food.)[501]

Among the shrimp fishermen looking to diversify into something that could be done during the off-season were brothers Oral and Al Burch, pioneers in Alaska's pink shrimp trawl fishery. The Burch brothers owned and operated the *Dawn* (86 feet) and *Dusk* (80 feet), and they began trawling for groundfish in February 1978. (Both of these vessels, since lengthened, are still active in the Gulf of Alaska bottomfish trawl fishery.) Another fisherman in the industry was Dave Harville, a California trawler who, according to an article in *Pacific Fishing*, was in 1977 "flush with Magnuson Act spirit and promise." Harville promptly built a 65-foot trawler, the *Linda Jeanne*, and began experimental trawling for hake along California's northern coast. The hake experiment met with only limited success, and in 1978 Harville headed for Alaska and began trawling for codfish for NEFCO. Harville recalled making three-day trips during which his vessel caught 60,000 pounds of codfish.[502]

In 1978 or 1979, NEFCO applied to the Alaska Fisheries Development Foundation (see chapter 15) for a grant of $2,183,000 that would be used to subsidize the establishment of a "model whitefish processing plant" at the company's Gibson Cove facility. But, despite the opportunities engendered in the Magnuson-Stevens Act, all was not well at NEFCO. Not well at all.

For the past decade, NEFCO's multifaceted operations had for a variety of reasons seldom shown a profitable year, and its debts were mounting. On April 23, 1980, after thirteen months of losing more than a million dollars a month, NEFCO filed for protection from its creditors under federal bankruptcy laws. Nine days later, the company refiled for complete liquidation. *Pacific Fishing* reported in late 1980 that one shore-based firm, likely NEFCO, had experimented with freezing cod fillets ashore but "ran into problems with European quality standards."[503]

Pete Harris, a former employee of NEFCO who was described in a *Pacific Fishing* article as the company's "chief bottomfish visionary," was among the five individuals who quickly organized the Alaska Food Company to purchase NEFCO's Gibson Cove plant. The new company, financed in large part by a $3.5 million loan from the Alaska Renewable Resources Corporation, a state-funded lending institution, promptly began processing codfish, though in small quantities. At the same time, the company received a $1.2 million Alaska Fisheries Development Foundation grant to establish a model whitefish-processing plant. In March 1981, the Alaska Food Company received its first substantial delivery of groundfish: 220,000 pounds of Pacific cod caught by the trawler *Storm Petrel*. The fish were caught in the Bering Sea and, despite at least four days in the vessel's refrigerated seawater holds, they were determined to be of sufficient quality to produce marketable fillets.

By early 1981, the Alaska Food Company had spent about $25,000 of its Alaska Fisheries Development Foundation grant on planning the facility, but later that year the foundation withdrew the remaining grant funds because of irreconcilable differences over the proposed plant design. The Alaska Food Company forged ahead despite that setback, but it was plagued by start-up problems and stung by unprofitable shrimp, crab, and salmon seasons. In October 1981, the company filed for protection from its creditors under federal bankruptcy laws to gain time to complete the installation of its groundfish-processing lines.

In January 1982, the Alaska Food Company again began processing groundfish, paying fishermen $0.19 per pound for head-on, gutted Pacific cod. The company's groundfish operation apparently began to show a profit, but Alaska Food remained financially overwhelmed by past debts and the costs of correcting problems in the original plant design. Its end came at the end of June 1982, when the Alaska Renewable Resources Corporation—stripped of its budget and ordered out of the venture capital business by Alaska's legislature—declined to finance further operations.[504] By then, however, Kodiak had two additional groundfish-processing operations: International Seafoods of Alaska and the Alaska Packers Association.[505]

Kodiak's first Pacific cod catcher-processor was likely the 66-foot steel trawler *Dominion*, owned and operated by Bernie Burkholder. In 1984, Burkholder built a shelter deck behind the vessel's wheelhouse. Beneath the shelter deck, he installed two plate freezers, a two-station fillet line, and a Baader skinning machine. He also refrigerated the *Dominion*'s hold, which enabled the vessel to carry about 60,000 pounds of frozen product.

With three crew members, Burkholder trawled in Shelikof Strait. Fishing was generally good, and a tow typically yielded 10,000 to 12,000 pounds of cod. And the fish were large, averaging between 16 and 32 pounds. Once the trawl was aboard, the fish were bled and then two crew members began hand filleting them and then passing them through the skinning machine. The third crew member packed the fish into 15-pound shatterpacks and put them into the plate freezers. Once the fish processing was complete, the trawl was put back in the water to begin the process anew.

Burkholder marketed his product to a seafood wholesaler in Seattle and to Favco, a seafood wholesaler in Anchorage. The Pacific cod catcher-processor venture lasted a couple years. Burkholder later reminisced that he had been "ahead of his time," that there were too many obstacles for an operation of his size. He later became an investor in Arctic Alaska Fisheries, which operated a fleet of factory trawlers.[506]

Codfish processing at Kodiak continued to grow, albeit slowly, until 1990. That year, a virtual doubling of the ex-vessel price of Pacific cod and a reduction in access to pollock sparked record participation in the industry. The Pacific cod fishery at Kodiak, wrote *Alaska Fisherman's Journal* reporter Donna Parker, was becoming "another derby fishery restrained only by bycatch caps."[507] (See chapter 24.)

The boom continued, and in early 1991, eleven fish plants were processing Pacific cod. By mid-March of that year, 163 vessels had delivered 22.5 million pounds of Pacific cod—about 25 percent more than the previous year. The dock price for longline-caught bled cod at Kodiak in early 1991 was $0.28 per pound.[508]

In the Bering Sea, the first U.S. post-Magnuson-Stevens Act effort to process codfish was in 1978 aboard the Universal Seafoods

Figure 58. Icicle Seafoods's floating processor *Arctic Star* processed king crab at Akutan in 1980. The *Bering Star*, an almost identical vessel, was anchored nearby and processed Pacific cod. (Courtesy of Kim Suelzle.)

processing ship *Viceroy* and the company's crab-processing barge *Vita*. The vessels were anchored at Dutch Harbor and primarily processed crab. Codfish processing was something of a sideshow.[509]

The first major effort by a domestic firm to process Pacific cod in the Bering Sea was early in 1980, when Icicle Seafoods began producing frozen H&G Pacific cod at Akutan, in the Aleutian Islands. Icicle's was a makeshift operation aboard the *Bering Star*, a 265-foot-by-55-foot barge that was originally designed to process and freeze salmon and crabs.

Icicle Seafoods brought the barge to Akutan in January 1980, and the following month the North Pacific Fishery Management Council granted the company a twelve-mile exclusive processing sanctuary around Akutan and Akun islands. In petitioning the council, Icicle had alleged that joint ventures constituted unfair competition because foreign processors were not required to comply with minimum wage laws and requirements of the Occupational Safety and Health Administration and the Food and Drug Administration. Though the council had given its approval, the secretary of commerce, who has the final word on federal fisheries management issues, vetoed the sanctuary.

Nevertheless, Icicle's venture proceeded, and the company purchased Pacific cod for $0.08 per pound from six crab boats that had been retrofitted with trawl gear. The boats ranged from 75 feet to 105 feet in length. Aboard the *Bering Star*, equipment designed to head, gut, and freeze salmon was used to do the same with Pacific cod. The frozen fish were sold at Akutan for $0.32 per pound to Joint Trawlers, Ltd., an Anglo-Swedish firm.

Icicle Seafoods had anticipated processing about 5 million pounds of cod, but severe weather constrained fishing, and only about 1.5 million pounds were processed before Icicle canceled the venture in mid-March. The tramp freighter contracted to carry the product to Europe was only partially loaded when it departed in late March after a longer than expected wait in Dutch Harbor. The less-than-full cargo and the extra time at Dutch Harbor added substantially to the cost of transportation. The worst, however, was yet to come: Joint Trawlers attempted to market the cod in Great Britain, but the British rejected it because of parasites (worms) and deterioration of the flesh. Parasites can be unavoidable, but the flesh deterioration was the result of the fish not being bled aboard the catcher boats and—because of equipment malfunctions aboard the *Bering Star*—the sometimes exceptionally long amount of time that elapsed between when the fish were caught and when they were processed. Most of the cod the British rejected was subsequently sold at a substantial discount to Norwegian salters, whose end product did not always require first-quality fish.[510]

Like NEFCO, Icicle Seafoods was at that time in a precarious financial positon, and its Bering Sea cod-processing venture was put on hold while the company strived to regain its financial footing.

Another early domestic codfish operation was the Alaska Brands Corporation's *Golden Alaska*, a 302-foot German-built fish-processing ship that had been purchased by U.S. interests and reflagged. The vessel began operations in the Bering Sea/Aleutian Islands area in the fall of 1982 and planned to take cod end deliveries of codfish year-round on the fishing grounds from five to eight trawlers (see chapter 23).[511] [512]

The first shore-based processing in the Bering Sea/Aleutian Islands area was a salting operation at Akutan, which is described in chapter 14.

ENDNOTES

497 "Bottomfish: The Fishery of the Future?" *Pacific Fishing* (May 1980): 18.

498 "Frozen Pacific Whitefish: Production Trails Market Potential," *Pacific Fishing* (March/April 1981): 53–58, 60.

499 Poor bottomfish market means cutbacks at plant," *Bering Sea Fisherman*, October 1980.

500 The $145,000 would cover the potential three-cents-per-pound losses on about 4.8 million pounds of fish.

501 Lynn Pistoll, "A Discussion of the Potential for Expanding into an Alaskan Bottomfish Industry," *Alaska Economic Trends* (August 1978): 1–5; Guide to the New England Fish Company Records, University of Washington Libraries, Seattle, Wash., accessed December 18, 2014, http://digital.lib.washington.edu/findingaids/view?docId=NewEnglandFishCompany3996.xml; Ward T. Bower, *Alaska Fisheries and Fur Industries in 1919* (Washington, DC: GPO, 1920), 57; "NEFCO begins bottomfish for food line," *Alaska Fisherman's Journal* (May 23, 1978): 78; Institute of Social and Economic Research, University of Alaska, "Prospects for

a Bottomfish Industry in Alaska," *Alaska Review of Social and Economic Conditions* (April 1980), 1–31.

502 Nancy Munro, "Alaskan Bottomfish Plan is Finally Submitted," *National Fisherman*, April 1978, 18; Al Burch, owner/captain of trawler *Dusk*, personal communications with author, December 2014; "'First' on West Coast: Sixty-five-foot Eureka vessel soon to fish for eatable hake," *Times Standard* (Eureka, Calif.), June 13, 1977; Patrick Reid, "Coast hake industry faces some obstacles," *Times Standard* (Eureka, Calif.), September 1, 1977; D. B. Pleshner, "An Hour with Dave Harville, *Pacific Fishing* (March 1989): 49–55.

503 Gene Viernes, "NEFCO Goes Under," *Alaskero News*, May 1980; Bruce Ramsey, "Life After Nefco," *Pacific Fishing* (July 1981): 55–62; "Born Again: A Resurrected Codfishery," *Pacific Fishing* (December 1980): 44–47.

504 "Grant to Alaska Food Company in jeopardy," *Alaska Fisherman's Journal* (March 1981): 4; Brad Matsen, "Alaska Food Co. Brings Bottomfish Dream 110 Tons Closer to Reality," *Alaska Fisherman's Journal* (April 1981): 8–9; Bruce Ramsey, "Life After Nefco," *Pacific Fishing* (July 1981): 55–62; Chris Blackburn, "Bottomfish lines start producing at Alaska Food," *Alaska Fisherman's Journal* (January 1982), 24–26; Chris Blackburn, "Alaska Food calls it quits," *Alaska Fisherman's Journal* (August 1982): 22–23.

505 Chris Blackburn, "Bottomfish fillets at second Kodiak plant," *Alaska Fisherman's Journal* (December 1978): 20; Nancy Freeman, "New Kodiak bottomfish plant modeled after 'factory trawler,'" *Alaska Fisherman's Journal* (February 1980): 49–51.

506 Bernie Burkholder, personal communication with author, October 18–19, 2018.

507 Donna Parker, "Codfish Booming in Kodiak," *Alaska Fisherman's Journal* (May 1991): 24.

508 Ibid.

509 UniSea, Inc., Aleutian Islands Risk Assessment report, Dutch Harbor, Alaska, September 1, 2009.

510 Nancy Hasselback, "Joint Ventures in Alaska Cry Foul as Icicle Gets Sanctuary," *National Fisherman*, March 1980, 114; A. D. Chandler, "Processing Limitations Curtail Icicle's Akutan Cod Operation, *National Fisherman*, June 1980, 14; John Sabella, "Bob Thorstenson, The Man Behind Icicle Seafoods, *Pacific Fishing* (June 1980): 23–27; Bruce Ramsey, "Life After Nefco," *Pacific Fishing* (July 1981): 55–62; Kim Suelzle, former manager, Icicle Seafoods, personal communication with author, March 10, 2008; Robert M. Thorstenson, Icicle Seafoods, to

Terry Leitzell, NMFS, March 20, 1980, reprinted in *American Fisheries Promotion, Hearing on H.R. 7039*, Ho. Subcommittee on Fisheries and Wildlife Conservation and the Environment, Committee on Merchant Marine and Fisheries, 96th Cong., 2d sess., May 6, 19, 20, 1980, Serial No. 96-44 (Washington, DC: GPO, 1980), 381, 388; A. D. Chandler, "Alaskan Bottomfish Shipment Runs Into European Resistance," *National Fisherman*, December 1980, 10–11.

511 The cod end is a cylindrical section of heavy netting at the trailing end of a trawl in which fish caught in the trawl accumulate. Typically, the cod end is hauled up a stern ramp onto the trawler's deck for unloading.

512 Doug McNair, "Joint Ventures—What are They? Who are They?" *Pacific Fishing* (December 1982): 50–53; "The Marathon Begins," *Pacific Fishing*, Yearbook (1983), 63.

Chapter 14
SALT COD REDUX

Salt cod: new venture into a forgotten fishery.

—*Alaska Fisherman's Journal*, September 1980[513]

Bale by bale, the Alaska salt cod and stockfish is finding its way into what was once an almost exclusively European and Icelandic controlled market.

—*Pacific Fishing*, Yearbook, 1983[514]

The final paragraph of part 1 of this book stated that the Salt Cod Era ended in 1950, with the last fishing voyage of the sailing schooner *C. A. Thayer*. That statement was premature. After the passage of the Magnuson-Stevens Act, the prospect of salting cod was alluring to Alaska fishermen and processors, in large part because initial start-up costs were relatively modest, the process was simple, and there appeared to be a good margin on the product. Additionally, codfish stocks along Norway's coast were in decline, which caused a reduction in the Norwegian production of salted cod and created an opportunity to use Alaska cod to fill the resultant market vacuum. Several Alaska operations sold salted cod to Norwegian firms, which then further processed it and sold it to their regular customers. Alaska fishermen without a domestic market for cod took advantage of a Magnuson-Stevens Act provision that gave the governor of Alaska authority to allow foreign processing vessels to operate in state waters if it could be shown that domestic processors could not handle fish being caught by U.S. fishermen. Relatively small trawlers delivered bled cod to foreign vessels, on which the cod were then processed by foreign crewmen.

Nigeria was Africa's largest consumer of salted cod and annually imported some 45,000 tons of stockfish, primarily from Norway and Iceland. Nigeria paid for its imports with revenue derived from oil exports, but those exports came to a dramatic halt in the early 1980s because of a glut of oil on world markets. The resultant shortage of oil revenue, combined with a lowered credit rating, inhibited greatly Nigeria's ability to import stockfish, and in 1982 Iceland cut off shipments of dried cod to Nigeria. *Pacific Fishing* wrote in early 1983 of Nigeria being "a very tentative market, at best." [515] The following year, *Pacific Fishing* noted the "virtual closing of the Nigerian stockfish markets."[516]

Figure 59. Crewmen with salted cod, schooner *Vansee*, 1980. (Courtesy of Per Odegaard.)

Nigeria's economy eventually recovered, and despite occasional interruptions caused by civil unrest, Nigeria has continued to be an avid purchaser of dried cod, including what has become an important ancillary product: dried cod heads (see chapter 20).

The first post-Magnuson-Stevens Act effort to salt Pacific cod was by Per Odegaard and Marvin Gjerde, captains, respectively, of the longline halibut schooners *Vansee* (87 feet long) and *Tordenskjold* (75 feet long). Faced with ever-shortening halibut seasons and sparked by rumors of a shortage of codfish in Norway, the men looked to catching and salting Pacific cod to supplement the employment of their vessels and crews. In early April 1980, after taking on salt in Seattle, the *Vansee*, with a crew of eight men, and the *Tordenskjold*, with a crew of seven men, departed for the Bering Sea fishing grounds. There, the vessels fished conventional longline gear, with number 10 hooks spaced at forty-two-inch intervals. To ensure their fish came aboard alive, the gear was allowed to *soak* (remain on the seafloor) for no longer than two hours. Severe weather that spring limited the schooners' fishing effort to just seventeen days—some of them partial days.

The cod caught by the schooners were bled, split, and salted conventionally in the vessels' holds. Then, after ten days curing in the salt, the fish were pulled out and resalted. Overall, salting cod proved to be hard and time-consuming work, especially given the schooners' limited work space.

The *Vansee*'s salted cargo totaled 26,000 pounds; the *Tordenskjold*'s totaled 20,000 pounds. Both vessels sold their cargos to Icicle Seafoods, which had a plant at Seward and marketed the fish in Italy. These deliveries marked the end of the schooners' codfish venture. In the words of *Pacific Fishing*, Odegaard and Gjerde had "found the hand labor involved in the many-step operation not worth the money."[517]

Interest in salting cod, however, remained, and in February 1980 several businessmen associated with Alaska's fisheries negotiated an agreement with a Norwegian firm for the purchase of salted cod. During the fall of that year, about a half-dozen king crab vessels that had been fitted with stern ramps, removable shelter decks, and German-made (Baader) heading-and-gutting, splitting, and washing machinery, began trawling for Pacific cod and salting their catch onboard. The fish they caught were large, averaging about twelve pounds. (The largest weighed seventy-three pounds.) The large size, however, was a mixed blessing because, while the market favored large fish, many were too large for machine processing and had to be processed by hand, a time-consuming job. The vessels' production was stored on Trident Seafoods' property at Akutan, and on Christmas Eve a tramp freighter loaded 250 tons of salted cod for delivery in Norway. It was an encouraging beginning.

The following year, the production of the six vessels involved in the salt cod fishery was stored in a building leased from Peter Pan Seafoods at Squaw Harbor, in the Shumagin Islands. The vessels' production in 1981 was an impressive 15 million pounds (round weight), but it was dwarfed by the 65 million pounds of cod that were frozen, mostly aboard floating processors.[518]

In February 1982, A. D. Chandler, writing in *National Fisherman*, called salting codfish in Alaska "a booming new game." At least six companies were involved, and together they were capable of processing a million pounds (round weight) of codfish per day. Technological developments facilitated production; in contrast to codfish being processed by a dress gang, as during the Salt Cod Era, now automated machinery headed and split the cod and—for better or for worse—reduced the amount of labor needed. (Gutting the cod was still done manually.) In 1981, there were six splitting machines in operation in the Aleutian Islands area; by the spring of 1982, there were eighteen. Nine were aboard vessels, six were at Dutch Harbor, and three were at Akutan. Each machine could handle approximately twenty-two cod per minute—almost eleven thousand fish per eight-hour shift.

In April 1982, Johansen/Sea Pro, a joint venture with the Norwegian firm Jan Magnus Johansen, began salting cod at Dutch Harbor.

The company took deliveries from five trawlers that in combination produced about 300,000 to 400,000 pounds of codfish per week. That same spring, Norwegian-owned Jangaard Alaska Fisheries announced its intention to construct a 25,000-square-foot salting plant at Dutch Harbor. The company apparently built a plant but left the industry that fall, apparently the victim of unfavorable exchange rates between the U.S. dollar and the Norwegian krone that pushed the price of salted codfish imported from Alaska to Norway beyond what the market could bear. Nevertheless, salted codfish production in Alaska increased to slightly more than 26 million pounds in 1982. Interest in salting cod, however, seemed to be waning, and some thought the number of companies involved might decrease.[519]

Salted codfish (*bacalhau* in Portuguese) "is to Christmas Eve in Portugal what turkey is to Thanksgiving in America," wrote the *New York Times* in 2008. Portugal had been a traditionally strong market for salted codfish, and in September 1982 a group of fishermen and businessmen from Kodiak and the Portuguese fish-processing company Pascoal & Filhos established the Alaskan/Portuguese Bottomfish Corporation. The following spring, the company brought to Kodiak the Portuguese floating processor *San Gabriel*. Alaska governor Jay Hammond, under the Magnuson-Stevens Act provision noted above, gave the Portuguese vessel a permit to process fish in Alaska state waters. A handful of local trawlers operating off the west shore of Kodiak Island and in Shelikof Strait supplied the *San Gabriel* with codfish. The vessels bled and gutted each cod they caught and delivered their catch daily to the processor.[520] The *San Gabriel* purchased the fish by the pound, and there was no minimum size limit. The *San Gabriel*'s state permit expired in June 1983, after which the vessel departed for Portugal. It was replaced later that year with the *Maria Ramos dos Pascoal*.

A provision of the state permit under which this vessel operated required the Alaskan/Portuguese Bottomfish Corporation to actively promote the development of shore-based fish processing in Alaska. The company, however, apparently made no effort to do so, and in February 1984, Alaska governor Bill Sheffield denied a request for an extension of the permit. Before its permit expired in February 1984, the *Maria Ramos dos Pascoal* had purchased about 5 million pounds of Pacific cod from five trawlers in the Akutan area. In the fall of that year, the *San Gabriel* returned to Alaska and, in exchange for its owner's (Pascoal & Filhos) purchase of a substantial quantity of salted cod from U.S. processors, was granted a joint-venture permit for up to 2,500 metric tons of Pacific cod, contingent upon the company purchasing additional quantities of salted cod from U.S. processors. During this venture, the *San Gabriel* operated in the Dutch Harbor/Akutan area.

One indirect benefit of the fishery at Kodiak was that it provided Alaska fishermen an opportunity to learn where the fish were and how best to trawl for them. Fortunately, the aforementioned strong 1977 year class provided an abundance of large cod in Shelikof Strait.[521]

In 1983, four trawlers—the *Gun-Mar*, *Royal Atlantic*, *Great Pacific*, and *Oceanic*—salted cod at sea.[522] The following year was different: *Pacific Fishing* described the Alaska salt cod industry as having gone "into hibernation." As had been the case with Jangaard Alaskan Fisheries, the strong U.S. dollar was to blame. Three years prior, when these vessels' owners had invested in the equipment to salt codfish, the Norwegian krone—the currency of their major competitors—was worth $0.22; by 1983 it had declined to about $0.13 and the U.S. producers could no longer compete on worldwide markets.[523]

There was, however, a foreign floating salt cod venture that year. It was the 200-foot Spanish processor *Mar de Labrador*. The vessel was operated by the Alaska Salt Fish Company as part of a joint venture with three U.S-owned trawlers that supplied it with cod for salting. Unfortunately, the *Mar de Labrador* sank west of Kodiak Island in November 1984. Fortunately, all thirty crew members, including two U.S. observers, were transferred to a trawler before the vessel sank. The three trawlers involved in the venture lost not only their market but also their cod ends, which had been on the deck of the Spanish vessel.[524]

A similar, later—and more successful—joint-venture operation was the *Estevao Gomes*, a 250-foot Puerto Rican floating processor. In the spring of 1987, the operation shipped 184,000 pounds of containerized salted cod from Dutch Harbor to Puerto Rico for additional processing.[525] By this time, however, domestic processing capacity had mostly replaced joint ventures, and the *Estevao Gomes* venture was likely the last of its kind.

SHORE-BASED COD SALTERS

The most ambitious shore-based effort to salt cod was the $12 million Trident Seafoods plant at Akutan, near Unimak Pass, which began operations in June 1982. The 49,000-square-foot plant, paid for in part by a $1.7 million grant from the Alaska Fisheries Development Foundation, was capable of processing more than 600,000 pounds (round weight) of codfish per day. As well as a freezing operation, the plant incorporated a 14,000-square-foot *cod house* in which cod were salted. The structure was air-conditioned and had sloped floors that

Figure 60. Stockfish drying structure at Trident Seafoods plant, Akutan, 1982. (Courtesy of Trident Seafoods.)

prevented pooling of the liquid that drained from the curing fish. It could accommodate 3.6 million pounds of split, salted cod stacked on wooden pallets for curing. In addition, Trident's Akutan plant had 100,000 square feet of stockfish drying racks that could accommodate about 260,000 cod.

Trident's plant marketed its salted cod and stockfish in Europe, northern South America, and the Caribbean. In the year beginning June 1, 1982, Trident purchased 35.4 million pounds of codfish, for which it paid fishermen an average price of about $0.125 per round (bled) pound. One of the trawlers that supplied Trident, the *Normar II*, fished mostly in Unimak Pass and itself delivered about 3 million pounds of codfish over the course of a year. Unfortunately, on June 9, 1983, a fire destroyed Trident's plant. By the time of the fire, the company had produced some 40 million pounds of salted cod. But though it had not realized a profit in doing so, Trident came close to breaking even—a significant accomplishment for an operation of this size and complexity in its first year of operation. Undeterred, Trident commenced rebuilding in January 1984. In August of that year, Trident began producing 15-pound shatterpacks (see chapter 20) of frozen cod and pollock fillets.[526] Salting cod would be left to others.

One of those others was Alyeska Seafoods—a Japanese-American joint-venture firm that operated a surimi plant at Dutch Harbor. The

Figure 61. Trident Seafoods advertisement, *Pacific Fishing*, Yearbook, 1983. (Image reproduced with permission of *Pacific Fishing*.)

company was not a regular producer of salted cod but, apparently sensing opportunity, shipped a million pounds of salted cod to Portugal in 1987 on a Japanese tramp freighter.[527] For the next few years, cod would be salted only on floating processors. And there weren't many, mostly due to low demand and difficulties in marketing the product.

The salt cod market eventually recovered, and in September 1995, *Pacific Fishing* reported that "a sizable chunk" of Alaska's cod production was salted in plants in Dutch Harbor, Sand Point, and King Cove for export to southern Europe. This market, however, had substantial pitfalls: European Union (EU) regulations subjected salted Pacific cod to quotas, duties, and minimum pricing that did not always apply to Atlantic cod. And Norway and Iceland, despite the fact that they are not members of the EU, had free access to the salt cod market in EU countries. On top of this was an influx into Europe of Barents Sea Atlantic cod from Russia.[528] To avoid the European Union limitations, Alyeska Seafoods, at Dutch Harbor, during 1994 exported a considerable amount of salted cod to eastern Canada.[529]

In 1999, Adak Fisheries, a company formed to process seafood at the abandoned U.S. Navy air station at Adak, in the central Aleutian Islands, purchased at least 6.5 million pounds of Pacific cod, which it salted. In 2002, Icicle Seafoods took over operation of the Adak seafood plant (see chapter 23) and, at least in 2004, produced salted cod.[530] Production of salted cod in Alaska, however, had been declining since the mid-1990s, and by 2008 only about 1.5 percent of Alaska's Pacific cod catch was salted. The production of salted cod declined again in 2009, then ended the following year.[531]

In recent years, however, Portuguese seafood processors have imported substantial quantities of frozen western-cut H&G cod for use in the production of salted cod.[532] Large-size fish are preferred and command the highest price. Much of the Portuguese salted cod is exported, with Brazil currently the largest market. Salted cod also remains popular in Europe, Latin America, and parts of Africa.[533]

ENDNOTES

513 "Salt cod: new venture into a forgotten fishery," *Alaska Fisherman's Journal* (September 1980): 62–63.

514 "Bottomfish," *Pacific Fishing*, Yearbook (1983): 63–87.

515 Ibid.

516 "Cod Update," *Pacific Fishing* (January 1984): 34–37, 40–41.

517 "Two schooners bring in salted Pacific cod," *Alaska Fisherman's Journal* (June 1980): 30; "Born Again: A Resurrected Codfishery," *Pacific Fishing* (December 1980): 44–47; Per Odegaard, personal communication with author, April 18, 2016.

518 "Salt cod: new venture into a forgotten fishery," *Alaska Fisherman's Journal* (September 1980): 62–63; A. D. Chandler, "Seven Hope to Resurrect Alaska's Salt Cod Fishery," *National Fisherman*, January 1981, 32; Roger Fitzgerald, "U.S. salt cod effort underway," *Alaska Fisherman's Journal* (February 1981): 18–19; A. D. Chandler, "Salt Cod Venture Prompts Conversion of Big Crabber, *National Fisherman*, May 1981, 68–70.

519 A. D. Chandler, "Alaska Salt Cod Industry Could Help Offset Crab Bust," *National Fisherman*, February 1982, 19; Kenneth J. Hilderbrand, *Model White Fish Project: Trident Seafoods, Akutan, Alaska, 1982/1985* (Anchorage: Alaska Fisheries Development Foundation, 1986), 7; Impact Assessment, Inc., *Unalaska: Ethnographic Study and Impact Analysis*, U.S. Department of the Interior, Minerals Management Service Tech. Rept. No. 92 (August 1983): 106; Mark Munro, "Dutch Harbor cod markets limited," *Alaska Fisherman's Journal* (May 1982): 40–41.

[520] Al Burch, captain of the trawler *Dawn* during the joint venture, once complained to the captain of the *San Gabriel* about the great amount of fishing time that was lost while his crew bled and gutted each codfish. The captain promptly provided the *Dawn* with a pair of what he termed "gutting machines"—Portuguese workers. Burch was amazed at how fast and efficiently the men worked and the quality of their work.

[521] Elaine Sciolino, "A Portuguese Tradition Faces a Frozen Future," *New York Times*, December 15, 2008; Chris Blackburn, "Portuguese arrive in Kodiak," *Alaska Fisherman's Journal* (May 1983): 36–37; "Portuguese J/V gets extension," *Alaska Fisherman's Journal* (July 1983): 52; "Sheffield bumps Portuguese from state waters," *Alaska Fisherman's Journal* (April 1984): 48; "Joint Ventures: 62 U.S. Vessels Net 275,000 mt in Alaska," *Alaska Fisherman's Journal* (July 1984): 16–17; "Portuguese Buy Salt Cod in Exchange for JV Permit," *Alaska Fisherman's Journal* (December 1984): 33; Al Burch, owner/captain of trawler *Dusk*, personal communications with author, December 2014 and April 2016; Dan Pratschner, personal communication with author, October 19, 2014; "What Is the Alaska Fisheries Development Foundation?" *Bering Sea Fisherman* (June 1980): 3; "Major Accomplishment, 1981–1983," Alaska Fisheries Development Foundation.

[522] "The Drive to Utilize Alaska Groundfish," *Pacific Fisherman*, Yearbook (1984): 125.

[523] "Alaska Groundfish," *Pacific Fishing*, Yearbook (1984): 126–135; A. D. Chandler, "Although Pacific Cod Is Abundant, There's No One Who Will Take It," *National Fisherman*, September 1983, 8; "Subsidizing Newfoundland," *Pacific Fishing*, Yearbook (1985): 62–63; A. D. Chandler, "Strength of U.S. Dollar is Problematic," *National Fisherman*, June 1984, 31.

[524] "JV Boat Rescues Spaniards," *Pacific Fishing* (January 1985): 20; "'Dona Genoveva' Rescues 30 from Spanish Processor," *Alaska Fisherman's Journal* (December 1984): 6.

[525] "First Salt Cod Out of Dutch," *Aleutian Eagle*, May 21, 1987.

[526] Alaska Fisheries Development Foundation, "Major Accomplishment, 1981–1983," 1984; Alaska Fisheries Development Foundation, "Interim Report, Model White Fish Project," 1983; Kenneth J. Hilderbrand, *Model White Fish Project: Trident Seafoods, Akutan, Alaska, 1982/1985* (Anchorage: Alaska Fisheries Development Foundation, 1986), 4, 6, 8, 30, 39–40, 42, 55, 81; Dan Pratschner, crewman aboard *Normar II*, personal communication with author, October 19, 2014; Glenn Boledovich, "Trident Akutan Plant Burns," *Aleutian Eagle*, June 15, 1983; J. P. Goforth, "Trident Seafoods Rebuilds Akutan Plant," *Aleutian Eagle*, February 1, 1984; Krys Holmes, "Lodestar Update," *Aleutian Eagle*, December 27, 1984; Doug McNair, "Codfish Update," *Pacific Fishing* (January 1984): 34–37, 40–41.

527 "Alyeska Ships Million Pounds," *Aleutian Eagle*, December 31, 1987.

528 "Seafood Report," *Pacific Fishing* (September 1995): 12.

529 Ann Touza, "Crabbers Turn to Cod in Dutch," *Alaska Fisherman's Journal* (December 1994): 10, 12.

530 Joel Gay, "Adak: Oasis or Mirage of the Aleutian Chain," *Pacific Fishing* (November 1999): 69–71; Wesley Loy, "Adak: Exit Icicle, Enter Swasand," *Pacific Fishing* (July 2004): 9.

531 Ben Fissel et al., *Stock Assessment and Fishery Evaluation Report for the Groundfish Fisheries of the Gulf of Alaska and Bering Sea/Aleutian Islands Area: Economic Status of the Groundfish Fisheries off Alaska, 2012* (Seattle: National Marine Fisheries Service, 2014), 250, accessed July 20, 2015, https://www.afsc.noaa.gov/refm/docs/2013/economic.pdf.

532 Joel Chetrick, "Sales of Premium Products to EU Drive Record U.S. Seafood Exports," *FAS Worldwide*, March 2007, accessed July 22, 2015, http://s3.amazonaws.com/zanran_storage/www.fas.usda.gov/ContentPages/1279531.pdf; David Little, Clipper Seafoods, personal communication with author, July 26, 2015.

533 Joel Chetrick, "Sales of Premium Products to EU Drive Record U.S. Seafood Exports," *FAS Worldwide*, March 2007, accessed July 22, 2015, http://s3.amazonaws.com/zanran_storage/www.fas.usda.gov/ContentPages/1279531.pdf; Ben Fissel et al., *Stock Assessment and Fishery Evaluation Report for the Groundfish Fisheries of the Gulf of Alaska and Bering Sea/Aleutian Islands Area: Economic Status of the Groundfish Fisheries off Alaska, 2012* (Seattle: National Marine Fisheries Service, 2014), 250, accessed July 20, 2015, https://www.afsc.noaa.gov/refm/docs/2013/economic.pdf.

Chapter 15

AMERICANIZATION OF ALASKA GROUNDFISH FISHERIES

PAYING FOR AMERICANIZATION

During the mid-1970s, those who followed the West Coast fisheries were well aware that the *Americanization* of the groundfish fisheries off Alaska's coast would be a high-risk venture. And it would be expensive. A study done for the National Marine Fisheries Service in 1978 estimated the cost of fishing vessels, including catcher-processors, would be $509 million (in 1979 dollars).[534] History showed this estimate to be low. There were at the time a number of suitable vessels that could have been purchased from European interests, but a provision of the Merchant Marine Act of 1920 (Jones Act) prohibited the use of foreign-built vessels of over 5 net tons capacity for commercial fishing in U.S. waters. The act, however, did not prohibit processing fish aboard foreign-built vessels until it was amended by the Commercial Fishing Industry Vessel Anti-Reflagging Act of 1987 (see below).[535] Additionally, there was the huge cost of developing the necessary shore-based infrastructure and facilities: harbors, fish-processing plants, docks, cold-storage facilities, fueling stations, and vessel-maintenance-and-repair facilities.

In the end, a combination of private money and government grants, subsidies, incentives, and loans over almost two decades financed the Americanization of Alaska's groundfish industry.

Private money

Alaska banks. Although Alaska banks theoretically knew potential borrowers best, they were too small to make the large loans necessary for building a fleet capable of replacing the foreign fleets that fished along Alaska's coast. (The conversion in 1987 of a surplus U.S. Navy tanker into the surimi factory trawler *Arctic Storm* cost $25 million. To fully develop the industry would require dozens of relatively large, sophisticated vessels, as well as shore plants.)[536]

Lower-48 banks. Even before the Magnuson-Stevens Act, Alaska fishermen and processors—especially those with large operations—depended to a large extent on lower-48 banks. In Washington, Seafirst Bank and Rainier Bank (both later absorbed into the Bank of America) had been heavily involved in Alaska's fisheries for many years and understood the industry, but the years following the passage of the Magnuson-Stevens Act were new territory for everyone and carried more risk than the banks were generally willing to expose themselves to. As one banker said, "When Marco [a now-defunct Seattle shipyard] was turning out a new crab boat every six weeks, they had it down to a science. It's a lot more difficult to predict when a shipyard is building just one boat and they haven't done it before." As well, since Washington banks served a large, diverse economy, they likely had less risky and shorter-term options for loaning money.[537]

Norwegian banks. Norway, with its seafaring and fishing heritage, has figured prominently in the development of Alaska's fisheries. Petersburg, Southeast Alaska's most successful fishing community, was founded by a Norwegian, Peter Buschmann. The community, which bills itself as Alaska's "Little Norway," continues to maintain and celebrate its Norwegian heritage. Norwegian immigrant fishermen commonly owned or crewed on deep-water halibut longline boats that fished along Alaska's coast. Moreover, by the late 1970s, Seattle-based first-generation Norwegian immigrants owned about half of the crab fleet that fished in the eastern Bering Sea.[538] These Norwegians were on the ground floor in the development of Alaska's groundfish industry, and their connection to their homeland helped.

Norway had a long history of fishing for groundfish, including codfish in the Atlantic, and its high-seas fleet was modern and efficient. Shipbuilding was big business in Norway, and Norwegian banks, particularly the government-controlled Christiana Bank, aggressively sought to finance the rebuilding of American vessels in Norway. Some Norwegian banks even opened branches in Seattle. Between 1987 and 1990 about fifteen factory trawlers were rebuilt in foreign shipyards, mostly in Norway, where the government offered significant inducements, and the Christiana Bank provided financing of up to 150 percent of a vessel's book value, according to some claims. The Norwegian banks expected some sort of catch-share program to be implemented in the not-too-distant future. Such a program would likely enhance the potential for profitability of the trawl fleet and, moreover, the shares could potentially be used as collateral for future loans.[539]

Government money

State government incentives. Not long after the passage of the Magnuson-Stevens Act, the Alaska Department of Commerce and Economic Development implemented a modest loss-guarantee program. In 1977, two firms, the New England Fish Company and Icicle Seafoods, signed contracts with the department for pilot groundfish-processing projects at their respective Kodiak and Petersburg plants. The department guaranteed each company up to three cents per pound against market losses for the first year of groundfish-processing operations, with a $145,000 ceiling. In exchange, the companies agreed to share with the department all information generated by the projects.[540]

In 1978, Alaska's legislature created the Alaska Renewable Resources Corporation, which undertook to expand the fishing, timber, and other renewable resource industries through state-subsidized venture-capital loans. The program was soon overwhelmed by bankrupt enterprises and loan defaults, and Alaska's legislature abolished it in 1982.[541]

In 1979, Alaska's legislature established the Alaska Commercial Fishing and Agriculture Bank as a private member-owned cooperative. The bank's mission was to provide financing to resident-owned businesses engaged in harvesting, processing, or marketing seafood or agricultural products in Alaska. Throughout its first two decades of existence, however, the bank was hard pressed to meet its own obligations, and it contributed little, if anything, to the development of the groundfish industry.[542]

Federal money. In addition to the loan guarantee from the State of Alaska noted above, the New England Fish Company obtained a $150,000 grant from the U.S. Economic Development Administration for fishermen and plant employee training programs.[543]

From 1977, the year the EEZ was officially established, until roughly 1990, Americanization of the fisheries was the national priority, and U.S. government policy fostered growth in fish-harvesting capacity in many ways. At the Internal Revenue Service, investment tax credit provisions in the tax code until 1986 stimulated spending on the construction of new vessels, while tax deferral programs stimulated vessel purchase, repair, and refitting. There were direct loans as well. The Fisheries Obligation Guarantee Program, now known as the Fisheries Financing Program, was created in 1971 to guarantee loans made by the private sector but evolved into a direct lending program. The National Marine Fisheries Service administers the program, and had by 1992 provided some $102 million in loans for thirty-three factory trawlers.[544] For all practical purposes, the U.S. EEZ was fully Americanized by the end of the 1980s.[545]

Direct grants from the federal Saltonstall-Kennedy Fund also helped facilitate the development of Alaska's groundfish industry. The origin of the Saltonstall-Kennedy Fund was Depression-era legislation that authorized the secretary of agriculture to purchase surplus agricultural commodities for distribution to the poor. In 1939, Congress broadened this legislation by directing the federal government to purchase up to $1.5 million annually in surplus domestically produced fishery products for distribution to the poor. Congress also directed the secretary of the interior to spend $75,000 annually "to promote the free flow of domestically produced fishery products in commerce by conducting a fishery educational service," and to spend $100,000 annually "to develop and increase markets for fishery products of domestic origin."[546]

The 1939 legislation retroactively became known as the Saltonstall-Kennedy Act in 1955, when it was amended by Congress under the leadership of Massachusetts senators Leverett Saltonstall and John Kennedy. The 1955 amendments mandated that a permanent appropriation equal to 30 percent of the customs duties collected by the U.S. government on imported fish and fish products be used to maintain what became known as the Saltonstall-Kennedy Fund (S-K Fund). Pursuant to the legislation, the secretary of the interior was directed to use this fund to:

- "promote the free flow of domestically produced fishery products in commerce by conducting a fishery educational service and fishery technological, biological and related research programs";
- "increase markets for fishery products of domestic origin"; and
- "conduct any biological, technological, or other research pertaining to American fisheries."

The legislation also directed the secretary to "cooperate with other appropriate agencies of the Federal Government, with State or local governmental agencies, private agencies, organizations, or individuals, having jurisdiction over or an interest in fish or fishery commodities."[547] The American Fisheries Promotion Act of 1980 amended the Saltonstall-Kennedy Act to make it more specific. The legislation directed the secretary of commerce to make grants to fishery development foundations or private nonprofit corporations in Alaska that were "carrying out research and development projects addressed to any aspect of United States fisheries, including, but not limited to, harvesting, processing, marketing, and associated infrastructures."[548]

The Bureau of Commercial Fisheries, an agency within the Department of the Interior, administered S-K Fund grants until 1970. That year, President Richard Nixon transferred almost all functions associated with the bureau to the Department of Commerce. The department then created the National Marine Fisheries Service, which began administering S-K Fund grants.[549]

The Alaska Fisheries Development Foundation was created in 1978 to serve as a conduit for S-K Fund grants in Alaska. The nonprofit foundation had a ten-person board—five fish-processing company representatives and five representatives of fishermen's organizations—and its primary focus was on undeveloped fisheries, especially pollock and cod.

During the calendar year 1979, the foundation awarded two grants: $475,000 for a conversion of the 124-foot crabber *Aleutian Mistress* into a catcher-processor that could longline for cod as well as fish for crab (see chapter 17) and $275,000 to convert a shrimp trawler into a groundfish trawler.[550]

LEGISLATING AMERICANIZATION

Processor Preference Amendment, 1978

Under the Magnuson-Stevens Act, the secretary of commerce did not have the authority to deny foreign processors permits to purchase U.S.-harvested fish. To overcome this impediment to Americanizing the fisheries, in 1978 Congress amended the Magnuson-Stevens Act to give domestic fishermen and processors prioritized access to fishery resources in the Fishery Conservation Zone through a three-tier system. The highest priority was *Domestic Annual Processing* (DAP)—fish reserved for U.S.-flag fishing vessels that sold their production to a domestic processor or processed their own catch. The second level of priority was reserved for *Joint Venture Processing* (JVP)—fish caught by U.S.-flag fishing vessels and delivered to foreign-flag motherships and factory trawlers. Of lowest priority was *Total Allowable Level of Foreign Fishing* (TALFF)—fish caught and processed by foreign-flag vessels.[551] To ensure compliance with regulations and catch limits, foreign harvests were monitored by onboard observers.

American Fisheries Promotion Act of 1980.

In early 1980, Congress was disappointed that the establishment of the Fishery Conservation Zone and the 1978 amendment to the

Magnuson-Stevens Act had not resulted in a rapid expansion of the domestic fishing industry. American fishermen in 1979 caught 33 percent of the fish landed in the Fishery Conservation Zone, an increase of only about 10 percent since the passage of the Magnuson-Stevens Act in 1976. Congress passed the American Fisheries Promotion Act in December 1980 to promote the development of the domestic fishing industry by increasing the American share of the total harvest in the Fishery Conservation Zone and by encouraging increased access of U.S. fish products in foreign markets.

The legislation, which was authored by Louisiana representative (later senator) John Breaux, a Democrat, included amendments to several provisions of the Magnuson-Stevens Act, most notably the establishment of a formula for reducing and ultimately eliminating the TALFF in the Fishery Conservation Zone. Under that formula, foreign allocations of groundfish in the Bering Sea and western Gulf of Alaska gradually decreased, and in 1989 they were totally eliminated. The formula depended only in part upon the U.S. capability to catch fish; the so-called Fish & Chips provision of the American Fisheries Promotion Act allocated portions of the TALFF to foreign nations based on their cooperation with and assistance to the U.S. fishing industry, as determined by the secretaries of state and commerce. Mostly, the allocations were related to how much U.S.-produced fish or fish products a country purchased and how willing a country was to participate in joint ventures with American fishermen.[552]

As mandated by the Magnuson-Stevens Act, foreign vessels fishing in the Fishery Conservation Zone were required to purchase a license and to pay a poundage fee and an observer fee, and were also subject to a surcharge fee. U.S. fishermen, on the other hand, paid nothing. As Kevin Bailey wrote in his book *Billion-Dollar Fish*, for U.S. fishermen, the fish in the Fishery Conservation Zone were "a public and common access resource, free for those with a means of accessing it."[553]

U.S. fishing vessels had another important advantage over their foreign counterparts: they could fish within the three-mile limit (state waters), such as in Aleutian Islands passes.

The legislation had the desired effect: in 1980, the joint-venture catch of Alaska groundfish was 34,500 metric tons; in 1982, the catch increased to 190,500 metric tons. This increase was, of course, offset by a reduction in the amount of fish foreign fishing vessels were allowed to harvest.[554]

In joint ventures, the cod ends of the trawl nets used by the American catcher boats were detachable from the main net, and fish caught in the trawls were usually transferred from the catcher boats to the processing vessels through a sometimes-tricky procedure known as a *cod end transfer*. During the transfer, the cod end was first tied off at its forward end to prevent fish from escaping. The cod end was then detached from the main net and then connected to a line from the processing vessel, which hauled the cod end onboard and unloaded it. The entire process could take as little as ten minutes. Cod ends full of cod floated (high enough for a fisherman to walk on) because the cods' air bladders inflated when the fish were brought up from the deep. Vessels making cod end transfers typically carried extra cod ends to prevent downtime after transferring the codend to a processing vessel.

> *In short, 1985 was the year the Americanization of the Alaska groundfish fisheries actually started to happen, the year the snowball started to roll.*
>
> —Chris Blackburn, *Alaska Fisherman's Journal*, January 1986[555]

By 1987, the accelerating Americanization of the Alaska groundfish fisheries made it apparent that joint ventures had little future. By 1989—the same year directed foreign fishing ended—the groundfish quota allocated to joint ventures had become insignificant, and the last joint-venture operation ended in 1990.[556]

To facilitate the management of the now fully Americanized fisheries, the National Marine Fisheries Service that same year implemented a comprehensive industry-funded onboard observer program to collect data on the total catch, including at-sea discards, and to accurately monitor bycatch. Currently, the observer fee is 1.25 percent of the ex-vessel value of the groundfish subject to the fee.[557]

Commercial Fishing Industry Vessel Anti-Reflagging Act of 1987

In 1986, just as the last of the foreign-flag fishing vessels operating in the U.S. EEZ were being replaced by U.S.-flag vessels, there remained four significant impediments to comprehensive Americanization of the fisheries:

1. It was still possible for corporations organized under U.S. laws to document vessels for use in U.S. fisheries, even if all the corporate stock was owned by foreign citizens.
2. Under the Merchant Marine Act of 1920 (Jones Act), foreign-built vessels could be used for fish processing within the EEZ.

3. Many of the workers on fishing and processing vessels were not U.S. citizens.

4. Although the Jones Act required U.S.-documented commercial fishing vessels measuring greater than five net tons that operated in U.S. waters to have been built in U.S. shipyards, fishing vessels that had been built in the United States could be rebuilt abroad. Several fishing companies had purchased U.S.-built vessels and had them completely rebuilt abroad, often retaining only a small portion of the original vessel.

The Commercial Fishing Industry Vessel Anti-Reflagging Act of 1987 was designed to reduce the level of foreign participation in the U.S. fisheries and to ensure that only U.S.-built vessels could participate in fishery-related activities. The legislation required:

- U.S. control of U.S.-flag fishing vessels;
- U.S.-flag fishing vessels to be crewed by U.S. citizens or lawfully admitted aliens;
- vessels processing fish or shellfish in the EEZ to have been U.S.-built (this provision amended the Jones Act); and
- ending what was essentially the construction in foreign shipyards of U.S.-flag vessels under the guise that they were being "rebuilt."[558]

The legislation, however, provided "grandfather" exemptions for rebuild projects that had already been planned. Speculators had anticipated this provision and took liberal advantage of it. Between 1987 and 1990 about fifteen factory trawlers were rebuilt in foreign shipyards, mostly in Norway. The 85-foot *Acona*, a former University of Alaska research vessel, was the most egregious example of the flouting of the intent of the Anti-Reflagging Act. The vessel—actually just part of its keel—was transformed in a Norwegian shipyard into the 286-foot *American Triumph*, a state-of-the-art factory trawler.[559] A 1990 article in *National Fisherman* derided the Anti-Reflagging Act as having loopholes "big enough to float a factory trawler through."[560]

The American Fisheries Act, 1998

The "rebuilt" trawlers contributed significantly to the excess capacity in the pollock industry, and, except for the law's vessel-manning provision, the Anti-Reflagging Act was considered a failure because of, according to Senator Stevens, "mistakes in, and misinterpretations of," the legislation.[561]

In 1998, about half of the Bering Sea pollock were harvested by, in the words of Stevens, "foreign interests on foreign-built vessels that are not subject to any U.S.-controlling interest standard."[562] Comprehensive Americanization of the fisheries was being thwarted, while overcapacity was growing. In 1998, Congress corrected some of the Anti-Reflagging Act's shortcomings by passing the American Fisheries Act.

The American Fisheries Act, which was introduced by Senator Stevens in September 1997 and passed by Congress in October of the following year, was an attempt to decapitalize and rationalize the Bering Sea pollock fishery, in part by fixing the mistakes in the Anti-Reflagging Act (see above). This complex legislation was designed to make the pollock fishery in the Bering Sea/Aleutian Islands area more efficient and flexible, but doing so had the potential to impact other fisheries, including the Pacific cod fishery. To minimize or eliminate these impacts, the NMFS added special provisions to regulations relating to this act.[563]

Fundamentally, the American Fisheries Act required the owners of all U.S.-flag fishing vessels to comply with a 75 percent U.S.-controlling interest standard. It also made provisions for the removal from U.S. fisheries at least half of the foreign-built factory trawlers that had entered the fisheries through the Anti-Reflagging Act foreign-rebuild grandfather provision and continued to be foreign owned as of September 25, 1997. The legislation listed nine factory trawlers as ineligible for continued participation in the industry.[564]

Importantly, the American Fisheries Act also contained major provisions—a vessel buyback program and a provision that allowed the formation of fishery cooperatives (including cooperatives tied to shore-based processors)—that fostered the *rationalization* and *Alaskanization* of Alaska's groundfish fisheries (see chapters 22 and 23).

NON-AMERICAN FISHERIES ACT BERING SEA TRAWLERS

About twenty Bering Sea trawlers historically targeted only Pacific cod and did not meet the eligibility requirements of the American Fisheries Act. Referred to as the *non-AFA Bering Sea trawlers*, these vessels now operate under a license limitation program but not as members of a cooperative or under an individual fishing quota program. In 2010, this fleet, which continues to harvest primarily Pacific cod, consisted of a dozen catcher vessels and one factory trawler.[565]

ENDNOTES

[534] Earl R. Combs, Inc., "Export and Domestic Market Opportunities for Underutilized Fish and Shellfish," U.S. Department of Commerce, National Marine Fisheries Service, December 1978; Kevin M. Bailey, *Billion-Dollar Fish: The Untold Story of Alaska Pollock* (Chicago: University of Chicago Press, 2013): 70.

[535] Merchant Marine Act of 1920, June 5, 1920 (41 Stat. 988); Commercial Fishing Industry Vessel Anti-Reflagging Act of 1987, January 11, 1988 (101 Stat. 1778); 46 U.S.C. § 12101(a)(1).

[536] Earl R. Combs, Inc., "Export and Domestic Market Opportunities for Underutilized Fish and Shellfish," U.S. Department of Commerce, National Marine Fisheries Service, December 1978; Kevin M. Bailey, *Billion-Dollar Fish: The Untold Story of Alaska Pollock* (Chicago: University of Chicago Press, 2013): 70.

[537] "Bankers Look at Overcapitalization," *Pacific Fishing* (March 1988): 45.

[538] James Strong and Keith R. Criddle, *Fishing for Pollock in a Sea of Change: A Historical Analysis of the Bering Sea Pollock Fishery* (Fairbanks, Alaska: Alaska Sea Grant, 2013), 21.

[539] Bill Saporito and Thomas J. Martin, "Most Dangerous Job in America," *Fortune*, May 31, 1993, accessed June 22, 2015, http://archive.fortune.com/magazines/fortune/fortune_archive/1993/05/31/77905/index.htm; James Strong and Keith R. Criddle, *Fishing for Pollock in a Sea of Change: A Historical Analysis of the Bering Sea Pollock Fishery* (Fairbanks, Alaska: Alaska Sea Grant, 2013), 37, 75.

[540] George W. Rogers, *Development of an Alaskan Bottomfish Industry and State Taxes* (Juneau, Alaska: Institute of Social and Economic Research, undated, but probably 1977), 38–39.

[541] Alaska Renewable Resources Corporation, *Session Laws of Alaska*, 1978, § 37.12.010–015; Gerald A. McBeath and Thomas A. Morehouse, *Alaska Politics and Government* (Lincoln: University of Nebraska Press, 1994), 67; Alaska Renewable Resources Corporation, *Session Laws of Alaska, 1982*, § 37.12.010–015.

[542] Alaska Commercial Fishing and Agriculture Bank, *Session Laws of Alaska*, *1979*, § 44.81.010.

[543] George W. Rogers, *Development of an Alaskan Bottomfish Industry and State Taxes* (Juneau, Alaska: Institute of Social and Economic Research, undated, but probably 1977), 39.

[544] The Fisheries Financing Program was originally created in 1971 as the Fishing Vessel Mortgage and Loan Insurance Program. It was renamed the Fishing

Vessel Obligation Guarantee Program in 1973. In 1998, it became the Fisheries Finance Program.

545 William T. Hogarth, Acting Asst. Administrator for Fisheries, National Marine Fisheries Service, in: *Reauthorization of the Magnuson-Stevens Fishery Conservation and Management Act, oversight hearing, May 10, 2001*, Serial No. 107-26, House Committee on Resources, 107th Cong., 1st sess. (Washington, DC: GPO, 2001), 20, 22; Donna Parker, "Alaska Onshore Wins in Washington," *Pacific Fishing* (May 1992): 28–30; 79 *Fed. Reg.* 36699, June 30, 2014.

546 Act of August 11, 1939 (53 Stat. 1411).

547 Act to Further Encourage the Distribution of Fishery Products, and For Other Purposes ("Saltonstall-Kennedy Act"), July 1, 1954 (68 Stat. 376), § 2(a), § 2(b).

548 American Fisheries Promotion Act of 1980, December 22, 1980 (94 Stat. 3287), § 2(a)(1)(B), § 2(c)(1).

549 AllGov, "National Marine Fisheries Service," accessed March 15, 2019, http://www.allgov.com/departments/department-of-commerce/national-marine-fisheries-service?agencyid=7135#historycont.

550 "What Is the Alaska Fisheries Development Foundation?" *Bering Sea Fisherman* (June 1980): 3.

551 An Act to authorize appropriations to carry out the Fishery Conservation and Management Act of 1976 during fiscal year 1979, to provide for the regulation of foreign fish processing vessels in the fishery conservation zone, and for other purposes, August 28, 1978 (P.L. 95-354).

552 Stephen C. Stanley, "Mare Clausum: The American Fisheries Promotion Act of 1980," *Syracuse Journal of International Law and Commerce*, vol. 9, no. 2 (1982): 403–426, accessed January 12, 2015, http://surface.syr.edu/jilc/vol9/iss2/16; American Fisheries Promotion Act of 1980, December 22, 1980 (94 Stat. 3287), §§ 230, 231; Rep. John B. Breaux, in: *American Fisheries Promotion, Hearing on H.R. 7039*, H. Subcommittee on Fisheries and Wildlife Conservation and the Environment, Committee on Merchant Marine and Fisheries, 96th Cong., 2d sess., May 6, 19, 20, 1980, Serial No. 96-44 (Washington, DC: GPO, 1980), 323; John Sabella, "Joint Ventures: Enormous Promise and Broken Promises," *Pacific Fishing* (January 1982): 33–35, 37; Beth A. McGinley, "Joint Ventures in Fisheries," *Pacific Fishing* (February 1989): 34–38.

553 Fishery Conservation and Management Act of 1976, April 13, 1976 (90 Stat. 331), § 204(b)(10); S. E. Jelley, "The Price of Fishing in American Waters," *Fishing News International*, 22, no. 5 (1983): 8–9; Kevin M. Bailey, *Billion-Dollar Fish: The Untold Story of Alaska Pollock* (Chicago: University of Chicago Press, 2013): 70.

554 John Sabella, "Joint Ventures: Enormous Promise and Broken Promises," *Pacific Fishing* (January 1982): 33–35, 37; Doug McNair, "Joint Ventures—What Are They? Who Are They?" *Pacific Fishing* (December 1982): 50–53; Kris Freeman, "The Logistics of Today's Joint Ventures," *Pacific Fishing* (December 1987): 24–26.

555 Chris Blackburn, "'85 a Landmark Year for Groundfish," *Alaska Fisherman's Journal* (January 1986): 50–52.

556 Beth A. McGinley, "Joint Ventures in Fisheries," *Pacific Fishing* (February 1989): 34–38; National Marine Fisheries Service, *Draft Environmental Assessment/Regulatory Impact Review for The Emergency Rule to Implement Reasonable and Prudent Steller Sea Lion Protection Measures in the Pollock Fisheries of the Bering Sea and Aleutian Islands Area and the Gulf of Alaska* (January 2000), 89.

557 U.S. Regional Fishery Management Councils, *Celebrating 40 Years of Regional Fisheries Management (2016)*, 34, accessed December 26, 2016, http://safmc.net/download/MSA40thAnniversary_May2016.pdf; NOAA Fisheries, North Pacific Halibut and Groundfish Observer Program, Observer Fee Collection, accessed December 26, 2016, https://alaskafisheries.noaa.gov/sites/default/files/observerfees.pdf.

558 Commercial Fishing Industry Vessel Anti-Reflagging Act of 1987 (101 Stat. 1778).

559 "Bering Cap: No Direct Foreign Fishing," *Pacific Fishing* (February 1988): 20–21; American Fisheries Act Coalition, "American Fisheries Act Coalition: Huge Factory Trawler, Rebuilt in Foreign Shipyard, Continues to Haunt Its Owners," news release, May 13, 1998, accessed March 7, 2015, http://www.prnewswire.com/news-releases/american-fisheries-act-coalition-huge-factory-trawler-rebuilt-in-foreign-shipyard-continues-to-haunt-its-owners-77892817.html.

560 Todd Campbell, "Vessel Anti-Reflagging Law Attacked as Weak," *National Fisherman*, May 1990, 22–23.

561 Sen. Ted Stevens, *Cong. Rec.* (October 21, 1998), S12777–S12778.

562 Ibid.

563 National Marine Fisheries Service, Alaska Region, *Final Environmental Impact Statement for American Fisheries Act Amendments 61/61/13/8* (February 2002), ES-1.

564 American Fisheries Act, October 21, 1998 (112 Stat. 2681-616).

565 David Witherell, Michael Fey, and Mark Fina (staff, North Pacific Fishery Management Council), *Fishing Fleet Profiles* (April 2012), 17–18.

SECTION B
Catching Codfish

Chapter 16
TRAWLING

Bottom trawling is the most common human-caused physical disturbance to the world's seabed.

—Aaron Orlowski, SeafoodSource News, August 2017[566]

TRAWL CATCHER VESSELS

The trawls (see chapter 5) codfishermen use are bottom (nonpelagic) trawls that have weighted footropes that are fitted with rubber disks or bobbins that roll along the seafloor to minimize contact with the seafloor and protect the net. Trawlers are commonly referred to as *draggers*, simply because they drag a net behind them.[567]

Originally, the vessels used in trawling were *side trawlers* that, as their name implies, set, towed, and hauled the trawl from the side of the vessel. The operation was an unwieldy, often dangerous process, especially in rough seas. Trawling became much more efficient and safer in the late 1960s, when stern ramp trawlers came into use. On these vessels, the trawl is set, towed, and hauled entirely over the stern of the vessel. The trawl is stored on a hydraulically powered reel mounted forward of an inclined ramp in the vessel's stern. The trawl doors are carried on stanchions on either side of the ramp. Each door is connected by a heavy steel cable to a hydraulically powered winch used, in coordination with its counterpart on the other side of the vessel, to deploy and retrieve the trawl, and to tow it.[568]

Modern bottom trawls that catcher vessels use to target Pacific cod typically have a front opening that, while being towed, is twelve to thirty feet deep and about sixty to seventy-five feet wide. The trawl is towed at a speed of 2.5 to 4 knots for, depending upon catch rates, from one to four hours.[569]

Beginning in the mid-1960s, trawls were often equipped with electronic monitors. Early monitors took snapshot photos at the headrope that showed only a small section of the entire net opening, but indicated the distance from the headrope to the footrope or to the ocean floor, and even showed fish entering the net. The next development was catch monitors that sent a signal when the cod end mesh stretched beyond a certain point due to the amount of fish that had accumulated in the trawl's cod end, indicating it was probably time to haul the net. Modern catch sensors—referred to as *eggs* by trawlermen

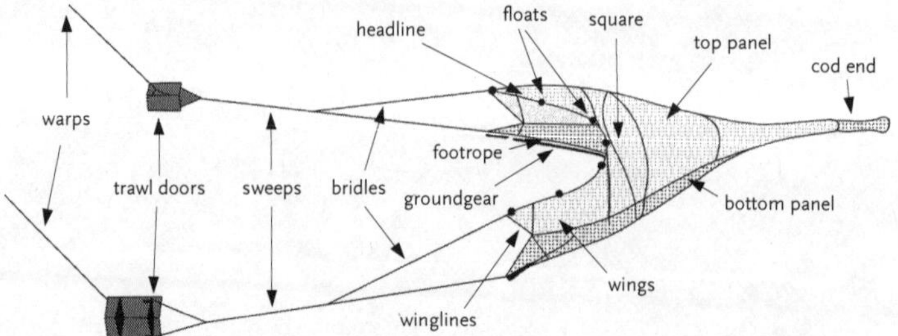

Figure 62. Modern otter bottom trawl. The footrope is weighted and is fitted with rubber discs, bobbins, or roller gear to shield the lower leading edge of the trawl from damage. (NOAA Fisheries, Alaska Fisheries Science Center, *Draft Programmatic Environmental Assessment for Fisheries and Ecosystem Research Conducted and Funded by the Alaska Fisheries Science Center*, June 2016, appendix A, A-6.)

Figure 63. Trawler *Tracy Anne* in rough seas near Kodiak. (Photographer unknown. Courtesy of Al Burch.)

because of their shape—provide a host of information, including the precise GPS location of the trawl and its doors and a sonar-generated view of the net, the bottom, and fish inside and outside the net. The sensors communicate with the vessel acoustically or via a *3rd wire* coaxial cable that is stored on its own reel.[570]

Bottom trawling is not without its critics: in 1995, a Canadian representative of inshore fishermen described bottom trawling as "like towing two D9 tractor blades linked with 150 feet of very heavy chain and rollers back and forth across every inch of the fishing grounds."[571]

Some forty to fifty catcher vessels target Pacific cod in the central Gulf of Alaska. Their effort is concentrated in the waters south of Kodiak Island, and they primarily deliver their catch to processing plants at Kodiak. In 2014, trawlers delivered some 29 million pounds of cod at Kodiak.[572] Typically, trawlers in the central Gulf of Alaska range from about 80 feet to about 120 feet long.

In the western Gulf of Alaska, the approximately dozen catcher vessels utilized in trawling tend to be smaller than those in the central gulf. These vessels typically participate in several other fisheries, including seining for salmon, a fishery in which vessels can be no longer than 58 feet overall. This fleet primarily fishes in the Shumagin Islands and delivers its catch to processing plants at Sand Point and King Cove.[573]

The Bering Sea/Aleutian Islands area catcher vessel trawl fleet consists of about a dozen vessels that tend to be from about 58 to 90 feet in length. The fleet's primary target is Pacific cod, and its effort for this species is concentrated immediately north of Cape Sarichef, on Unimak Island, in an area known to fishermen as "Cod Alley." More than half of the fleet's catch is delivered in Dutch Harbor and Akutan.[574]

The trawl fishery is typically conducted mostly during the first few months of the year.

Trawl catcher vessels that operate in the central and western Gulf of Alaska and in the Bering Sea are represented by the Alaska Whitefish Trawlers Association and by United Catcher Boats. Trawl catcher vessels homeported in Kodiak are also represented by the Alaska Groundfish Data Bank.

FACTORY TRAWLERS

As for destroying the resource, everybody who's got a factory trawler has got something up to fifty or sixty million dollars invested in each one. It's not in our best interest, even if we're as venal as

Figure 64. Codfish in trawler's *alleyway*. (Courtesy of Alaska Groundfish Data Bank.)

> *some people think we are, to suck up all the fish. We're going to take years and years to recover that money. We have a vested interest more than anybody to assure that resource continues.*
>
> —Bob Morgan, president, Alaska Factory Trawler Association, 1990[575] [576]

As the term implies, a factory trawler is a trawler that incorporates a factory in which the vessel's catch is processed, usually by freezing, though some are equipped as well with machinery to convert fish waste into fish meal.

Factory trawlers are the largest class of vessels currently fishing in Alaska waters. The 1991 Alaska factory trawler fleet, which mostly fished pollock but caught cod as well, was entirely based in Puget Sound. It consisted of sixty vessels that ranged from 104 feet long (*Golden Fleece*) to 376 feet long (*Alaska Ocean*) and averaged about 225 feet long. Pulling a trawl requires a lot of power, and the drive engine horsepower on factory trawlers generally ranges from about 1,500 horsepower to about 3,000 horsepower. The trawls typically employed today are about 300 feet wide and have a vertical opening of from 12 feet to 30 feet. A single *drag* with a trawl usually lasts about two hours and can yield more than 100 tons of groundfish. The vessels usually carry from twenty to forty crewmen.[577] Factory trawlers are today represented by the At-Sea Processors Association and the Groundfish Forum.[578]

The story of factory trawlers in Alaska began in the U.S. Congress in the late 1950s. Congress had become concerned that the high cost of constructing fishing vessels in domestic shipyards had put U.S. fishermen at a competitive disadvantage vis-à-vis their foreign counterparts. To put U.S. fishermen on a more equal footing, Congress in 1960 passed the U.S. Fishing Fleet Improvement Act, which provided a construction subsidy to offset the difference between the cost of constructing the vessel in a U.S. shipyard and the estimated cost of constructing it in a foreign yard. The subsidy maximum was established at one-third of the domestic cost of an eligible vessel's construction.[579]

Eligibility, according to the U.S. Bureau of Commercial Fisheries, which administered the construction subsidy program, was "restricted to vessels of advanced design, capable of fishing in expanded areas (fishing grounds not usually fished by the majority of vessels working in a particular fishery), equipped with newly developed gear, and scheduled for operation in a fishery where such use will not cause economic hardship to other operators now in that fishery."[580] Because of U.S. shipyard cost increases, Congress raised the subsidy maximum to one-half in 1964. As well, Congress authorized an annual appropriation of $10 million to fund the construction subsidy program. The U.S. Fishing Fleet Improvement Act expired in 1969, but before doing so, the legislation helped pay for the construction of the first U.S. factory trawlers.[581]

During the mid-1960s, about half of the fish consumed in the United States was imported. In 1965, as an effort to pioneer the development of large-scale U.S. groundfish trawling and at the same time help the United States catch up to the foreign—particularly Japanese and Soviet—competition, the firm American Stern Trawlers was incorporated. The company contracted for the construction in a Baltimore shipyard of two virtually identical *factory-supertrawlers*, the *Seafreeze Atlantic* and the *Seafreeze Pacific*. The vessels were launched in 1968.

The 295-foot vessels—the first U.S.-flag factory trawlers—were the largest and most costly vessels in the entire U.S. fishing fleet. Each was equipped with automated filleting machinery, freezers that were capable of freezing 40,000 pounds per day, a 2-million-pound-capacity cold storage, a fish meal plant, and accommodations for sixty crew members. The cost of each vessel was close to $6 million, about $2.6 million of which was provided by the U.S. Fishing Fleet Improvement Act subsidy.[582]

The first deployments of these vessels were financial disasters, mostly because of the inability of the vessels' owners to hire skilled crewmen who had a demeanor that allowed them to function and remain civil during months at sea. The *Seafreeze Atlantic* fished in the North Atlantic Ocean for three trying years until it was laid up in Norfolk, Virginia, and offered for sale. The *Seafreeze Pacific* made its debut in the North Pacific Ocean in November 1969 and fished along the coasts of Washington, British Columbia, and Southeast Alaska on and off until December 1970, when, because of unprofitable operations, it was laid up in Seattle and offered for sale.

In the spring of 1973, Pan-Alaska Fisheries, a successful Seattle-based king crab processing company, purchased the mothballed *Seafreeze Pacific*. The company also had an option to purchase the *Seafreeze Atlantic*, but chose not to do so. The *Seafreeze Pacific* was renamed *Royal Sea* and that summer sent to the Bering Sea. Ronald Jensen, president of Pan-Alaska Fisheries, explained the reason for the purchase: "We regard the area of the Bering Sea as having the greatest fisheries potential of any place in the world." Jensen added: "This is the only way we are going to get these guys (foreign fleets) out of there before the resources are depleted and it's too late to do anything."[583]

Though its commitment was admirable, Pan-Alaska Fisheries was ahead of its time, and the venture did not work out well. Once on the fishing grounds, the *Royal Sea* was plagued by mechanical problems that required it to return to Seattle before any serious fishing or processing occurred.[584] The vessel was then converted into a crab processor. In 1986, under new owners, the *Royal Sea* was converted into a pollock factory trawler in Norway.[585]

ARCTIC TRAWLER

> *We feel that enough has been said about developing the North Pacific bottomfisheries without much being done about it. Now we are going to do something about it.*
>
> —Konrad Uri, part owner, *Arctic Trawler*, 1980[586]

In 1979, four experienced Seattle-based fisherman—Konrad Uri, John Sjong, John Boggs, and Richard Hastings—incorporated as Trans-Pacific Industries to purchase the *Seafreeze Atlantic*, which had been laid up on the East Coast for almost a decade. The partnership, together with Erik Breivik, a Norwegian national, paid $2 million for the vessel and then spent an additional $4 million to retrofit it in Seattle to trawl for cod and to produce frozen fillets onboard. Interest rates then were in the 20 percent range. The two German-made Baader filleting lines

Figure 65. *Arctic Trawler.* (Courtesy of Konrad Uri.)

included in the retrofit were capable of producing thirty tons of fillets per day—either in 15-pound shatterpacks or in 16.5-pound blocks—with a catch-to-frozen cycle of five hours. Once in operation, the system's recovery rate from round fish to fillets ranged from 16 to 20 percent, depending on the size and conformation of the fish processed.

The vessel was rechristened *Arctic Trawler*. Only one of the Trans-Pacific partners, Konrad Uri, however, had any trawling experience, so they arranged for Erik Breivik, who had extensive experience on a codfish factory trawler in the Atlantic and had designed the *Arctic Trawler*'s nets, to sail aboard the vessel as *fishing master*.[587]

With John Boggs at the helm, the *Arctic Trawler* departed Seattle on its maiden voyage to the Aleutian Islands in mid-May 1980. The crew comprised mostly young and inexperienced but enthusiastic Americans and a small group of experienced Norwegian factory trawlermen who had been granted temporary work visas and who would train the Americans.

This crewing strategy worked well; Erik Breivik and his fellow Norwegians, according to Konrad Uri, did "a fantastic job." Under Norwegian tutelage, the Americans learned their jobs well and, after that first voyage, returned to crew on future voyages. (The decent pay helped.) Some of the *Arctic Trawler*'s original crew subsequently became officers on other factory trawlers.

On its first voyage, the *Arctic Trawler* began fishing in the Bering Sea, but during the first couple of months of fishing, harvestable schools of cod were scarce. In mid-July, however, Konrad Uri replaced John Boggs as the vessel's captain, and he convinced Breivik to go to Seguam Pass (between Seguam and Amlia islands, in the

Aleutian Islands), where they located substantial concentrations of cod. Uri described the fishing there as "tremendous," and the *Arctic Trawler* returned to Seattle in late August carrying something over a million pounds of frozen cod fillets. (This was less than a full load; the *Arctic Trawler*'s holds had a total capacity of 1.8 million pounds of frozen fillets.) Samples of the catch, under the brand name Arctic Fresh, were shipped to test kitchens of fast-food and retail sales chains. According to *Pacific Fishing*, the fish was rated "superior to most Canadian production and equal to or better than most European production."[588]

Nevertheless, the traditional markets for cod were accustomed to Atlantic cod and were somewhat reluctant to purchase the Pacific variety. Because Icelandic, Norwegian, and Canadian companies dominated the U.S. East Coast market, the early post-Magnuson-Stevens Act Alaska industry mainly focused on selling its product in California and Nevada, though Trans-Pacific's marketing effort was nationwide. Touting the American origin of their product was part of the company's marketing strategy, but it took a year before sales reached a significant volume.[589]

By the fall of 1982, Trans-Pacific's *Arctic Trawler* venture was considered an unqualified success.[590] Mike Nordby, general manager of Trans-Pacific, noted that his operation had "borrowed a lot from the Norwegians in what we've done, but there's a lot more that we do differently, with as good or better results."[591]

Don Bevan, fisheries scientist and North Pacific Fishery Management Council member, praised Konrad Uri for having "done more toward developing the Alaska groundfish industry than anyone else." And Ted Evans, of the Alaska Factory Trawlers Association, later praised Erik Breivik for having "set the standard for factory trawler operations." Under Uri and Breivik, the *Arctic Trawler*'s catch generally ranged from five tons to twenty-five tons per tow, and fillet production ranged from five tons to thirty tons per day. During 1982, the *Arctic Trawler* landed some 23 million pounds of Pacific cod. Reminiscing years later, Konrad Uri said he had greatly enjoyed prospecting for codfish, but that they tore up a lot of trawl gear in the process. (Mike Petersen, captain of the factory trawler *American No. 1* from 1985 through 2008, recalled occasionally dragging up old anchors that he assumed had been lost by cod-fishing schooners during the Salt Cod Era.) Today, most of the hazards to trawls—rock piles and ledges, for example—are well known.

Trans-Pacific's product was well received, and, more importantly, the venture proved that trawling for cod in the Bering Sea was economically viable. Yet the question remained regarding whether others could match the results of the savvy operators of the *Arctic Trawler*. As Doug McNair, the editor of *Pacific Fishing*, pointed out in January 1984, "Promotors of the Pacific coast industry are fond of saying that Alaskan producers have proven their product is as good or better than Icelandic. But that's a claim based almost entirely on the performance of one company, Trans-Pacific. No other company has really been around long enough or marketed enough cod to be judged."[592]

Erik Breivik left Trans-Pacific, and in 1982, with several fishermen friends, he incorporated the Glacier Fish Company to build a factory trawler. The company's 201-foot *Northern Glacier* was the first large factory trawler built in the U.S. since the *Seafreeze Atlantic* and *Seafreeze Pacific* were launched in 1968, the first built on the U.S. West Coast, and the first built entirely with private money. The vessel was designed specifically for fishing off Alaska's coast and was a model of fish-catching and fish-processing technology. J. M. Martinac Shipbuilding, in Tacoma, Washington, built it, and it was launched in mid-1983. Like the *Arctic Trawler*, the *Northern Glacier* was equipped to produce primarily frozen cod fillets.[593]

This vessel joined two other pioneer factory trawlers: Arctic Alaska Fisheries's 162-foot *Northwest Enterprise*, which had begun producing Pacific cod fillets in 1982, and North Pacific Fishing's 160-foot *American No. 1*, which had begun producing H&G Pacific cod in 1983. "We would like to put the foreign fleets on notice that this is the beginning of their end," said Ken Peterson, part owner of the *American No. 1* since 1979, when the vessel was launched.[594]

In 1984, the capacity to harvest Pacific cod expanded dramatically with the entry into the industry of five new factory trawlers: Trans-Pacific's second vessel, the 160-foot *Aleutian Trawler*, the Glacier Fish Company's aforementioned 201-foot *Northern Glacier*, American Fish Ventures's 133-foot vessels *Ocean Bounty* and *Aleutian Bounty*, and Arctic Alaska Fisheries's 162-foot *Aleutian Enterprise*. But there was a problem: the large 1977 year class of cod (see chapter 20) had by this time, in the words of *Pacific Fishing*, "succumb[ed] to cod ends and old age," and by the fall of 1984 the vessels found it difficult to locate commercial concentrations of Pacific cod. Moreover, the price that year was low, in part due to Canadian government–subsidized Atlantic cod being placed on the U.S. market at artificially low prices. In an attempt to make up the difference, the vessels began targeting Alaska pollock, which was abundant, though its price, too, was low. By year-end, there were reported to be 2.5 million pounds of layer-pack frozen pollock fillets in storage.[595]

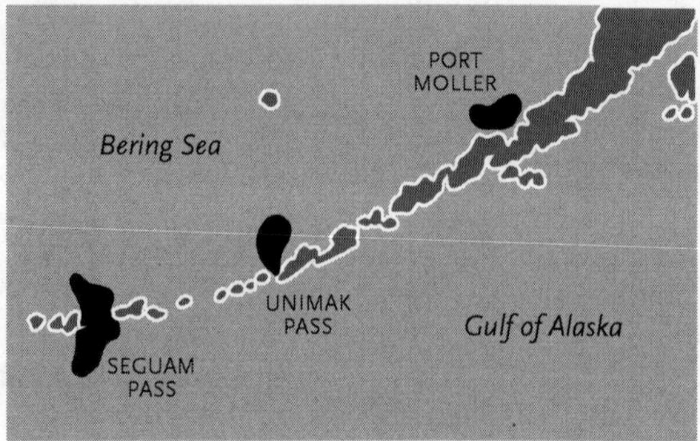

Figure 66. *Pacific Fishing*, 1984: "Virtually all of the Alaska cod trawlers fish three areas. Port Moller, where the old-time cod schooners used to work, is an early summer fishery. Seguam Pass, with its strong tides and terrible weather, is avoided during the winter. The Unimak fishery is concentrated along the 100-fathom curve that runs from the northwest corner of Unimak Island toward the Pribilofs. The boats work bottom trawls in 50–200 fathoms." (Image reproduced with permission of *Pacific Fishing*. "Alaska Groundfish," *Pacific Fishing*, Yearbook (1984), 126–35.)

In early 1982, *Pacific Fishing* had praised Alaska pollock as "the billion-dollar-a-year fish," based almost completely on the value of the almost 10-billion-pounds-per-year catch by foreign vessels.[596] But Americanization of the fishery was proceeding, and increasing amounts of the catch were being allocated to U.S. fishermen. According to *Pacific Fishing* in early 1985, "the future of the Bering fleet lies with pollock." It was a matter of volume: the North Pacific Fishery Management Council, which regulated the groundfish fisheries in federal waters off Alaska's coast, set the allowable catch for pollock for 1985 at 1.6 million metric tons, while the allowable catch of Pacific cod was about one-sixth of that amount—270,000 metric tons. Not only were there too many boats chasing too few cod, there was also a marked seasonality in cod availability to trawl gear. High catch rates were reliably experienced for only about six months of the year, February through July.[597]

Early in 1985, the four companies that operated factory trawlers in Alaska (nine factory trawlers fished that year) organized the Alaska Factory Trawlers Association to promote their frozen-at-sea products and to lobby for their common interests. During 1985, factory trawlers caught almost the entire harvest of Pacific cod in the Bering Sea. In the fall of that year, the Alaska Seafood Marketing Institute (see chapter 21), a state agency, inaugurated a $600,000 campaign to promote pollock as an inexpensive alternative to cod.[598]

Coincidentally, it was about this time that Pacific cod gained broad acceptance in the general market. Emblematic of this acceptance was the switch by the national Skippers Seafood & Chowder House chain and Seattle's Ivar's seafood restaurant chain to exclusively serving frozen-at-sea Pacific cod from Alaska. The year 1985 was also characterized by a shortage of Canadian and European Atlantic cod, which increased the demand for and price of cod produced in Alaska. According to *Pacific Fishing*, processors reported they could "sell all the fish they could fillet."[599]

Nevertheless, beginning in 1985, the *Arctic Trawler* began exclusively targeting pollock. It was a fortuitous time to do so because a sharp decline in Atlantic cod production in 1986 caused the price of Alaska pollock to rise substantially.[600] Pacific cod prices, of course, rose as well, but for the owners of the *Arctic Trawler*, the math was clear: although pollock, on a per-pound basis, was less valuable than Pacific cod, the great volume of pollock that could be caught more than offset its lower price. By 1991, there were fully sixty factory trawlers working off Alaska's coast. Their primary target was pollock.[601]

The development of the pollock fishery represented a diversification and great expansion of the fishing industry off Alaska's coast, but it did not come at the expense of the Pacific cod fishery. Instead, the Pacific cod fishery became a distinct and important component of an array of groundfish fisheries. By species and in terms of total poundage caught, Pacific cod became Alaska's second-largest fishery, behind pollock.

The *Arctic Trawler*, now a half-century old, has been renamed *Seafreeze Alaska* and remains engaged in the pollock fishery. Its equally aged sister ship, formerly the *Seafreeze Pacific* and *Royal Sea*, is now the *Katie Ann* and likewise remains engaged in the pollock fishery and processes Pacific cod (see chapter 22).[602] Today, the factory trawlers that target Pacific cod in the North Pacific—commonly referred to as the *head & gut fleet* (*H&G fleet*) because of the product they typically produce—are primarily represented by the trade association Groundfish Forum.

In 2017, Konrad Uri reminisced fondly about his fishing experience:

> Fishing in Alaska in the 1980s was a special time and place—we were in this together (processors and fishermen) and worked as if our lives depended on it. Which they did—this was serious business. When we look back on the experience

> of fishing together as a family and in a fleet that treated each other like family, I have nothing but the fondest memories, no matter how hard we were actually working.[603]

The newest vessel in the Alaska factory trawler fleet is the 194-foot ice-hardened *Araho*, which is owned by the O'Hara Corporation, a Maine-based seafood company that in recent years has operated four factory trawlers in the Bering Sea. Built in Panama City, Florida, and launched in July 2016, the *Araho* is the first new U.S.-flag factory trawler launched in three decades. The O'Hara Corporation characterized its new vessel as the "most technologically advanced fishing boat in the Bering Sea." The *Araho*'s primary power is a 4,000-horsepower diesel engine, and its accommodations are designed for fifty-four persons and include a hospital room. The vessel joined the Bering Sea factory trawler fleet in the spring of 2017 and produces nineteen-kilogram blocks of frozen headed-and-gutted groundfish, Pacific cod among them.[604]

In 2017, two additional factory trawlers were under construction. The 262-foot *America's Finest* was being built at the Dakota Creek Industries shipyard in Anacortes, Washington, and the 261-foot *North Star* was being constructed at Eastern Shipbuilding, the same Panama City, Florida, shipyard that had constructed the *Araho*. But for both vessels, all did not go according to plans.

Figure 67. O'Hara Corporation's factory trawler *Araho*. (Courtesy of O'Hara Corporation. Casey O'Hara, O'Hara Corporation, personal communication with author, August 1, 2017.)

Owned by Seattle-based Fishermen's Finest, which currently operates two factory trawlers, *America's Finest* incorporates fish meal and fish oil plants and had a projected total cost of $75 million. It was scheduled to be delivered within the year and soon after enter Alaska's groundfish fishery, replacing the company's thirty-eight-year-old *American No. 1.*

Relatively late in the construction process, however, the quantity of modified foreign steel used in the *America's Finest* construction was found to be very substantially in excess of that authorized under the Jones Act (see chapter 15). Fishermen's Finest sought a Congressional waiver to the act—such waivers are exceptionally rare—but in March 2018, Congress declined to grant the waiver, and it seemed likely that the company was going to have to sell the vessel—at a very substantial loss—to a foreign buyer. Fortunately for Fishermen's Finest, in November 2018 Congress reversed itself and granted the *America's Finest* a waiver, though with temporary restrictions that limited the amount of fish that Fishermen's Finest's factory trawlers can harvest and the amount of catcher-vessel fish that can be processed aboard them.[605]

The *North Star*, owned by the Seattle-based Glacier Fish Company, which currently operates three factory trawlers in Alaska, was in early October 2018 near completion and expected to depart for Alaska the following month. On October 10, however, Hurricane Michael, the largest hurricane ever to hit the Florida Panhandle, changed everything. The storm ripped the *North Star* from its moorings and left it lying on its starboard side in Saint Andrews Bay's shallow waters. The vessel remained stranded in that position for nearly two months before a salvage crew righted it. The next step will be a major overhaul of the *North Star's* interior and systems.[606]

During the 2016 Pacific cod federal season in the Bering Sea/Aleutian Islands area, factory trawlers caught 2,306 metric tons of Pacific cod, less than half of what trawl catcher vessels caught.[607]

ENDNOTES

[566] Aaron Orlowski, "Global impacts of trawling quantified in new study," *Seafood-Source News*, August 8, 2017, accessed August 8, 2017, https://www.seafood-source.com/commentary/global-impacts-of-trawling-quantified-in-new-study.

[567] See chapter 5, for a description of a trawl and its operation.

[568] Robert J. Browning, *Fisheries of the North Pacific: History, Species, Gear and Processes*, revised edition (Anchorage: Alaska Northwest Publishing Company, 1980), 195–196.

[569] David Witherell, Michael Fey, and Mark Fina (staff, North Pacific Fishery Management Council), *Fishing Fleet Profiles* (April 2012), 26.

[570] Mike Hillers, "Trawling, Trawl Monitoring and the Future of Trawl Monitoring," *Pacific Fisheries Review*, a supplement to *Fishermen's News* (December 2013): 23–27.

[571] Cabot Martin, "The Collapse of the Northern Cod Stocks: Whatever Happened to 86/25?" *Fisheries* (May 1995): 6–8.

[572] McDowell Group, *Economic Impact of the Seafood Industry on the Kodiak Island Borough* (2014), 38; North Pacific Fishery Management Council, Fishery Management Plan for Groundfish of the Gulf of Alaska, appendix D, *Life History Features and Habitat Requirements of Fishery Management Plan Species* (November 2016), D-16, accessed July 15, 2017, https://www.npfmc.org/wp-content/PDFdocuments/fmp/GOA/GOAfmpAppendix.pdf.

[573] David Witherell, Michael Fey, and Mark Fina (staff, North Pacific Fishery Management Council), *Fishing Fleet Profiles* (April 2012), 24; North Pacific Fishery Management Council, *Fishery Management Plan for Groundfish of the Gulf of Alaska*, appendix D, *Life History Features and Habitat Requirements of Fishery Management Plan Species* (November 2016), D-16, accessed July 15, 2017, https://www.npfmc.org/wp-content/PDFdocuments/fmp/GOA/GOAfmpAppendix.pdf.

[574] David Witherell, Michael Fey, and Mark Fina (staff, North Pacific Fishery Management Council), *Fishing Fleet Profiles* (April 2012), 18; North Pacific Fishery Management Council, *Fishery Management Plan for Groundfish of the Gulf of Alaska*, appendix D, *Life History Features and Habitat Requirements of Fishery Management Plan Species* (November 2016), D-16, accessed July 15, 2017, https://www.npfmc.org/wp-content/PDFdocuments/fmp/GOA/GOAfmpAppendix.pdf.

575 The Alaska Factory Trawlers Association later became the American Factory Trawlers Association and then the At-Sea Processors Association.

576 "Bob Morgan On Trawler Trashing: AFTA President Speaks His Piece," *Alaska Fisherman's Journal* (July 1990): 10–15, 38.

577 "1991 Factory Trawler Directory," *Pacific Fishing* (September 1991): 64–71; "Typical Alaska Catcher-Processor," *Pacific Fishing*, Yearbook (1984): 127; Division of Commercial Fisheries, Alaska Department of Fish and Game, "What Kind of Fishing Boat is That?" brochure, April 2009; Marine Stewardship Council, *Alaska Pacific Cod—Bering Sea and Aleutian Islands, January 2015*, accessed February 11, 2015, http://www.msc.org/track-a-fishery/fisheries-in-the-program/certified/pacific/bering-sea-and-aleutian-islands-pacific-cod/fishery-name.

578 Josh Keaton (NOAA Fisheries, Alaska Region), *BSAI Inseason Management Report* (December 2016), 11; Josh Keaton (NOAA Fisheries, Alaska Region), *GOA Inseason Management Report* (December 2016), 8.

579 United States Fishing Fleet Improvement Act, June 12, 1960 (74 Stat. 212).

580 U.S. Bureau of Commercial Fisheries, "Fishing Vessel Construction Subsidy Regulations Adopted," news release, December 10, 1964.

581 United States Fishing Fleet Improvement Act, June 12, 1960 (74 Stat. 212), amended August 30, 1964 (78 Stat. 614).

582 John Wiese, "New Operation May Harvest Alaska's Groundfish," *Pacific Fishing* (May 1973): 24, 54, 58–60.

583 Richard Adams Carey, *Against the Tide: The Fate of the New England Fisherman* (Boston: Houghton Mifflin, 2000), 271–274; *23rd Annual Report of the Pacific Marine Fisheries Commission for the Year 1970*, 10; "Alaska Fisheries Purchase New Boat," *Fairbanks Daily News-Miner*, April 12, 1973; "Ship to Compete for Bering Fish," *Fairbanks Daily News-Miner*, April 13, 1973.

584 *Report of the Technical Sub-Committee of the International Groundfish Committee Appointed by the Second Conference on Coordination of Fisheries Regulations Between Canada and the United States* (June 19–21, 1974), 18.

585 Kevin M. Bailey, *Billion-Dollar Fish: The Untold Story of Alaska Pollock* (Chicago: University of Chicago Press, 2013): 94, Beth McGinley, "1988 Factory Trawler Directory," *Pacific Fishing* (September 1988): 59–64.

586 Robert Browning, "*Seafreeze Atlantic* is Reincarnated for Pacific Groundfishery," *National Fisherman*, February 1980, 19.

587 Kevin M. Bailey, *Billion-Dollar Fish: The Untold Story of Alaska Pollock* (Chicago: University of Chicago Press, 2013): 94; Robert Browning, "Seafreeze Atlantic is Reincarnated for Pacific Groundfishery, *National Fisherman*, February 1980, 19; Bruce J. Cole, "Arctic Trawler is off to Alaska on Maiden Trip," *National Fisherman*, July 1980, 33; John Sabella, "Arctic Trawler Brings Home the Bacon: 1.5 Million Pounds of Cod Fillets and Insight into Bottomfish," *Pacific Fishing* (December 1981): 35–36, 38, 65.

588 "Arctic Trawler: Freezer-Trawler Brings Home First Load of Fillets," *Pacific Fishing* (September 1980): 29; Konrad Uri, personal communication with author, September 9, 2016.

589 Kevin M. Bailey, *Billion-Dollar Fish: The Untold Story of Alaska Pollock* (Chicago: University of Chicago Press, 2013): 94; Doug McNair, "Cod Update," *Pacific Fishing* (January 1984): 34–37, 40–41; Jon Sabella, "Arctic Trawler Brings Home the Bacon: 1.5 Million Pounds of Cod Fillets and Insight into Bottomfish," *Pacific Fishing* (December 1981): 35–36, 38, 65.

590 Not so successful was the 160-foot factory trawler *Blue Ocean*, which fished for cod for only a short time in the Gulf of Alaska before giving up in 1982.

591 Doug McNair, "Bottomfish Update," *Pacific Fishing* (November 1982): 65–71, 119.

592 Terry Johnson, "Highliner Award Given to Veteran Fishermen," *National Fisherman*, November 1981, 8, 97; Beth A. McGinley, "Erik Breivik," *Pacific Fishing* (December 1988): 50–56; "Bottomfish," *Pacific Fishing*, Yearbook (1983): 63–87; A. D. Chandler, "Although Pacific Cod is Abundant, There's No One Who Will Take It," *National Fisherman*, September 1983, 8; Konrad Uri, personal communication with author, September 9, 2016, and September 15, 2016; Mike Petersen, personal communication with author, November 20, 2016; Doug McNair, "Cod Update," *Pacific Fishing* (January 1984): 34–37, 40–41.

593 Kevin M. Bailey, *Billion-Dollar Fish: The Untold Story of Alaska Pollock* (Chicago: University of Chicago Press, 2013): 95; Glacier Fish Company, accessed March 9, 2015, http://www.glacierfish.com/departments.htm; Glacier Fish Company, accessed March 9, 2015, http://www.glacierfish.com/ng.htm; "The Marathon Begins," *Pacific Fishing*, Yearbook (1983): 63.

594 "American No. 1 Gears Up for Cod," *National Fisherman*, May 1983, 57; Joseph Plesha (Trident Seafoods Corp.), "The In-shore/Off-shore Dispute: Impact of Factory Trawlers on Fisheries in the North Pacific and Proposals to Regulate the Fleet," Eighth Annual National Fishery Law Symposium, University of Washington School of Law, Seattle, Wash. (October 1990), 22; "American No. 1 launched as challenge to foreign fleet," *Alaska Fisherman's Journal* (September 1979): 1, 6.

[595] "Slicing a New Fillet Market," *Pacific Fishing*, Yearbook (1985): 61–62.

[596] Ken Talley, "Alaska Pollock: The Billion-Dollar-A-Year Fish," *Pacific Fishing* (February 1982): 50–52, 54.

[597] "Pollock: Fish of the Future for a Growing Fleet," *Pacific Fishing* (January 1985): 26; Alaska Groundfish," *Pacific Fishing*, Yearbook (1984): 126–135.

[598] "John Sabella, "New Wave of Domestic Factory Ships Wants to Play Ball, *National Fisherman*, July 1985, 5, 36.

[599] Chris Blackburn, "'85 a Landmark Year for Groundfish," *Alaska Fisherman's Journal* (January 1986): 50–52; "Cod Market Improves," *Pacific Fishing*, Yearbook (1986): 80–81.

[600] The price per pound of a standard three- to five-ounce shatterpack of Alaska pollock fillets rose from $0.85 in early 1986 to $1.50 in early 1987. By year's end, however, the price had retreated about 20 percent.

[601] R. G. Bakkala and J. W. Balsinger, eds., *Condition of Groundfish Resources of the Eastern Bering Sea and Aleutian Islands Region in 1986*, NOAA Tech. Mem. NMFS F/NWC-117 (June 1987), 41; Daniel D. Huppert, *Managing Alaska Groundfish: Current Problems and Management Alternatives* (Seattle: Fisheries Management Foundation and Fisheries Research Institute, 1988), 2–3; Kris Freeman, "Is Alaska Groundfish Overcapitalized?" *Pacific Fishing* (March 1988): 42–48; "1991 Factory Trawler Directory," *Pacific Fishing* (September 1991): 64, 66–67, 69–71.

[602] Michael Crowley, "Staying Power," *National Fisherman*, September 2013, 28–31.

[603] Konrad Uri, "One Last Haul," *Alaska Fisherman's Journal* (North Pacific Focus) (Fall 2017): 22.

[604] Eastern Shipbuilding Group, Inc., "Eastern Shipbuilding Group and O'Hara Corporation Launch the F/T *ARAHO*," news release, September 8, 2015, accessed December 19, 2016, https://www.google.com/search?q=Eastern+Shipbuilding+Group+and+O%E2%80%99Hara+Corporation+Launch+the+F%2FT+ARAHO&oq=Eastern+Shipbuilding+Group+and+O%E2%80%99Hara+Corporation+Launch+the+F%2FT+ARAHO&aqs=chrome..69i57.4735j0j4&sourceid=chrome&ie=UTF-8; O'Hara Corporation Fleet, accessed December 19, 2016, https://www.oharacorporation.com/alaskan-fishing/our-fleet.html; Jessica Hathaway, "Newbuild factory trawler Araho en route to Alaska," *National Fisherman*, February 8, 2017, accessed February 11, 2017, https://www.nationalfisherman.com/boats-gear/newbuild-factory-trawler-araho-alaska/?utm_source=informz&utm_medium=email&utm_campaign=newsletter&utm_content=newsletter; Jean Paul Vellotti, "Building boom: North Pacif-

ic fleet makeover will generate billions," *National Fisherman*, March 24, 2017, accessed March 30, 2017, https://www.nationalfisherman.com/boats-gear/boatbuilding-boom-north-pacific-fleet-makeover-billions/?utm_source=informz&utm_medium=email&utm_campaign=newsletter&utm_content=newsletter; Bruce Buls, "Araho Arrives," *National Fisherman*, May 2017, 28–31.

605 "Fishermen's Finest to build advanced trawler," *Deckboss* (blog) November 24, 2014, accessed December 19, 2016, http://deckboss.blogspot.com/2014/11/fishermens-finest-to-build-advanced.html; Kara Carlson, "New Anacortes-built trawler could be grounded by old law, endangering two local firms," *Seattle Times*, June 3, 2017; Samuel Hill, "Jones Act requirement grounds America's Finest," *National Fisherman*, August 2017, 14; Jessica Hathaway, "America's Finest fails to win waiver," *National Fisherman*, March 22, 2018, accessed March 22, 2018, https://www.nationalfisherman.com/boats-gear/americas-finest-fails-to-win waiver/?utm_source=informz&utm_medium=email&utm_campaign=newsletter&utm_content=newsletter); Steve Bittenbender, "Trump signs Coast Guard bill into law, includes Jones Act waiver for America's Finest," *Seafood-Source News*, December 6, 2018, accessed December 6, 2018, https://www.seafoodsource.com/news/supply-trade/trump-signs-coast-guard-bill-into-law-includes-jones-act-waiver-for-america-s-finest; Frank LoBiondo Coast Guard Authorization Act of 2018, December 4, 2018, §§ 835–836.

606 McDowell Group, *Modernization of the North Pacific Fishing Fleet: Economic Opportunity Analysis* (November 2016), 15, 47, accessed December 18, 2016, https://www.portseattle.org/Supporting-Our-Community/Economic-Development/Documents/Fleet%20Modernization%20Final%2011_11.pdf; Hal Bernton, "Hurricane damaged Seattle company's Alaska-bound factory trawler in Florida shipyard," *Seattle Times*, October 12, 2018; Jessica Hathaway, "North Star rising: Storm-tossed trawler righted in Florida," *National Fisherman*, December 7, 2018, accessed December 7, 2018, https://www.nationalfisherman.com/alaska/north-star-rising-storm-tossed-trawler-righted-in-florida/?utm_source=marketo&utm_medium=email&utm_campaign=newsletter&utm_content=newsletter&mkt_tok=eyJpIjoiTlRoalltWmpNVE5pWVRSbSIsInQiOil1WDVNMFVZNFhQbW9EZDBkWGVNVElXMGlWNmw1em5MUVY1UCtZco5KdU9lMzNtWkRpWkltRFl4ZkEyTW1uVStsbjV2WERFUnpYK2JsQWtUQjR4bUJ3ZWxpOHMzZXUxKoowb2hHVkkxZWcwdUkxaERLRXJNZXNocDBzMjFSelVlVyJ9.

607 Josh Keaton (NOAA Fisheries, Alaska Region), *BSAI Inseason Management Report* (December 2016), 15.

Chapter 17
LONGLINING

LONGLINE CATCHER VESSELS

As their technical name implies, longline catcher vessels are vessels that catch fish with longline gear and deliver their catch to shore-based or floating processors. The longline catcher-vessel fleet that targets Pacific cod consists of about seventy-five vessels that range in length from about 30 feet to almost 90 feet, with most being less than 60 feet. Their catch is usually carried in refrigerated seawater. A large portion of the longline catcher-vessel fleet that fishes for Pacific cod is also active in the halibut and sablefish ("black cod") fisheries and derives most of its income from one or both of these fisheries. The halibut and sablefish fisheries are managed under an individual fishing quota (IFQ) program in which individual fishermen own the right to catch a percentage of the total allowable catch in a regulatory area or areas.

The Pacific cod longline catcher-vessel fishery is centered in the central Gulf of Alaska, particularly along the east shore of Kodiak Island. The fishery also occurs, though less intensively, along the west shore of Kodiak Island and throughout the western Gulf of Alaska. Beginning in 2012, Pacific cod longline catcher vessels in the central Gulf of Alaska less than 50 feet in length overall have been allocated 14.6 percent of the total allowable catch, while vessels greater than 50 feet in length overall are allocated 6.7 percent. Pacific cod longline catcher vessels in the western Gulf of Alaska are allocated 1.4 percent of the area's total allowable catch.[608] These allocations are further apportioned between A and B seasons, which begin in January and September, respectively.

Most Pacific cod longline catcher vessels today use modern versions of the stationary longlines described in chapter 5 of this book. The differences are materials: the use of nylon and polypropylene lines instead of tarred cotton and hemp lines, the replacement of wooden buoy kegs with plastic floats, and the replacement of J hooks with circle hooks, as well as the fact that the gear is hauled by hydraulic power rather than by hand. Some vessels employ *snap-on* gear in which the gangion (snood), rather than being permanently spliced into the groundline, is fastened with a metal clip, usually of galvanized or stainless steel. This allows the gangion, with its baited hook, to be snapped onto the groundline as it is paid out—usually from a hydraulically driven drum—and snapped off as the gear is retrieved. Some longline catcher vessels today are equipped with automated longline systems that incorporate baiting machines (*autobaiters*). Baiting hooks by hand is a time-consuming process that limits a vessel's potential. For example, crewmen aboard the 58-foot cod longliner *Tribute*, based in Seward, Alaska, once spent up to twenty hours each day baiting perhaps 12,000 hooks. A Mustad baiting machine later installed on the vessel was capable of baiting 20,000 hooks per day, an increase of almost 70 percent.[609]

As noted in chapter 14, the 1980 endeavor by the halibut schooners *Tordenskjold* and *Vansee* to produce salted cod was the first significant post-Magnuson-Stevens Act domestic effort to longline for Pacific cod in Alaska. Among those that followed was a joint venture during the summer of 1986 between the Japanese-owned North Pacific Co-operative Fisheries Company and the Alaska Longline Fisherman's Association. Under the agreement—which was intended "[t]o promote mutual understanding between U.S. and Japanese longline fishermen"—ten U.S. longliners agreed to deliver 2,000 metric tons of bled Pacific cod to two approximately 160-foot-long Japanese longline vessels that did not fish but acted as motherships to the American vessels.[610] The Japanese paid the Americans $0.15 per pound for their catch. In the end, only six or seven U.S. longliners participated in the venture.

The boats began fishing in Shelikof Strait but found few cod and quickly moved to the Shumagin Islands. Fishing was better there, but still not as productive as the fishermen had expected. A good day's catch for a vessel was 12,000 pounds, and within a couple of weeks, the venture was over. Deliveries to the motherships totaled only 300 metric tons. Jana Suchy, writing in *Alaska Fisherman's Journal*, summed up the venture from the fishermen's perspective: "They worked long and hard for little."[611]

Meanwhile, Pacific cod had become, according to *Alaska Fisherman's Journal*, "the latest hot item along the waterfront" at Kodiak. In the fall of 1986, five plants there were processing codfish, but all complained that they needed more fish and even advertised in the local paper for fishermen to deliver fish. Most of the cod delivered to the plants was caught offshore by trawlers, but a number of small vessels, including several salmon seiners, began longlining Pacific cod in nearby bays.

Processors paid the longliners $0.15 per pound for bled, iced fish, but it wasn't enough for many fishermen.

Chez Matosuka, captain of the 50-foot seiner *Homeward*, said that with two crew members he usually caught between 14,000 and 22,000 pounds during a three-day trip, but that "[w]e are just making wages." Another fishermen said he enjoyed longlining codfish but "They don't pay us enough." Both processors and fishermen agreed that the price of cod would have to increase before enough fishermen would be attracted into the industry to enable the fish plants to operate at full capacity. It took a few years, but to almost everyone's delight, in 1990 the dock price for bled cod at Kodiak increased to $0.23, making longlining profitable. By this time, however, many local longliners had begun converting their vessels to catch cod with pots (see chapter 18).[612]

The dock price for Pacific cod at Kodiak averaged a little less than $0.37 per pound during the years 2005 to 2014. The price during that period peaked at $0.65 per pound in 2008, and sank to a low of $0.26 per pound in 2010 and 2013. In 2014, the dock price for Pacific cod at Kodiak was $0.32 per pound, and longline vessels delivered nearly 13 million pounds of Pacific cod. Many of the vessels were under 58 feet long and also participated in the halibut and sablefish longline fisheries.[613]

FREEZER-LONGLINERS

> *[T]he growing number of hook-and-line boats is making its presence felt on the groundfish scene. Most are involved in catching Pacific cod, and with cod stocks in the North Atlantic severely depleted, these longliners are having a significant impact on world markets. . . . The future will likely see increased entries into this fleet.*
>
> —*Pacific Fishing*, October 1990[614]

Freezer-longliners use longline gear to catch Pacific cod that are then frozen onboard, mostly as a headed and gutted (H&G) product. Although the term *factory longliner* was used early on and better describes this type of vessel, the owners apparently wanted to avoid the less-than-popular "factory" label and coined the term *freezer-longliner*. Congressional legislation specific to the Bering Sea/Aleutian Islands fishery refers to freezer-longliners as *longline catcher processors*.[615] This gear group was represented originally by the trade group Pacific Cod Freezer Longliners, then by the Alaska Longline Cod Commission, and is currently represented by the Freezer Longline Coalition.

Because fish caught by freezer-longliners are handled individually, bled while still alive (which results in better flesh color), not subject to scale loss, skin damage, and damage to the flesh by crushing, such as may occur in a trawl net's cod end, and are promptly processed and frozen, they command a higher price than trawl-caught fish or fish delivered to shore plants.[616]

The primary markets for H&G Pacific cod in the early 1990s were Norway and Japan. Cargoes were delivered dockside in Dutch Harbor or Kodiak for containerization or were loaded aboard Japan-bound tramp freighters. Today, a considerable portion of the H&G cod production is marketed domestically. The primary export markets are Europe, Japan, and China (see chapter 21).[617]

Aleutian Mistress

> *The first automated longliner-processor of her kind to operate in the North Pacific, the ALEUTIAN MISTRESS, has initiated a ground swell in the Alaska cod fishery. Since December [1980], she has fished in the Aleutians, proving that the vessel and crew can handle a winter fishery, that the Mustad Autoline system can be operated by the crew of six to nine men to produce an average of almost a pound of cod per hook, and that the longline-caught product, bled and processed immediately after catching, is of superior quality and commands top prices.*
>
> —Alaska Fisheries Development Foundation, 1981[618]

The first domestic freezer-longliner vessel to target Pacific cod in Alaska waters after the passage of the Magnuson-Stevens Act was the *Aleutian Mistress*. The 124-foot, wheelhouse-forward vessel had been constructed originally to fish for king crabs in Alaska and was launched at Tacoma, Washington, in 1979. It was owned by a vertically integrated partnership that included SeaWest Industries, an Edmonds, Washington–based seafood processor. Following its first year of fishing, the *Aleutian Mistress* underwent a $915,000 remodel to enable it to catch and process both crabs and codfish. A little more than half the cost of the remodel, which was completed in August 1980, was funded by a $475,000 grant from the Alaska Fisheries Development Foundation. The foundation was interested in the economics of adapting a crabber to longline and process Pacific cod.

The remodel included the installation of a removable shelter deck over the *Aleutian Mistress*'s aft deck. The structure afforded a degree of safety and comfort for the crew, which had to handle gear and fish during often-harsh conditions. The remodel also included what might have been the first automated longline system to be employed on the U.S. West Coast.

The system, which was capable of fishing thirty-five thousand J-style hooks per day, was built by Mustad Autoline, a Norwegian firm. (The much more efficient circle hooks came into widespread use several years later.) Once fish were removed from gear being hauled, the Autoline machinery sequentially cleaned hooks of unused bait, untwisted gangions, and separated hooks from the groundline. The equipment then racked hooks onto magazines, with the groundline draped between the hooks, and, as the gear was reset, baited each hook with a piece of bait that had been previously machine cut and fed into position. Fishermen tending the magazines replaced bent or broken hooks.

Also installed were automated fish-processing machinery (including a fillet machine made by Baader, a German firm), blast freezers capable of freezing 18,000 pounds per day, and 2,290 cubic feet of frozen storage capacity—enough to hold about 115,000 pounds of frozen product. The venture hoped to catch 20,000 pounds (round weight) of codfish per day, which would yield perhaps 5,000 pounds of skinless/boneless fillets. The fillets would be packaged in 15-pound-capacity plastic-lined cardboard boxes. Fish carcasses and offal would be discarded overboard.

In early August 1980, the remodeled *Aleutian Mistress* departed Seattle. The first part of its voyage, per contractual obligations with the Alaska Fisheries Development Foundation, included spending ten days catching and processing cod in the Gulf of Alaska and along the Aleutian Chain. The fillets produced were shipped to Bellingham, Washington, for secondary processing and then used as samples in potential markets. The *Aleutian Mistress* was then converted to crabbing. In late November, the crab season having been concluded, the *Aleutian Mistress* was converted from a fully rigged crabber back into a freezer-longliner. The conversion, including the reinstallation of the shelter deck, took just three days despite Dutch Harbor's typically inclement weather.

During 1981, the *Aleutian Mistress*, equipped with the latest electronic fish-locating technology, explored the waters along both sides of the Aleutian Chain. The vessel found a lot of cod. Guidelines provided by Mustad indicated that a catch of a half pound per hook was "good fishing," but the *Aleutian Mistress* averaged closer to one pound per hook, and the daily catch slightly exceeded the hoped-for catch of 20,000 pounds per day. Many of the cod, however, were larger than the vessel's filleting machine could accommodate and had to be filleted by hand. Jerry Tilley, a part owner of the *Aleutian Mistress* who was aboard during some of the fishing, recalled cod coming aboard so fast at times that the processing crew could not keep up. Codfish were at times three feet deep on the deck. "More than a few times we had the boat so loaded beyond her processing capacity that we were forced to sell tons as [crab] bait," said Bill Peck, one of the *Aleutian Mistress*'s crew members.[619]

For bait, the crew used a mixture of squid and herring. Of the two, herring was generally considered the better bait, but the herring's smooth, relatively hard scales as often as not prevented the Autoline system hooks from properly snaring it, and about half the herring-baited hooks threw off their bait as they went overboard. On the other hand, squid, with its soft skin, were efficiently baited by the system. (A useful characteristic of automated longline systems is that the number of hooks set can be easily increased to compensate for the number of hooks that might not be baited.)

In addition to being pleased with the amount and quality of the cod caught, Jerry Tilley was very satisfied with the quality of *Aleutian Mistress*'s frozen fillets. But marketing them proved a challenge.

The target market was Norway, and Tilley made two trips there to promote his product. The Norwegians liked it, but offered only $0.70 per pound, about $0.40 less per pound than the fillets had cost to produce. The market situation did not improve, and in 1982, the owners of the *Aleutian Mistress* gave up their cod-fishing ventures. Don Barton, part owner of the *Aleutian Mistress*, summed up their experience: "You don't longline for cod in the North Pacific and make it—no matter how good you are. We were associated with the best people around, and we lost in seven figures."[620]

The *Aleutian Mistress* subsequently went through several ownerships and name changes. It is today the factory trawler *Vaerdal* and fishes in the Bering Sea/Aleutian Islands area.

Seattle Star

A second freezer-longliner, the *Seattle Star*, followed close on the *Aleutian Mistress*'s heels and had a cod-fishing career that spanned almost three decades. Unlike the *Aleutian Mistress*, the *Seattle Star* was a privately financed venture.

The prime mover behind the *Seattle Star* was Erling Skaar, who had emigrated from Norway in 1962, begun commercial fishing as a deckhand on the king crabber *Foremost* in 1967, and had crewed on three groundfish trawlers.[621]

Skaar was ambitious and was early on critical of trawling, saying, "It is plain to me that if the fishing industry continues to build big

Figure 68. *Seattle Star* christening, Sunnmore, Norway, July 1980. (Courtesy of Erling Skaar.)

trawlers, it is going to run into problems. Longlining is much more efficient on fuel, offers a product of consistently higher quality, and does not disturb the ocean floor."[622]

In late 1979, a naval architect colleague informed Skaar of a vessel that would be a good candidate for conversion into a freezer-longliner. The 138-foot, wheelhouse-aft *Austholm* had been built in the United States in 1946 as the *Pan Trades Andros*, one of a dozen trawlers purchased by the U.S. Army and chartered to German interests in 1949 to help rebuild Germany's commercial fishing industry. In 1956, Norwegian interests purchased the *Pan Trades Andros*, converted it to a trawler/seiner, and renamed it *Polar Tral*. The vessel was later renamed *Austholm*.

Although the *Austholm*'s hull was sound, the vessel was outdated and scheduled to be scrapped under Norway's fleet reduction program. As such, it was available at a bargain price. Moreover, Norway's export program would finance its rebuilding and conversion in a Norwegian shipyard. Skaar and two partners—one of them a Norwegian—formed the Nordic Seafood Company, which in December 1979 purchased the vessel and rechristened it *Seattle Star*. As half-owner, Skaar would be its captain, and he oversaw its rebuilding and conversion into a freezer-longliner at the Myklebust Shipyard in Sunnmøre, Norway. The work, which began almost immediately, cost $3.1 million, and the *Seattle Star*—basically a new vessel from just above the waterline—was launched in July 1980. Among its newly installed equipment was a Mustad Autoline system, a Baader heading and gutting machine, freezers capable of freezing 40-pound blocks of cod in three hours, and 10,600 cubic feet of cold storage—enough to store more than 350,000 pounds of finished product.

Because of processing space limitations and financial considerations, the *Seattle Star* produced H&G cod rather than fillets, as were produced by the *Aleutian Mistress* and *Arctic Trawler*, both of which had begun fishing only months before. Some retail experts warned that thawing the H&G cod for reprocessing and then freezing the product manufactured from it (e.g., fillets or portions) might result in measurable damage that could diminish the final product's marketability. Skaar, however, was confident the high quality of his fish would offset any problems with thawing and refreezing. His reasoning proved correct, and H&G cod became the standard product produced by freezer-longliners.

With a crew of eight, Skaar promptly set a course for Seattle via the Panama Canal. At sea, Skaar appreciated the efficiency of the *Seattle Star*'s long, deep, and narrow hull, as well as the stability afforded by the 200,000 pounds of pig iron ballast in its keel.[623]

The *Seattle Star* arrived in Seattle in mid-September. Skaar and his partners understood that longlining for Pacific cod would be, in his words, "no bed of roses," but they hoped markets would open up. An article in *Alaska Fisherman's Journal* introduced the *Seattle Star* as "the latest entry in the Great American Cod Game to determine who can harvest bottomfish and make it pay."[624]

With a fishing crew of twelve, the *Seattle Star* departed Seattle on October 16, and about a week later, near Raspberry Island, in Shelikof Strait, it began its career as a freezer-longliner. But there was considerable gear conflict with trawlers working the area, so the *Seattle Star* moved to the Bering Sea side of the Aleutian Islands. Not long after, the vessel's baiting machine broke and the crew began baiting hooks by hand. Some days they baited 18,000 hooks.

But there were cod, big ones, with some in the seventy- to eighty-pound range. On one trip, the fish averaged forty-eight pounds. The large fish had to be gaffed to be brought aboard, lest their weight cause the hooks to tear out.

In December 1980, Skaar shipped four containers—168,000 pounds—of frozen H&G Pacific cod from Dutch Harbor to Seattle. Not long after, he flew to Seattle to begin marketing the fish.

Skaar and his partners had hoped to market their product to the Safeway grocery chain. Carrying two seafood dishes his wife had

Figure 69. Nordic Seafood Company poster. (Courtesy of Erling Skaar.)

carefully prepared from cod caught aboard the *Seattle Star*, Skaar met with the chain's seafood buyer. The buyer—who obviously had little experience with seafood—had two problems with the product: cod, he said, wasn't an attractive name for marketing fish, and "fish is supposed to smell and this fish does not smell." Safeway was a no-sell. Skaar, however, was able to persuade Ivar's, a local chain of seafood restaurants, to put Pacific cod on its menu, where it remains to this day. Over the course of the next two months, Skaar managed to sell the four containers of fish, but at a price of $0.40 per pound, he and his partners realized little profit. Like the *Aleutian Mistress*, the *Seattle Star* exited the Pacific cod fishery in 1982. To make ends meet, the vessel pot fished for sablefish, longlined for halibut, and, as opportunities arose, did various types of charter work.

Markets for Pacific cod gradually improved, and the *Seattle Star* reentered the fishery in 1988. In 1991, Skaar began shipping his product to Norway, where it was well received and sold for remunerative prices. The Norwegians split and salted the fish and marketed it worldwide. (The Vatican was a customer.) Norway became the primary market for the *Seattle Star*'s production.

Erling Skaar and his partners left the codfish business on Christmas Eve in 2002, when they sold the *Seattle Star* —including the federal License Limitation Program (LLP) license it had been awarded in

2000—to the Seattle Cod Company. In 2005, this company sold the *Seattle Star* and its LLP license to Seattle-based Blue North Fisheries, which had been operating freezer-longliners since 1994. Blue North renamed the vessel *Blue Star* and sent it to sea. In 2010, the *Blue Star*'s nearly three decades as a freezer-longliner ended when Blue North sold it to an Alaska construction company that converted it into a floating bunkhouse.[625]

Growth of the freezer-longliner fleet

Despite its inauspicious beginning, there was a recognized opportunity in longlining for Pacific cod, and the freezer-longliner fleet grew quickly beginning in about 1986. As of 1991, the Pacific cod quota in the Bering Sea had never been reached, but with the additional vessels due to enter the fishery, there were concerns that the freezer-longliner fleet might be the next victim of overcapitalization. And it was (see chapter 22).

In early 1991, the freezer-longliner fleet comprised twenty-four vessels, and seven more were under construction. The vessels ranged from 76 feet to 188 feet long, and most had been converted to freezer-longliners from other uses, especially oil rig supply. Typically, a freezer-longliner then carried a crew of from six to eighteen and fished about 20,000 hooks per day. Daily cod catches ranged from about 15,000 pounds to 20,000 pounds (round weight), although the most efficient vessels were capable of catching and processing as much as 40,000 pounds per day.[626]

The first purpose-built freezer-longliners in the Alaska codfish industry were the 123-foot *Aleutian Chalice* and the 135-foot *Frontier Spirit*. The *Aleutian Chalice* was constructed by Atlantic Marine, in Jacksonville, Florida, and delivered to Bonar Peterson, an Edmonds, Washington, fisherman, in July 1989. The vessel was equipped with a U.S.-built (Marco) automated longline system and could carry 400,000 pounds of frozen product.[627]

The *Aleutian Chalice* was a first-rate vessel, but all was not aboveboard with Bonar Peterson. The following October, Alaska Division of Fish and Wildlife Protection officers were alerted that the *Aleutian Chalice* had been fishing for Pacific cod in closed waters near Sand Point, in the Shumagin Islands. Investigating officers quickly determined that the *Aleutian Chalice* had been fishing in the closed waters for a couple of weeks, and they subsequently found approximately 200,000 pounds of illegally caught, frozen headed-and-gutted Pacific cod aboard the vessel. For his illegal fishing and a licensing infraction, Peterson was fined $20,000 and placed on two years' probation. Additionally, he paid $160,000—the approximate value of the illegally taken fish—to the State of Alaska. And there was more to come: in mid-December, a state trooper informed the National Oceanic and Atmospheric Administration that the *Aleutian Chalice* had delivered a significant amount of frozen Greenland turbot, a prohibited species, to a Japanese tramp freighter earlier in the month. A subsequent investigation showed the amount to be 26.5 metric tons. For this and several lesser infractions, Peterson was fined $33,500.[628]

Not long after this incident, the *Aleutian Chalice* was renamed *Maria N.* The vessel was renamed *North Cape* in 1993 and was at some point acquired by American Seafoods. In a 2010 exchange, American Seafoods transferred ownership of the vessel to Coastal Villages Region Fund, a community development quota (CDQ) group (see chapter 23). Coastal Villages Region Fund operated the *North Cape* until 2013, after which it was retired from the fishery.[629]

Not far behind the *Aleutian Chalice* was the *Frontier Spirit*, which was built at the now-defunct Marine Construction & Design Company (Marco) shipyard, in Seattle, and was launched in August 1989. The vessel's owner was the Alaska Frontier Company, a Japanese-American joint-venture firm that planned to sell the *Frontier Spirit*'s production of frozen headed-and-gutted Pacific cod in Japan. The vessel, which carried a crew of twenty-three, was equipped with a Marco autoline system and had 14,000 cubic feet of frozen storage that could hold about 650,000 pounds of frozen product.

A second Alaska Frontier Company freezer-longliner, the *Frontier Mariner*, was constructed at the Marco shipyard and entered service in 1990. The vessel was essentially identical to the *Frontier Spirit*, as was the third Marco-constructed freezer-longliner in the Alaska Frontier Company fleet, the *Frontier Explorer*, which entered service in 1991. All three vessels were purchased by Seattle-based Clipper Seafoods in 2008 and remain employed in the Pacific cod fishery.[630]

The first freezer-longliner to be built for Alaska fishermen was the 150-foot *Alaskan Leader*, which was built in Coden, Alabama, for a group of Kodiak longliners and launched in the fall of 1991.

Unconventionally, the owners—who incorporated as Alaskan Leader Fisheries in 1990—chose to eschew automatic baiting in favor of hand baiting. According to Nick Delaney, one of the vessel's owners, "We opted for hand-baiting because we [thought we could] keep more hooks in the water that way."[631]

Alaskan Leader Fisheries subsequently partnered half-and-half with the Bristol Bay Economic Development Corporation, a CDQ group, to build and operate three additional freezer-longliners. The partnership's first vessel, the 167-foot *Bristol Leader*, was launched in 1998. As with its predecessor, hooks on this vessel were hand baited. In 2004, the *Bristol Leader* underwent a major retrofit to accommodate an automated baiting system. The *Alaskan Leader* underwent a similar retrofit in 2006, the year the Bristol Bay Economic Development Corporation obtained half ownership of the vessel. The second vessel built by the partnership, the 124-foot *Bering Leader*, incorporated an automated baiting system when it was launched in 2005. The partnership's third vessel, the 184-foot *Northern Leader*, was launched in 2013 (see below).[632] All four Alaskan Leader Fisheries freezer-longliners are homeported in Kodiak.

Freezer-longliner operations

Although several freezer-longliners have been equipped with machinery to produce frozen cod fillets,[633] the U.S. Pacific cod fleet's production was and remains almost completely frozen headed-and-gutted (H&G) and frozen whole gutted fish that is subsequently processed into a variety of consumer products, mostly in foreign plants.

Industry professionals generally estimate that to turn a profit, a freezer-longliner has to fish ten months per year and average twenty days fishing during each month. One advantage longliners have over trawlers is that longlines can be fished during stormy weather. The gear can be set as a storm approaches, allowed to soak on the seafloor while the vessel seeks shelter or rides out the weather at sea, then retrieved when the weather abates. One highliner longline fisherman's advice: "When weather threatens or picks up, set the gear out."[634] Another freezer-longliner advantage over trawlers was that bycatch, especially of crab and halibut, was not nearly as large a problem. Bycatch of halibut is capped by the North Pacific Fishery Management Council, and once the cap is reached, an offending fishery can be closed even if the allowable catch of the target species has not yet been reached.

Almost all freezer-longliners today are schooner-style (wheelhouse-aft) vessels in the 120- to 180-foot range. Fish are processed and frozen in a factory located in the *tween deck*—the enclosed space between the main deck and a shelter deck that extends the length of the main deck. Freezer-longliners have minimal capacity to store freshly caught fish, and ideally a vessel's processing capability slightly exceeds its potential catch rate.

Longline gear is serviced and readied for setting in an enclosed space in the aft section of the upper deck. Freezer-longliners have two enclosed setting stations at the stern, one for hooks and groundline

Figure 70. Placing headed-and-gutted cod in freezer trays aboard freezer-longliner. (Courtesy of Alaska Seafood Marketing Institute.)

and another for anchors and buoys. The gear is set through closeable openings.

Circle hooks

The widespread use of circle hooks, in which the hook's point is turned perpendicularly back toward the hook's shank to form an approximately circular shape (fig. 71), began in the mid-1980s. The timing was fortuitous for Pacific cod fishermen because circle hooks catch substantially more fish than conventional J hooks, an important benefit in a developing but potentially marginally profitable fishery.

The main reason for circle hooks' high "catching power" is that they are more efficient than J hooks at both hooking and retaining fish. The hooks almost uniformly catch fish in the gristly part of the jaw at the corner of the mouth, and gut-hooked fish are rare. Some fishermen believe fish hooked in this manner do not fight as hard to rid themselves of the hook and stay alive longer and are less likely to be attacked by sand fleas—the previoiusly discussed voracious amphipod crustaceans that live on the ocean floor and can render a fish worthless in a matter of hours. Moreover, there may be a multiplier effect: hooked fish are thought to attract other fish, which may themselves take a baited hook. Another significant advantage of circle hooks is a lower release mortality for unwanted fish (bycatch) because the hooks are less likely to fatally damage the fish. In 1986, Marco introduced its CircleMatic system, the first system that automatically baited circle hooks. The semi-circle "EZ-Baiter" design, which Mustad introduced in 1987, is also effective and has become the industry standard in the freezer-longliner fleet. Today, Mustad supplies 95 percent of worldwide market for mechanical line fishing gear. The company's hooks are primarily manufactured of Swedish steel that is drawn into wire in Norway. The wire is then shipped to China for manufacture into hooks.[635]

Today, most longline cod vessels use Mustad-brand circle or semi-circle hooks on ten- to fourteen-inch-long gangions. The gangions are tied to stainless steel swivels that are in turn fastened to stainless steel *swivel clips*—barrel-like fittings that encircle the groundline at forty-two- to forty-eight-inch intervals. The swivel clips are able to freely rotate around the groundline, which is usually 7/16 inch in diameter. Stainless steel stops clamped to the groundline on either side of each swivel clip prevent the clip from sliding along the groundline.

Favored baits are chopped squid, herring, mackerel, and Pacific saury. Typically, a freezer-longliner consumes about 4,000 pounds of bait daily. Modern autoline systems are capable of baiting hooks at a rate that allows vessels to set gear at speeds of up to 9 knots, but most vessels set gear at about four knots—about three hooks per second.

Catching codfish

For most freezer-longliners, fishing is largely a numbers game: get as many baited hooks in the water as possible. A typical set is in a straight line about seven or eight nautical miles long and is made in water from 30 to 100 fathoms deep. The groundline is anchored and buoyed at each end and may itself be weighted with lead strands or have removable weights attached at intervals along its length. With a forty-two-inch spacing between hooks, a seven-mile-long set would deploy approximately 15,000 hooks. Typically, about two or three sets—usually parallel to one another and one-half mile to three-fourths mile apart—are made each day, with the total number of hooks set ranging from about 40,000 to 50,000. (To put this figure in perspective, there are 86,400 seconds in a day.) The total time the gear is in the water (*soaking*) ranges from about four to twenty hours.

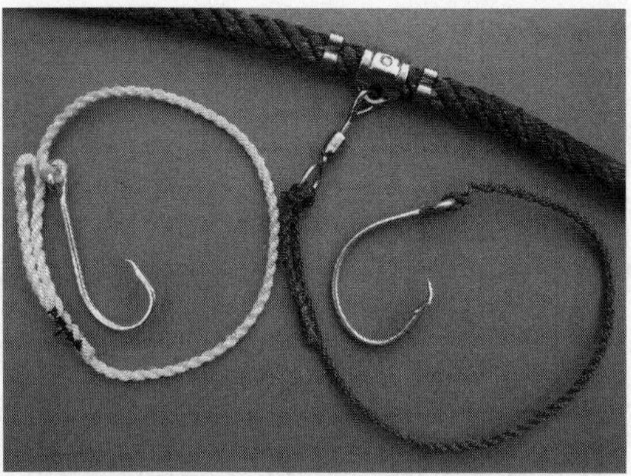

Figure 71. Modern longline gear: hooks, gangions (snoods), swivel, swivel clip, stops, and groundline. The hook on the left is a Mustad "EZ-Baiter" semi-circle design that is favored by the freezer-longliner fleet. The hook on the right is a true circle hook such as are commonly used on small ("coastal") codfish longliners. Photo by the author.

However, at least one company, Cape Romanzof Fisheries, chooses to fish in more difficult grounds that do not accommodate long sets but where the cod tend to be above average in size. These fish command a premium price in certain markets. The company's two freezer-longliners, *Baranof* and *Courageous*, often fish the nearshore Aleutian Islands waters.[636]

Some freezer-longliners are equipped with a computerized longline control and data storage system that is connected to the vessel's GPS system, depth sounder, baiting machine, and line hauler. The system records the time, position, and depth at which the gear is set, as well as the setting speed and the percentage of hooks that are successfully baited. While the gear is being hauled, the system records the hauling speed, hooks per minute, line tension, and the catch rate and weight.

The groundline, hopefully laden with codfish, is pulled aboard by a hydraulically driven line hauler over a roller mounted on the lower edge of a closeable opening that is typically located several feet above the waterline about amidships on a vessel's starboard side. The groundline is typically retrieved at a rate of about thirty to forty hooks per minute. The fisherman working at the roller (the *roller man*) wears a harness that is tethered to the vessel to prevent him from falling overboard or being washed overboard.[637] He lifts especially large cod

Figure 72. Roller man Brad Lambert brings a Pacific cod aboard freezer-longliner *Frontier Explorer*. (Courtesy of Clipper Seafoods.)

aboard with a gaff hook to prevent their weight as they leave the water from causing the hook to pull out. The roller man is also responsible for the careful release of untargeted fish, though the speed at which the gear is retrieved can make doing so challenging.

The line hauler then pulls the fish into a dehooking device fittingly called a *crucifier*, which consists of a pair of approximately four-inch-diameter metal pipes mounted vertically about 3/4-inch apart on a solid frame. The groundline and hooks can pass between the pipes, but a fish's head cannot. Upon having its head pulled against the pipes, the fish has the hook instantly torn from its mouth.

The fish then slides down a chute into a bin, from which it is carried on a conveyor belt into the vessel's factory and past an observer station, where a fisheries observer may monitor the catch. The conveyor belt then deposits the fish on the sorting table, where a crewman cuts its throat and then, depending on the fish's size, slides it or sends it via a conveyor belt into one of several bleeding tanks that contain fresh running seawater in which the fish is allowed to bleed out. Small cod may require only ten minutes bleeding time; large cod may require thirty to forty-five minutes. The bleeding tanks are worked in succession so fish can be bleeding in one tank while already-bled fish in another tank can be fed by conveyor belt into automated heading-and-gutting machinery. Heading and gutting proceeds at a rate of up to about thirty-five fish per minute. The machinery cannot handle especially large cod, which must be headed and gutted by hand.

The H&G fish are sorted by size into tubs, where they are rinsed in fresh running seawater. Rinsed fish are then placed in metal trays that each hold about forty pounds (twenty kilograms). The trays are then transferred into plate freezers, where they are frozen over the course of about four to five hours to around -13 to -22 degrees Fahrenheit (-25 to -30 Celsius). This allows for up to three full freezing cycles per day. Once frozen, each block of fish is removed from its tray, glazed with saltwater, and slid into a plastic-lined paper bag. The bag is then tied off at its open end, weighed, and transferred to the vessel's storage freezer. Modern freezer-longliners are typically capable of producing up to about 1,500 bags—equal to about 66,000 pounds—of frozen H&G Pacific cod per day.

Fishing trips usually range in length from two to four weeks. Catch rates are generally highest during the winter, and vessels normally make shorter trips at this time. Frozen product is offloaded at cold storage facilities in port (primarily at Dutch Harbor and Kodiak), where it is sorted and palletized for shipment via container ship or freighter.[638] [639]

The freezer-longliner fleet today

From 1991 through 1999, trawl gear accounted for 52 percent of the catch of Pacific cod in Alaska, while longline gear accounted for 37 percent. (Pot and jig gear accounted for the remaining 11 percent.) In 2000, however, the catch of Pacific cod by longliners eclipsed for the first time the catch by trawlers. It has continued to do so each year since. On average, during the years 2003 through 2015, longline gear accounted for 45 percent of the catch, while trawl gear accounted for 34 percent.[640]

There were thirty-nine vessels in the freezer-longliner fleet in 2010. The fleet comprised both newer vessels that were designed and built specifically as freezer-longliners and older vessels that had been converted from other uses. The average vessel was close to forty years old, making the freezer-longliner fleet one of the oldest offshore fleets in the United States.

The oldest vessels operating in the freezer-longliner fleet are the Seattle-based Cape Romanzof Fisheries's vessels *Courageous* and *Baranof*. These nearly identical 180-foot vessels were built in Duluth, Minnesota, as the U.S. Coast Guard icebreaking buoy tenders *Tupelo* and *Balsam*, respectively. The vessels were launched in 1942, decommissioned in 1975, and converted into crab catcher-processors in

1978–1979. Beginning in 1984 or 1985, both vessels began using pots on longlines to catch sablefish.

In 1990, Cape Romanzof Fisheries installed a Mustad automated longline system on the *Courageous* and then sent the vessel north to engage in the Pacific cod fishery seasonally as a freezer-longliner. The following year, the company installed an identical longline system on the *Baranof* and sent it north to fish alongside the *Courageous*. In addition to fishing for Pacific cod, both vessels continue to participate in the sablefish and crab fisheries.

The fact that the *Courageous* and *Baranof*—both now more than seventy-five years old—continue to operate in one of the harshest marine environments on the planet is a testament to the high standards under which they were constructed and to a rigorous maintenance program.[641]

Eventually—and perhaps sooner rather than later—these vessels, as well as other aging vessels in the freezer-longliner fleet, will need to be replaced. Anticipating this need, in November 2016 Vigor Industrial, a firm that operates several shipyards in the Pacific Northwest and Alaska, released the design for an "affordable" 144-foot freezer-longliner that—once buyers are found—it intends to serially fabricate to hold the price down. A Vigor Industrial official commented: "We didn't set out to design a Lamborghini. . . . This design is more like a heavy duty pick-up truck."[642]

Two of the newer vessels in the freezer-longliner fleet are the 184-foot *Northern Leader*, which was launched in January 2013 at the J. M. Martinac Shipbuilding Corporation yard, in Tacoma, Washington, and the 136-foot *Arctic Prowler*—the first large commercial fishing vessel ever built in Alaska—which was launched in May 2014 at the Vigor Industrial shipyard (formerly Alaska Ship & Drydock) at Ketchikan, Alaska. The *Arctic Prowler* is owned by Prowler Fisheries, which is in turn owned by the Aleutian Pribilof Island Community Development Association, a community development quota (CDQ) group, and several individuals with long ties to the Alaska fishing industry. Prowler Fisheries owns four additional freezer-longliners: *Prowler*, *Bering Prowler*, *Gulf Prowler*, and *Ocean Prowler*. In a February 2016 consolidation, Prowler Fisheries contracted Seattle-based Blue North Fisheries, which itself operates five freezer-longliners, to manage the operations of its five freezer-longliners.[643]

Built for Alaskan Leader Fisheries at a cost reported to be nearly $35 million, the *Northern Leader* was the largest freezer-longliner ever built and the largest commercial fishing vessel built in the United States in more than two decades. It was also, according to Robin

Figure 73. Freezer-longliner *Northern Leader* (Courtesy of Alaskan Leader Fisheries.)

Samuelsen, chairman of Alaskan Leader Fisheries, one of the "most technologically advanced and innovative commercial fishing vessels ever built." Among the innovations incorporated in the *Northern Leader* is a diesel-electric propulsion system that incorporates a pair of pod-mounted 360-degree directional propellers (*z-drives*). This system enables the vessel to be maneuvered with a single joystick. Another innovation is processing equipment that allows nearly full utilization of the targeted species. (Fish parts that cannot be processed into more valuable products are ground and frozen onboard and then processed into animal feed at shore plants.) The vessel is capable of freezing 153,000 pounds of H&G Pacific cod daily and has freezer space for more than 1.8 million pounds of product.[644]

Pioneering a new era in marine communications, in July 2016, Alaskan Leader Fisheries signed a contract with Inmarsat, a British satellite telecommunications company, for high-speed broadband internet service on its freezer-longliner *Alaskan Leader*. The service—its first use by a fishing vessel in the Bering Sea—will enable the reliable transmission of up-to-date catch information to the company's sales office as well as allow the vessel's crew to communicate with family and friends and to browse the internet—considerable amenities during fishing trips that typically last for weeks.[645]

Nearly all freezer-longliner vessels fish in both the Bering Sea and Gulf of Alaska; just one or two vessels fish exclusively in the

Gulf of Alaska. In 2013, twenty-nine freezer-longliners fished for Pacific cod in the Bering Sea/Aleutian Islands area.[646] About half of the freezer-longliner fleet has some level of community development quota (CDQ) group ownership. The average annual gross revenue for a freezer-longliner is $5.8 million.[647] [648]

In the Bering Sea, the freezer-longliners' fishing season for cod is calendar-year based and is divided into two subseasons: The A season begins January 1 and ends when that season's quota is harvested or on June 10, whichever occurs first. The B season begins June 10 and runs until that season's quota is harvested or on December 31, whichever occurs first. In the central and western Gulf of Alaska, the A season begins on January 1 and ends when that season's quota is harvested or on June 10, whichever occurs first. The B season begins on September 1 and ends when that season's quota is harvested or on December 31, whichever occurs first. Seasons can be closed prematurely if bycatch limits are exceeded.

ENDNOTES

[608] David Witherell, Michael Fey, and Mark Fina (staff, North Pacific Fishery Management Council), *Fishing Fleet Profiles* (April 2012), 45.

[609] "Cod longliner ecstatic over his new automatic baiting machine," *Pacific Fishing* (April 2013): 34, 36–38.

[610] Steve Fish, captain of the longliner *Lualda*, one of the vessels that participated in the venture, recalled that the Japanese processors headed Pacific cod on what was essentially a large table saw without a blade guard. It is doubtful that the U.S. Occupational Safety and Health Administration would have approved such a device on an American vessel.

[611] Steve Fish, F/V *Lualda*, personal communication with author, October 14, 2016; Jana M. Suchy, "Longliners Share Cod Joint-Venture Adventures," *Alaska Fisherman's Journal* (October 1986): 26–28.

[612] Donna Parker, "Codfish Booming in Kodiak," *Alaska Fisherman's Journal* (May 1991), 24; Cecil Ranney, "Kodiak Processors Cry for Cod," *Alaska Fisherman's Journal* (December 1986): 41.

[613] McDowell Group, DRAFT *Economic Impact of the Seafood Industry on the Kodiak Island Borough* (April 2016), 36, 39; Darius Kasprzak, Alaska Jig Association, personal communication with author, July 29, 2018.

[614] "Factory Longline Directory," *Pacific Fishing* (October 1990): 55–56, 58–59.

[615] John Bragg, "Freezer Longliners: A Healthy Fleet in a Hot Cod Market," *Pacific Fishing* (May 1991): 53–59; Longline Catcher Processor Subsector Single Fishery Conservation Act, December 22, 2010 (124 Stat. 3583).

[616] Ben Fissel et al., *Stock Assessment and Fishery Evaluation Report for the Groundfish Fisheries of the Gulf of Alaska and Bering Sea/Aleutian Islands Area: Economic Status of the Groundfish Fisheries off Alaska, 2012* (Seattle: National Marine Fisheries Service, 2014), 253.

[617] John Bragg, "Freezer Longliners: A Healthy Fleet in a Hot Cod Market," *Pacific Fishing* (May 1991): 53–59; David Little, Clipper Seafoods, personal communication with author, January 25, 2015.

[618] Alaska Fisheries Development Foundation, Inc., bulletin, June 1981.

[619] Neil Rabinowitz, "Aleutian Mistress Pioneers Alaska Groundfish Project," *National Fisherman*, November 1981, 110–112, 114.

[620] "Aleutian Mistress: Domestic Vessel Sets Out to Longline for Pacific Cod," *Pacific Fishing* (September 1980): 38; Robert Browning, "Crabber Gets a Shelterdeck for Longline Fishing," *National Fisherman*, November 1980, 99–101; "What Is the Alaska Fisheries Development Foundation?" *Bering Sea Fisherman* (June 1980): 3; "Born Again: A Resurrected Codfishery," *Pacific Fishing* (December 1980): 44–47; Neil Rabinowitz, "Aleutian Mistress Pioneers Alaska Groundfish Project," *National Fisherman*, November 1981, 110–112, 114; C. Sheldon, "Mechanized Longline Fishing—A Technique with a Future," *National Fisherman*, November 1981, 106–109; *Conversion of a Crabber to an Auto-Longlining Catcher/Processor, Executive Summary*, Alaska Fisheries Development Foundation, November 1981; "Major Accomplishment, 1981–1983," Alaska Fisheries Development Foundation; David L. Abbott, "Color Sounders for Longlining: An Idea Whose Time has Come," *Fishermen's News* (July 1981): 21–22; Jerry Tilley, personal communication with author, January 27, 2015; "Bottomfish," *Pacific Fishing*, Yearbook (1983): 63–87; Roger Fitzgerald, "The Highliner and The Mistress," *Alaska Fisherman's Journal* (August 1983): 14–18.

[621] Erling Skaar, personal communication with author, January 6, 2016.

[622] Roger Fitzgerald, "Clean fishing 'Seattle Star' enters cod fishery," *Alaska Fisherman's Journal* (November 1980): 8–9.

[623] A. D. Chandler, "Seattle Star Tries Out Longlining on Alaska Bottomfish," *National Fisherman*, December 1980, 68–69; Erling Skaar, personal communication with author, January 6, 2016.

[624] "Seattle Star," *Alaska Fisherman's Journal* (August 1983): 11.

[625] Erling Skaar, personal communication with author, January 6, 2016; Patrick Burns, Blue North Fisheries, personal communication with author, January 25, 2016.

[626] John Bragg, "Freezer Longliners: A Healthy Fleet in a Hot Cod Market," *Pacific Fishing* (May 1991): 53–57, 59–60; "Factory Longline Directory, 1991," *Pacific Fishing* (May 1991): 38–42. Note: Not all of the vessels listed in the directory fished for Pacific cod.

[627] Charles Piatt, "New Combination Boat for Alaska," *National Fisherman*, November 1989, 62–63; Bruce Buls, "Factory Longliners for the North Pacific," *National Fisherman*, January 1990, 44–45.

[628] *State of Alaska v. Bonar D. Peterson*, District Court for the State of Alaska, Unalaska, 3UN-89-245, October 23, 1989; "Fisherman Fined," *Daily Sitka Sentinel*, October 27, 1989.

[629] State of Alaska, Dept. of Public Safety, Case No. 89-68450, November 7, 1989; In the Matter of: Bonar D. Peterson Chalice Trawler Corp., Respondent, 6 O.R.W. 774 (N.O.A.A.), 1992 WL 347569, Docket No. 116-238, 116-239, June 2, 1992, at 3, 7; Coastal Villages Region Fund newsletter, Spring 2010, 2; Coastal Villages Region Fund newsletter, Summer 2014, 5.

[630] John Bragg, "Freezer Longliners: A Healthy Fleet in a Hot Cod Market," *Pacific Fishing* (May 1991): 53–59; "Frontier Spirit Enters Longline Fishery," *Pacific Fishing* (November 1989): 21–23; "MARCO Launches New 'Spirit,'" *Alaska Fisherman's Journal* (October 1989): 70–71; "Factory Longline Directory," *Pacific Fishing* (October 1990): 55–56, 58–59; "Factory Longline Directory, 1991," *Pacific Fishing* (May 1991), 38–41; "Pot and Longline At-Sea Processors," *Pacific Fishing* (September 1993): 43–49; "Clipper Seafoods Buys Alaska Frontier," archive, *Pacific Fishing* (September 8, 2008), accessed March 6, 2015, http://www.pacificfishing.com/Archives/week_of_090808pf.html; Clipper Seafoods, accessed March 6, 2015, http://www.clipperseafoods.com/vessels/.

[631] Charles Piatt, "150' Factory Longliner for Kodiak Fishermen," *National Fisherman*, December 1991, 40–41.

[632] Jennifer Karuza, "Staying Out of Trouble," *National Fisherman*, September 2002, 26–27; Alaskan Leader Fisheries, "*Bristol Leader*," accessed March 11, 2016, http://www.alaskanleaderfisheries.com/index.php?page=fv_bristol_leader; Alaskan Leader Fisheries, "*Alaskan Leader*," accessed March 11, 2016, http://www.alaskanleaderfisheries.com/index.php?page=fv_alaskan_leader); Alaskan Leader Fisheries, "*Bering Leader*," accessed March 11, 2016, http://www.alas-

kanleaderfisheries.com/index.php?page=fv_bering_leader; Bristol Bay Economic Development Corporation, *Annual Report 2006*, 13.

633 Vessels equipped to produce frozen fillets are: the *Aleutian Mistress*, which fished in 1980 and 1981 (see above); the ill-fated *Galaxy*, a converted U.S. Coast Guard buoy tender that began codfishing in January 1998 but burned and sank in the Bering Sea in October 2002 (see chapter 26); the *Glacier Bay*, a crab catcher-processor that was converted into a freezer-longliner in 2002, and the *Blue North*, which entered service in the fall of 2016 (see afterword).

634 Jeb Wyman, "Cod's Next Step," *Pacific Fishing* (April/May 2003): 32–33, 36; Brad Warren, "Longliner Highlights Efficiency, Comfort," *Pacific Fishing* (November 2001): 37, 100; Sig Jaeger, "Fishing Objectives and Longline Techniques," Alaska Fisheries Development Foundation (April 1983).

635 Doug McNair, "Circle Hooks," *Pacific Fishing* (December 1983): 60–63; Brad Matsen, "Marco's New Circle-Hook Baiter Could Be a Longliner's Best Friend," *National Fisherman*, April 1986, 30; "Marco Keeps on Trucking with New Circle Hook Baiter," *Alaska Fisherman's Journal* (October 1986): 41–42; Terje Paulsberg, *Mustad: Fish Hooks for the World* (Gjøvik, Norway: Alfa Forlag As, 2007), 147–149, 161.

636 Chuck Hosmer, Cape Romanzof Fisheries, personal communication with author, January 6, 2016.

637 Of the 158 Alaska fishermen fatalities between 2000 and 2012, 49 were caused by fishermen being washed overboard or falling overboard.

638 Seattle-based Coastal Transportation, Inc., operates a fleet of small refrigerated freighters that primarily serve western Alaska's groundfish industry. The company has a cold storage at Dutch Harbor that provides a host of services to the Bering Sea freezer-longliner fleet.

639 David Witherell, Michael Fey, and Mark Fina (staff, North Pacific Fishery Management Council), *Fishing Fleet Profiles* (April 2012), 30; Sig Jaeger, "Fishing Objectives and Longline Techniques," Alaska Fisheries Development Foundation (April 1983); Alan Haig-Brown, "Prowler Tools Up," *Pacific Fishing* (October 2003): 24–26; Jeb Wyman, "Cod's Next Step," *Pacific Fishing* (April/May 2003): 32–33, 36; David Little, Clipper Seafoods, personal communication with author, January 26 and January 28, 2015, and April 15, 2015; Alaska Longline Cod Commission, "Quality," accessed February 17, 2015, http://alaskalonglinecod.com/quality/; Manuel Valdes, "New Boat Seeks Safer Fishing on Deadly Bering Sea," *Huffington Post*, December 13, 2013, accessed February 18, 2015, http://www.huffingtonpost.com/huff-wires/20131213/us--less-deadly-catch/?utm_hp_ref=green&ir=green.

[640] NOAA Fisheries Service, Alaska Fisheries Science Center, "Pacific Cod Fact Sheet," 2010; NOAA Fisheries, Pacific cod landings database, provided by Mary Furuness (NOAA Fisheries), December 18, 2016.

[641] Cape Romanzof Fisheries, "About Us," accessed January 9, 2016, http://www.baranofcourageous.com/#!copy-of-about-me/cyjn; "Pot and Longline At-Sea Processors, *Pacific Fishing* (September 1993): 43–49.

[642] "Vigor to unveil new freezer longliner design," MarineLog.com, September 14, 2016, accessed November 22, 2016, http://www.marinelog.com/index.php?option=com_k2&view=item&id=23089:vigor-to-unveil-new-freezer-long-liner-design&Itemid=223; Michael Crowley, "Designed to fish for the next 40 years; 57-foot salmon tender built in Ore.," *National Fisherman*, March 2017, 35, 37.

[643] Tom Seaman, "US cod longline catcher Blue North inks deal to manage Prowler fleet," *Undercurrent News*, February 18, 2016, accessed August 27, 2016, https://www.undercurrentnews.com/2016/02/18/us-cod-longline-catcher-blue-north-inks-deal-to-manage-prowler-fleet/.

[644] "Lt. Gov. Treadwell Speaks at Christening of F/V 'Northern Leader,'" *Alaska Business Monthly* (August 2, 2013), accessed April 10, 2015, http://www.akbizmag.com/Alaska-Business-Monthly/August-2013/Lt-Gov-Treadwell-Speaks-at-Christening-of-F-V-Northern-Leader/; "Freezer longliners modernize fleet," *Deckboss* (blog), May 11, 2013, accessed April 10, 2015, http://deckboss.blogspot.com/2013/05/freezer-longliners-modernize-their-fleet.html; Alaskan Leader Fisheries, "Alaskan Fishing Company Contracts Large Eco-friendly Commercial Fishing Vessel with Tacoma Shipyard," media release, February 14, 2012; Kathleen Gleaves, "Northern Leader," *Fishermen's News Online*, October 14, 2013, accessed June 4, 2016, http://fnonline-news.blogspot.com/2013/10/northern-leader.html; Kathleen Gleaves, "Northern Leader," *Pacific Fisheries Review*, special supplement to *Fishermen's News* (July 2012): 10–14.

[645] Ashley Herriman, "Alaskan Leader signs broadband deal with Inmarsat," *National Fisherman*, July 19, 2016, accessed July 19, 2016, http://www.nationalfisherman.com/news-events/inside-the-industry/6898-alaskan-leader-signs-broadband-deal-with-inmarsat.

[646] David Witherell and Jim Armstrong (staff, North Pacific Fishery Management Council), *Groundfish Species Profiles, 2015*, 4, accessed December 2, 2015, www.npfmc.org/wp-content/PDFdocuments/resources/SpeciesProfiles2015.pdf.

[647] Consumer Price Index–adjusted from 2010 values.

[648] McDowell Group, *Modernization of the North Pacific Fishing Fleet: Economic Opportunity Analysis* (November 2016), 11, 21, accessed December 18, 2016, https://www.portseattle.org/Supporting-Our-Community/Economic-Development/Documents/Fleet%20Modernization%20Final%2011_11.pdf.

Chapter 18
POT FISHING

> *In only a few years, cod pot fishing has risen from relative obscurity to a modest popularity, especially among those concerned with halibut bycatch.*
>
> —*Alaska Fisherman's Journal*, October 1991[649]

Relatively late in development, the pot cod fishery provides an opportunity for vessels to diversify and to augment income from other fisheries. This virtue is especially relevant for crab vessels during the years that crabs are scarce. Importantly, pots experience very low levels of halibut bycatch.

Not infrequently, groundfish, such as Pacific cod, find their way into king and Tanner crab pots.[650] Crab fishermen welcome the fish as free, fresh hanging bait. By the mid-1960s, a few king crab fishermen in Kodiak began targeting cod during the off-season, using the same side-entry pots that they had long used to catch king crabs. The fishermen didn't catch a lot of cod, but they sold what they did catch to a local crab processor, who presumably froze the fish and later sold it as bait.[651]

As noted in chapter 13, several companies that were experimenting with processing and marketing groundfish in 1978 purchased some Pacific cod from fishermen who had begun using king and Tanner crab pots to target cod.[652]

> *By late 1989, the local [Kodiak] crab fleet began looking at the pot cod fishery as a legitimate target fishery that could supplement the declining king and Tanner crab fisheries.*
>
> —Alaska Department of Fish and Game, 1991[653]

The serious use of pots to catch Pacific cod, however, began modestly in 1985 in the waters around Kodiak Island, and the following year six seine boats using pots there made eleven deliveries of Pacific cod that totaled 330,000 pounds. In 1987, about ten vessels pot fished for cod near Kodiak, and by mid-November they had landed a little over a million pounds. In 1988, however, pot fishermen caught more than 3 million pounds of Pacific cod in the Gulf of Alaska. The catch dropped to around 1 million pounds in 1989 but increased dramatically to 13 million pounds in 1990 and then to nearly 20 million pounds—fully 20 percent of the central Gulf of Alaska total allowable Pacific cod catch—in 1991. Part of the catch increase was driven by decreased landings on the East Coast of Atlantic cod, which caused the prices for Pacific cod to surge. Between 1989 and early 1991, the per-pound price processors at Kodiak paid for pot-caught Pacific cod more than doubled, from $0.12 to $0.29. Almost sixty vessels at Kodiak engaged in the three-month pot cod fishery that ended April 29, 1991. One thing was clear: the pot cod fishery had gained recognition at Kodiak as a viable supplement to declining landings of king and Tanner crabs. Writing for the *Anchorage Daily News*, reporter Hal Bernton commented that "the once-lowly cod are now a bread-and-butter fishery" that enabled crab boats to work through the lean winter months.[654]

> **1990:** *In a year swamped with high bycatch and a lot of finger-pointing at dirty fishermen, the groundfish pot fishery has ridden into the North Pacific like a white knight.*
>
> —Donna Parker, *Alaska Fisherman's Journal*, August 1990[655]

Figure 74. Crew members of F/V *Oracle* pose with pot cod in state waters near Unimak Island. (Photo by Pete Neaton. Courtesy of Buck Laukitis.)

Moreover, many fishermen liked the simplicity of fishing for Pacific cod with pots rather than longlines. There were no hooks to bait, no tubs of gear to coil, no snarls of groundline to untangle, and no snags on the seafloor.[656] Not to say it was easy work: "These are 18- to 20-hour days of turning over up to 240 pots a day," said Kodiak fisherman Ron Eads. Pot fishing from his eighty-six-foot crab boat *Rough & Reddy*, Eads delivered more than two million pounds of codfish during a three-month period beginning in March 1990.[657]

A cod pot, like a king crab pot or a Tanner crab pot, is constructed of a steel frame that ranges from about five to eight feet square and from two to three feet deep. The frame is covered with tarred nylon netting. Located on each of two opposing sides of the pot is a narrowing tunnel that leads into the pot. At the inner end of each tunnel is a rectangular *eye* made of steel rod. Mounted in the eye is a *trigger*—a one-way gate that closes from the inside to prevent cod that have entered a pot from escaping.

To prevent a pot from *ghost fishing* if the line connecting the pot to its buoy should break or if the pot becomes derelict, federal and state regulations require each cod pot to be fitted with an escape panel—a section of netting held in place by decomposable twine—to allow fish or crabs to eventually escape from the pot. Fishermen have an alternative name for escape panels: "rotten cotton."

While early pot cod fishermen often converted their pots back and forth between cod and crabs, many pot cod fishermen today use purpose-built pots. Modifying a crab pot for fishing cod, however, is simple: replacement of the pot's crab triggers with triggers designed specifically for cod.

Cod triggers were originally developed in the late 1970s for crab fishermen who wanted to catch their own bait. These triggers, however, were relatively crude devices. During the mid-1980s, Seattle fishing equipment manufacturers Bob and Ed Wyman developed an effective cod trigger design, which they patented in 1989. The Wymans' cod trigger has not significantly changed since that time and is the industry standard today.

The trigger consists of rows of flexible plastic strips (*fingers*) that are fitted to the top and bottom of a rectangular plastic frame. The fingers are angled toward the center of the pot. The upper row fingers interlock with those of the lower row, and both rows of fingers converge at their ends. Fitted with its fingers, the trigger frame is fastened flush to the inside of the tunnel eye with plastic wire ties. A cod swimming into a tunnel can easily push its way through the fingers, but a cod

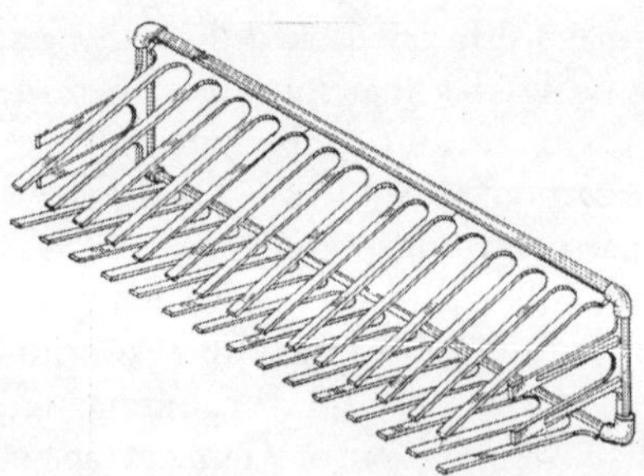

Figure 75. Neptune Marine Products cod trigger. (Neptune Marine Products.)

trying to escape the pot encounters the converged ends of the fingers, which effectively block the exit.

Federal regulations prohibit the size of an entry tunnel eye opening on cod pots from being greater than nine inches in any one dimension. The State of Alaska entry tunnel eye opening regulation is more flexible, stating only that the perimeter of an opening cannot exceed thirty-six inches. In their own self-interest, however, state-waters pot cod fishermen adhere to the federal regulation because it prevents all but the smallest halibut from entering cod pots. And even if they do find their way into a pot, halibut can generally be released unharmed. This makes pot cod fishing, along with jigging, the "cleanest" of the methods used to catch Pacific cod. For this reason, in 1993 the National Marine Fisheries Service, after a series of temporary exemptions, instituted a permanent exemption of the pot cod fishery (and the jig fishery) from the halibut bycatch limits that can close trawl and longline cod fisheries (see chapter 24).[658] Pot gear is classified as fixed gear, along with longline gear and jig gear.

Typically, a tunnel eye is nine inches high and either eighteen inches or twenty-seven inches wide, and the dimensional limitation is achieved by welding vertical uprights at nine-inch intervals in the tunnel eye or by mounting a purpose-built halibut excluder on the outer face of the cod trigger. A simple halibut excluder is a nine-inch square made of plastic-coated 10-gauge steel wire. The excluder is mounted on one side of a nine-by-eighteen-inch cod trigger to create two side-by-side openings, or it can be centered on a nine-by-twenty-seven-inch cod trigger to provide three openings. Additionally, the netting used on cod pots can be selectively sized to allow small cod and other fish to escape.

Some fishermen who target cod add one or two additional tunnels. In recent years, three-tunnel pots are probably the most common among cod fishermen. In 2000, a cod pot, including buoy line and buoys, cost about $600.[659]

Today, vessels in federal waters typically fish seventy-five to one hundred pots, while those fishing in state waters are, except under special circumstances, limited to sixty pots per vessel (see chapter 23). Cod pots are usually baited with cloth bags filled with chopped Pacific mackerel or Pacific saury and are set singly, usually in water forty to fifty fathoms deep. A buoy line connects the pot to a plastic buoy and a smaller auxiliary buoy at the surface. Once set, the pots are allowed to *soak*—sit on the seafloor—for about six to eight hours before being hauled. Once aboard, the pots are opened and the catch unloaded and sorted. If the catch is satisfactory, the pot is quickly rebaited and set again. Weather and sea conditions permitting, this schedule allows a vessel to theoretically haul each pot three times per day. Typically, however, pots are hauled twice daily.[660]

Some vessels, such as the *Katmai*, a catcher-processor that sank in 2008, deployed multiple pots off longlines approximately 2.5 nautical miles in length. The longlines were weighted at each end with anchor chain and were also buoyed at each end.[661]

It is thought that most cod that enter a pot do so in the first hours after it is set, and that their effort to get at the bait attracts other cod. The effort of these secondary arrivals to get at the bait, in turn, further increases the activity in the pot and attracts even more cod.

Because pot-caught cod come aboard alive, are promptly bled, and are not subject to crushing (as in the cod end of a trawl), they are of very high quality. Processing plants normally require delivered fish to have been out of the water no more than two or, in some cases, three days. Almost all the vessels that fish for cod with pots are 58-foot *limit seiners* (vessels of the maximum overall length allowed to seine for salmon in Alaska), crab boats, or salmon tenders that fish for cod when not engaged in other fisheries. The vessels are equipped with refrigerated seawater (RSW) systems to keep their catch cold. The ex-vessel price for pot-caught Pacific cod is typically two or three cents per pound higher than for trawl-caught fish.

The waters around Kodiak Island, especially Shelikof Strait, can be very stormy during the fall and winter months, making it difficult and sometimes dangerous for relatively small vessels, such as limit seiners, to deliver their catch to processors at Kodiak on a regular schedule. To rectify this situation, Kodiak seafood processors sometimes send tenders to the fishing grounds to purchase Pacific cod from pot boats.

The tenders are usually about 100 feet long and can carry about 150,000 pounds of cod in refrigerated seawater. The price paid on the fishing grounds is usually about three cents less than is paid dockside.

In the Bering Sea, pot cod fishing became an attractive option in the fall of 1994, when crab fishermen were facing a drastic reduction in Tanner crab quotas and grim prospects for king crab fishing. For some crab fishermen, pot cod fishing was a way to keep up with their vessel mortgage payments. At least nineteen vessels, including two catcher-processors, fished for cod. The ex-vessel price at Dutch Harbor was $0.15 to $0.18 per pound during 1993 and 1994.[662]

In 1998, eighty-five vessels participated in the Bering Sea pot cod fishery and caught some 57 million pounds of Pacific cod, about 12.4 percent of the total allowable catch for all gear sectors.[663]

Several catcher-processors are also involved in the pot cod fishery and produce frozen headed-and-gutted (H&G) product that rivals the quality produced by freezer-longliners.

According to Bob Wyman, one of the designers of the modern cod trigger, the average per-pull catch by cod pots during the late 1980s ranged from 200 to 500 pounds. In 1987, one fisherman reported catching more than 2,000 pounds in a single pot after a twenty-four-hour soak, while another fisherman reported taking 4,500 pounds from eight pots fished near Dutch Harbor. These fishermen likely used the cod as bait to catch king crab. A crab boat fishing, say, two hundred to three hundred pots might outfit fifteen or twenty of them to catch bait cod.[664]

A vessel fishing in the Bering Sea in 1993 reported catching from five to about forty cod per pot, with the fish averaging from eight to ten pounds. The cod in some deliveries from the Unimak Pass area to the Alyeska Seafoods plant in Dutch Harbor that year, however, averaged twenty pounds.[665]

Pot vessel landings are often very substantial. In early 2015, the *Aleutian Mariner*, a 117-foot vessel that alternated between fishing for cod and fishing for crab, made a delivery of 280,000 pounds of cod to Dutch Harbor. And the 100-foot *Farrar Sea* alone in early 2015 caught more than 3 million pounds of cod in federal waters near Dutch Harbor.[666]

Historically, the North Pacific Fishery Management Council allocated fish quota among competing gear groups based upon a policy of "protect[ing] those who have made significant long-term investments and are dependent on the fishery." Cod pot gear, however, was combined with longline gear in the *fixed gear* allocation category. In the spring of 1993, as the council prepared to debate how cod in the

Bering Sea would be allocated, pot cod interests lobbied for a separate and increased allocation to their fishery based on merit—the fact that pots caught less bycatch than the other gear. They characterized their bycatch as *de minimis*—too trivial to warrant consideration.

The effort that year was to no avail, but in 1999 the council, to "stabilize the growing hook-and-line and pot fleets," passed BSAI[667] Amendment 64. The amendment was implemented the following year. One of its provisions allocated 18.3 percent of the non-CDQ Bering Sea total allowable catch (TAC) to the pot fleet. Another provision, designed to provide fishing opportunities for small vessels, allocated an additional 0.7 percent of the non-CDQ TAC to pot vessels and hook-and-line vessels less than 60 feet long. These vessels would first utilize the 18.3 percent allocation open to all length classes and, when that fishery closed, utilize the 0.7 percent allocation. The pot cod allocations reflected the approximate average harvest of these gear sectors since the mid-1990s.[668]

Amendment 64 expired at the end of 2003 and was replaced by Amendment 77. This amendment, which was implemented for the 2004 season, continued the same overall allocations as under Amendment 64, but divided the 18.3 percent pot gear allocation between pot catcher vessels, which received 15 percent of the non-CDQ TAC, and pot catcher-processor vessels, which received 3.3 percent of the non-CDQ TAC. The council also increased the allocation to pot vessels and hook-and-line vessels less than 60 feet long to 1.4 percent to allow for growth in this mainly shore-based sector.[669]

As circumstances change, reallocations continue. In 2008, the NMFS implemented Amendment 85, which reduced the allocations for pot catcher vessels and pot catcher-processors to 8.4 percent and 1.5 percent, respectively. The allocation to pot vessels and hook-and-line vessels less than 60 feet long, however, was increased to 2 percent.[670]

Allocating Pacific cod to the pot fleet in the Gulf of Alaska took until 2012, when the NMFS implemented GOA[671] Amendment 83, which provided for gear sector groundfish allocations in the gulf based primarily on historical catches.[672] Pot vessels in the central Gulf of Alaska were allocated 27.8 percent of the allowable catch, while in the western Gulf, pot vessels were allocated 38 percent of the allowable catch.[673] Relatively few catcher-processors utilize pot gear in the Gulf, and these vessels work off the same allocation as the catcher vessels.

The pot cod fishery in state waters is open entry, but vessels in the Chignik and South Alaska Peninsula areas, as noted above, may not exceed 58 feet in length overall and are limited to 60 pots each. (Statewide, there can be temporary exceptions to the 60-pot limit.

The commissioner of the Alaska Department of Fish and Game may temporarily suspend the limit in an area if the commissioner determines the area's Pacific cod guideline harvest level will not be reached without doing so.[674])

In the Kodiak area in 2015, state regulations limit pot cod vessels greater than 58 feet in length to harvesting no more than 25 percent of the total guideline harvest level for Pacific cod. The 2015 combined guideline harvest level for Pacific cod in the Kodiak, Chignik, and South Alaska Peninsula areas was 41 million pounds.[675]

In 2015, halibut bycatch mortality—halibut that are killed or are statistically expected to die after being returned to the water—in the Pacific cod pot fishery was estimated to be just 5,000 pounds, and the North Pacific Fishery Management Council, in an effort to reduce bycatch of halibut as well as Chinook salmon by trawlers in the Gulf of Alaska, began consideration of a plan to give trawlers the opportunity to voluntarily convert to pot gear to catch Pacific cod.[676] Today, in terms of Pacific cod production, pot fishermen are not far behind trawlers. Statewide landings by pot fishermen in 2016 totaled 83,281 metric tons, while trawlers that same year landed 97,875 metric tons.[677]

Codfish report, 1993: *Cod product mix. The product mixes differ substantially between the cod trawl fishery and the other two cod fisheries for 1990–92. Cod taken with longline or pot gear is used principally to produce headed-and-gutted (H&G) cod. For example, H&G cod accounted for between 93% and 96% of the product weight of cod in the cod longline fishery and for between 78% and 84% of the product weight of cod in the cod pot fishery. H&G cod accounted for only between 29% and 41% of the product weight of cod in the cod trawl fishery. Fillets, which accounted for from 0% to 2% of the cod product weight in the cod longline and for from 2% to 6% in the pot fisheries, accounted for from 19% to 31% of the cod product weight in the cod trawl fishery. Both whole fish and salted and split cod accounted for larger shares of the cod product weight in the cod trawl fishery than in the other two cod fisheries.*

—National Marine Fisheries Service, 1993[678]

ENDNOTES

649 "Cod Pots on the Rise," *Alaska Fisherman's Journal* (October 1991): 94.

650 The Alaska Department of Fish and Game used a formula of ten pounds of cod per pot pull. In setting cod quotas, federal fisheries managers used this formula, at least for a while, to estimate the crab fleet's incidental catch of cod.

651 Sue Jeffrey, "Kodiak Crab Boats Lose BSAI Cod," *Pacific Fishing* (January 2000): 16; Al Burch, owner/captain of trawler *Dusk*, personal communication with author, April 6, 2015.

652 Lynn Pistoll, "A Discussion of the Potential for Expanding into an Alaskan Bottomfish Industry," *Alaska Economic Trends* (August 1978): 1–5.

653 David Carlile, Tom Dinnocenzo, and Leslie Watson (Alaska Department of Fish and Game), *An Evaluation of the Effectiveness of Modified Crab Pots for Increasing Catch of Pacific Cod and Decreasing Catches of Halibut and Crab* (March 1991), 1.

654 Hal Bernton, "'A Real Clean Fishery,'" *Anchorage Daily News*, May 5, 1991.

655 Donna Parker, "Council Gives Pot Gear a Boost," *Alaska Fisherman's Journal* (August 1990): 18.

656 Alaska Fisheries Development Foundation, "Getting Potted for Cod," *Lodestar* (Spring 1991): 1–2; Donna Parker, "Pots May Save Cod Fishery," *Alaska Fisherman's Journal* (July 1990): 127–129; John van Amerongen, "Crab Pots for P-Cod," *Alaska Fisherman's Journal* (February 1988): 44; "Cod Pots on the Rise," *Alaska Fisherman's Journal* (October 1991): 94; Tom Wray, "Irish Research Creates Interest," *Fishing News* (March 19, 1993): 45; Hal Bernton, "'A Real Clean Fishery,'" *Anchorage Daily News*, May 5, 1991; Joel Gay, "In Alaska, Everything's Coming Up Cod," *National Fisherman*, August 1991, 28–30.

657 Donna Parker, "Pots May Save Cod Fishery," *Alaska Fisherman's Journal* (July 1990): 127–129.

658 David Witherell, Michael Fey, and Mark Fina (staff, North Pacific Fishery Management Council), *Fishing Fleet Profiles* (April 2012), 52; Alaska Administrative Code, 5 AAC 28.050(e); 58 *Fed. Reg.* 14524, March 18, 1993; 59 *Fed. Reg.* 7647, February 16, 1994.

659 "'Canada Should Go Cod Potting,'" *Fishing News International* (October 1992): 43; Charlie Ess, "Pacific (Cod) Time," *National Fisherman*, August 2000, 18–21; Wyman et al., U.S. Patent No. 4,843,756, July 4, 1989; Edward Wyman, Neptune Marine Products, personal communication with author, April 29, 2016.

[660] Ann Touza, "Crabbers Turn to Cod in Dutch," *Alaska Fisherman's Journal* (December 1994) 10, 12; Jim Paulin, "Pot cod fishery numbers all in," *Bristol Bay Times–Dutch Harbor Fisherman* (April 28, 2017), accessed May 1, 2017, http://www.thebristolbaytimes.com/article/1717pot_cod_fishery_numbers_all_in.

[661] Marine Board of Investigation, U.S. Coast Guard, *Sinking of the F/V Katmai in the Amchitka Pass, North Pacific Ocean, on October 22, 2008, with Multiple Loss of Life*, MISLE Activity No. 3351236 (April 26, 2010), 44.

[662] Ann Touza, "Crabbers Turn to Cod in Dutch," *Alaska Fisherman's Journal* (December 1994) 10, 12.

[663] Michel Drouin, "Cod Wars," *Pacific Fishing* (August 1999), 20–22.

[664] John van Amerongen, "Crab Pots for P-Cod," *Alaska Fisherman's Journal* (February 1988): 44; Imre Nemeth, "Insert Turns Crab Pot Into Cod Pot," *Alaska Journal of Commerce* (March 14, 1988): 6; Masahiko Takeuchi (Minato Tsukiji), "US Pacific cod suppliers selling more to home market," *Undercurrent News*, April 7, 2014, accessed December 10, 2015, https://www.undercurrentnews.com/2014/04/07/alaskan-pacific-cod-suppliers-selling-more-to-north-america/.

[665] John van Amerongen, "Going to Pot," *Alaska Fisherman's Journal* (June 1993): 6; Ann Touza, "Crabbers Turn to Cod in Dutch," *Alaska Fisherman's Journal* (December 1994) 10, 12.

[666] Annie Ropeik, "Double Boiler," *National Fisherman*, May 2015, 26–27, 30.

[667] BSAI is an initialism for the Bering Sea/Aleutian Islands area.

[668] National Marine Fisheries Service, Alaska Region, Alaska Groundfish Fisheries, *Final Programmatic Supplemental Environmental Impact Statement* (June 2004), C-88; Ed Wyman, "Preferential Allocation or Preferential Management," *Alaska Fisherman's Journal* (April 1993): 19; 65 *Fed. Reg.* 51553, August 24, 2000.

[669] North Pacific Fishery Management Council, *BSAI Amendment 85, Environmental Assessment/Regulatory Impact Review/Initial Regulatory Flexibility Analysis, Secretarial Review Draft* (October 11, 2006), 7.

[670] 72 *Fed. Reg.* 50788, September 4, 2007.

[671] GOA is an initialism for the Gulf of Alaska area.

[672] 76 *Fed. Reg.* 37763, June 28, 2011.

[673] North Pacific Fishery Management Council, *Fishery Management Plan for Groundfish of the Gulf of Alaska* (November 2016), 22.

[674] Alaska Department of Fish and Game, *2013–2014 Statewide Commercial Groundfish Fishing Regulations*, 30, 37, 44, 50, 56.

[675] Division of Commercial Fisheries, Alaska Department of Fish and Game, "Kodiak, Chignik, and South Alaska Peninsula Areas State-Waters Pacific Cod Guideline Harvest Levels and Season Opening Dates," news release, January 6, 2015, accessed November 8, 2015, http://www.adfg.alaska.gov/static/applications/dcfnewsrelease/508203729.pdf; Alaska Department of Fish and Game, *2015–2016 Statewide Groundfish Commercial Fishing Regulations*, 44.

[676] International Pacific Halibut Commission, *Annual Report, 2015* (Seattle: International Pacific Halibut Commission, 2016), 30; Laine Welch, "Trawlers may convert to pot gear for cod catches," *Alaska Fish Radio*, October 28, 2015, accessed November 1, 2015, http://www.alaskafishradio.com/trawlers-may-convert-to-pot-gear-for-cod-catches/.

[677] Mary Furuness (NOAA Fisheries), personal communication with author, March 1, 2017.

[678] Staff (Seattle, Wash.), National Marine Fisheries Service, *Draft for Secretarial Review, Environmental Assessment/Regulatory Impact Review/Initial Regulatory Flexibility Analysis of Alternatives to Allocate the Pacific Cod Total Allowable Catch by Gear and/or Directly Change the Seasonality of the Cod Fisheries, Amendment 24 to the Fishery Management Plan for the Groundfish Fishery of the Bering Sea and Aleutian Islands Area* (October 5, 1993), 21.

Chapter 19
JIGGING[679] [680]

Modern jigging is merely a continuation of the hand lining of yore from dories, but the line is now spooled on a motorized reel, which allows for more hooks to be fished deeper and retrieved faster. The basic concept of hand-tended, vertical hook-and-line fishing remains unchanged.

—Darius Kasprzak, Alaska Jig Association, January 2017[681]

Bycatch is minimal, gear conflicts with other boats are nil, and operating costs are low.

—Joel Gay, *National Fisherman*, May 1992[682]

The jig fishery is among those with the lowest barriers to entry, as it can be prosecuted with a small vessel and has very modest gear requirements.

—Dory Associates, 2009[683]

The jig fishery, though relatively small, is a key fishery in the Gulf, providing entry-level opportunity into area fisheries and contributes to a diversified fishing portfolio for combination fishing vessels throughout GOA coastal Alaskan communities.

—North Pacific Fishery Management Council, Alaska Department of Fish and Game, and National Marine Fisheries Service, 2012[684]

The jig sector also serves as a truly entry-level opportunity for local and especially young fishermen in Alaska.

—Theresa Peterson, Alaska Marine Conservation Council, Alaska Jig Association, 2018[685]

Alaska's Pacific cod jig fisheries today are relatively low-production affairs that were devised primarily to provide opportunities for local small-boat operators. As such, in the federal and parallel Pacific cod fisheries of the Gulf of Alaska, vessels using five or fewer jigging machines (one line per machine, thirty hooks per line) are exempt from the federal license limitation program. In the federal and parallel Pacific cod fisheries of the Bering Sea, catcher vessels less than 60 feet in overall length using jig gear are also exempt from the license limitation program.[686] The state-waters jig fisheries in both the Gulf of Alaska and Bering Sea are open access. In combination, the seasons for the federal, parallel, and state-waters fisheries provide very liberal opportunities for jig fishermen to ply their trade.[687]

In Alaska, jig gear is classified as fixed gear, along with longlines and pots. Because the few halibut that jig fishermen catch can usually be released with minimal injury, the jig fishery is exempt from halibut bycatch-induced closures.

Jig-caught cod spend little time on the hook, can be bled and iced promptly, and are typically delivered to a processor within three days of being caught. These factors warrant jig-caught cod's reputation for high quality. Reflecting this, fish processors typically pay several cents more per pound for jig-caught cod than for cod caught by trawlers. (As noted above, processors pay a similar premium for cod caught by pot and longline gear.)[688]

Jig fishermen operate primarily from the ports of Kodiak and Homer in the central Gulf of Alaska, and from Sand Point in the western Gulf. All but a few jig for cod to supplement their income from other fisheries, such as salmon and halibut. They typically fish from one to five jigging machines and generally utilize either of two methods to fish for Pacific cod: (1) using baited hooks fished at or near the seafloor from an anchored vessel; or (2) drift *fishing with rubber*—actively jigging lures that are basically a short length of colored flexible plastic tubing that is slid over a slightly bent long-shank hook. The lures may be augmented with bait.

A jigging machine weighs about thirty-five pounds and is mounted on a pipe stanchion, at the top of which is an outward-pointing rigid metal or flexible fiberglass arm about thirty inches long with a pulley at the end.

The mainline, which is usually 200- to 500-pound-test monofilament or something of similar strength (usually braided Spectra- or Dynex-brand polyethylene), runs from the reel up through a pulley near the top of the stanchion, then outward to the pulley at the end of the fiberglass arm, then downward. Fastened to the mainline's end is a nylon or steel ring to which is clipped a *setup*—a length of leader line, typically monofilament, fitted at intervals with swivels and/or leaders and their respective lures or baited hooks. A setup typically consists of about ten or fifteen lures or baited hooks spaced one to three feet apart. Clipped to the end of the setup is a lead, steel, or iron sinker that weighs about five to ten pounds (see fig. 77). If a setup is damaged or if fishing circumstances require, it can be quickly unclipped and replaced.

The majority of jig-caught Pacific cod are caught using the baited-hook method. With this method, many fishermen use relatively simple

Figure 76. Greg Perkins jigging for cod near Kodiak Island, 2009. Note Kachemak bottom reels. (Courtesy of Darius Kasprzak.)

bottom reels in which the reel is belt driven by an electric motor that can operate on 12- or 24-volt direct current. These reels, which retail for about $1,300, are known locally as Kachemaks, after the Kachemak Gear Shed, the Homer, Alaska, company that sells them. (Waterman Industries, in Florida, manufactures Kachemaks.)

Bottom reels do not automatically jig the gear. However, when the reels are used with a flexible fiberglass arm, wave action causes the arm to flex and rebound, which imparts a jigging motion on the gear hanging below. Jig fishermen use either circle hooks or semi-circle hooks that are typically baited with squid, octopus, herring, or mackerel. Many jig fishermen also attach a mesh chum bag filled with chopped mackerel or herring to a mainline leader or on a separate line to help attract cod to the baited hooks. At times, jig fishermen drift while fishing with baited hooks and may use a sea anchor to slow their drift.

Fishing with rubber is most often done with computerized jigging machines while drifting. Like bottom reels, either 12- or 24-volt direct current powers the machines. Some are very sophisticated, incorporating built-in as well as user-definable programs that control the jigging speed, frequency, and type. When multiple jigging

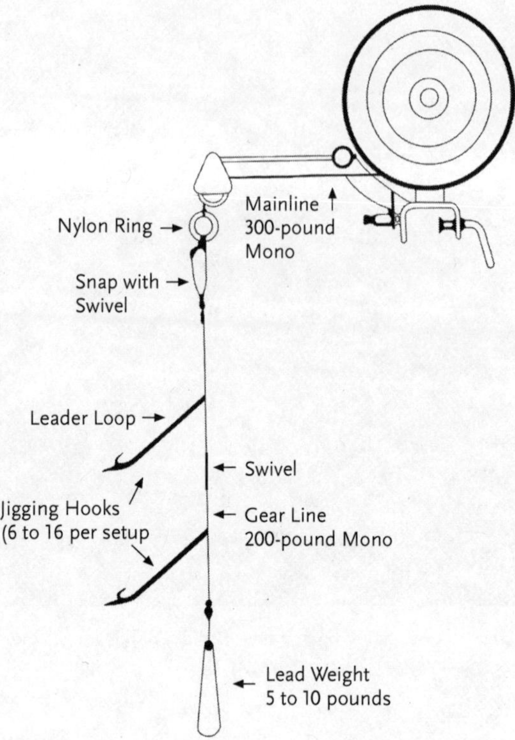

Figure 77. Jigging machine/gear diagram. (Courtesy of Kachemak Gear Shed.)

machines are employed on a single vessel, they can be programmed to communicate with each other to keep fish beneath the vessel for as long as possible by ensuring that at least one machine continues jigging—and attracting fish—while the others haul fish to the surface.

The basic function of a computerized jigging machine can be divided into four parts:

- pay out the mainline and detect when the sinker hits the bottom
- jig the line to attract fish
- detect when fish bite the hooks
- haul the fish to the surface when a preset line tension is reached[689]

DNG-brand jigging machines, which are manufactured in Iceland and cost about $4,600, are the most commonly employed brand in Alaska waters.[690]

Figure 78. Codfish jig advertisement, circa 1980. (Courtesy of Ed Wyman.)

On the fishing grounds, jig fishermen use electronic fish-finding technology to locate concentrations of Pacific cod, and then anchor their vessels or position them to drift over the fish. Generally, Pacific cod are found nearshore in relatively shallow water (five to forty fathoms) during the spring and summer, but are deeper (forty to sixty fathoms) during the winter. Fish hauled up by the machines are removed one by one.

The use of jigging machines in Alaska began in the late 1970s. It was a modest beginning: the harvest of Pacific cod by fishermen using the machines in the nearshore waters of the central and western Gulf of Alaska during the 1980s was considered too trivial to document. Eventually, however, a few fishermen who wanted to supplement their income from other fisheries began outfitting their boats with jigging machines. For the years 1993 through 1995, the Alaska Department of Fish and Game reported that the small Gulf of Alaska jig fleet had accumulated landings of 917,000 pounds.[691]

The jigging effort increased substantially beginning in 1997, when the Alaska Board of Fisheries established five state-waters groundfish fisheries. For jig fishermen targeting Pacific cod, the Kodiak and South Alaska Peninsula areas are by far the most important. In the

Kodiak area that year, jig fishermen received an allocation of 50 percent of the 8.5 million pound Pacific cod guideline harvest level. The seventy-three vessels that participated in the fishery landed nearly 2 million pounds of Pacific cod. The effort and the catch in the South Alaska Peninsula fishery—where jig fishermen did not receive a sector allocation—was considerably smaller: forty-three jig vessels landed just shy of 350,000 pounds.[692]

In the Kodiak state-waters management area during the years 1997 through 2015, an annual average of ninety-three vessels jigged for Pacific cod. Their total annual catch averaged 3.27 million pounds. The trend, however, has been boom-and-bust cycles, such as low catch rates during the years 2006 through 2008 and high catch rates during the years 2009 through 2012. The high year in the Kodiak area was 2012, when 145 jiggers caught nearly 8 million pounds of Pacific cod. The following year, however, the effort and catch plummeted. The total catch of the fifty-five jiggers who participated in the fishery landed slightly less than 600,000 pounds—the smallest catch in any of the years from 1997 through 2015.[693]

In the South Alaska Peninsula state-waters management area during the years 1997 through 2016, an annual average of thirty-nine jig vessels had a total annual average catch of 1.59 million pounds of Pacific cod. As with the Kodiak area fishery, there has been an element of boom and bust, and landings vary considerably. The high year was 2003, when sixty-five jiggers caught 3.63 million pounds. The low year was 2006, when the dozen jiggers who participated in the fishery landed only 99,552 pounds of Pacific cod.[694]

Farther west, at Unalaska/Dutch Harbor, three jig vessels fished for groundfish—almost exclusively Pacific cod—in 1992. Only two jig vessels fished in 1993, but participation in the fishery surged to twelve vessels the following year. Participation peaked in 1996, when eighteen jig vessels were active. Markets at Unalaska/Dutch Harbor for what were by necessity small and sporadic deliveries of cod were initially somewhat tentative, but local processors soon managed to accommodate jig fishermen.

During the summer of 1994, one Unalaska/Dutch Harbor fisherman using a single jigging machine mounted on a 20-foot aluminum skiff delivered 60,000 pounds of Pacific cod. The following summer, the fisherman, now fishing from a 30-foot aluminum bowpicker, single-handedly caught 100,000 pounds of Pacific cod. That year, the Unalaska Native Fishermen's Association's fleet comprised fifteen jig vessels.[695]

The Unalaska/Dutch Harbor jig fleet fished mostly west of Unalaska Bay, on the north side of Unalaska Island. Perhaps due to local depletion of the cod resource, the number of active jig vessels at Unalaska/Dutch Harbor declined to nine in 1996, and in 2000 there were just seven.

In 2007, because of a desire of the North Pacific Fishery Management Council to "maintain and expand entry-level, local opportunities" in the Bering Sea/Aleutian Islands area, the cod allocation to the jig sector (and to hook-and-line and pot catcher vessels less than 60 feet long) was increased beyond those sectors' catch histories. During the years 2004 and 2005, jig vessels accounted for only 0.1 percent of the retained harvest of Pacific cod, but in 2007 the sector was allocated 1.4 percent of the non-CDQ total allowable catch. Similarly, the council in 2011 increased the central and western Gulf of Alaska jig sector's allocation beyond its catch history, again to enhance entry-level opportunities and to "maximize seasonal access to Federal waters for jig vessels in conjunction with State waters jig fisheries, thereby increasing jig vessel fishing opportunities."[696]

During the years 1995 through 2014, the average annual catch by jig fishermen in the waters around Unalaska/Dutch Harbor was a little less than 300,000 pounds. The catch peaked in 2011 at nearly 950,000 pounds, but the fishery since then has almost disappeared.

Jigging is also permitted in the state's Aleutian Islands District (in the vicinity of Adak Island), but the effort there, too, is minimal. During the years 2006 through 2014, jig fishermen participated in the district's Pacific cod fishery only four years. It likely had more to do with Adak's remoteness rather than a lack of fish, because in 2018, the lone jig fisherman at Adak delivered about 100,000 pounds of Pacific cod to the Golden Harvest Alaska Seafood plant during a month of fishing in nearby waters during the B season. The fish were especially large and were filleted, but the pin bones were left in and the skin left on. These fillets were frozen and then sold in Portugal, where they were salted.[697]

A 2009 study on restricted-access fisheries by Kodiak-based Dory Associates implied that the efforts to provide access to the cod fishery—at least in the Gulf of Alaska—through jigging might have been "too small to be meaningful." The study assumed that a vessel with appropriate jig gear could be purchased for $60,000, which, based on a six-year loan, would require two annual loan payments of $7,539 each. If the ex-vessel price of Pacific cod was $0.30 per pound, a fisherman would need to land 25,130 pounds of cod just to make one loan payment. And fishermen had other major expenses, such as fuel and insurance. The study pointed out that in 2007, the average Pacific cod jig fisherman at Kodiak landed 16,400 pounds of cod; on the Alaska Peninsula, the average catch was 8,900 pounds.[698]

Figure 79. The late Lisa Malutin jigging cod near Kodiak Island, 1999. (Courtesy of Darius Kasprzak.)

The year 2007, however, was not representative: Pacific cod landing by jig fishermen in both the Kodiak and South Alaska Peninsula state-waters areas that year were far below the average for the years 1997 through 2015. (On occasion, individual jig vessels in these areas have landed more than a half-million pounds in a single year.) Moreover, the math is considerably better for fishermen who already own their vessels and jig for cod to supplement their income from other fisheries.[699]

Coastwise, the jig fleet in 2010 consisted of one catcher-processor and seventy-six catcher vessels. The fleet caught 8.5 million pounds of Pacific cod that year, with an ex-vessel value of $2.4 million. For jig fishermen, the average ex-vessel price per pound was $0.284.[700] In recent years, however, participation in the Pacific cod jig fishery has declined significantly. In 2016, the coastwise jig fishery landed 4.6 million pounds of Pacific cod.[701]

Nevertheless, jig-caught Pacific cod and the concept of local, small-boat fisheries have a certain allure. In 2014, the Kodiak-based Alaska Jig Association, which was established in 2004 to represent jig fishermen, partnered with the Alaska Marine Conservation Council, an organization that works to ensure, in its own words, the "long-term health of Alaska's coastal and marine ecosystems and the communities that rely on them," to form Kodiak Jig Seafoods. The new company worked and is continuing to work to develop new markets for Pacific cod as well as rockfish by emphasizing quality, conservation, and community. The U.S. Department of Agriculture has funded an effort by Kodiak Jig Seafoods to expand in-state markets.

The fishermen who supply Kodiak Jig Seafoods are almost all local, small-boat fishermen who, to ensure a top-quality product, make short fishing trips, bleed their fish, and chill them in slush ice (a slurry of ice and seawater). For this effort, Kodiak Jig Seafoods pays the fishermen about $0.15 per pound above *dock price*—the price large processors at Kodiak pay for dockside delivery of Pacific cod. Kodiak Jig Seafoods, however, has no processing facility of its own and contracts to have its fish processed and packaged.[702]

ENDNOTES

679 In some locations, regulations also allow the use of dinglebar troll and hand troll gear to catch Pacific cod, but the use of this gear is essentially nonexistent. Dinglebar troll gear is a variant of salmon power troll gear that utilizes a long, heavy iron bar attached to the bottom end of the line to keep lures or baited hooks attached to the line close to the ocean floor. While the gear is pulled through the water by a vessel making way, the iron bar frequently contacts the ocean floor. Mostly, dinglebar troll gear is used in Southeast Alaska to target lingcod. Hand troll gear is similar to dinglebar troll gear, except that the gear is raised and lowered manually rather than hydraulically. According to Clem Tillion, the inclusion of hand troll gear in the state-waters fisheries was apparently Alaska state senator Bill Ray's price for supporting the establishment of state-waters groundfish fisheries. Ray served in Alaska's Senate from 1970 until 1986.

680 Clem Tillion, personal communication with author, April 8, 2016.

681 Darius Kasprzak, Alaska Jig Association, personal communication with author, January 24, 2017.

682 Joel Gay, "Cod jiggers like their gear but face a management Catch 22," *National Fisherman*, May 1992, 40–41.

683 Dory Associates, *Access Restrictions in Alaska's Commercial Fisheries: Trends and Considerations* (January 2009), 12, accessed May 12, 2015, http://www.akmarine.org/wp-content/uploads/2014/06/AMCC_access-restrictions-dory-report-06-01-09.pdf.

684 North Pacific Fishery Management Council, Alaska Department of Fish and Game, and National Marine Fisheries Service, *Gulf of Alaska Pacific Cod Jig Fish-*

ery—Management Report and Update, North Pacific Fishery Management Council and Alaska Board of Fisheries, March 2012, 1.

[685] Theresa Peterson, "Jigged for success," *National Fisherman*, August 2018, 5.

[686] David Witherell, Michael Fey, and Mark Fina (staff, North Pacific Fishery Management Council), *Fishing Fleet Profiles* (April 2012), 47.

[687] National Marine Fisheries Service, *Fishery Management Plan for Groundfish of the Bering Sea and Aleutian Islands Management Area, 2015*, appendix, D-16.

[688] City of False Pass, Alaska, "Pacific Cod," accessed April 7, 2015, http://home.gci.net/~cityoffalsepass/cod.htm; Darius Kasprzak, Alaska Jig Association, personal communication with author, January 30, 2017.

[689] DNG jigging machine brochure (acquired at Pacific Marine Expo, November 2015).

[690] Ibid.

[691] Charlie Ess, "Jigging Machines: has their moment arrived?" *Pacific Fishing* (March 1997): 37–42.

[692] David Jackson and Dan Urban (Alaska Department of Fish and Game), *Western Region Report on 1997 State Managed Pacific Cod Fishery*, ADF&G Regional Information Rept. No. 4K98-2 (January 1998), 4–6.

[693] Alaska Department of Fish and Game, Staff Comments on Bering Sea Tanner Crab and Kodiak Groundfish and Finfish Proposals, Alaska Board of Fisheries Meeting, Kodiak, Alaska, January 10–13, 2017, table 50-2.

[694] Nathaniel Nichols, Paul Converse, and Kim Phillips (Alaska Department of Fish and Game), *Annual Management Report for Groundfish Fisheries in the Kodiak, Chignik, and South Alaska Peninsula Management Areas, 2014*, ADF&G Fishery Management Rept. No. 15-41 (October 2015), 39–40; Natura Richardson (Alaska Department of Fish and Game), personal communication with author, January 2, 2018.

[695] Charlie Ess, "Jigging Machines: has their moment arrived?" *Pacific Fishing* (March 1997): 37–42.

[696] Joel Gay, "Cod Jiggers Like Their Gear But Face a Management Catch 22," *National Fisherman*, May 1992, 40–41; 72 *Fed. Reg.* 5660, February 7, 2007; 72 *Fed. Reg.* 50789, September 4, 2007; 76 *Fed. Reg.* 74673, December 1, 2011.

[697] Heather Fitch and Janis Shaishnikoff (Alaska Department of Fish and Game), *Annual Management Report for Groundfish Fisheries in the Bering Sea–Aleutian Islands Management Area, 2013–2014*, ADF&G Fishery Management Rept. No. 15-43 (November 2015), 9–10, 21, 23; Adam Lalich, F/V *Yorjim*, personal communication with author, October 18, 2018, and October 24, 2018.

[698] Dory Associates, *Access Restrictions in Alaska's Commercial Fisheries: Trends and Considerations* (January 2009), 12, accessed May 12, 2015, http://www.akmarine.org/wp-content/uploads/2014/06/AMCC_access-restrictions-dory-report-06-01-09.pdf.

[699] Darius Kasprzak, Alaska Jig Association, personal communication with author, January 24, 2017, and July 29, 2018.

[700] Charlie Ess, "Jigging Machines: has their moment arrived?" *Pacific Fishing* (March 1997): 37–42; Alaska Regional Office, National Marine Fisheries Service, *Steller Sea Lion Protection Measures, Final Supplemental Environmental Impact Statement*, appendix F1, *Groundfish Community Socioeconomic Profiles* (November 2001), 21, 23–24; David Witherell, Michael Fey, and Mark Fina (staff, North Pacific Fishery Management Council), *Fishing Fleet Profiles* (April 2012), 47–48.

[701] Mary Furuness (NOAA Fisheries), personal communication with author, February 1, 2017; Natura Richardson (Alaska Department of Fish and Game), personal communication with author, February 7, 2017.

[702] Alaska Marine Conservation Council, accessed January 27, 2017, http://www.akmarine.org/fisheries-conservation/; "Alaska firm looks to promote hook and line-caught cod, rockfish," *Undercurrent News*, April 28, 2014, accessed December 11, 2015, https://www.undercurrentnews.com/2014/04/28/alaska-firm-looks-to-promote-hook-and-line-caught-cod-rockfish/; Theresa Peterson (Alaska Marine Conservation Council), personal communication with author, November 25, 2015 and December 14, 2015.

SECTION C

Processing and Marketing Pacific Cod

Chapter 20

GETTING THE MOST FROM CODFISH: MODERN PRODUCTS

Total utilization of the retained catch is an economically and socially responsible goal that both generates revenue and enhances the public's image of and acceptance of a fishery.

The trend since the implementation of the Magnuson-Stevens Act has been toward total utilization of each fish caught, and beginning in 1998 the North Pacific Fishery Management Council required full retention of all Pacific cod caught in the Bering Sea and Gulf of Alaska (see chapter 22). Manufacturing fish meal from discards is a relatively easy way to fully utilize the catch, but fish meal has comparatively low value. If the Pacific cod fishery is to thrive, its goal must always be to maximize the value of each and every component of the catch.

Iceland has led the world in increasing the value of codfish. Between 1981 and 2011, the Icelandic codfish catch declined 60 percent, yet the total export value of cod products doubled. Icelanders attribute

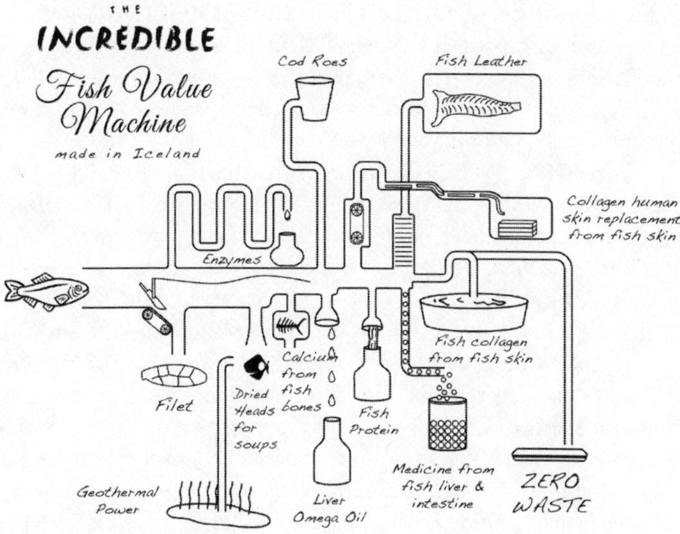

Figure 80. Icelandic "Incredible Fish Value Machine," 2016. (Iceland Ocean Cluster. Thor Sigfusson, "Introducing the Incredible Fish Value Machine," Iceland Ocean Cluster, August 2016, accessed October 12, 2016, http://www.sjavarklasinn.is/en/introducing-the-incredible-fish-value-machine.)

their success to product diversification and increased raw material utilization.[703] Specific codfish parts and organs—milt and stomachs, for example—that were formerly discarded or manufactured into fish meal have become valuable products that have dramatically increased the revenue derived from the county's codfish catch.

Following Iceland's lead, Alaska's Pacific cod industry has made substantial strides in producing products from fish parts that were often manufactured into fish meal or even discarded overboard. Cod milt, cod stomachs, and cod skins are examples. The industry is also moving away from producing commodities such as headed-and-gutted (H&G) fish and moving toward producing retail products. The individually vacuum-packed fillets produced aboard Alaskan Leader Seafoods' vessels are an example of this heartening trend, as are the cod liver oil produced at Dutch Harbor by Bering Select (see below) and the dog treats recently introduced by Alaskan Leader Seafoods.

WHOLE (RELATIVELY) FISH

The primary domestic frozen Pacific cod production today comprises headed-and-gutted fish, whole gutted fish (head on, gills intact), and skinless/boneless fillets. The head & gut fleet (H&G fleet) of catcher-processor vessels—trawlers, longliners, and pot boats—is so named because it produces primarily frozen headed-and-gutted fish, which, in turn, is raw material to secondary processors that is not contaminated by blood and viscera.

Frozen H&G Pacific cod, which represents more than half of the Alaska industry's production, comes in two forms: *western cut*, in which the head has been cut off forward of the pectoral girdle/collar, or *eastern cut* (*J-cut*), in which the pectoral girdle/collar has been cut off with the head. The recovery rate—the weight of finished product as a percentage of the live weight of the fish it was produced from—for western cut Pacific cod is 64 percent; for eastern cut, it is 58 percent.

H&G fish are reprocessed into consumer-ready products in domestic plants or are exported to Europe, Japan, or China for similar reprocessing. In fishing industry parlance, a fish that has been thawed, reprocessed, and then refrozen is referred to as *twice-frozen* or *double-frozen*. Some of the Chinese products are re-exported to the United States. (Given its carbon footprint, one has to question the wisdom of shipping a frozen product to China and back.)

Frozen whole gutted Pacific cod (WGC) is, by volume, the industry's second most important product, and it is a relative newcomer in the Pacific cod product line. It is utilized in a manner similar to H&G fish, but allows for fuller utilization of the fish. For this product, which is mostly sold in China, the recovery rate from live fish is 85 percent.

FILLETS

> *[Codfish fillets are] a staple, not a specialty item, and the whole marketing effort has to reflect this. . . . [Y]ou're not selling fish but a bland white slab of protein for which moisture content, texture, color and shelf life are all important.*
>
> —Trans-Pacific Industries, Inc., 1981[704]

> *Pacific cod is equal in quality to the Atlantic variety, and because the fish tend to be bigger, yields a thicker, more appealing fillet.*
>
> —John Sabella, *Pacific Fishing*, citing Palmi Ingvarsson, international seafood consultant, 1982[705]

> *Cod fillets are typically 2-3 times more valuable than H&G product (the primary alternative cod product) and can require 2-4 times as much labor input per pound, according to processing plant managers.*
>
> —McDowell Group, 2018[706]

Pacific cod fillets are primarily sold in domestic markets and are usually packed in one of five manners:

- **Block.** Fillets are frozen together into blocks from which portions of uniform size and weight can be cut at a secondary processing facility. These blocks, which typically weigh fifteen pounds, are primarily used in the manufacture of breaded fish portions such as are common in fish sandwiches and at casual "fish-and-chips" restaurants.
- **Cellopack.** One to three fillets are wrapped together in a thin plastic film, then frozen. Often sold in five-pound boxes containing six packets of fillets.
- **Layerpack.** Layers of fillets, with each layer separated from adjoining layers by thin plastic film, are frozen, typically in fifteen-pound boxes. This method allows the layers to be easily separated and removed while still frozen.

- **Shatterpack.** Individual fillets wrapped in plastic film in a manner that prevents the fish from sticking together and then frozen into a block that can be struck against a hard surface to break loose individual fillets. This method provides more control than layerpacks over the number of fillets that can be removed at one time.
- **IQF (individually quick frozen).** Individual size-graded fillets that are quick frozen and then glazed. IQF fillets are packed loosely in a polyethylene-lined box.

Layerpacks, shatterpacks, and individually frozen fillets are used by upscale restaurants, institutional food service companies, and retail fish markets.

The recovery rate for skinless/boneless Pacific cod fillets is 25 percent.[707] There is, however, an unsavory method of increasing the yield: soaking the fillets in an aqueous solution containing sodium tripolyphosphate. The chemical causes the flesh to absorb water, increasing its weight and thereby substantially lowering the net cost to the processor. Some reprocessors—including in the U.S.—treat cod fillets cut from thawed H&G fish with it. Commonly, the added moisture increases the fillets' weight by 15 percent, but in some cases the solution is set for a water uptake of as much as 25 percent. (The water is released when the fish is cooked.) The fillets are subsequently sold in retail establishments. Although many nations regulate the amount of sodium tripolyphosphate allowed in food products, the U.S. Food and Drug Administration generally recognizes the chemical as "safe when used in accordance with good manufacturing practice" and provides little oversight of its use.[708]

> *Alaska's cod industry has arrived at the threshold of the world cod markets.*
>
> —Doug McNair, *Pacific Fishing*, January 1984[709]

In the early 1980s, the U.S. market for frozen cod fillets was approximately 175 million pounds annually. Most of the fillets were imported from Iceland, Norway, and Canada. Of these countries, Iceland had the best reputation for quality and possessed a more-or-less monopoly in the high-end, cellophane-wrapped (cellopack) frozen cod fillet market. Norway, too, had a good reputation for quality and had no problem selling its fish. The Canadian industry, however, was in disarray because the quality of the cod it produced was inconsistent and because the Canadian industry had not developed steady markets and had made a practice of dumping large quantities of cod on the market at discount prices. Hence, Canadian cod did not usually command as high a price as Icelandic or Norwegian cod. (By 1983, the Canadian government had begun purchasing frozen cod to keep it off the market and prevent price decreases. When market inventories fell, the government would sell the cod back to the producers. And there was even talk then of nationalizing part of the industry.)[710]

The developing Alaska frozen cod industry's strategy for marketing its product was straightforward, at least in theory: "Maintain higher quality than the Canadians and a lower price than Iceland and Norway," wrote Doug McNair, the editor of *Pacific Fishing*, in January 1984. In addition to quality and price, there was a geographic component to marketing frozen cod fillets in the United States. Icelandic, Norwegian, and Canadian companies had, in the words of McNair, "sewn up" the U.S. East Coast market, but breaking into West Coast markets was less of a challenge, and the Alaska industry focused on California and Nevada. The expected frozen fillet production in 1984 in Alaska by the seven factory trawlers that planned to fish was about 15 million pounds—less than 9 percent of what the domestic market could absorb.[711]

Codfish report, summer 1986: *Abundance of Pacific cod remains high at 1.1 million [metric tons]. A substantial increase in cod abundance occurred in the late 1970s due to recruitment of exceptionally strong 1977 and 1978 year classes. These year classes, although relatively old at ages 8 and 9 years in 1986, still made up a substantial portion of the population biomass. Contributions of moderately strong year classes spawned in 1982 and 1984 have maintained the overall biomass at a high level.*

—National Marine Fisheries Service[712]

Because of a decline in the cod fishery off the east coasts of the United States and Canada, Alyeska Seafoods, at Dutch Harbor, during the 1994 season was able to successfully invade the U.S. East Coast market and sold about 15 percent of its cod production as fresh fillets, mostly in Chicago, New York, and Boston. The fish was shipped via scheduled air carriers.[713]

Inshore processors in the Bering Sea/Aleutian Islands area have in recent years invested substantially in increasing their capability to produce Pacific cod fillets. Although the amount of Pacific cod fillets produced there fluctuates year to year, 2016 was a record. The 29.3

million pounds of fillets produced that year represented 33 percent of the area's total Pacific cod product volume.[714]

ANCILLARY PRODUCTS

Fish meal/fish oil

Fish meal and fish oil are important by-products of Alaska's groundfish industry and are produced by cooking and pressing fish waste. Fish meal is used for livestock/animal feed, aquaculture feed, and fertilizer. Fish oil is primarily used as feed in agriculture and aquaculture but also as a dietary supplement for humans and for the manufacture of pharmaceuticals and cosmetics. (Cod liver oil is a separate product.) There are about ten onshore fish meal plants in Alaska, and several catcher-processors are equipped with fish meal plants. Alaska's largest onshore fish meal plants are at Dutch Harbor, Akutan, Sand Point, and Kodiak. About 80 percent of the fish meal produced from groundfish is exported, primarily to China, Korea, and Japan. (Fish meal produced from salmon is primarily sold on the domestic market for use in pet food.) During the years 2005–2014, annual groundfish-derived fish meal production averaged about 57,000 metric tons. Of this, an average of 1,700 metric tons, or 3 percent, was derived from Pacific cod. Most—an average of more than 90 percent—was derived from pollock.[715] Fish meal is typically packaged in twenty-five-kilogram-capacity plastic-lined paper bags.

The first domestic operations to produce fish meal and fish oil in the Bering Sea/Aleutian Islands area began in 1991, when Alyeska Seafoods and Westward Seafoods began operating fish meal plants at Unalaska/Dutch Harbor. Fish oil produced in the plants was used to fuel the plants' boilers.[716]

By 2005, Alaska's groundfish industry produced an estimated 8 million gallons of fish oil annually, mostly in seafood plants at Unalaska/Dutch Harbor and at Akutan. At the UniSea Corporation's Dutch Harbor seafood processing facility, the fish meal plant is capable of producing 20,000 gallons of fish oil per day. As with the Alyeska Seafoods and Westward Seafoods plants, much of the oil, some of which is manufactured from Pacific cod offal and carcasses, was used to fuel the plant's boilers, but UniSea also blended fish oil with diesel fuel to power the company's diesel-electric generators.[717] Fish oil not utilized on-site is sold outside Alaska.

The Kodiak Fishmeal Company, at Kodiak, is Alaska's only independent fish meal plant. Its 1,100-metric-ton-per-day-capacity plant operates year round, but during the months when the local Pacific cod fisheries are open (typically January, February, March, November, and December), it produces fish meal and fish oil primarily from the remains of Pacific cod processed at local seafood plants. According to Gary Anthony, the company's plant manager, the fact that the Marine Stewardship Council has certified the Pacific cod fisheries as sustainable (see chapter 21) enhances the marketability of fish meal derived from the species.[718]

Cod liver oil

The therapeutic value of cod liver oil has been long recognized, and Alaska codfishermen during the Salt Cod Era regularly retained livers for rendering into cod liver oil. This practice was discontinued when the final voyage of the schooner *C. A. Thayer*, in 1950, marked the end of the Salt Cod Era (see chapter 1). During the following sixty-plus years, Alaska's cod-fishing industry discharged cod livers overboard with other fish offal or, at best, used them with other fish waste as raw material in the manufacture of fish meal and fish oil (as described above).

The utilization of cod livers in Alaska took a great stride forward in 2014. Early that year, Clipper Seafoods and Siu Alaska Corporation,[719] which together operate eight freezer-longliners, partnered with Marine Ingredients,[720] a Pennsylvania-based manufacturer of fish oil products, and Marine Therapeutics, a scientific advisor, to form a company, Bering Select. The new company promptly constructed a plant at Dutch Harbor to produce high-quality cod liver oil—the first such plant in North America. Bering Select's plant began preliminary operations during the winter of 2014–2015, processing frozen-at-sea codfish livers delivered by the partners' vessels.[721]

During that winter, the plant produced about seventy-five tons of oil, which was shipped to several lower-48 factories and to a factory in Iceland for further refining into medicinal-quality oil that is sold as a health supplement. According to Bering Select, manufacturing cod liver oil from livers frozen at sea provides the company with "the purest, most natural cod liver oil in the world." Bering Select is the only company worldwide that manufactures cod liver oil from frozen livers, which eliminates the rancid taste associated with cod liver oil produced from old livers that have not been frozen. The company's cod liver oil is minimally processed to preserve its natural nutrient profile.[722]

To date, Bering Select, which plans to eventually manufacture customer-ready cod liver oil products at Dutch Harbor, is still in the process of developing and refining its cod liver oil products. So far, the company has received livers from only the partners' vessels, but its plant has the capacity to process livers from a considerable number of vessels in both the freezer-longliner and factory trawler fleets. The Marine Stewardship Council's certification of Alaska's Pacific cod fishery adds to the marketability of the company's products.[723]

Surimi

Surimi, a staple of the Japanese diet, is a fish paste made from the minced flesh of white-meated fish. It is the basic ingredient in a variety of traditional products, including *kamaboko*, which is surimi to which starch and seasonings have been added that is then shaped into loaves and steamed. Sliced *kamaboko* is eaten with a dipping sauce. Surimi is also the basic ingredient in *seafood analogs*: imitation crab, shrimp, and scallops. Surimi is primarily manufactured from pollock, but Pacific cod and other groundfish are used as well. Surimi manufactured from pollock, however, is considered the highest quality.[724] Surimi is frozen into blocks for shipping and storage. In 2013, surimi accounted for 11 percent, by value, of all Alaska seafood exports.[725] The first surimi plant in Alaska was a pilot project subsidized by the Alaska Fisheries Development Foundation at the Alaska Pacific Seafoods plant, at Kodiak. The plant began operating in January 1985.[726]

Milt

Cod milt is the sperm-filled reproductive organ of male fish and has a mild, creamy flavor. The product is a delicacy in Japan, where it is known as *shirako*, which translates to "white children," and is typically served tempura fried or in soup dishes. Cod milt is also appreciated in Korea and is consumed in Europe as an ingredient in soups, dips, and smoked products.

During the spawning season, typically the months of February and March, fresh cod milt, which is highly perishable, is shipped by air from western Alaska fishing ports via Anchorage to Japan, where it commands a premium price (up to $9 per pound in 2001). This market, however, can quickly become saturated. Frozen cod milt is far less valuable (about $1.20 per pound in 2015), but, as one processor said, "it's better than throwing it away." The milt is usually frozen in fifteen-kilogram blocks, much of which is shipped to China for further processing and later resale in Japan and Europe. During the spawning season, cod milt can account for from 1 to 3 percent of cod production weight. Average Alaska Pacific cod milt production during the years 2011–2015 was 2,900 metric tons, of which fully 93 percent was produced at shore plants.[727]

Roe

Pacific cod roe (eggs) is frozen and primarily sold in Japan and Korea, where it is salted and used in a variety of dishes. Cod roe is not as valuable as milt. Annual production during the years 2011–2015 averaged about 4,000 metric tons, with an average wholesale value of about $1,900 per metric ton—less than a dollar per pound.[728]

Other ancillary products

> **Iceland, 2016:** *Cod byproducts are being used for a variety of different consumer goods, including lucrative cod oil supplements, in order to generate new profitability for the fishery in Iceland. Companies in the cluster are making cosmetics from cod enzymes and cod skin is being made into clothing leather and bandages.*
>
> —*SeafoodSource News*, September 2016[729]

Cod heads are often ground to make fish meal and fish oil, but some are ground, then frozen and sold to pet food manufacturers. Cod heads are also frozen whole, either with the gills in ("green") or with the gills removed, depending on the market. While most frozen heads are sold to manufacturers of animal food, some heads are exported for human consumption, primarily to Asia. Frozen heads (and cheeks) are also exported to Norway, where they are salted and dried and then sold on traditional salt cod markets. Annual production of cod heads in Alaska during the years 2011–2015 averaged 5.9 million pounds, with a wholesale value of $3.6 million.[730]

Cod stomachs are frozen and exported to Korea, where they are used as an ingredient in soup. The stomachs are also a source of the enzyme trypsin, which is used in food preparation, including for processing salmon roe.[731] (In Iceland, a factory under construction will produce collagen, a protein that in purified form is widely used in cosmetic surgery, from cod skins.)

Currently, the market for cod tongues is limited, and no Alaska processors market them.[732]

Cod skins and trimmings are frozen and sold to manufacturers of pet treats. In October 2016, Alaskan Leader Seafoods began marketing its own brand of dog treats—Wild Alaskan Cod Crunchies—made from cod skins and trimmings produced in its onboard fillet operations. A year later, the product won an award at the Alaska Fisheries Development Foundation's annual Symphony of Seafood competition, which honors product innovation in Alaska's seafood industry.[733]

ENDNOTES

[703] Iceland Ocean Cluster, *Ocean Cluster Analysis, April 2013*, accessed October 15, 2015, http://sjavarklasinn.is/en/wp-content/uploads/2014/11/OceanCluster-AnalysisApril2013-2.pdf.

[704] Jon Sabella, "Arctic Trawler Brings Home the Bacon: 1.5 Million Pounds of Cod Fillets and Insight into Bottomfish," *Pacific Fishing* (December 1981): 35–36, 38, 65.

[705] John Sabella, "American Fish and Food Service," *Pacific Fishing* (June 1982): 53–61.

[706] McDowell Group, *Economic Impact of Inshore Seafood Processing in the Bering Sea/Aleutian Islands Region* (2018), 31.

[707] Ben Fissel et al., *Stock Assessment and Fishery Evaluation Report for the Groundfish Fisheries of the Gulf of Alaska and Bering Sea/Aleutian Islands Area: Economic Status of the Groundfish Fisheries off Alaska, 2012* (Seattle: National Marine Fisheries Service, 2014), 249, 252, accessed July 20, 2015, https://www.afsc.noaa.gov/refm/docs/2013/economic.pdf; Loh-Lee Low et al. (Alaska Fisheries Science Center), *A Review of Product Recovery Rates for Alaska Groundfish* (November 1989), 10; David Little, Clipper Seafoods, personal communication with author, July 26, 2015.

[708] Paul Gilliland, Clipper Seafoods, personal communication with author, January 13, 2016; *Electronic Code of Federal Regulations* (January 6, 2016), Title 21, Substances Generally Recognized as Safe, § 182.6810.

[709] Doug McNair, "Codfish Update," *Pacific Fishing* (January 1984): 34–37, 40–41.

[710] Ibid.

[711] Ibid; Doug McNair, "Bottomfish Update," *Pacific Fishing* (November 1982): 65–71, 119.

[712] Gary D. Stauffer, "Bering Sea Groundfish Subtask," *Northwest and Alaska Fisheries Center (National Marine Fisheries Service) Quarterly Report* (July–September 1986): 22–24.

[713] Ann Touza, "Crabbers Turn to Cod in Dutch," *Alaska Fisherman's Journal* (December 1994) 10, 12.

[714] McDowell Group, *Economic Impact of Inshore Seafood Processing in the Bering Sea/Aleutian Islands Region* (2018), 31.

[715] Alaska Office of International Trade, *Alaska Export Report, 2014 Alaska Export Update*, accessed October 16, 2015, http://gov.alaska.gov/Walker_media/documents/alaskaexportcharts2014.pdf; Alaska Fisheries Science Center, *Wholesale Market Profiles for Alaska Groundfish and Crab Fisheries* (May 2016), 101–104; Dan Lesh (McDowell Group), personal communication with author, November 28, 2018.

[716] Yereth Rosen, "An Aleutian Island Booms and Busts Over Bottomfish," *Christian Science Monitor*, April 24, 1991; "Pair of Plant Openings," *Alaska Business Monthly* (September 1, 1991), accessed October 17, 2015, http://www.thefreelibrary.com/Pair+of+plant+openings.-a011368798.

[717] Wesley Loy, "Fish oil gains as alternative fuel," *Pacific Fishing* (July 2005): 13, 15; UniSea, Inc., *Aleutian Islands Risk Assessment Report*, Dutch Harbor, Alaska, September 1, 2009.

[718] Kodiak Fishmeal Company, accessed October 17, 2015, http://www.kodiakfishmealcompany.com/production.html; Gary Anthony, plant manager (Kodiak Fish Meal Company), personal communication with author, October 16, 2015.

[719] Siu Alaska Corporation is a wholly owned for-profit subsidiary of Nome, Alaska–based Norton Sound Economic Development Corporation, a Community Development Quota group.

[720] In late 2016, Marine Ingredients merged with KD Pharma, a German manufacturer of nutraceutical fish oil.

[721] "Bering Select Opens First Ever Omega-3 Plant in Dutch Harbor," Business Wire, accessed October 5, 2016, accessed February 9, 2019,http://www.businesswire.com/news/home/20150114006340/en/Bering-Select-Opens-Omega-3-Plant-Dutch-Harbor.

[722] Alex DeMarban, "Not your parents' cod-liver oil: New Alaska plant looks to rejigger old-school product," *Alaska Dispatch News*, June 7, 2015.

[723] Diana Haecker, "NSEDC board contemplates expanding bulk fuel program," *Nome Nugget*, February 13, 2014; Diana Haecker, "Siu Alaska to sell defunct Dutch Harbor fish plant," *Nome Nugget* (May 14, 2015); Bill Giebler, "Dutch Harbor's Bering Select Opens First Cod Liver Oil Plant in US," *Seafood News*, January 8, 2018, accessed January 9, 2018, http://seafood.com/Story/1088128/Dutch-Harbors-Bering-Select-Opens-First-Cod-Liver-Oil-Plant-in US.

[724] Trans-Ocean Products (surimi products manufacturer), "Where T.O.P. Quality Begins," accessed October 13, 2015, http://www.trans-ocean.com/production.html.

[725] Alaska Seafood Marketing Institute, *Seafood Market Bulletin, Spring 2014*, accessed October 13, 2015, http://www.alaskaseafood.org/industry/market/seafood_spring14/2014-alaska-seafood-market-outlook.html.

[726] "Alaska Pacific Nets Pilot Surimi Project," *Alaska Fisherman's Journal* (November 1984): 5; "Alaska Pacific Fires Up Surimi Line," *Alaska Fisherman's Journal* (February 1985): 9.

[727] Jeb Wyman, "Can Freezing Preserve Cod Milt's Value?" *Pacific Fishing* (July 2001): 32; Masahiko Takeuchi (Minato Tsukiji), "Pacific cod roe prices soar," *Undercurrent News*, January 24, 2014, accessed December 11, 2015, http://www.undercurrentnews.com/2014/01/24/alaskan-pacific-cod-roe-prices-soar/; Michelle Theriault Boots, "Alaska fish processors chase Japanese market for an unusual product—cod semen," *Alaska Dispatch News*, February 2, 2015, accessed October 10, 2015, http://www.adn.com/article/20150202/alaska-fish-processors-chase-japanese-market-unusual-product-cod-semen; Steve Melchert, "Pacific Grey Cod Milt," *Ace Flyer* (ACE Air Cargo Co. newsletter), April 2005; McDowell Group, *Analysis of Specialty Seafood Products* (November 2017), 17; NOAA Fisheries, *Fisheries Catch and Landings Reports, Pacific Cod and Pollock Products* (Codes), accessed January 16, 2018, https://alaskafisheries.noaa.gov/fisheries-catch-landings.

[728] McDowell Group, *Analysis of Specialty Seafood Products* (November 2017), 35; Harry Yoshimura (Mutual Fish Company, Seattle, Wash.), personal communication with author, January 15, 2018.

[729] Christine Blank, "Entrepreneurs getting creative with seafood byproducts," *SeafoodSource News*, September 26, 2016, accessed September 27, 2016, http://www.seafoodsource.com/news/supply-trade/entrepreneurs-getting-creative-with-seafood-byproducts.

[730] McDowell Group, *Analysis of Specialty Seafood Products* (November 2017), 6–7.

[731] Yereth Rosen, "An Aleutian Island Booms and Busts Over Bottomfish," *Christian Science Monitor*, April 24, 1991.

[732] Jason Holland, *SeafoodSource News*, January 20, 2017, accessed January 24, 2017, http://www.seafoodsource.com/news/supply-trade/icelandic-collaboration-looks-to-add-more-value-to-cod-skin; Laine Welch, "Tongues and Cheeks," *Alaska Journal of Commerce* (June 2007), accessed October 11, 2016, http://www.alaskajournal.com/community/2007-06-10/tongues-and-cheeks#.V_oeYOArKM8.

[733] Michael Ramsingh, "Alaskan Leader Seafood Providing Pet Food Makers with MSC Certified Bering Sea Cod Raw Materials," *Seafood News*, March 16, 2016, accessed October 11, 2016, http://www.seafoodnews.com/Story/1011674/Alaskan-Leader-Seafood-Providing-Pet-Food-Makers-with-MSC-Certified-Bering-Sea-Cod-Raw-Materials; Laine Welch, "Dog treats made from cod trimmings

hit stores," *National Fisherman*, October 3, 2016, accessed October 4, 2016, https://www.nationalfisherman.com/alaska/dog-treats-made-from-cod-trimmings-hit-stores/; Brian Hagenbuch, "Newcomer Alaskan Leader Seafoods wins two Symphony of Seafood awards," *SeafoodSource News*, November 21, 2017, accessed November 21, 2017, https://www.seafoodsource.com/news/foodservice-retail/newcomer-alaskan-leader-seafoods-wins-two-symphony-of-seafood-awards.

Chapter 21
MARKETING PACIFIC COD

It's no fish ye're buying, it's men's lives.

—Sir Walter Scott, Scottish historical novelist, playwright, poet, and historian, 1816[734]

Alaska cod producers continue to enjoy one of the world's few strong cod resources, which doesn't hurt the price. In Europe, scientists are calling for a total closure on North Sea cod stocks, saying nothing less will give the fish a chance to regenerate. New England fishery managers were grappling toward yet another major reduction in catches at press time. Atlantic Canada's cod stocks are still staggering. The only serious competition at present comes from the Barents Sea, where the joint Norwegian-Russian Fishery Commission set a quota of 506,000 [metric] tons for 2004.

—*Pacific Fishing*, December 2003[735]

Cod is the big winner in the per capita consumption numbers as it has taken an increased share of the whitefish market over the past five years.

—*Seafood News*, November 2017[736]

U.S. retailers [are] requesting more and more fish that is a product of the U.S.A.

—Bill Weed, Blue North Fisheries, September 2018[737]

Among the challenges facing Alaska cod producers in the years following the passage of the Magnuson-Stevens Act were the need to dispel suspicions in the national marketplace that Pacific cod was inferior to its Atlantic cousin and to show that the industry could reliably produce a high-quality product. By the spring of 1981, the factory trawler *Arctic Trawler* (see chapter 16) had made a pair of million-pound deliveries of frozen Pacific cod fillets to Seattle. The high quality of the fillets largely put market suspicions to rest. As *Pacific Fishing* wrote in early 1981, "[T]he *Arctic Trawler* has demonstrated that both retail and institutional markets will accept Pacific cod, even at premium prices."[738]

The state agency tasked with marketing Alaska's seafood is the Alaska Seafood Marketing Institute (ASMI), which was established by

Figure 81. In the years following the passage of the Magnuson-Stevens Act, there was uncertainty regarding the marketability of Alaska groundfish. (Image reproduced with permission of *Pacific Fishing*. "Frozen Pacific Whitefish: Production Trails Market Potential," *Pacific Fishing* [March/April 1981]: 53–58, 60.)

Alaska's legislature in 1980. With an annual budget of about $25 million, the institute is a public-private partnership between the State of Alaska and the commercial fishing industry. Its self-proclaimed mission is "to foster the economic development of a renewable economic resource."[739]

In partnership with the Alaska Seafood Marketing Institute, some domestic companies began marketing Pacific cod as *Alaska cod* or *Alaskan cod* to tie the product to Alaska's well-known reputation for a clean environment and sustainable fisheries. In 1984, the Alaska Seafood Marketing Institute published a handsome four-page brochure that discussed the characteristics, handling, and preparation of Alaska cod and included eight recipes.[740]

Recently, large supplies of competing aquaculture products—tilapia and pangasius, in particular—combined with a rebound in Atlantic cod production, have presented a marketing challenge to producers of Pacific cod.[741] The fish, nevertheless, remains very popular, and some producers have successfully responded to the weakened international markets by focusing on products that cater to the domestic retail and

Figure 82. Alaska Seafood Marketing Institute Pacific cod advertisement, 2014. Image courtesy of the Alaska Seafood Marketing Institute.

foodservice sectors.[742] In 2015, Wendy's, the fast-food chain, touted "wild-caught North Pacific Cod" as "the gold standard for fried fish."[743] For the record, reporters for *Business Insider* in March 2019 sampled fried-fish sandwiches from every major U.S. fast-food chain and determined Wendy's North Pacific Cod Sandwich to be the best.[744]

On a broader scale, there is an ongoing trend of domestic retailers increasingly requesting domestically produced fish. Domestic sales of Pacific cod at Blue North Fisheries, a company that operates a small fleet of freezer longliners, increased 60 percent from 2015 to 2017. It's likely that Blue North's competitors have experienced a similar increase.[745]

SUSTAINABILITY CERTIFICATION

In an effort to do what some governments were failing to do—protect fish stocks from overfishing—some conservation groups developed eco-certification and seafood guide schemes, like the MSC and Seafood Watch. In doing so, they introduced into fisheries management a new player—the consumer. . . . Pressure was brought to bear on retailers and restaurants to sell or serve

Figure 83. Dairy Queen, a fast-food chain, began advertising its Alaskan Pacific Cod sandwich in 2016. (Courtesy of American Dairy Queen Corp.)

only certified seafood, which in turn put pressure on the fishing industry to apply for sustainable seafood certification.

—Nelson Bennett, *Richmond News* (British Columbia), 2018[746]

New survey sees seafood consumers placing sustainability before price and brand.

—Madelyn Kearns, *SeafoodSource News*, July 2016[747]

The record of fishery development is replete with examples of overexploited fish stocks. Many consumers today want to enjoy the flavor of fish and to benefit from its nutritional qualities, but they want also to know the fish they purchase is harvested sustainably, that the process of catching the fish is not ecologically harmful, and that the fishery itself is conducted in a socially responsible manner. Seafood certification programs and seafood advisory lists are designed to increase consumer awareness of these factors and to foster ocean-friendly, socially responsible purchases. Certification also provides a significant marketing advantage. In Europe and increasingly in the United States, eco-labeled (certified) products enjoy broader market acceptance than noncertified products and can command a premium price.

Several organizations have developed third-party seafood certification programs. The standards measured vary among the organizations,

Figure 84. Marine Stewardship Council label. (Courtesy of Marine Stewardship Council.)

but they mostly involve fishery sustainability. Also, some certifying organizations have more stringent requirements than others.

Among the most prominent seafood certification organizations is the Marine Stewardship Council (MSC), an independent, global, nonprofit organization founded in 1997 that is striving to reverse the decline of the world's fisheries primarily through its fishery certification program. In their book *Fishing for Pollock in a Sea of Change*, which was published in 2013, James Strong and Keith Criddle praised MSC certification as "the premier fisheries eco-label."[748]

To obtain MSC certification, a fishery must meet three criteria. First, it must be sustainable—"fishing activity must be at a level which ensures it can continue indefinitely." Second, it must minimize environmental impact—"managed to maintain the structure, productivity, function and diversity of the ecosystem." Third, the fishery must be effectively managed—"comply with relevant laws and have a management system that is responsive to changing circumstances." Additionally, the MSC requires compliance with national and international forced-labor laws.[749] The MSC authorizes companies that participate in its certification program to place the organization's distinctive eco-label on their packaged fish. MSC-certified fisheries are audited annually and are reassessed every five years. In 2002, Brad Warren, *Pacific Fishing*'s editor, praised the MSC as "the only outfit in the eco-labeling arena that has developed genuinely credible standards and criteria for evaluating fisheries."[750] In 2016, *SeafoodSource News*, characterized MSC representative An Yan as the "most influential seafood executive in China."[751]

The first Alaska fisheries to be certified by the MSC were the salmon fisheries (all gear types, statewide), which were certified in 2000.[752] The first Alaska groundfish fisheries to obtain MSC certification were the pollock fisheries in both the Gulf of Alaska and Aleutian Island/Bering Sea areas. The pollock certification process began in early 2001 when, at the urging of European seafood buyers, the At-Sea Processors Association, a seafood trade group that then comprised seven member companies that together operated nineteen catcher-processor vessels and accounted for approximately 40 percent of the annual pollock harvest, applied to the MSC for an assessment of the pollock fisheries. The assessment began that same year, and the fisheries were certified in 2005.[753] U.S. seafood buyers, too, had begun shifting their purchases toward fish that were considered sustainably harvested.

Regarding Pacific cod, at least one industry professional recognized that MSC sustainability certification could provide a marketing advantage. That person was Paul Gilliland, of the Bering Select Seafoods Company, the wholly owned marketing subsidiary of Clipper Seafoods, a Seattle-based company that at that time operated four freezer-longliners and accounted for about 23 percent of the freezer-longliner sector's annual Pacific cod harvest in the Bering Sea.[754] Gilliland recognized that the collapse of Atlantic cod stocks along the east coasts of the United States and Canada, overfishing in the North Sea and the Baltic Sea, and projections of overfishing in the Barents Sea had left many consumers reluctant to purchase cod because they feared doing so would contribute to the problem. He sensed an opportunity to enhance his company's marketing opportunities by distinguishing Pacific cod caught in Alaska's well-managed fisheries from generic cod. Certification was a means of doing so.

In 2004, Bering Select, together with one other freezer-longliner company, hired a consulting firm to perform a preassessment of the Bering Sea/Aleutian Islands freezer-longliner Pacific cod fishery. This study concluded that the fishery would easily meet MSC's rigorous assessment standards. After failing to interest other companies that operated freezer-longliner vessels to share in the cost of certifying their fishery, Bering Select made the commitment to proceed as the sole sponsor.[755]

The assessment began the following year, and in February 2006, the fishery was granted certification—the first cod fishery in the world to be certified by the council. Certification, however, applied only to vessels and companies that opted to pay a share of the assessment costs Bering Select had fronted. According to Gilliland, the premium received for his MSC-certified Pacific cod was as much as 2 or 3 percent, but he thought the most significant benefit of certification was access to new markets.[756]

In 2007, the Alaska Fisheries Development Foundation (see chapter 15), representing more than twenty companies and organizations involved in the Pacific cod fishery in Alaska, applied to the MSC for blanket certification—all gear groups in all locations—of Alaska's cod fisheries.

Following a thorough assessment of the fisheries, MSC certification was granted in January 2010.[757] Several months later, companies representing about 70 percent of the Alaska freezer-longliner fleet created the Alaska Longline Cod Commission to promote Pacific cod from what it termed "one of the world's most eco-friendly fisheries." On average, about half of the commission members' production is sold in Europe, with the remainder sold in Japan, the United States, and China.[758]

Alaska's Pacific cod fishery was recertified in 2015.[759] By that time, about a dozen cod fisheries worldwide—including cod fisheries off the coasts of Iceland and Norway and in the Barents Sea—had been certified against the MSC standard. Alaska's cod fishery remains the only MSC-certified cod fishery in the Pacific Ocean.[760]

Also in 2010, the Alaska Seafood Marketing Institute applied on behalf of the Pacific cod industry for blanket certification of Alaska's Pacific cod fisheries by Ireland-based Global Trust, a private company whose Pacific cod certification program is based on the United Nations Food and Agriculture Organization's Code of Conduct for Responsible Fisheries. The Global Trust certification was granted in April 2013.[761]

The reason for this second certification was that ASMI, as well as some others within and outside the commercial fishing industry, felt the MSC had become dominated by the environmental non-governmental organization (ENGO) community, which, according to ASMI, "drives [the MSC's] agenda and ever-changing criteria." ASMI complained that "[i]n addition to eroding the Alaska brand, the ENGO sustainability movement creates the additional challenge of defending market access for Alaska seafood and the state's right to govern its own resources."[762] Some within Alaska's commercial fishing industry, however, have questioned the wisdom of abandoning MSC's globally recognized and respected certification program, which the Alaska Longline Cod Commission praised in 2015 as the "the leading eco-rating program for wild-caught seafood," for an alternative program few have ever heard of.[763]

The Monterey Bay Aquarium, in California, does not certify fisheries but maintains the highly regarded Seafood Watch, a program that, in the aquarium's words, "creates science-based recommendations that help consumers and businesses make ocean-friendly seafood choices." The recommendations are published in consumer guides that rate seafood as Best Choice, Good Alternative, or Avoid. Seafood Watch rates all Pacific cod caught in Alaska as Best Choice.[764] Note, however, that seafood guides generally reflect the publishing organization's policy stand on fishery issues, and, as a result, these guides sometimes contradict each other. Moreover, guides draw more of the public's attention by reporting bad news than good news, which is perhaps an incentive for some organizations to report bad news.

NOAA's FishWatch program, which was initiated in 2007, does not rank or rate one species over another, but provides up-to-date, relatively technical information on the status, science, and enforcement of U.S. commercial fisheries.[765]

Codfish report, 2007: *In the Bering Sea-Aleutian Island fisheries this year, Pacific cod has been the star, with market conditions driving ex-vessel prices to all-time highs. Ex-vessel price in Dutch Harbor for trawl cod was 47–48 cents a pound, and for fixed gear as high as 51 cents.*

—*Pacific Fishing*, May 2007[766]

Codfish report, 2009: *Recession, exchange rates, thump Pacific cod markets; will other fisheries follow? As the new Alaska groundfish season opened in early January, a sense of dread gripped many producers. Said one industry player: "It's panic city in P-cod."*

Indeed, the Pacific cod party seems decidedly over, the music drowning in a sea of economic gloom. Fish that had been commanding as much as 75 cents a pound dockside had plunged [to] as low as 30 cents by late last year, and lots of fish reportedly languishing in cold storage.

—*Pacific Fishing*, February 2009[767]

Codfish report, 2013: *What may look good for the cod resource globally might not be so good for Alaska's Pacific cod market. Harvest levels for 2013 in the Bering Sea and Gulf of Alaska are around 320,000 metric tons and will most likely be around the same for the upcoming year. The European producers—namely Russia, Iceland, and Norway—are expected to increase production by up to 10 percent in 2014. . . . The oversupply of cod worldwide will likely have a negative effect on prices seen at the dock. It is a simple case of supply and demand. To make matters worse, there are high transportation costs from Alaska to China.*

—Laine Welch, *Pacific Fishing*, November 2013[768]

THE CHINA CONNECTION: "CAUGHT IN ALASKA, PACKED IN CHINA"[769]

Until the late 1990s, about 80 percent of the fish products leaving Dutch Harbor, which, in terms of volume of product moved, is the United States' largest fishing port, was shipped to Puget Sound for distribution in domestic and international markets. Mostly tramp freighters carried the remaining approximately 20 percent to foreign ports, mostly in Asia and Europe. By 2003, the ratio had been essentially reversed, with the bulk of the exported product going to Asia. Cheap Asian labor was the reason, and this labor attracted capital that built, in the words of *Pacific Fishing*, "gleaming modern fish plants, some of which outclass many U.S. seafood facilities." The Chinese exported a significant quantity of the fish they processed back to the United States. Europe, too, imported processed-in-China fish, causing a loss of seafood processing jobs in fish-dependent communities in Norway.[770]

Cheap labor and modern seafood processing plants notwithstanding, consistent and uniform quality control was lacking in China, and the quality of the seafood products varied. Markets place a very high value on consistency, and increasingly less Pacific cod is being sent to China for reprocessing.

EXCHANGE RATES AND TARIFFS

Given that a large percentage of the Pacific cod produced in Alaska is exported and that profit margins are often thin, exchange rates play an important part in determining the fish's price. A weak dollar generally bodes well for U.S. exports. For example, in mid-2008, the U.S. dollar was weak vis-à-vis the euro, and it took about $1.59 to buy one euro. The U.S. Pacific cod industry received record prices (in dollars) that year for Pacific cod sold on the European market.[771] In late 2015, however, the dollar was considerably stronger—it took only about $1.08 to buy one euro—which represented about a 32 percent increase in the value of the dollar. The result was that if Europeans in 2015 were paying the same price (in euros) for Pacific cod as they did in 2008, U.S. producers were receiving about 32 percent less (in dollars) for their product. The story is similar for the Japanese yen.

Tariffs also figure into the equation. The European Union allowed the tariff-free importation of up to 50,000 metric tons of headed-and-gutted cod (*Gadus morhua*, *Gadus macrocephalus*, and

Gadus ogac[772]) annually. The limit was increased to 75,000 metric tons in 2016, and to 90,000 metric tons in 2017. Imports beyond this limit are subject to a 12 percent tariff.

Japan imposes a 6 percent tariff on all imports of frozen Pacific cod. While China did not impose a tariff on Pacific cod that will be reprocessed and exported, it does impose a tariff on Pacific cod destined for domestic consumption. The tariff on frozen Pacific cod for domestic consumption had been 10 percent, but it was scheduled to be reduced to 7 percent in July 2018.

The trade war between the United States and China in 2018 changed everything. In retaliation for tariffs levied by the United States on more than a thousand categories of Chinese manufactured goods, the Chinese government levied tariffs on an array of U.S. goods, including a 25 percent tariff on a host of seafood products, Pacific cod among them. The seafood tariff became effective on July 6, 2018, and is in addition to any previous tariffs. This tariff, however, does not apply to Pacific cod that will be reprocessed and exported. For Pacific cod that will be consumed domestically, the new total tariff is a hefty 32 percent.[773]

Escalating the trade war, the United States imposed a 25 percent tax on seafood sent to China for processing. The tax became effective on January 1, 2019. The United States also threatened to levy a 10 percent tariff on all seafood sent from China to the United States, including U.S. products (such as Pacific cod) that are reprocessed in China. In what was considered a victory for Alaska's fisheries, the At-Sea Processors Association, which represents six companies that operate factory trawlers that fish for and process cod and pollock, successfully lobbied to have those species removed from the list of seafood items targeted for the tariff.[774]

Fishmonger's maxim: *Retail customers quickly forget paying "too much" for fish, but they never forget poor quality.*

ENDNOTES

[734] Sir Walter Scott, *The Antiquary* (1816), vol. 1, chap. 11.

[735] Jeb Wyman, Susan Chambers, and Brad Warren, "Seafood Report," *Pacific Fishing* (December 2003): 7.

[736] John Sackton, "Cod, Shrimp and Pangasius See Gains in US Consumption, Canned Tuna Continues to Slide," *Seafood News*, November 3, 2017, accessed November 3, 2017, http://www.seafoodnews.com/Story/1081582/Cod-Shrimp-and-Pangasius-See-Gains-in-US-Consumption-Canned-Tuna-Continues-to-Slide.

[737] Brian Hagenbuch, "Pacific cod fillets from Alaska gain traction in domestic markets," *SeafoodSource News*, September 26, 2018, accessed September 26, 2018, https://www.seafoodsource.com/premium/global-bulletin/pacific-cod-fillets-from-alaska-gain-traction-in-domestic-market.

[738] "Frozen Pacific Whitefish: Production Trails Market Potential," *Pacific Fishing* (March/April 1981): 53–58, 60.

[739] Alaska Department of Revenue, Tax Division, *Seafood Marketing Assessment, Historical Overview*, accessed July 22, 2017, http://www.tax.alaska.gov/programs/programs/reports/Historical.aspx?60636; Alaska Seafood Marketing Institute, *2014 Annual Report*, 25, 30; Alaska Seafood Marketing Institute, *2013 Annual Report*, 1.

[740] Alaska Seafood Marketing Institute, *Alaska Cod: Handling and Preparation of Alaska Cod*, 1984; Alaska Seafood Marketing Institute, "Wild Alaska Cod Fact Sheet," accessed December 16, 2015, http://www.alaskaseafood.org/retailers/pdfs/Cod/Wild%20Alaska%20Cod%20Fact%20Sheet.pdf; Paul Gilliland, Clipper Seafoods, personal communication with author, December 16, 2015.

[741] Ben Fissel et al., *NPFMC Draft Stock Assessment and Fishery Evaluation Report for the Groundfish Fisheries of the Gulf of Alaska and Bering Sea/Aleutian Islands Area: Economic Status of the Groundfish Fisheries off Alaska, 2014* (Seattle: National Marine Fisheries Service, 2015), 191, accessed December 9, 2015, http://www.afsc.noaa.gov/REFM/stocks/plan_team/economic.pdf.

[742] Masahiko Takeuchi (Minato Tsukiji), "US Pacific cod suppliers selling more to home market," *Undercurrent News*, April 7, 2014, accessed December 10, 2015, https://www.undercurrentnews.com/2014/04/07/alaskan-pacific-cod-suppliers-selling-more-to-north-america/.

[743] Wendy's Company, "Wendy's North Pacific Cod Sandwich is Back on the Line," *PR Newswire*, February 9, 2015, accessed August 26, 2016, http://www.prnewswire.com/news-releases/wendys-north-pacific-cod-sandwich-is-back-on-the-line-300032525.html.

[744] Hollis Johnson, "We tried fried-fish sandwiches from every major fast-food chain — and the winner is clear," *Business Insider*, March 6, 2019, accessed March 8, 2019, https://www.msn.com/en-us/foodanddrink/tipsandtricks/we-tried-fried-fish-sandwiches-from-every-major-fast-food-chain-%E2%80%94-and-the-winner-is-clear/ss-BBUrB5S?li=BBnb7Kz#image=9.

[745] Brian Hagenbuch, "Pacific cod fillets from Alaska gain traction in domestic markets," *SeafoodSource News*, September 26, 2018, accessed September 26, 2018, https://www.seafoodsource.com/premium/global-bulletin/pacific-cod-fillets-from-alaska-gain-traction-in-domestic-market.

[746] Nelson Bennett, "Consumers seek certainty about sustainable seafood labelling," *Richmond News* (Richmond, British Columbia), November 7, 2018, accessed November 9, 2019, https://www.richmond-news.com/news/consumers-seek-certainty-about-sustainable-seafood-labelling-1.23489869.

[747] Madelyn Kearns, "New survey sees seafood consumers placing sustainability before price and brand," *SeafoodSource News*, July 13, 2016, accessed July 13, 2016, http://www.seafoodsource.com/news/foodservice-retail/new-survey-sees-seafood-consumers-placing-sustainability-before-price-and-brand.

[748] James Strong and Keith R. Criddle, *Fishing for Pollock in a Sea of Change: A Historical Analysis of the Bering Sea Pollock Fishery* (Fairbanks, Alaska: Alaska Sea Grant, 2013), 122.

[749] Marine Stewardship Council, "MSC Fisheries Standard," accessed September 19, 2015, https://www.msc.org/about-us/standards/fisheries-standard; Marine Stewardship Council, MSC Chain of Custody Certification Requirements, v.2.0, February 2015, 23, accessed September 23, 2015, https://www.msc.org/documents/scheme-documents/msc-scheme-requirements/msc-coc-certification-requirements-v2.0/; Madelyn Kearns, "Urged by industry, MSC weighs adding social sustainability standards," *SeafoodSource News*, April 27, 2017, accessed April 27, 2017, https://www.seafoodsource.com/news/environment-sustainability/how-should-certifiers-approach-social-sustainability--sanford-ceo-top-industry-panel-weighs-in.

[750] Brad Warren, "Can Eco-labeling Work? Watch Pollock," *Pacific Fishing* (July 2002): 4.

751 Mark Godfrey, "10 most influential seafood executives in China," *SeafoodSource News*, November 30, 2016, accessed March 15, 2019, https://www.seafoodsource.com/features/10-most-influential-seafood-executives-in-china.

752 Marine Stewardship Council, "Alaska Salmon," accessed September 19, 2015, https://www.msc.org/track-a-fishery/fisheries-in-the-program/certified/pacific/alaska-salmon/alaska-salmon.

753 Mary Pemberton, "Pollock fishery awarded eco-label," *Anchorage Daily News*, June 16, 2004; Mary Pemberton, "Alaska Bering Sea Pollock Gets Eco-label," *Associated Press*, October 5, 2004, accessed September 19, 2015, http://www.enn.com/top_stories/article/13696; At-Sea Processors Association, "Green Label of Sustainability Sought for Largest U.S. Fishery," January 4, 2001, accessed September 19, 2015, http://www.atsea.org/press/MSC_Green_Label_JAN2001.htm; Marine Stewardship Council, "Formal Notification of the Intent to Proceed with an Assessment of the U.S. Alaska Pollock Fisheries," January 4, 2001, accessed September 19, 2015, https://www.msc.org/track-a-fishery/fisheries-in-the-program/certified/pacific/bsai-pollock/assessment-downloads-1/BSAI_formal_notification_MSC.pdf.

754 Bering Select Seafoods Company was merged into Clipper Seafoods in 2011.

755 Paul Gilliland (Clipper Seafoods), personal communication with author, September 27, 2015.

756 Marine Stewardship Council, "Bering Sea and Aleutian Islands Alaska (Pacific) Cod—Freezer Longline Fact Sheet," accessed September 20, 2015, https://www.msc.org/documents/fisheries-factsheets/net-benefits-report/Bering-Sea-Aleutian-Is-cod.pdf/view; Paul Gilliland (Clipper Seafoods), personal communication with author, September 27, 2015.

757 Moody Marine, Ltd., to Alaska Fisheries Development Foundation, Inc., re Bering Sea and Aleutian Islands (BS/AI) Pacific Cod Fisheries (Trawl, Longline, Jig, Pot) Marine Stewardship Council Certification, November 19, 2007; Moody Marine, Ltd. to Alaska Fisheries Development Foundation, Inc., re Gulf of Alaska Pacific Cod Fisheries (Trawl, Longline, Jig, Pot) Marine Stewardship Council Certification, November 15, 2007; Marine Stewardship Council, *Alaska Pacific Cod—Bering Sea and Aleutian Islands*, July 16, 2015, accessed September 21, 2015, https://www.msc.org/track-a-fishery/fisheries-in-the-program/certified/pacific/bering-sea-and-aleutian-islands-pacific-cod.

758 Freezer Longline Coalition, "Alaska fishermen form export trade group to promote sustainable cod fishery," news release, April 6, 2010, accessed September 23, 2015, http://www.freezerlonglinecoalition.com/Alaska-fishermen-form-export-trade-

group-to-promote-sustainable-cod-fishery.html; Paul Gilliland (Clipper Seafoods), personal communication with author, September 27, 2015.

[759] Marne Stewardship Council, *Alaska Pacific Cod—Gulf of Alaska*, September 3, 2015, accessed September 21, 2015, https://www.msc.org/track-a-fishery/fisheries-in-the-program/certified/pacific/gulf-of-alaska-pacific-cod.

[760] Madelyn Kearns, "Cod caught from the eastern Baltic Sea no longer MSC-certified," *SeafoodSource News*, December 17, 2015, accessed December 22, 2015, http://www.seafoodsource.com/news/environment-sustainability/cod-caught-from-the-eastern-baltic-sea-no-longer-msc-certified.

[761] Global Trust Certification, Ltd., *FAO-Based Responsible Fisheries Management Certification Full Assessment and Certification Report for the Alaska Pacific Cod Commercial Fisheries (200 mile EEZ)*, April 2013.

[762] Alaska Seafood Marketing Institute, *2014 Annual Report*, 10, accessed September 21, 2015, http://ebooks.alaskaseafood.org/ASMI_Annual_Report/#/12/.

[763] Alaska Longline Cod Commission, "Sustainability," accessed September 23, 2015, http://alaskalonglinecod.com/sustainability/.

[764] Monterey Bay Aquarium, *Seafood Watch West Coast Consumer Guide*, July–December 2015.

[765] NOAA Fisheries, "FishWatch Gets a Fresh Look," fact sheet, accessed October 1, 2015, http://www.nmfs.noaa.gov/ocs/mafac/meetings/2012_05/docs/2012_fishwatch_factsheet_final.pdf.

[766] Ann Touza, "Pacific cod big news in Aleutians, Bering," *Pacific Fishing* (May 2007): 30.

[767] "Recession, exchange rates, thump Pacific cod markets; will other fisheries follow?" *Pacific Fishing* (February 2009): 4.

[768] Laine Welch, "Alaska cod struggles in classic supply-demand squeeze," *Pacific Fishing* (November 2013): 29.

[769] Brad Warren and Michel Drouin, "Caught in Alaska, Packed in China," *Pacific Fishing* (October 2003): 18–20.

[770] Ibid.

[771] Wesley Loy, "Recession, exchange rates, thump Pacific cod markets; will other fisheries follow?" *Pacific Fishing* (February 2009): 4–5.

772 *Gadus ogac* is Greenland cod, which may be the same species as Pacific cod (*Gadus macrocephalus*).

773 Paul Gilliland, Clipper Seafoods, personal communications with author, December 2015 and July 2018; Peggy Parker, "China to Cut Import Tariffs on More Than 200 Seafood Products on July 1, 2018," *Seafood News*, June 1, 2018, accessed June 1, 2018, https://www.seafoodnews.com/Story/1105434/China-to-Cut-Import-Tariffs-on-More-Than-200-Seafood-Products-on-July-1-2018; Ana Swanson, "U.S. and China Expand Trade War as Beijing Matches Trump's Tariffs," *New York Times*, June 15, 2018; John Sackton, "China Slams Alaska and US Seafood Industry with 25% Tariff on $1 Billion in Exports," *Seafood News*, June 15, 2018, accessed June 15, 2018, https://www.seafoodnews.com/Story/1107044/BREAKING-NEWS-China-Slams-Alaska-and-US-Seafood-Industry-with-25-percent-Tariff-on-1-Billion-in-Exports.

774 Laine Welch, "Analysis: Here's how the trade war with China is affecting the outlook for Alaska seafood," *Anchorage Daily News*, July 16, 2018, accessed July 17, 2018, https://www.adn.com/business-economy/2018/07/16/heres-how-the-trade-war-with-china-is-affecting-the-outlook-for-alaskas-seafood-industry/; John Sackton, "10% Tariffs on $200 Billion Chinese Exports Take Effect; Some Pollock and Cod Products May be Exempt," *Seafood News*, September 18, 2018, accessed September 18, 2018, https://www.seafoodnews.com/Story/1116969/10-percent-Tariffs-on-200-Billion-Chinese-Exports-Take-Effect-Some-Pollock-and-Cod-Products-May-be-Exempt; Cliff White, "Trump hits China with another USD 200 billion in tariffs, but Alaska gets a break," *SeafoodSource News*, September 18, 2018, accessed September 18, 2018, https://www.seafoodsource.com/news/supply-trade/trump-hits-china-with-another-usd-200-billion-in-tariffs-but-alaska-gets-a-break.

SECTION D

Challenges, Development, and the Future of Alaska's Pacific Cod Fishery

Chapter 22
RATIONALIZATION OF ALASKA'S GROUNDFISH FISHERIES

The real problem with groundfish [is that] it's difficult to sustain a high-volume, low-margin industry on shortened seasons. Huge boats and plants can't afford to operate a few months a year. The fishery must stretch over most of the year—fish and process continuously. We can't do that with present management.

—Jim Branson, Executive Director, North Pacific Fishery Management Council, 1987[775]

The American trawl and longline fleets on the offshore grounds of the Bering Sea moved from the adolescence of a decade of unrestricted development into tough, competitive adulthood as the 1990s began. The capacities of the factory trawler, freezer-longliner and catcher boat navies far outstrip stock abundance for year-round operations, and the consequences of overcapitalization are hitting home.

—Brad Matsen, *National Fisherman*, Yearbook, 1991[776]

In 1992, the Council committed to rationalize the groundfish and crab fisheries and begin development of a Comprehensive Rationalization Plan (CRP). The CRP was prompted by concerns that expansion of the domestic harvesting fleet, in excess of that needed to efficiently harvest the optimum yield, was burdening compliance with the Magnuson-Stevens Act and severely deteriorating the economic benefits derived from the crab and groundfish fisheries.

—National Marine Fisheries Service, 2004[777]

OVERCAPITALIZATION

Overcapitalization reduces the potential net value that could be derived from the non-pollock groundfish resource by dissipating rents, driving variable operating costs up, and imposing economic externalities. At the same time, excess capacity and effort diminish the effectiveness of current management measures (e.g., landing limits and seasons, bycatch reduction measures). Overcapitalization has diminished the economic viability of members of the

> *fleet and increased the economic and social burden on fishery dependent communities.*
>
> —National Marine Fisheries Service, 2012[778]

In 1980, only one factory trawler, the *Arctic Trawler*, operated in Alaska. By 1987, however, the factory trawler fleet had grown dramatically. And it was continuing to grow. There were at least nineteen factory trawlers on the Alaska fishing grounds in 1987, and as many as a dozen more—at a cost of about $15 million each—were expected to join the fleet in 1988.[779] Konrad Uri, one of the original partners in the *Arctic Trawler*, described the situation as being "like the king crab boom only on a much larger scale financially."[780] The main target species of the factory trawler fleet was pollock, but it caught Pacific cod as well.

The rapidly increasing size of the factory trawler fleet raised the concern that it would soon be capable of catching and processing the entire domestic allocation of groundfish in Alaska, to the detriment of joint-venture vessels and shore-based processors. The fleet's ability to rapidly catch fish was also of concern to fisheries managers because monitoring the catch would be a challenge and overharvesting might occur. There was concern also over the industry's ability to profitably market all the fish it might catch.

"Is Alaska Groundfish Overcapitalized?" was the title of a March 1988 *Pacific Fishing* feature article by Kris Freeman, a contributing editor to the journal. According to Freeman, "As the Bering Sea and Gulf of Alaska become 'Americanized,' fishermen are facing the prospect of too many expensive new boats chasing too few cheap fish." Freeman wrote of "a veritable steamroller of investment" in the Bering Sea fisheries that had shifted fish allocations so quickly that some joint-venture vessels became obsolete within months of being launched.[781]

Well aware of the potential problems faced by the industry, in September 1987, the North Pacific Fishery Management Council issued a carefully worded, yet vague, statement regarding a possible moratorium on the entry of new vessels into the groundfish fisheries. Potential cutoff dates for entry into the fisheries were December 31, 1987, and December 31, 1988. This moratorium was implemented, but not until 1995, and it incorporated February 9, 1992, as the cutoff date (see below).

Also in September 1987, the council established the Future of Groundfish Committee to determine whether the amount of investment in the fisheries was "rational" and, if not, what should be done to correct the problem. In December 1987, the committee reported that "[t]here is more harvesting capacity than needed to harvest the resource over a year-round season, but it's not necessarily the right mix to harvest and process the resource most efficiently."[782] Moreover, the committee concluded that overcapitalization and its associated problems would worsen under the open-access system in effect at that time, and it recommended limiting access to the groundfish fisheries (as well as the crab and halibut fisheries).[783]

The Future of Groundfish Committee was disbanded in 1988, but "comprehensive rationalization" of Alaska's groundfish fisheries—basically matching the industry's fish-catching and fish-processing capabilities with the amount of fish that could be harvested efficiently and in a manner that maximized its value, reduced bycatch, and minimized waste—became, with the establishment of a complementary management regime to ensure that fishing was conducted in a sustainable and ecologically sound manner, the primary goals of the council.[784]

But the concept of comprehensive rationalization itself was fraught with uncertainty because, among other considerations, fish populations—even unexploited fish populations—and markets for fish products are dynamic and sometimes fluctuate dramatically. By definition, a comprehensively rationalized industry would not be burdened by significant overcapitalization. The NPFMC never developed a comprehensive rationalization plan. Rather, the mostly rationalized industry that harvests Pacific cod today along Alaska's coast was developed piecemeal over the course of more than two decades through management actions, private initiative, congressional legislation, and actions by the State of Alaska.

Despite the concerns of the North Pacific Fishery Management Council and others regarding overcapitalization and the issues associated with it, Bob Morgan, formerly the president of the Pacific Seafood Processors Association, was of the opinion in 1987 that keeping cod and pollock stocks healthy was key to a prosperous, sustainable industry: "We need to protect the renewability of the resource. As long as we have a stable resource, everything else will fall into line," he said.[785]

Codfish report, 1990: *The worldwide cod market grows stronger as each day passes. The shortage of cod is pushing wholesale prices up quickly now . . . The real shortage is in Atlantic cod, but that makes Pacific cod more attractive. In fact, Norway has reportedly bought substantial amounts of Pacific cod to use as fill-in to customers who can't get Atlantic cod.*

—*Pacific Fishing*, July 1990[786]

A 1988 report by the Fisheries Management Foundation and the Fisheries Research Institute was far less sanguine. The report noted the concern that "continuation of the current management regime will encourage too much investment in fishing capacity" and warned that overcapitalization would lead to "much reduced profitability, short and disorderly fishing seasons, and intense pressure on conservation agencies to slacken fishing regulations."[787]

As was the pattern during the Salt Cod Era, shortages of Atlantic cod spurred interest in Alaska's Pacific cod resource. In 2016, *National Fisherman* reporter Ashley Herriman described the early 1990s collapse of cod stocks in the eastern Atlantic Ocean:

> Forever, it seems, Newfoundland was cod heaven—or more correctly, cod fishermen's heaven—and by the 1960s, nearly a millennium after Leif Erikson and his cohort first caught cod off Labrador, the Newfies were landing upward of 800,000 tons a year. It all came to an end in the early 1990s, with a crash that made headlines around the world.[788]

Reporter Alan Harman, writing in *National Fisherman* in 1992, provided some details:

> The Canadian Atlantic Fisheries Scientific Advisory Committee has found that northern cod has declined dramatically in the past year and a half, and the spawning stock is about 10% of the long-term average. . . . More than 500,000 tons of northern cod have been lost in Canadian waters in the past two years. The committee says the decline is "consistent with extreme environmental conditions in 1991." These include particularly cold water temperatures off Labrador and Newfoundland, greater than normal ice coverage and low ocean temperatures.[789]

A 1995 assessment of the situation by the president of the Newfoundland Inshore Fisheries Association was considerably harsher:

> Canada's North Atlantic cod fishery closed down due to chronic overfishing borne of greed, poor science, and political expediency. The moratorium on directed fishing for cod remains in place. The rape of the northern cod stocks ranks with the greatest of human insults against this Earth—not to mention the loss of the annual production of up to a billion pounds of protein in a hungry world.[790]

The upside of the collapse of cod stocks in the western Atlantic—if there indeed was one—was that it fostered what appeared to be reasonably long-term opportunities in Alaska's Pacific cod fishery and stimulated the growth of the fishery. It was also a warning of what could happen if cod stocks weren't prudently managed and that changes in sea temperature could dramatically affect cod stocks.

An assessment of the Bering Sea/Aleutian Islands Pacific cod fishery by the North Pacific Fishery Management Council in April 1993 showed, however, that the overcapitalization problem was worsening. The fishery, according to the council, "exhibits numerous problems which include: compressed fishing seasons, periods of high bycatch, waste of resource, gear conflicts and an overall reduction in benefit from the fishery."[791] Fishing seasons were characterized by some in the industry as "Olympic-style derbies" in which each fishermen raced to get as much of the allowable catch as possible before the season closed.[792] In their book *Fishing for Pollock in a Sea of Change*, James Strong and Keith Criddle characterized the derbies as "economically perverse."[793]

Ironically, the National Marine Fisheries Service—the federal agency responsible for implementing groundfish management plans in the EEZ off Alaska—had contributed to the overcapitalization problem. Through its Fisheries Obligation Guarantee Program, the agency had by 1992 provided some $102 million in loans for thirty-three factory trawlers.[794] Moreover, the threat of a moratorium on entry into the groundfish fisheries spurred the construction of vessels by individuals and companies who feared being frozen out of fisheries.[795]

The rationalization of Alaska's groundfish industry wasn't a clear, step-by-step process. Rather, it was a series of actions—punctuated by several detours—over nearly two decades by the NPFMC, Congress, and the industry. And it still isn't complete.

Codfish report, 1994: *The world cod market is still recovering from the blow dealt last year by the Russians, who dumped an enormous amount of cod, some good, some bad, all cheap.*

—Clark Miller, *National Fisherman*, July 1994[796]

Codfish report, 1999: *Cod is changing from a low-cost staple to a high-end seafood.*

—Brad Warren, *Pacific Fishing*, February 1999[797]

RATIONALIZATION, ACTION 1: GEAR SECTOR ALLOCATIONS, SEASONS

> *Fisheries off Alaska officially came of age in December [1985] when the North Pacific Fishery Management Council (NPFMC) found itself for the first time with more requests for fish than there were fish to allocate.*
>
> —Chris Blackburn, *National Fisherman*, 1986[798]

The first action toward rationalizing the Pacific cod fishery was the allocation of percentages of the total allowable catch (TAC) in the Bering Sea/Aleutian Islands area to the various fishing sectors. Prior to these allocations, the whole of the figurative harvest "pie" was available to all sectors, which intensified the race to fish. Under the allocation system, each sector's potential catch was limited, but, in exchange, each sector was guaranteed a percentage of the total allowable catch. This fostered efficiency and helped stabilize the fishery.

The first allocation of Pacific cod in Alaska was a somewhat belated, though probably more comprehensive than expected, response to a proposal by the North Pacific Fixed Gear Coalition, an organization that was established in 1987 mostly to oppose bottom trawling. In 1991, the coalition proposed to the North Pacific Fishery Management Council that fixed-gear fishermen be given preferential access to certain groundfish species in the Bering Sea/Aleutian Islands areas.

Three years later, in 1994, the council provided what it termed "a bridge to comprehensive rationalization" that would provide at least a temporary measure of stability in the industry. The council's Amendment 24 to the Bering Sea/Aleutian Islands fishery management plan allocated the total allowable catch of Pacific cod in Bering Sea/Aleutian Islands area among the three existing gear sectors: trawl gear, fixed gear (longline and pot), and jig gear. (Jig gear is classified as fixed gear but usually receives a separate allocation.) No similar action, however, was taken for the Gulf of Alaska groundfish fisheries.

Amendment 24 was implemented in January 1994 and was scheduled to expire at the end of 1996. The percentage allocations established under Amendment 24 for the 1994 through 1996 fishing seasons were: trawl gear, 54 percent; fixed gear, 44 percent; and jig gear, 2 percent. The percentages for the trawl sector and the fixed-gear sector reflected the annual harvests in those sectors from 1991 to 1993. The 2 percent allocation to jig gear exceeded the recent harvest percentage taken by jig fishermen and was intended to enable an increase in the participation of small locally based vessels.

To provide management flexibility, the amendment authorized the NMFS regional director to reallocate Pacific cod among the gear sectors if the director determined that a gear sector would not be able to harvest its allocation.

Amendment 24 also authorized the NMFS to divide each year into three seasons, each of four months duration, for the fixed-gear sector, and to divide that sector's Pacific cod allocation among the three seasons in proportions recommended by the North Pacific Fishery Management Council. This action was in response to longline industry representatives who argued for a seasonal allowance of the fixed-gear allocation of Pacific cod to provide for a first- and third-season fishery, when halibut bycatch rates were lowest, product quality was high, and markets were most advantageous.[799]

In 1996, citing "significant regulatory, economic, and biological changes" that had occurred in the Pacific cod fishery since the implementation of Amendment 24, the council passed Amendment 46, which revised the sector allocations as follows: fixed gear, 51 percent; trawl gear, 47 percent; and jig gear, 2 percent. The amendment also divided the trawl gear allocation equally between catcher vessels and catcher-processors.

Amendment 46 was implemented in January 1997 and was followed by three increasingly specific sector allocation amendments, as shown in the table below.[800] Note that the community development quota (CDQ) program currently receives a 10.7 percent *off-the-top* allocation of the total allowable catch. The remainder is divided among the various sectors, as is shown in the table below.

Pacific cod allocation in the central and western Gulf of Alaska areas had been structured since 1992 around the requirement that 90 percent of the groundfish catch in the areas had to be delivered to shore-based processors (see chapter 23). Historically, gear sectors in these areas did not receive allocations of the total allowable catch, but were restricted primarily by seasonal catch limits designed to provide a relatively steady flow of fish to mostly shore-based processors. In the late 2000s, competition among the gear sectors intensified, resulting in derby-style races for fish, with the associated efficiency, bycatch, safety, and wastage problems. This untenable situation was improved in 2012, with the implementation of Amendment 83 of the NMFS's Gulf of Alaska fishery management plan, which was designed to "reduce the uncertainty regarding the distribution of Pacific cod catch, enhance stability among the sectors, [and] maintain processing limits to protect coastal fishing communities" by establishing gear sector groundfish allocations based primarily on historical catches.[801]

Table 1. Non-CDQ Bering Sea Pacific cod allocations, percentage by sector, 1994–present.

Sector	Amend. 24 (1994)	Amend. 46 (1997)	Amend. 64 (2000)	Amend. 77 (2004)	Amend. 85 (2007)
Jig	2.0	2.0	2.0	2.0	1.4
Hook-&-Line/Pot CV <60 ft. LOA	44.0	51.0	0.7	0.7	2.0
Hook-&-Line/Pot CV ≥60 ft. LOA			0.2	0.2	0.2
Hook-&-Line CP			40.8	40.8	48.7
Pot CV ≥60 ft. LOA			9.3	7.6	8.4
Pot CP				1.7	1.5
AFA trawl CP	54.0	23.5	23.5	23.5	2.3
Non-AFA trawl CP					13.4
Trawl CV		23.5	23.5	23.5	22.1

Abbreviations: CV = catcher vessel; CP = catcher-processor vessel; LOA = length overall; AFA = American Fisheries Act (1998).
Data source: 72 *Fed. Reg.* 5655, February 7, 2007; 72 *Fed. Reg.* 50789, September 4, 2007.

The amendment also provided incentives for new entrants into the jig sector through an initial allocation that was above the sector's historical catch in the fishery and a potential performance-based increase in this sector's allocation. It also superseded the 1992 requirement that 90 percent of the Pacific cod caught in the Gulf of Alaska be delivered to shore-based processors.[802] In 2011, the central Gulf of Alaska area, which included the waters around Kodiak Island and south along the eastern shore of the Alaska Peninsula approximately to Chignik, accounted for 62 percent of the Gulf of Alaska's 66,100-metric-ton total allowable catch. The western Gulf area accounted for 35 percent, and the eastern Gulf area just 3 percent.[803]

Pursuant to Amendment 83, the jig fishery was treated special. In the central Gulf of Alaska, it was allocated—off the top—1 percent of the total allowable non-CDQ catch of Pacific cod. The remainder was allocated as follows:

- trawl catcher vessels: 41.6 percent
- trawl catcher-processors: 4.2 percent
- hook-and-line catcher vessels less than 50 feet long overall: 14.6 percent
- hook-and-line catcher vessels more than 50 feet long overall: 6.7 percent
- hook-and-line catcher-processors: 5.1 percent
- pot (catcher vessels and catcher-processors): 27.8 percent

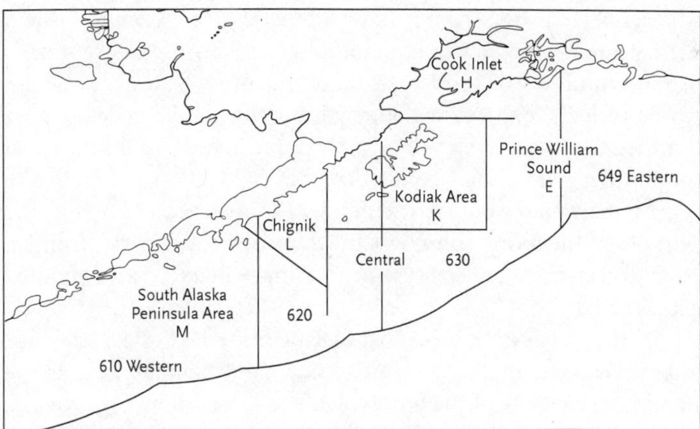

Figure 85. Chart of state Pacific cod management areas (South Alaska Peninsula, Chignik, Kodiak, Cook Inlet, and Prince William Sound) and federal regulatory areas (Western, Central, and Eastern) in the Gulf of Alaska, 2011. (National Marine Fisheries Service. 76 *Fed. Reg.* 44700, July 26, 2011.)

In the western Gulf of Alaska area in 2012, the jig fishery likewise received special treatment—an off-the-top allocation of 1.5 percent of the total allowable catch of Pacific cod. The remainder was allocated as follows:

- trawl catcher vessels: 38.4 percent
- trawl catcher-processors: 2.4 percent
- hook-and-line catcher vessels: 1.4 percent
- hook-and-line catcher-processors: 19.8 percent
- pot (catcher vessels and catcher-processors): 38 percent[804]

RATIONALIZATION, ACTION 2: MORATORIUM

> *In 12 years of Olympic-style scrambling to outcatch one another, fishermen have built an armada that handily exceeds what's needed to haul about 4 billion of 5 billion lbs. of food from the Bering Sea and the Gulf of Alaska each year. The excess of fishing power has fueled a war over allocations that drags the industry into bad blood and legal fees. No one wants this to continue, but the pressure driving overcapitalization is difficult to ease.*
>
> —Brad Warren, *National Fisherman*, 1992[805]

The next action in the comprehensive rationalization of the cod fishery was to prevent new entrants into the fishery. By 1989, there was widespread fear in the fishing industry that the North Pacific Fishery Management Council would not limit access to Alaska's groundfish fisheries until it was too late to be of much use, and bankruptcies would result.[806] The council adopted moratorium language in 1992, but the moratorium was not actually implemented until September 1995, when the council concluded that "allocation conflicts and overcapitalization would worsen under the open access system." The purpose of the moratorium was to "transition the fisheries from an open-access management system to a more market-based, limited access system."[807]

In the words of the National Marine Fisheries Service, the moratorium was "an interim measure to slow significant increases in the harvesting capacity of the groundfish and crab fishing fleets until a Comprehensive Rationalization Plan (CRP) could be implemented." The moratorium was scheduled to expire in three years (it was later extended), and it applied to the groundfish fisheries as well as the crab fisheries in both the Bering Sea/Aleutian Islands area and the Gulf of Alaska.[808]

Concerned that there had been speculative entry into the fisheries during its multiyear discussion of management alternatives, the council established a control date of February 9, 1992. Any vessel that had not participated in the fisheries before that date would not be guaranteed future access to the fisheries. Moratorium permits were granted to vessels that made at least one legal landing of any moratorium species during the qualifying period of January 1, 1988, through February 9, 1992.[809]

Because the permits were granted to vessels, the only means of gaining entry into the fishery was to purchase a permitted vessel. One criticism of the retroactive nature of the moratorium was its potential to "resurrect the dead"—to grant permits to vessels that had a history of participation in the fisheries but had been absent from the fishery for a long while.[810]

> **Codfish report, 1995:** *Cod fishermen off Alaska had more fish to catch in 1995, as the North Pacific Fishery Management Council increased the Bering Sea quota to 250,000 tons, up from 191,000 in 1994. The Gulf of Alaska quota also grew last year from 50,400 to 69,200 tons. While improved stocks and quotas are always good news, an additional benefit to producers is that the market for cod was also reasonably healthy. Eastern Canadian and New England production continued to be dismal, but increased production from Europe and competition from Russia kept the lid on cod prices.*
>
> —*Pacific Fishing*, March 1996[811]

RATIONALIZATION, SIDESTEP 3: INDIVIDUAL TRANSFERABLE QUOTA ("CATCH SHARES")—AN IDEA THAT DIDN'T FLOAT[812]

> *By allocating the catch to individuals, cooperatives, communities, or other entities catch share programs are intended to promote sustainability and increase economic benefits.*
>
> —National Marine Fisheries Service, 2015[813]

> *Catch shares: a polite-sounding term that implied things got shared when what they really got was privatized.*
>
> —Lee van der Voo, author of *The Fish Market*, 2016[814]

Once the moratorium was in place, the next logical step in rationalizing the groundfish fisheries would have been to devise and institute an individual transferable quota (ITQ) share program such as the North Pacific Fishery Management Council instituted in the spring of 1995 for the halibut and sablefish fisheries. Under an ITQ program, the National Marine Fisheries Service would allocate a fixed percentage of each season's total allowable catch in a fishery to a vessel (or its owners), based on a vessel's participation and fishing success in that fishery during a qualifying period. In possession of individual transferable quota—essentially an exclusive right to harvest a public resource—fishermen would have little incentive to race to fish. To foster efficiency in the fleet, ITQs would be transferable among vessels. As well, they could be sold or leased or inherited or even used as collateral. ITQs would also be a conservation incentive because fishermen would have a vested interest in maintaining—or increasing—the stocks of the species.

The concept of ITQs for Alaska's groundfish fisheries, however, engendered considerable criticism. As a market-based approach, ITQs would institutionalize a regime that would favor those of great financial means, especially large food-manufacturing firms. In the crosshairs of critics was Arkansas-based Tyson Foods, America's largest producer of poultry, which had purchased the operations of the Arctic Alaska Fisheries Corporation in 1992 and stood to gain fishing rights conceivably worth many millions of dollars. Some individuals familiar with the industry thought the aggregate value of Alaska groundfish ITQs would be worth on the order of a billion dollars.[815] At the same time, many Alaskans worried that an ITQ system would eventually result in a fishery where only wealthy individuals and corporations—mostly from the lower-48 states—would hold the available permits.

In February 1994, Congress's House Subcommittee on Fisheries Management held a hearing on the potential issuance of individual transferable quota. A wealth of information was presented, and a variety of often-conflicting opinions were voiced. In favor of an ITQ program was the Puget Sound–based factory trawler fleet, which was at that time severely overcapitalized. Owners of financially troubled vessels hoped they would be able to cash in on their vessels' catch histories. Among those opposed to ITQs was the Pacific cod fishery's fixed-gear sector (longline and pot gear fishermen). This sector opposed ITQs as well as other market-based allocation schemes because its share of the Pacific cod catch was growing. An ITQ or similar program had the potential to prevent further growth in the fixed-gear sector. Shore-based processors opposed any program that would favor factory trawlers, their historic rivals. In the end, the issue engendered so much controversy that the hearing spawned no congressional action.[816]

In early 1996, Alaska senator Ted Stevens characterized ITQs as "the most contentious issue that [he had] faced in the process of the Magnuson Act." Until ITQs could be further studied, Stevens opposed the development of an ITQ system, and, with several like-minded Senate colleagues, he called for a moratorium on the development of any new licensing systems.[817]

In accordance with the wishes of Stevens and his colleagues, the October 1996 Sustainable Fisheries Act, which reauthorized the Magnuson-Stevens Act, imposed a four-year moratorium on the creation of new individual fishing quota programs. The legislation also required the National Academy of Sciences, in consultation with the secretary of commerce and regional fishery management councils, to prepare a comprehensive report on individual fishing quotas.[818]

Despite the positive effects ITQs might afford a fishery, the idea of the direct awarding of exclusive fishery harvest rights in the waters off Alaska's coast to primarily lower-48 corporations remained controversial, and no ITQ program, per se, has yet been implemented in Alaska's groundfish fisheries. However, quota shares of restricted transferability were issued to trawl catcher-processors in the Bering Sea/Aleutian Islands (see Amendment 80, below), and the fishery cooperatives established in Alaska's groundfish fisheries, as will also be discussed below, function much as an ITQ program.

RATIONALIZATION, ACTION 4: AMERICAN FISHERIES ACT, 1998

In addition to correcting shortcomings in the Commercial Fishing Industry Vessel Anti-Reflagging Act of 1987, the American Fisheries Act incorporated provisions to further rationalize Alaska's groundfish fisheries. (This legislation was intended primarily for pollock, but, as noted below, it created a template that was later applied to cod and other groundfish species.) It prohibited the entry into the Bering Sea pollock fishery of any new fishing vessel that was greater than 165 feet in registered length or of more than 750 gross registered tons or that had an engine or engines capable of producing a total of more than 3,000 shaft horsepower.

The act also authorized $75 million in loans to further reduce excess capacity in the fishery through a vessel buyout program. The loans would be repaid through a small levy on each pound of pollock caught by the vessels that remained in the fleet.[819]

Very importantly, the American Fisheries Act, by allocating percentages of the total allowable catch of pollock in the Bering Sea/ Aleutian Islands area (after subtracting the Community Development Quota Program allocation and a bycatch allowance) among three sectors—inshore catcher vessels, offshore factory trawlers (catcher-processors), and mothership catcher vessels—fostered the establishment of the first *catch share* program in the North Pacific groundfish fisheries. The sector allocations enabled the formation of fishery cooperatives in each sector.[820]

The Fishermen's Collective Marketing Act of 1934, a law designed to allow fishermen to bargain collectively with processors, provides limited antitrust exemptions that enable members of a fishery cooperative to operate collectively rather than in competition, provided the market price of its production is not "unduly enhanced."[821]

Cooperatives, wrote the National Marine Fisheries Service in 2002, "eliminate problems flowing from the common property status of fisheries resources."[822] Members of a cooperative could divide the available quota for their sector among themselves in a manner that facilitated efficient fishing, maximized productivity, enabled the production of higher quality (and more valuable) products, and minimized bycatch. These *quota shares* could be leased, which enabled some vessels to profit from the fishery without even putting a net in the water.[823] Because shares can be exchanged among a cooperative's members, the National Marine Fisheries Service referred to the cooperative program as a "quasi-Individual Fishing Quota program."[824]

Quota shares, however, are not permanent property rights. Rather, they are *harvest rights* that exist subject to a private contractual agreement and are automatically canceled if the cooperative dissolves. As well, the quota shares are subject to being rescinded without compensation by the National Marine Fisheries Service as a penalty for violation of fisheries regulations.

Thanks to provisions in the American Fisheries Act of 1998, the open-access race for pollock was a thing of the past, but not everyone was pleased with the new order. In an article titled "This Is Americanization?" Brad Warren, editor of *Pacific Fishing*, wrote of what he considered the shortcomings of the American Fisheries Act. First among them was that the legislation granted "virtual private-property rights to Bering Sea pollock . . . to a select group of industry players" and had locked out all their competitors.[825] Little could Warren have suspected that the American Fisheries Act was actually a template for the rationalization of other groundfish fisheries.

The rationalization of the pollock fishery created the potential for pollock vessels to use their operational advantages to expand into other fisheries, some of which were already fully utilized. The cod and crab fisheries were of particular concern. This potential, however, was recognized in the American Fisheries Act, which contained language that provided—at least until superseded by NPFMC regulations—protection for other fisheries from adverse impacts caused by the pollock fishery cooperatives. Specifically, the American Fisheries Act limited the catcher-processor sector in the Bering Sea/Aleutian Islands area to landing the same percentage of non-pollock groundfish as it had landed during the years 1995 through 1997. Catcher-processors using trawl gear were banned from fishing in the Gulf of Alaska, but they could process, in aggregate, up to 10 percent of the cod harvested from a limited area of the gulf.[826] (Such restrictions are known as *sideboards*, which the NMFS defines as "restrictions placed on vessels and/or [License Limitation Program] licenses to prevent 'spillover' effects from potential increases in effort when management programs free up vessels from participation in other fisheries."[827])

In 1999, the NPFMC passed sideboard regulations that allowed some pollock vessels to expand their cod-fishing effort in the Bering Sea/Aleutian Islands area. Catcher vessels that delivered to motherships were restricted to their historical catch level only until March 1 of each year. After that date, they could fish as hard as they desired. In addition, the council eliminated any restriction in the cod catch of pollock vessels that landed less than 1,700 metric tons of pollock per year.[828]

In 2002, the NPFMC refined the sideboard regulations to protect small trawlers that operated in "Cod Alley"—a cod-rich area north of Cape Sarichef, on Unimak Island, in the Aleutian Islands—from excessive competition by pollock trawlers rationalized under the American Fisheries Act.[829]

RATIONALIZATION, ACTION 5: WASTE MANAGEMENT: "YOU CATCH IT, YOU KEEP IT."[830]

> **The goal:** *[P]rovide for the rational and optimal use, in a biological and socioeconomic sense, of the region's fisheries resources as a whole.*
>
> *—Bering Sea/Aleutian Islands Area Fishery Management Plan, 1998*[831]

In early 1992, the North Pacific Fishery Management Council established the Discard Committee to address bycatch. The committee's goal was to

> [i]ncrease the quantity and quality of food and byproducts produced from the fishery resources harvested in the BS/AI and GOA by reducing the amount of harvest discarded to the maximum extent practicable while recognizing the contributions of these fishery resources to our marine ecosystems and the economic and social realities of our fisheries.[832]

In practical terms, the desired goal at the time was full retention—to somehow utilize every fish that was caught. During 1992, Alaska's groundfish industry was reported to have discarded some 507 million pounds of bycatch—equal to about 12 percent of the total groundfish harvest—for various reasons, including wrong species, wrong size, and wrong sex.[833]

In 1993, Alaska senator Frank Murkowski, calling the level of fish discards in the commercial fisheries "a disgrace of unparalleled proportions," introduced legislation to limit bycatch and encourage the processing of waste fish into meal or other products.[834] Senator Murkowski's well-intended legislation did not become law, and bycatch continued while options for reducing it were contemplated both administratively and in Congress.

Echoing Senator Murkowski's earlier condemnation of bycatch, in September 1996 Alaska senator Ted Stevens called the waste by distant-water fishing vessels of 500 million to 700 million pounds each year "unacceptable, totally unacceptable."[835] That same month, the North Pacific Fishery Management Council adopted BSAI Amendment 49, which required full retention of pollock and Pacific cod—no matter how or where the fish were caught—beginning in 1998. (Rock sole and yellowfin sole were to follow in 2003.) Fish not suitable for human consumption, however, could be legally discarded. The amendment also established a 15 percent minimum utilization standard for all at-sea processors.[836]

The council had targeted the pollock and cod vessels first because of the sheer volume of their discards, which were mostly fish that were of the right species but were considered too small for utilization. The volumes were staggering: had the regulations been in effect in 1995, some 220 million pounds of pollock and 95 million pounds of Pacific cod would have stayed aboard the vessels.[837]

Waste in the Gulf of Alaska groundfish fishery was similarly severe, and was addressed in GOA Amendment 49, which the council adopted in June 1997. This amendment required all vessels fishing for groundfish to retain all pollock and Pacific cod beginning in 1998, and all shallow-water flatfish beginning in 2003. Like its counterpart amendment in the Bering Sea/Aleutian Islands area, it established a 15 percent minimum utilization standard for all at-sea processors.[838]

Together, these parallel measures dramatically reduced overall discard of groundfish and fostered fuller utilization of groundfish by at-sea processors. Additionally, as will be discussed in chapter 24, under the federal Prohibited Species Donation Program, halibut and salmon caught by trawl catcher vessels may be retained if the fish will be delivered to an authorized shore-based processor or stationary floating processor for donation to hunger relief agencies, food bank networks, or food bank distributors.[839]

RATIONALIZATION, ACTION 6: LICENSE LIMITATION PROGRAM (LLP)

The NPFMC had early on identified an individual transferable quota (ITQ) program as having the greatest potential for resolving most of the issues surrounding overcapitalization, but questions remained regarding what specific form of ITQ program would be most appropriate and how long it might take to develop and implement such a program. And, as noted above, the idea of instituting an ITQ program in Alaska's groundfish fisheries engendered considerable controversy. A simpler, less controversial, and more expeditious alternative was a license limitation program (LLP), an interim action that would provide stability in the fishing industry while an ITQ program was developed and implemented. The LLP would limit the number, size, and specific operation of the vessels in the Bering Sea/Aleutian Islands and Gulf of Alaska groundfish and crab fisheries. In doing so, the program would establish something of an upper limit on the amount of capitalization that could occur in those fisheries.

In 2000, the North Pacific Fishery Management Council replaced the moratorium it had established in 1995 with an LLP. Based on a vessel's past participation in the fisheries, including the area fished and if processing was done aboard the vessel, an LLP license was issued to the vessel's owners. The licenses:

1. endorsed fishing activities in specific regulatory areas in the Bering Sea/Aleutian Islands area and Gulf of Alaska;
2. restricted the length of the vessel on which the LLP license may be used;

3. designated the fishing gear that may be used on the vessel; and
4. designated the type of vessel operation permitted (catcher vessel or catcher-processor).

An LLP license is transferable, but only to a vessel with a length overall that is no greater than the maximum length overall stated on the license, but the endorsements for specific regulatory areas, gear designations, and vessel operational types are nonseverable from the LLP license.

For the Bering Sea/Aleutian Islands area and the Gulf of Alaska combined, the National Marine Fisheries Service subsequently issued more than three hundred LLP licenses endorsed for trawl gear and more than one thousand licenses endorsed for non-trawl gear.

To foster entry-level opportunities in the Pacific cod fishery, jig gear was later exempted from the license limitation program.[840] A shortcoming of the license limitation program was that although it capped the number of participants in the fisheries, it did not limit the amount of fishing effort. Regulation of the fishing effort was subsequently accomplished through amendments to fisheries management plans and congressional action.

RATIONALIZATION, ACTION 7: FISHING CAPACITY REDUCTION PROGRAM

The first fishing capacity reduction program in Alaska waters was incorporated in the American Fisheries Act in 1998, which through a buyout program removed nine Seattle-based pollock factory trawlers from the fleet. (The legislation permanently removed the vessels from all fisheries conducted in U.S. waters. Eight were scrapped in a San Francisco shipyard.)[841]

In 2004, Congress authorized up to $75 million to finance fishing capacity reduction programs (buyouts) within the Alaska non-pollock fisheries. However, because industry players wanted to await the results of a major effort underway to reallocate Pacific cod and other groundfish species among the gear sectors, no action was taken immediately.[842] In 2007, at the request of the Freezer Longline Conservation Cooperative, an association formed in 2006 to work toward a buyout, the National Marine Fisheries Service implemented a voluntary fishing capacity reduction loan program for the freezer-longliner sector, which at the time comprised thirty-nine vessels. Members of the cooperative subsequently voted to accept four offers. At a cost of $35 million, the program removed three active freezer-longliner vessels, the *Northern Aurora*, *Horizon*, and *Western Queen*, including their associated licenses, as well as an inactive LLP license, itself valued at $1.5 million. Based on their production during the 2003 through 2005 seasons, the removal of the three vessels reduced the freezer-longliner fleet capacity by about 7.5 percent and was expected to result in an average annual net revenue gain of $302,560 for each vessel remaining in the fishery.

The interest rate on the 30-year loan was fixed at 6.84 percent, and the loan is being repaid through a fee levied on each pound of cod landed by the remaining freezer-longliner vessels. The original fee was $0.02 per pound, but it has subsequently been reduced several times, and the rate since January 1, 2013, has been $0.0111 per pound.

In 2012, to reduce potential competition in the freezer-longliner sector, a second fishing capacity reduction program retired an inactive LLP license at a cost of $2.7 million. As with the previous program, the loan is being repaid through a small fee on the freezer-longliner fleet's cod landings, in this case $0.001 per pound. The freezer-longliner sector—the dominant player in the Alaska codfish industry—is the only sector of the Alaska codfish industry that has requested fishing capacity reduction program funds.[843]

RATIONALIZATION, ACTION 8: AMENDMENT 80

The American Fisheries Act in 1998 contributed to the rationalization of the pollock fishery in the Bering Sea/Aleutian Islands area in part by restricting entry into the fishery and by authorizing the formation of fishery cooperatives in which members divided the available pollock quota among themselves in a manner that facilitated efficient fishing, maximized productivity, reduced bycatch, and enabled the production of a higher-quality product.

While not without its critics, this aspect of the legislation accomplished its objectives, and in 2006 the North Pacific Fishery Management Council, in adopting Amendment 80 to its Bering Sea/ Aleutian Islands area management plan, expanded rationalization of the fisheries.

Amendment 80, which was implemented in 2008, was directed at the non-pollock trawl fisheries, which during the years leading up to the amendment had remained classic races to fish, with the associated safety, inefficiency, and waste issues. Discard rates reached up to 30 percent of the catch limit and premature season closures became a regular occurrence as trawlers reached bycatch limits before they could harvest all their target species.[844]

The direct goals of Amendment 80 were to minimize waste, reduce bycatch, improve utilization, and minimize impacts on other fisheries.[845] Amendment 80 assigned quota shares for six non-pollock trawl groundfish species—Pacific cod among them—to owners of eligible vessels in the non–American Fisheries Act trawl catcher-processor sector.[846] The National Marine Fisheries Service determined that twenty-eight vessels qualified for the program. One vessel, the *Golden Fleece*, opted out of the program, so the National Marine Fisheries Service allocated to owners of the twenty-seven individual eligible vessels quota shares based on their vessels' catch histories for the listed species during the years 1998 through 2004. Amendment 80 quota shares are nonseverable from the vessel permit to which they were assigned.

Of the qualified vessels, twenty-four participated in the Bering Sea/Aleutian Islands fisheries under Amendment 80. These vessels ranged in length from just over 100 feet to just under 300 feet.

Under the Magnuson-Stevens Act, the Amendment 80 fleet is considered a limited-access privilege program, and participating vessels are subject to cost-recovery fees of up to 3 percent of the ex-vessel value of their catch to cover the federal government's costs of management, data collection and analysis, and enforcement.

The amendment also provided the non-pollock trawl fisheries sector with a specific allocation of crab and halibut bycatch (prohibited species catch) and authorized the formation of harvesting cooperatives. Amendment 80 vessels can vest their shares either in one or more cooperatives formed by participating members or in the limited-access common-pool fishery. The Alaska Seafood Cooperative (formerly the Best Use Cooperative) was formed in 2008 and included sixteen of twenty-four participating vessels. The remaining vessels participated in the Amendment 80 limited-access common-pool fishery until 2011, when they formed a second cooperative, the Alaska Groundfish Cooperative.[847]

With the race to fish ended, the Amendment 80 fleet was amenable to gear and operational changes that helped reduce the effects of bottom trawling on the marine environment. The fleet's focus, in the words of the Groundfish Forum, an industry advocacy group, changed from "dollars per day" to "dollars per fish."[848] Bycatch and discards were reduced, while the fleet's efficiency—and the quality of its products—increased. Moreover, as James Strong and Keith Criddle wrote in their book *Fishing for Pollock in a Sea of Change*, cooperatives enabled "[c]ompanies that were once looking to make it through the season . . . to plan for the future, [and] . . . to justify purchases of equipment that may take several years to amortize."[849]

To prevent the Amendment 80 fleet from increasing its participation in other fisheries as a result of its increased efficiency, sideboard regulations developed by the North Pacific Fishery Management Council limit participation of Amendment 80 vessels in Gulf of Alaska fisheries to historical levels. The annual average catch of Pacific cod in the Gulf of Alaska by the Amendment 80 fleet during the years 2008–2012 was approximately 1,000 metric tons.

Consolidation in the Amendment 80 fleet has gradually reduced the number of active vessels. In 2008, the active fleet consisted of twenty-two vessels; in 2015, there were eighteen. The number of fishing companies has also diminished. Originally, nine companies owned vessels in the Amendment 80 fleet, but by 2013 consolidation had reduced this number to six.[850]

Amendment 97, passed by the North Pacific Fishery Management Council in 2012, allows owners of Amendment 80 vessels, to "for any reason at any time," replace their vessels with "newer, larger, safer, and more efficient vessels" albeit with some provisions, including an length overall limit of 295 feet.[851] So far, the amendment has resulted in the construction of three new vessels, the 194-foot *Araho*, the 264-foot *America's Finest*, and an as-yet-unnamed vessel being constructed in a Florida shipyard (see chapter 16).

Despite the operational advantages engendered in Amendment 80, the vast majority of the fleet's Pacific cod allocation is harvested not in the directed fishery, but as bycatch. In 2014, for example, fully 90 percent of the Alaska Seafood Cooperative's 22,370-metric-ton Pacific cod allocation was caught while targeting other species. The cooperative refers to Pacific cod as a "constraining species."[852]

In addition to fishing themselves, some Amendment 80 factory trawlers are simultaneously employed in the Bering Sea and Aleutian Islands as motherships for Pacific cod trawl catcher vessels. (The practice is known as *mothershipping*.) During the years 2012–2015, only one Amendment 80 factory trawler did so, but in 2016 it was joined by five others, and in 2017 one more joined. In 2017, the ad hoc Amendment 80 mothership fleet (along with one American Fisheries Act factory trawler) processed Pacific cod from seventeen trawl catcher vessels that each delivered at least part of their annual catch to the motherships. The motherships' success came at a price: a commensurate decline in deliveries to the area's shore-based processors. This affected not only the shore-based processors, but also the communities in which the processing plants were located: they were losing jobs, general economic activity, and tax revenues. The fact that there are no regulations to prevent further growth of the mothership fleet aggravated the situation. Some fishermen, however, countered that the motherships provided needed competition. In December 2017, the North Pacific Fishery Management Council voted to study this issue, which is on a smaller scale reminiscent of the inshore/offshore conflict that occurred over pollock in the early 1990s.[853]

In April 2019, the council, to help return the fishery to, in the words of its advisory panel, "long-term stable processing patterns," reduced the number of factory trawlers that can act as motherships in the Bering Sea/Aleutian Islands area trawl cod fishery to two: the *Seafreeze Alaska*, an Amendment 80 vessel owned by United States Seafood, and the *Katie Ann*, an American Fisheries Act vessel owned by American Seafoods.[854] Both vessels had histories as motherships that predated the implementation of Amendment 80. (The *Seafreeze Alaska* and the *Katie Ann* are sister ships, built in 1968 as the *Seafreeze Atlantic* and the *Seafreeze Pacific*, respectively [see chapter 16]).

RATIONALIZATION, ACTION 9: REMOVING LATENT LLP LICENSES

In 2000, as noted above, the National Marine Fisheries Service issued some 1,300 LLP licenses. In the Gulf of Alaska fixed-gear Pacific cod fishery, however, a substantial number of the vessels to which those licenses were assigned had not made any landings of Pacific cod since their licenses were issued. These licenses were referred to as *latent* licenses. By about 2005, only about one-fourth of the fixed-gear endorsed LLP licenses were in active use. In 2007, at the request of active participants in the Gulf of Alaska fixed-gear Pacific cod fishery—fishermen concerned that holders of latent licenses could resume fishing and potentially adversely affect their fishing operations—the North Pacific Fishery Management Council initiated a review of LLP licenses endorsed for fixed gear in the gulf and subsequently recommended establishing a fixed-gear LLP endorsement for Pacific cod there. In GOA Amendment 86, which became informally known as the "GOA fixed gear recency action," the council made the issuance of LLP Pacific cod fixed-gear endorsements in the Gulf of Alaska contingent upon landing thresholds during the years 2002–2008. To receive an endorsement, longline and pot vessels less than 60 feet in length were required to have landed at least ten metric tons of

Pacific cod during the period, while longline and pot vessels 60 feet or longer and catcher-processors were required to have landed at least fifty metric tons. Jig fishermen were required to have made at least one landing of Pacific cod. Amendment 86 was implemented for the 2011 fishing season.[855]

In 2010, the NMFS issued 1,226 non-trawl bottomfish LLPs for the central and western Gulf of Alaska areas. In those areas in 2012, the agency issued fixed-gear Pacific cod endorsements to just 378 LLPs. Natural attrition likely contributed in a small way to this very substantial reduction.[856]

(SELF-)RATIONALIZATION, ACTION 10: FREEZER LONGLINE CONSERVATION COOPERATIVE

> *Without a quota system, you could never make this kind of investment.*
>
> —Kenny Down, Blue North Fisheries, on the launching of the company's freezer-longliner *Blue North*, which cost approximately $40 million, 2016[857]

Fishery cooperatives had been organized for the pollock fisheries in the Bering Sea/Aleutian Islands area under the American Fisheries Act (1998) and in non-pollock trawl fisheries in the same area under the North Pacific Fishery Management Council's Amendment 80 (2008). The trawler cooperatives had the desired effect, and in 2010 Congress authorized the formation of a cooperative within the freezer-longliner fleet with the passage of the Longline Catcher Processor Subsector Single Fishery Cooperative Act. The legislation authorizes the secretary of commerce, at the request of eligible members of the Bering Sea/Aleutian Islands freezer-longliner fleet holding at least 80 percent of the licenses issued for their fishery, to approve the organization of a single fishery cooperative.[858]

David Little, president of Clipper Seafoods, which in 2010 operated four freezer-longliners, characterized the legislation as a "win-win for vessel operators, crew, and the environment." In his July 2010 testimony before the U.S. House of Representatives Subcommittee on Insular Affairs, Oceans, and Wildlife, Little described a few of the most valuable benefits of the cooperative:

> Eliminating the race for fish will allow vessel operators to safely slow down with the knowledge that they will still catch their

> share of the resource. A slower pace will improve safety, allow the crew more time to rest and provide greater job stability by extending the fishery over the entire year. Fishermen will have greater flexibility to maximize the value of their catch, they will increase product yield and quality, and will become even better stewards of the resource.[859]

This legislation, however, has not been utilized. Partly, this is because the cooperative it authorized would have been considered a limited-access privilege program under the Magnuson-Stevens Act. As such, the cooperative would have been subject to cost-recovery fees of up to 3 percent of the ex-vessel value of the cooperative's catch to cover the federal government's costs of management, data collection and analysis, and enforcement. By contrast, a voluntary cooperative is not required to pay cost-recovery fees.[860]

Since 2006, most of the Bering Sea/Aleutian Islands area freezer-longliner fleet had been members of the Freezer Longline Conservation Cooperative, a voluntary organization formed in 2004. In the fall of 2010, participation in the cooperative reached 100 percent.[861]

The Freezer Longline Conservation Cooperative operates pursuant to a formal agreement among its members that imposes heavy financial penalties for noncompliance. Each year, the National Marine Fisheries Service allocates quota of Pacific cod to the Bering Sea/Aleutian Islands area freezer-longliner sector. Members of the cooperative then subdivide the quota among themselves in proportion to each member's historical Pacific cod-fishing activity in the Bering Sea/Aleutian Islands area. Members are free to exchange their quota shares among themselves. A private company monitors compliance with the agreement.

Among the benefits of a safer, stable, and more efficient fishery are that it is conducive to new investment, including the construction of new, more efficient, and safer vessels by the cooperative's members, and it provides more opportunities for the retention of by-products.

Freezer-longliners in the Gulf of Alaska operate as a similar, though smaller, voluntary cooperative, the Gulf of Alaska Freezer Longliner Conservation Cooperative. Participation is 100 percent and membership overlaps with its counterpart in the Bering Sea/Aleutian Islands area.

RATIONALIZATION STALLED: GULF OF ALASKA TRAWLERS

> *The greatest challenge facing fishery managers and communities to date has been how to adequately protect communities and working fishermen from the effects of fisheries privatization, notably excessive consolidation and concentration of fishing privileges, crew job loss, rising entry costs, absentee ownership of quota and high leasing fees, and the flight of fishing rights and wealth from fishery dependent communities.*
>
> —Alaska Marine Conservation Council, 2016[862]

Amendment 80 (see above), which applies to the Bering Sea and Aleutian Islands, has no parallel in the Gulf of Alaska except for the Central Gulf of Alaska Rockfish Program (which incorporates a small Pacific cod allocation to trawl catcher vessels). Outside of this program, the trawlers that fish for groundfish in the federal waters of the central and western gulf and in the West Yakutat District of the eastern gulf (trawling is not permitted in the southeast districts of the eastern gulf) are required to have a license limitation program (LLP) license—which can be worth several hundred thousand dollars—and applicable endorsements, but they are essentially free to fish at their own pace within their sector. The situation is an anomaly: except for the approximately twelve to seventeen non–American Fisheries Act trawlers that typically operate in the Bering Sea and Aleutian Islands and the trawlers that operate in the central and western Aleutian Islands state-waters fisheries, all of the West Coast trawl fisheries, including in Canada, operate as cooperatives or under individual fishing quota (IFQ) programs.

Nevertheless, the groundfish trawl fisheries in the Gulf of Alaska work pretty well: absent bycatch-induced closures, vessels sometimes fish for fully eleven months and even then don't always harvest their sector's entire Pacific cod allocation. Contributing to this success is the fact that in 2016 there were forty-nine trawlers that fished for Pacific cod in the gulf, fifteen fewer than in 2008, the year Amendment 80 was implemented in the Bering Sea.863 Perhaps more important are the constraints trawlers have voluntarily placed upon themselves.

For example, the Pacific cod season for trawlers in the gulf typically begins on January 20, but trawlers agree to wait a while—even until March—to start fishing because the cod milt by then has matured and reached its greatest value, and, not insignificantly, the weather tends to be less inclement. Based on the number of vessels and the available quota, the trawlers that deliver their catch at Kodiak and other Gulf of Alaska ports may also impose per-vessel landing limits—as well as bycatch limits—upon themselves. Fishery managers refer to these landing limits as "voluntary catch shares." A potential problem with the voluntary constraints is that if even one or two fishermen chose to operate outside of the agreements, the entire effort at self-regulation would collapse.[864]

According to Clem Tillion—a former commercial fisherman who had been an Alaska legislator, chaired the North Pacific Fishery Management Council, and had served as a fisheries advisor ("Fish Czar") to three Alaska governors—most of the owners of trawl catcher vessels in the Gulf prefer status quo management. Nevertheless, if rationalization is going to happen, the trawlers don't want be left out. For them, the positive side of rationalization is that when it comes time for them to end their fishing careers, they'll have additional fishing rights to sell, which may also enhance their ability to sell their boats.[865] Others look more broadly at the potential negative effects of rationalization: fleet consolidation, loss of crew jobs, and the erosion of the coastal Alaska tradition of fishing vessel owner-operators, the latter of which is especially robust in Kodiak, one of the top commercial fishing ports in the United States.[866]

An amendment passed in 2012 by the North Pacific Fishery Management Council that required trawlers and longline vessels in the gulf to reduce their halibut bycatch triggered the most recent effort to rationalize the gulf's trawl fisheries. The amendment required trawlers to reduce their halibut bycatch by 15 percent over three years, beginning in 2014.[867]

In the course of deliberating the reduction of bycatch limits, the council acknowledged what processors and some trawlers had been saying for years: rationalizing the gulf trawl fleet could aid in reducing bycatch while mitigating the potential for adverse impacts that result from fishing seasons that are altered or curtailed by bycatch "hard caps." At the urging of the Alaska Groundfish Data Bank, a Kodiak-based organization that represents trawlers and shore-based processors, the council then began the process of developing a rationalization plan for the trawlers in the gulf. Trawlers, it must be noted, are responsible for only about a quarter of the Pacific cod catch in the Gulf of Alaska federal- and parallel-waters fisheries.

Cora Campbell, commissioner of the Alaska Department of Fish and Game (2011–2014) and a member of the council, proposed an individual quota system, but trawlers opposed the idea because they feared that companies that operated factory trawlers would buy up and consolidate quota and leave the Alaska vessels "rusting in port."[868]

As an alternative, in 2014 the council began discussing a plan that would apportion shares of the total allowable catch of groundfish in the central and western Gulf of Alaska to cooperatives that comprised trawl catcher vessels and shore-based processors. This alternative had broad support among the Alaska-based trawl fleet as well as shore-based processors.

In November 2014, however, Alaskans elected Bill Walker as their governor. Alaska's economy was suffering severely from low oil prices, and the Walker administration—well aware that the commercial fishing sector is the state's largest private source of employment—wanted to ensure that coastal communities' economic prospects were maximized. Walker replaced Cora Campbell with Sam Cotten, a former state legislator, as the Department of Fish and Game's commissioner.

Cotten opposed the harvester/processor cooperative alternative, claiming it would inevitably result in consolidation of the trawl fleet and would separate the rights to fish from Gulf of Alaska communities that are dependent on groundfish. He pointed out that "[t]he majority of the owners of the fishing rights that would have enjoyed these benefits don't live here. They've fished here but they don't live here," adding that some of the advocates for fishing rights would "like to monetize these fishing rights and sell them," and that the buyers would likely be corporations.

In October 2015, Cotten introduced a proposal that would give individual fishermen quota for bycatch rather than for the target species. At the council's February 2016 meeting, trawlers—both catcher vessels and factory trawlers—opposed the proposal, arguing that it would do nothing to end the race for fish because vessels would simply fish up to their individual bycatch limit. (A benefit of the measure, however, was that it would curtail vessels that fished "dirty" and were responsible for a disproportionate amount of bycatch.) The trawlers were also concerned that, depending on what eligibility criterion was used, the distribution of individual bycatch quota, potentially a marketable item, could result in a redistribution of wealth to relatively new entrants into the fishery.[869] Cotten's bycatch quota alternative nevertheless remained on the table.

It was a contentious exercise, but the council continued to explore ways to rationalize the Gulf of Alaska trawl fisheries. During its June 2016 meeting, the council stated its desire to minimize "economic barriers for new participants by limiting harvest privileges that may be allocated (target species and/or prohibited species) in order to maintain opportunity for entry into GOA trawl fisheries."[870] The concern of particularly the Alaska council members was that the leasing of fishing rights obtained under a rationalization program would become standard practice and would constrain the ability of future generations of Alaska fishermen to enter the groundfish fisheries.

Compromise has proved elusive. In the council's December 2016 meeting, Cotten's rationalization proposal was stridently opposed by both trawlers and processors, who preferred status quo management over what they portrayed as an "experiment." Led by its Alaska members (six of the council's eleven voting members), the council then determined the issue was too divisive and decided to postpone work on it indefinitely.[871]

RATIONALIZATION CONSIDERED: BERING SEA/ALEUTIAN ISLANDS AREA TRAWLERS

In December 2017, the council began discussing the potential rationalization of the Bering Sea/Aleutian Islands area trawl catcher vessel Pacific cod fishery. The council was concerned about the growth in offshore deliveries of Pacific cod to Amendment 80 vessels (see above) operating as motherships and the impacts of these deliveries on shore-based processors and the communities in which they are located. The council was also concerned about the shortening duration of the fishery, which makes it more difficult to maximize its value.

This concern was highlighted in short order: the 2018 federal Pacific cod trawl fishery A season in the Bering Sea (a management subarea of the Bering Sea/Aleutian Islands area[872]), which opened on January 20, was the shortest on record—twenty-two days. Two factors were responsible: the quota was 20 percent lower than in 2017, and there was an increase in effort. Three factors were at play in the increased effort: the price of Pacific cod was high, several eligible fishermen chose to fish in the Bering Sea rather than the Gulf of Alaska, where the Pacific cod quota had been cut dramatically (see afterword), and some fishermen were building catch histories in case a rationalization program is adopted.[873]

Commenting on the short season, Brent Paine, executive director of United Catcher Boats, which represents trawlers in the Bering Sea, said something needed to be done to regulate fishing in the congested "Cod Alley" (just north of Cape Sarichef, on Unimak Island). "If you don't do anything," Paine said, "we're all going to be losers."[874]

ENDNOTES

[775] D. B. Pleschner, "Alaska," *Pacific Fishing* (March 1987): 34–41.

[776] Brad Matsen, "Alaska," *National Fisherman*, Yearbook (1991): 18–20.

[777] National Marine Fisheries Service, *Alaska Groundfish Fisheries, Final Programmatic Supplemental Environmental Impact Statement* (June 2004), C-56.

[778] 77 *Fed. Reg.* 58778, September 24, 2012.

[779] Kris Freeman, "Factory Trawlers Take Off," *Pacific Fishing* (September 1987): 38–41, 48–49.

[780] Konrad Uri, personal communication with author, March 17, 2019.

[781] Kris Freeman, "Is Alaska Groundfish Overcapitalized?" *Pacific Fishing* (March 1988): 42–48.

[782] Inshore-Offshore Analytical Team (North Pacific Fishery Management Council), *Final Supplemental Environmental Impact Statement and Regulatory Impact Review/Final Regulatory Flexibility Analysis of Proposed Inshore/Offshore Allocation Alternatives (Amendment 18/23) to the Fishery Management Plans for the Groundfish Fishery of the Bering Sea and Aleutian Islands and the Gulf of Alaska* (March 5, 1992), chap. 1, p. 17; Kris Freeman, "The Logistics of Today's Joint Ventures," *Pacific Fishing* (December 1987): 24–26.

[783] National Marine Fisheries Service, Alaska Region, *Alaska Groundfish Fisheries, Final Programmatic Supplemental Environmental Impact Statement* (June 2004), C-36, D-33.

[784] Staff, North Pacific Fishery Management Council, *Allocation of Pacific Cod Among Sectors and Apportionment of Sector Allocations between Bering Sea and Aleutian Islands Subareas, Initial Review Draft Environmental Assessment/Regulatory Impact Review/Initial Regulatory Flexibility Analysis for Proposed Amendment 85 to the Fishery Management Plan for Groundfish of the Bering Sea/Aleutian Islands Management Area* (January 13, 2006), 2–7; 63 *Fed. Reg.* 52642, October 1, 1998.

[785] Kris Freeman, "Factory Trawlers Take Off," *Pacific Fishing* (September 1987): 38–41, 48–49.

[786] "Seafood Report," *Pacific Fishing* (July 1990): 12.

[787] Daniel D. Huppert, *Managing Alaska Groundfish: Current Problems and Management Alternatives* (Seattle: Fisheries Management Foundation and Fisheries Research Institute, 1988), vi, 3.

[788] Ashley Herriman, "Cod Opera," *National Fisherman*, May 26, 2016, accessed May 26, 2016, http://www.nationalfisherman.com/blogs/mixed-catch/6685-cod-opera?utm_source=informz&utm_medium=email&utm_campaign=newsletter&utm_content=newsletter.

[789] Alan Harman, "Newfoundland cod fishery closed for two years," *National Fisherman*, September 1992, 19.

[790] Cabot Martin, "The Collapse of the Northern Cod Stocks: Whatever Happened to 86/25?" *Fisheries* (May 1995): 6–8.

[791] Staff, North Pacific Fishery Management Council, *Initial Review Draft Environmental Assessment/Regulatory Impact Review/Initial Regulatory Flexibility Analysis for Proposed Amendment 85 to the Fishery Management Plan for Groundfish of the Bering Sea/Aleutian Islands Management Area, Allocation of Pacific Cod Among Sectors and Apportionment of Sector Allocations Between Bering Sea and Aleutian Islands Subareas* (January 13, 2006), 4.

[792] Staff (Seattle, Wash.), National Marine Fisheries Service, Draft for Secretarial Review, *Environmental Assessment/Regulatory Impact Review/Initial Regulatory Flexibility Analysis of Alternatives to Allocate the Pacific Cod Total Allowable Catch by Gear and/or Directly Change the Seasonality of the Cod Fisheries, Amendment 24 to the Fishery Management Plan for the Groundfish Fishery of the Bering Sea and Aleutian Islands Area* (October 5, 1993), E-1; Scott C. Matulich, "IFQ Ownership: The Rationale for Including Processors," *Pacific Fishing* (March 1996): 47–53; At-Sea Processors Association, glossary, accessed February 20, 2015, http://www.atsea.org/glossary.php.

[793] James Strong and Keith R. Criddle, *Fishing for Pollock in a Sea of Change: A Historical Analysis of the Bering Sea Pollock Fishery* (Fairbanks, Alaska: Alaska Sea Grant, 2013), 73.

[794] Now known as the Fisheries Financing Program, the Fisheries Obligation Guarantee Program was originally created to guarantee loans made by the private sector, but it evolved into a direct-lending program.

[795] Donna Parker, "Alaska Onshore Wins in Washington," *Pacific Fishing* (May 1992): 28–30; 79 *Fed. Reg.* 36699, June 30, 2014; "Big Longliners On Their Way," *Alaska Fisherman's Journal* (December 1990): 80.

[796] Clark Miller, "Proposed Changes in the Management of Alaska's Cod Fishery Could Leave this Small, Clean-fishing Fleet Out in the Cold," *National Fisherman*, July 1994, 18–19.

[797] Brad Warren, "Pollock and Cod," *Pacific Fishing* (February 1999): 9.

[798] Chris Blackburn, "Alaska's Fisheries Grapple with Allocation Realities, *National Fisherman*, February 1986, 5.

[799] Susan Froetschnel, "W. Coast Fishermen May Form Coalition," *Sitka Sentinel* (Sitka, Alaska), January 17, 1986; National Marine Fisheries (staff, Seattle, Wash.), National Marine Fisheries Service Draft for Secretarial Review, *Environmental Assessment/Regulatory Impact Review/Initial Regulatory Flexibility Analysis of Alternatives to Allocate the Pacific Cod Total Allowable Catch by Gear and/or Directly Change the Seasonality of the Cod Fisheries, Amendment 24 to the Fishery Management Plan for the Groundfish Fishery of the Bering Sea and Aleutian Islands Area* (October 5, 1993), E-1; 61 *Fed. Reg.* 43325, August 22, 1996.

[800] National Marine Fisheries Service, Alaska Region, *Alaska Groundfish Fisheries, Final Programmatic Supplemental Environmental Impact Statement* (June 2004), C-37, C-66, C-88.

[801] 76 *Fed. Reg.* 37763, June 28, 2011.

[802] 76 *Fed. Reg.* 74670, December 1, 2011.

[803] 76 *Fed. Reg.* 11111, March 1, 2011.

[804] David Witherell, Michael Fey, and Mark Fina (staff, North Pacific Fishery Management Council), *Fishing Fleet Profiles* (April 2012), 23, 25, 45.

[805] Brad Warren, "Moratorium Points to Tough Choices," *National Fisherman*, September 1992, 6.

[806] Beth A. McGinley, "Joint Ventures in Fisheries," *Pacific Fishing* (February 1989): 34–38.

[807] North Pacific Fishery Management Council, fisheries management plan amendments: BSAI Amendment 23 and GOA Amendment 28.

[808] Ibid.

[809] Ibid; 60 *Fed. Reg.* 40763–40775, August 10, 1995.

[810] Laine Welch, "Council 'Freaks' Over ITQs," *Pacific Fishing*, Yearbook (1994): 16, 18.

[811] "Groundfish, 1995 Review," *Pacific Fishing* (March 1996): 85–86.

[812] The term *Individual Transferable Quota* (ITQ) is often used interchangeably with *Individual Fishing Quota* (IFQ).

[813] Ben Fissel et al., *NPFMC Draft Stock Assessment and Fishery Evaluation Report for the Groundfish Fisheries of the Gulf of Alaska and Bering Sea/Aleutian Islands Area: Economic Status of the Groundfish Fisheries off Alaska, 2014* (Seattle: National Marine Fisheries Service, 2015), 11, accessed December 9, 2015, http://www.afsc.noaa.gov/REFM/stocks/plan_team/economic.pdf.

[814] Lee van der Voo, *The Fish Market: Inside the Big-Money Battle for the Ocean and Your Dinner Plate* (New York: St. Martin's Press, 2016), 8.

[815] Hal Bernton, "Battle for the Deep," *Mother Jones* (July/August 1994), accessed March 10, 2015, http://www.motherjones.com/politics/1994/07/battle-deep.

[816] *Transferable Quotas Under the Magnuson Act*, Hearing Before the House Subcommittee on Fisheries Management, Serial No. 103-82, 103rd Cong., 1st sess., February 9, 1994; 72 *Fed. Reg.* 5655, February 7, 2007.

[817] Charlie Ess, "Stevens Slows ITQs," *Pacific Fishing* (March 1996): 18.

[818] Sustainable Fisheries Act, October 11, 1996 (110 Stat. 3559), §§ 106(d), 106(f).

[819] American Fisheries Act, October 21, 1998 (112 Stat. 2681-616).

[820] The inshore catcher-vessel sector comprises multiple cooperatives, each formed around a specific shore-based processor. The offshore factory trawlers formed the Pollock Conservation Cooperative, while catcher vessels that deliver pollock to the factory trawlers formed the High Seas Catcher's Cooperative. Catcher vessels that deliver to motherships formed the Mothership Fleet Cooperative.

[821] Fishermen's Collective Marketing Act of 1934, June 25, 1934 (48 Stat. 1213); National Marine Fisheries Service, Alaska Region, *Final Environmental Impact Statement for American Fisheries Act Amendments 61/61/13/8* (February 2002), 2–13.

[822] National Marine Fisheries Service, Alaska Region, *Final Environmental Impact Statement for American Fisheries Act Amendments 61/61/13/8* (February 2002), ES-9.

[823] American Fisheries Act, October 21, 1998 (112 Stat. 2681-616), § 210; National Marine Fisheries Service, Alaska Region, *Final Environmental Impact Statement for American Fisheries Act Amendments 61/61/13/8* (February 2002), ES-2; Kevin M. Bailey, *Billion-Dollar Fish: The Untold Story of Alaska Pollock* (Chicago: University of Chicago Press, 2013), 169.

[824] National Marine Fisheries Service, Alaska Region, *Final Environmental Impact Statement for American Fisheries Act Amendments 61/61/13/8* (February 2002), ch. 1, p. 4.

[825] Brad Warren, "This Is Americanization?" *Pacific Fishing* (September 1999), 5.

[826] American Fisheries Act, October 21, 1998 (112 Stat. 2681-616), § 211.

[827] NMFS, Alaska Region, "North Pacific License Limitation Program (LLP)," accessed July 5, 2015, https://alaskafisheries.noaa.gov/ram/llp.htm.

[828] Brad Warren and Joe Childers, "Sideboards Limit Crab, Cod Take By Pollock Boats," *Pacific Fishing* (August 1999): 17–18.

[829] Jim Paulin, "No More Brawls in Cod Alley," *Alaska Fisherman's Journal* (February 2003): 6–7.

[830] U.S. Regional Fishery Management Councils, *Celebrating 40 Years of Regional Fisheries Management* (2016): 34.

[831] 62 *Fed. Reg.* 34429–34438, June 26, 1997.

[832] Dave Witherell, North Pacific Fishery Management Council, email message to author, April 10, 2015.

[833] Joel Gay, "Fleet Heads Toward Full Retention," *Pacific Fishing* (June 1996): 14; Joel Gay, "Council Says Keep What You Catch," *Pacific Fishing* (November 1996): 20.

[834] Hal Bernton "Murkowski Introduces Bill to Reel In Fish Waste," *Anchorage Daily News*, November 23, 1993.

[835] Sen. Ted Stevens, *Cong. Rec.* S10933 (September 19, 1996).

[836] National Marine Fisheries Service, *Alaska Groundfish Fisheries, Final Programmatic Supplemental Environmental Impact Statement* (2004), C-69.

[837] Joel Gay, "Council Says Keep What You Catch," *Pacific Fishing* (November 1996): 20.

[838] National Marine Fisheries Service, *Alaska Groundfish Fisheries, Final Programmatic Supplemental Environmental Impact Statement* (1994), D-57.

[839] 50 CFR, Federal Fishing Regulations for Fisheries of the Exclusive Economic Zone Off Alaska, § 679.26, Prohibited Species Donation Program.

[840] 63 *Fed. Reg.* 52642–52652, October 1, 1998; National Marine Fisheries Service, Alaska Region, *Alaska Groundfish Fisheries, Final Programmatic Supplemental Environmental Impact Statement* (June 2004), C-85, D-33; *NOAA Fisheries, North Pacific License Limitation Program*, accessed May 4, 2015, http://alaskafisheries.noaa.gov/ram/llp.htm#transfer; 75 *Fed. Reg.* 43118–43136.

[841] American Fisheries Act, October 21, 1998 (112 Stat. 2681-616), § 207; National Marine Fisheries Service, *United States Plan of Action for the Management of Fishing Capacity* (2004), 20, 25, accessed January 22, 2018, http://www.nmfs.noaa.gov/op/pds/documents/01/113/01-113-01.pdf; North Pacific Fishery Management Council, *Impacts of American Fisheries Act* (February 2002), vii, accessed January 22, 2018, https://alaskafisheries.noaa.gov/sites/default/files/congress202.pdf.

[842] Wesley Loy, "Congress approves $75 million for Alaska groundfish vessel buyout," *Pacific Fishing* (January 2005): 16.

[843] "Cod freezer longliners work toward fleet buyout," *Pacific Fishing* (November 2006): 20; FY 2005 Appropriations Act, December 8, 2004 (118 Stat. 2888), § 219; Leo Irwin, National Marine Fisheries Service, ballot transmittal letter, March 21, 2007, accessed May 12, 2015, http://www.nmfs.noaa.gov/mb/financial_services/buyback_docs/03212007_Ballot_Transmittal_Letter.pdf; National Marine Fisheries Service, Longline Catcher Processor (CP) Subsector of the Bering Sea and Aleutian Islands (BSAI) "Non-Pollock Groundfish Fishery Buyback Program," accessed May 12, 2015, http://www.nmfs.noaa.gov/mb/financial_services/longline_cp_non-pollock_groundfish_buyback.html; 72 *Fed. Reg.* 20836–20837, April 26, 2007; Wesley Loy, "Fewer cod boats, new Yukon cold storage, charter halibut, and more," *Pacific Fishing* (June 2007): 21; 77 *Fed. Reg.* 58775–58776, September 24, 2012; 77 *Fed. Reg.* 68106–68107, November 15, 2012.

[844] Karly McIlwain and Jos Hill, *Catch Shares in Action: United States Bering Sea and Aleutian Islands Non-Pollock (Amendment 80) Cooperative Program*, Environmental Defense Fund (2003), 4.

[845] Northern Economics, Inc., *Five-year Review of the Effects of Amendment 80, Prepared for the North Pacific Fishery Management Council* (October 2014), ES-3, accessed October 9, 2015, https://alaskafisheries.noaa.gov/sustainablefisheries/amds/80/.

846 The other species listed in Amendment 80 are yellowfin sole (*Pleuronectes asper*), rock sole (*Lepidopsetta bilineata*), flathead sole (*Hippoglossoides elassodon*), Atka mackerel (*Pleurogrammus monopterygius*), and Pacific Ocean perch (*Sebastes alutus*).

847 Jason Anderson and Beth Concepcion, *Alaska Seafood Cooperative Report to the North Pacific Fishery Management Council for the 2013 Fishery* (March 26, 2013), 3; Ben Fissel et al., *NPFMC Draft Stock Assessment and Fishery Evaluation Report for the Groundfish Fisheries of the Gulf of Alaska and Bering Sea/Aleutian Islands Area: Economic Status of the Groundfish Fisheries off Alaska, 2014* (Seattle: National Marine Fisheries Service, 2015), 266, accessed December 9, 2015, http://www.afsc.noaa.gov/REFM/stocks/plan_team/economic.pdf; Magnuson-Stevens Fishery Management and Conservation Act, § 303A(e).

848 Groundfish Forum, *Management Programs and Industry Initiatives to Reduce Halibut Mortality Rates and Amounts in North Pacific Groundfish Fisheries*, accessed October 9, 2015, http://www.iphc.int/documents/2012bycatch/3bSwansonBonneyDownbackground1.pdf.

849 James Strong and Keith R. Criddle, *Fishing for Pollock in a Sea of Change: A Historical Analysis of the Bering Sea Pollock Fishery* (Fairbanks, Alaska: Alaska Sea Grant, 2013), 119

850 72 *Fed. Reg.* 21198, April 30, 2007; North Pacific Fishery Management Council, *Trawl Catcher/Processor Vessel Replacement (AFA and Amendment 80 vessels)*, discussion paper (October 2012), 2–3; Northern Economics, Inc., *Five-year Review of the Effects of Amendment 80, Prepared for the North Pacific Fishery Management Council* (October 2014), 19, 22, 31, accessed October 9, 2015, https://alaskafisheries.noaa.gov/sustainablefisheries/amds/80/; North Pacific Fishery Management Council, *Gulf of Alaska Trawl Bycatch Management*, discussion paper (June 2016), 95–96.

851 77 *Fed. Reg.* 13253–13256, March 6, 2012.

852 Jason Anderson, Beth Concepcion, and Mark Fina, *Alaska Seafood Cooperative Report to the North Pacific Fishery Management Council for the 2014 Fishery* (March 27, 2015), 16, accessed October 1, 2016, https://alaskafisheries.noaa.gov/sites/default/files/reports/asc14.pdf.

853 Jim Paulin, "Council to address cod conflict," *Bristol Bay Times–Dutch Harbor Fisherman*, December 15, 2017, accessed December 18, 2017, http://www.thebristolbaytimes.com/article/1750council_to_address_cod_conflict; North Pacific Fishery Management Council, *Participation and Effort in the BS Trawl CV Pacific Cod Fishery*, discussion paper (December 2017), 13–21; Jim Paulin, "Fight over America's Finest vessel part of bigger processor battle," *Alaska Journal of Commerce*, May 23, 2018, accessed May 24, 2018, http://www.alaskajournal.

com/2018-05-23/fight-over-americas-finest-vessel-part-bigger-processor-battle#.Wwb4akgvyM8.

[854] Advisory Panel, North Pacific Fishery Management Council, "Motions and Rationale," (C3 Motion), April 2–5, 2019, accessed April 9, 2019, https://meetings.npfmc.org/CommentReview/DownloadFile?p=5f08c3f4-006d-4e8b-88d1-bab52dad9f53.pdf&fileName=C3%20AP%20Motion.pdf; Peggy Parker, "North Pacific Council Restricts Motherships in Bering Sea Trawl Cod Fishery," *Seafood News* (April 5, 2019), accessed April 9, 2019, https://www.seafoodnews.com/Story/1137362/North-Pacific-Council-Restricts-Motherships-in-Bering-Sea-Trawl-Cod-Fishery.

[855] 75 *Fed. Reg.* 43118–43136, July 23, 2010; Dave Witherell, North Pacific Fishery Management Council, personal communication with author, September 15, 2016.

[856] NOAA Fisheries, Alaska Regional Office, "Permits and Licenses," accessed July 17, 2017, https://alaskafisheries.noaa.gov/permits-licenses?field_fishery_pm_value=License+Limitation+Program+%28LLP%29.

[857] "Outlook Alaska," *Alaska Fisherman's Journal: North Pacific Focus* (Fall 2016): 10.

[858] Longline Catcher Processor Subsector Single Fishery Cooperative Act, December 22, 2010 (124 Stat. 3583).

[859] Freezer Longline Coalition, "Alaska Freezer Longline Fleet Receives Congressional Approval to Form Bering Sea Fishery Cooperative," news release, December 10, 2010, accessed February 17, 2015, http://alaskalonglinecod.com/news/.

[860] Magnuson-Stevens Fishery Management and Conservation Act, § 303A(e).

[861] National Marine Fisheries Service, Alaska Region and North Pacific Fishery Management Council, *Final Regulatory Impact Review/Initial Regulatory Flexibility Analysis for Amendment 45 to the Fishery Management Plan for Bering Sea and Aleutian Islands King and Tanner Crabs Revising Freezer Longline GOA Pacific Cod Sideboards* (April 2015), 29; David Witherell, Michael Fey, and Mark Fina (staff, North Pacific Fishery Management Council) *Fishing Fleet Profiles* (April 2012), 29.

[862] Alaska Marine Conservation Council, reprinted in: North Pacific Fishery Management Council, *Gulf of Alaska Trawl Bycatch Management*, discussion paper (June 2016), 178.

[863] North Pacific Fishery Management Council, *Participation and Effort in the BS Trawl CV Pacific Cod Fishery*, discussion paper (December 2017), 13–21; D. J. Summers, "Cotten, council get earful from trawlers," *Alaska Journal of Commerce* (February 10, 2016), accessed December 16, 2016, http://www.alaskajournal.com/2016-02-10/cotten-council-get-earful-trawlers#.WFTPtvkrKM8; D. J.

Summers, "Gulf of Alaska trawlers call for four-day stand down," *Alaska Journal of Commerce* (January 28, 2016), accessed December 17, 2016, http://www.alaskajournal.com/2016-01-28/gulf-alaska-trawlers-call-four-day-stand-down#.WFXWOPkrKM8; NOAA Fisheries, Pacific cod landings database, provided by Mary Furuness (NOAA Fisheries), January 9, 2017; Jim Paulin, "Council to address cod conflict," *Bristol Bay Times–Dutch Harbor Fisherman*, December 15, 2017, accessed December 18, 2017, http://www.thebristolbaytimes.com/article/1750council_to_address_cod_conflict.

[864] NOAA Fisheries, *Central Gulf of Alaska Rockfish Program Informational Guide* (November 2015), accessed January 29, 2018, https://alaskafisheries.noaa.gov/sites/default/files/rockfish-faq.pdf; Al Burch, owner/captain of trawler *Dusk*, personal communications with author, December 2014; Cristy Fry, "IFQs for cod, pollock trawl fishery on hold," *Homer News* (Homer, Alaska), December 21, 2016, accessed December 23, 2016, http://homernews.com/homer-features/seawatch/2016-12-21/ifqs-for-cod-pollock-trawl-fishery-on-hold; Julie Bonnie, Alaska Groundfish Data Bank, to Glenn Merrill, NMFS, August 28, 2015; 76 *Fed. Reg.* 74670, December 1, 2011.

[865] Clem Tillion, personal communication with author, January 2, 2017.

[866] D. J Summers, "Council Cracks over Catch Shares," *Alaska Journal of Commerce* (January 5, 2017), accessed January 5, 2017, http://www.alaskajournal.com/2017-01-05/council-cracks-over-catch-shares#.WG9KVfkrKM8.

[867] 79 *Fed. Reg.* 9625–9642, February 20, 2014.

[868] Cristy Fry, "IFQs for cod, pollock trawl fishery on hold," *Homer News*, December 21, 2016, accessed December 29, 2016, http://homernews.com/homer-features/seawatch/2016-12-21/ifqs-for-cod-pollock-trawl-fishery-on-hold.

[869] North Pacific Fishery Management Council, *Gulf of Alaska Trawl Bycatch Management*, discussion paper (June 2016), 129; Alaska Groundfish Data Bank to Glenn Merrill, NMFS, August 28, 2015, accessed January 1, 2017, https://alaskafisheries.noaa.gov/sites/default/files/goatrawl-comments092915.pdf; D. J. Summers, "Cotten, council get earful from trawlers," *Alaska Journal of Commerce* (February 10, 2016), accessed December 17, 2016, http://www.alaskajournal.com/2016-02-10/cotten-council-get-earful-trawlers#.WFWnIPkrKM8; D. J. Summers, "Council convenes in Kodiak with Gulf catch shares in focus," *Alaska Journal of Commerce* (June 1, 2016), accessed January 1, 2017, http://www.alaskajournal.com/2016-06-01/council-convenes-kodiak-gulf-catch-shares-focus#.WGnmo_krKCg, Peggy Parker and John Sackton, "Gulf Rationalization: Two Different Points of View," *Seafood News*, June 10, 2016, accessed June 13, 2016, http://seafoodnews.com/Story/1022202/Gulf-Rationalization-Two-Different-Points-of-View.

[870] North Pacific Fishery Management Council, *GOA Trawl Bycatch Management—Preliminary Analysis*, BCY 16-010 (December 6, 2016), unpaged.

[871] North Pacific Fishery Management Council, *GOA Trawl Bycatch Management*, accessed December 17, 2016, http://www.npfmc.org/goa-trawl-bycatch-management/; "Gulf of Alaska Trawl Fishermen Cease Fishing," *Alaska Business Monthly* (January 28, 2016), accessed December 17, 2016, http://www.akbizmag.com/Fisheries/Gulf-of-Alaska-Trawl-Fishermen-Cease-Fishing; Laine Welch, "Alaska commercial fishing picks and pans for 2016," *Alaska Dispatch News*, December 30, 2016, accessed December 31, 2016, https://www.adn.com/business-economy/2016/12/30/alaska-commercial-fishing-picks-and-pans-for-2016/.

[872] In recent years, about 95 percent of Pacific cod caught by all gear groups in the Bering Sea/Aleutian Islands area is caught in the Bering Sea subarea.

[873] North Pacific Fishery Management Council, "Bering Sea Pacific cod Trawl CV Participation" (newsletter), December 14, 2017, accessed February 20, 2018, https://www.npfmc.org/bering-sea-pacific-cod-trawl-cv-participation/; North Pacific Fishery Management Council, *Participation and Effort in the BS Trawl CV Pacific Cod Fishery, discussion paper* (December 2017), 4.

[874] Jim Paulin, "Cod caught quickly," *Bristol Bay Times–Dutch Harbor Fisherman*, February 16, 2018, accessed February 20, 2018, http://www.thebristolbaytimes.com/article/1807cod_caught_quickly.

Chapter 23
ALASKANIZATION OF THE PACIFIC COD FISHERIES

The legislature shall provide for the utilization, development, and conservation of all natural resources belonging to the State, including land and waters, for the maximum benefit of its people.

—Alaska Constitution, 1959[875]

The State shall: support the development of a resident Alaskan bottomfishery by encouraging the utilization, development, conservation and rehabilitation of existing resources.

—State of Alaska, *Program for the Development of the Bottomfish Industry*, 1979[876]

While the bottomfish fishery may provide up to several thousand jobs in the long run, it is questionable whether many of these will be held by Alaskan residents.

—Institute of Social and Economic Research, University of Alaska, 1980[877]

This year, only one of the 20 U.S. processing vessels that cruised Alaska's offshore waters berthed in Alaska.

—Chris Blackburn, Alaska Groundfish Data Bank, 1986.[878]

I just come up here, kill fish, make money, go home.

—Deckhand, *Bering Sea* trawler, January 2019[879]

Now that the 200-mile EEZ off Alaska is fully "Americanized" and the entire groundfish pie sits on the table before domestic processing interests, the fight for the North Pacific whitefish has evolved from an international conflict between American and international harvesters to a civil war between at-sea and shore-based processors.

—Paul Drummond, *Alaska Fisherman's Journal*, November 1989[880]

Alaskanization—garnering benefits for the state and its residents from the groundfish fisheries—incorporated four components that were addressed separately and at different times. They are:

- allocations of fish between the shore-based (inshore) and offshore processing sectors;
- community development quota;
- state management of groundfish within state waters;
- taxes.

INSHORE/OFFSHORE ALLOCATIONS

If the Golden Alaska [a 302-foot floating pollock and Pacific cod processor] symbolizes things to come, then shore-based plants might become a thing of the past.

—Robert Mann, *Alaska Fisherman's Journal*, November 1982[881]

In the fall of 1989, the Americanization process was almost complete in the EEZ along Alaska's coast. In fact, in the Gulf of Alaska, domestic fishing and processing capacity for all groundfish species already exceeded the availability of fish. The Bering Sea was not far behind: fishing and processing capacity there for most groundfish species—including Pacific cod—exceeded the availability of fish. In 1990, there were some fifty-five factory trawlers working Alaska's coast, and about ten or fifteen more were expected the following year.[882] Floating processing ships, such as the *Golden Alaska*, were another, though lesser, concern.

Overcapacity—overcapitalization—was the essence of the problem, but there was a very real threat that the still-growing factory trawler (offshore) sector of the industry would preempt the inshore sector, which included shore-based processors as well as "stationary floating processors"—processing vessels that operated solely within Alaska State waters. Moreover, joint ventures were on the cusp of being phased out, and a lot of catcher vessels were about to lose their markets. Concerned about their future, a group of fishermen, seafood processors, and municipalities formed the Coastal Coalition and asked the NPFMC to ban factory trawlers from the Gulf of Alaska and from "shore-based fishing areas" of the Bering Sea.

In October 1989, the NPFMC recognized the "possible pre-emption of one industry component by another with the attendant social and economic disruptions."[883] The factory trawler owners—who were winning the race for fish against the inshore sector—argued that they had made solid business decisions when they elected to build their vessels, and they shouldn't now be penalized. They asserted, too, that the free market should be allowed to determine by whom fish were caught and processed, and that social and economic disruptions were simply a cost of doing business in a free-market economy. And they portrayed inshore processing preferences as "a social welfare issue for western Alaska."[884]

Shore-based interests countered that they could more fully utilize fish that were caught because it was relatively easy for a shore-based plant to add equipment when an opportunity arose. They also had the law on their side: the Magnuson-Stevens Act required management plans produced by the regional fishery management councils to "take into account the social and economic needs of the States." Regarding the economic well-being of the State of Alaska, fish landed at shore-based plants and floating processors in state waters would be subject to the state's fisheries business tax, which at the time was 1 percent of the ex-vessel value on species such as pollock and Pacific cod. Additionally—and very importantly—Alaska's senior senator, Ted Stevens, apparently favored special consideration for inshore operations.[885]

A shortcoming of shore-based fish-processing plants is that heavy fishing in nearby waters has the potential to deplete local stocks of fish, forcing fishermen to travel farther afield to find economically viable fishing. Especially in this situation, the fish caught might be held aboard fishing vessels for several days before being delivered to a processing plant. Fish are usually held in refrigerated seawater—ideally just above freezing—but deterioration of the flesh and other parts still occurs and can lower the quality of the final products. This is in contrast to catcher-processors and motherships, which can move with the fish and typically process them as they are caught.

An article in *Pacific Fishing* in March 1990 spoke of an ongoing "civil war" over inshore/offshore allocations that pitted fishermen against fishermen, processors against processors, and even state against state.[886]

Nevertheless, in December 1991, the North Pacific Fishery Management Council approved GOA Amendment 23, which allocated 90 percent of the Pacific cod catch in the Gulf of Alaska to the inshore sector and 10 percent to the offshore sector. The Department of Commerce, apparently under pressure from Senator Stevens, approved the amendment as an interim measure to maintain the status quo until a long-term solution to over-capitalization could be devised.[887] Incorporating a three-year sunset provision, Amendment 23 became effective in June 1992. It was subsequently extended twice and became permanent in 2002, when GOA Amendment 62 removed the sunset provision.[888]

The council addressed inshore/offshore allocations of pollock in the Bering Sea/Aleutian Islands area (BSAI Amendment 18, implemented in 1992), but it has not addressed Pacific cod allocations in that area.

During the three-year period 2015–2017, the inshore sector in the

Bering Sea/Aleutian Islands area made capital improvements totaling $175 million. A portion of this investment was focused on expanding Pacific cod processing capacity, especially for the production of fillets.[889]

> *Inshore/offshore was a big one. . . . Otherwise Alaska would have gotten nothing. We'd have been a distant-water fishery.*
>
> —Clem Tillion, 2009[890]

COMMUNITY DEVELOPMENT QUOTA PROGRAM

> *When it was created in [1991], Alaska's Community Development Quota Program stood as one of the most revolutionary fishery management initiatives in the country. In short, it set aside a lucrative share of the enormous Bering Sea commercial fisheries for the benefit of disadvantaged villages along the state's western coastline.*
>
> —Wesley Loy, *Pacific Fishing*, September 2006.[891]

> *The motivation and means for placing a portion of the pollock under the control of the people of western Alaska were purely political, a matter of vote bargaining in the broader debate on permanent allocation of the billion-dollar fishery. For better or worse, the council, congressional delegation, state officials and interested parties are slowly carving out a complicated deal to transfer ownership of the resource to the private sector. Without CDQs, the eventual privatization of the groundfisheries, in all likelihood, would have been slowed or sunk by congressional pressure or lengthy judicial challenges.*
>
> —Brad Matsen, *National Fisherman*, March 1993[892]

A component of the Alaskanization of the federal fisheries off Alaska—to the extent it has occurred—is the Harold Sparck Memorial Bering Sea Community Development Quota Program, which allocates a percentage of the total allowable catch of all the commercial fisheries in the U.S. Bering Sea Exclusive Economic Zone to sixty-five villages in western Alaska. Village eligibility is based on proximity to the coast and historic involvement in Bering Sea/Aleutian Islands fisheries.

Not linked by roads and isolated from Anchorage, Alaska's major population center, these villages are some of the nation's most geographically remote communities. They were also economically depressed; in 1990, a fourth of the region's population of about twenty-three thousand was living below the poverty level, a situation exacerbated by substandard infrastructure and the high cost of living. The price of gasoline, for example, could be triple the national average. Unemployment in some locations exceeded 40 percent. Meanwhile, along the same coast are some of the most valuable fisheries in the world.[893]

Harold Sparck is considered the father of the Alaska community development quota (CDQ) program. Sparck was a native of Baltimore, Maryland, and, upon the recommendation of Alaska senator Ernest Gruening, came to Alaska in 1968. He became a longtime resident of Bethel, on the Kuskokwim River, in Southwest Alaska. In 1973, Sparck founded Nunam Kitlutsisti, ("protectors of the land," in Yu'pik), a nonprofit organization that represented some fifty-six Native villages in the Yukon-Kuskokwim Delta.

The people of the delta relied heavily on salmon, and Sparck was part of a successful effort during the 1980s to eliminate Japanese high-seas salmon gillnetting in the Bering Sea. About this time, Sparck began advocating for a program to provide residents of the villages along the Bering Sea coast an opportunity to participate in the Bering Sea's offshore fisheries, which were dominated by Seattle corporate interests.[894] Of concern, too, were the Aleut people on the Pribilof Islands. The harvest of fur seals (*Callorhinus ursinus*) in the Pribilofs had since the Russian occupancy been the islands' economic mainstay, but it was terminated in 1985, when the United States refused to ratify an extension of the international treaty that governed the harvest of fur seals in the North Pacific. Providing the people of the Pribilofs with an opportunity to participate in the Bering Sea groundfish fisheries would help mitigate the economic damage that the loss of the fur seal industry had caused.

Sparck had an important ally: Henry Mitchell, who was an attorney and the director of the Bering Sea Fishermen's Association (in which Sparck was very active), and who represented western Alaska's interests on the North Pacific Fishery Management Council.

It was Sparck and Mitchell's good fortune to have the support of Alaska senator Ted Stevens, who in 1989 introduced legislation that incorporated a CDQ program for Bering Sea villages. To no one's surprise, the Seattle-based fishing companies that operated in the Bering Sea opposed Stevens's proposed legislation. And the companies had the support of their representatives in Congress. Washington senator Slade Gorton derided the proposed CDQ program as "an entitlement program that will be paid for largely by the Washington fishing industry." Stevens's CDQ proposal failed to become law.[895]

Enter, in 1990, Alaska governor Walter Hickel. Years earlier, Hickel had stood one night with several Native leaders on a beach on St. Lawrence Island, in the Bering Sea. The lights of foreign fishing vessels filled the horizon. "They fish all the time, and we don't even know what's there," said one of Hickel's companions.[896] In December 1990, Hickel began his second term as Alaska's governor, and shortly thereafter, he gave the aforementioned Clem Tillion, his "Fish Czar" (see chapter 22), marching orders. The governor, who owned Anchorage's Hotel Captain Cook, where the dining room served fresh Atlantic cod imported from Canada, told Tillion that he wanted his hotel to be able to serve fresh Alaska fish year-round. And he wanted Tillion to do something to help the Native villagers along Alaska's Bering Sea coast.[897] Tillion, who had been chairman of the North Pacific Fishery Management Council from October 1978 through August 1983, used his influence with the council, and in 1991, as part of a regulatory package (Amendment 18) that divided the Bering Sea pollock fishery between the offshore and on-shore sectors, the council created the Western Alaska Community Development Quota (CDQ) program. Henry Mitchell wrote the CDQ provision. Under this program, 7.5 percent of the annual total allowable catch of Bering Sea/Aleutian Islands area pollock was allocated to western Alaska communities. The council's allocation was implemented in the spring of 1992.

An article in *Pacific Fishing* characterized the CDQ program as a "bold new allocation scheme" that was "nothing ever seen in American fishing."[898] Chronically poor communities along the Bering Sea coast would now have a slice of a burgeoning industry that had developed, literally, on their doorstep. Fifty-six eligible communities—all required to be within fifty miles of the Bering Sea coastline—were organized into six CDQ groups that divided an allocation of 101,000 metric tons of pollock among themselves. The groups were then able to lease their individual allocations to existing players in the industry, with the caveat that the money received from the leases be used, in the words of a National Academy of Sciences study, "to provide the participating communities with the means to develop ongoing commercial fishing activities, create employment opportunities, attract capital, develop infrastructure, and generally promote positive social and economic conditions."[899]

The CDQ program was originally scheduled to sunset in 1995, but it was institutionalized by Congress in 1996 in the Sustainable Fisheries Act, which reauthorized the Magnuson-Stevens Act. In 1998, the American Fisheries Act increased the CDQ allocation of Bering Sea/Aleutian Islands pollock to 10 percent.[900]

Pacific cod, along with a number of other species of groundfish, became part of the CDQ program in 1997 as part of the North Pacific Fishery Management Council's Amendment 39 to the Bering Sea and Aleutian Islands Fisheries Management Plan, which was approved by the National Marine Fisheries Service in 1998. Under Amendment 39, the CDQ groups received an off-the-top *reserve* allocation of 7.5 percent of the Bering Sea/Aleutian Islands area total allowable catch (TAC) of Pacific cod and other groundfish species. In 2007, Amendment 85 to the plan increased the Pacific cod reserve to 10 percent of the total allowable catch for the species.[901]

Under the Magnuson-Stevens Act, the CDQ program is considered a limited-access privilege program. As such, the CDQ groups are subject to cost-recovery fees of up to 3 percent of the ex-vessel value of their catch to cover the federal government's costs of management, data collection and analysis, and enforcement.[902]

Over the years, the groups, which now represent about twenty-seven thousand people in sixty-five communities, have acquired ownership throughout the commercial fishing industry and have become among the largest private-sector employers in western Alaska. Today, companies that are fully or partially owned by CDQ groups harvest and process more than three-fourths of CDQ fish allocations. For example, the CDQ group Coastal Villages Region Fund owns a number of vessels. Among them are three freezer-longliners: *Lilli Ann*, *North Pacific*, and *North Cape*.[903] Another CDQ group, the Bristol Bay Economic Development Corporation, is part owner of Alaskan Leader Seafoods, which operates four freezer-longliners: *Alaskan Leader*, *Bering Leader*, *Bristol Leader*, and *Northern Leader*. And the Aleutian Pribilof Island Community Development Association, another CDQ group, is part owner of Prowler Fisheries, which operates five freezer-longliners: *Arctic Prowler*, *Prowler*, *Bering Prowler*, *Gulf Prowler*, and *Ocean Prowler* (see chapter 17). Adding to this incomplete list, the CDQ group Norton Sound Economic Development Corporation is part owner of the Glacier Fish Company, which operates three factory trawlers: *Alaska Ocean*, *Pacific Glacier*, and *Northern Glacier*. The trawlers catch and process a variety of groundfish species, Pacific cod among them.

By 2003, the six CDQ groups had combined assets estimated at $227 million. In 2013, the CDQ groups reported gross revenues of $318 million, and their combined net assets totaled $899 million.[904]

State-waters Pacific cod fisheries

> *The Board's goal was to provide a slower-paced, low-bycatch fishery at a time of year more adaptable to the nearshore, small-boat fleets.*
>
> —Alaska Department of Fish and Game, 1999[905]

> *State-waters Pacific cod fisheries are open access fisheries and are prosecuted independent of the federal/parallel fisheries.*
>
> —Alaska Department of Fish and Game, 2011[906]

> *Cod is one of the top-three pillars that resident fishermen build their success upon—alongside salmon and halibut. This species feeds a range of user groups and communities spanning the coastline, offering important winter opportunities and bolstering the statewide economy.*
>
> —Trever Shaishnikoff, Unalaska, Alaska, 2018[907]

Two distinct Pacific cod fisheries occur within designated Alaska state waters of both the Gulf of Alaska and in the Bering Sea/Aleutian Islands. The first of these are *parallel fisheries*, which mirror the season, gear, and bycatch limits in the federal season in adjacent federal waters (EEZ) unless superseded by other Alaska Board of Fisheries regulations.

The second fisheries are *state-waters fisheries*, which are independent of the parallel fisheries and are managed exclusively by the Alaska Department of Fish and Game under guidelines developed by the Alaska Board of Fisheries. Typically, state-waters fisheries are open when federal and parallel fisheries are closed. Pots and jigging machines are legal in most state-waters Pacific cod fisheries. Longline gear is permitted in Prince William Sound and Southeast Alaska. All state-waters Pacific cod fisheries are open access, and collectively they afford fishermen the most extensive open-access opportunities in the state. To foster participation by Alaska fishermen, the overall vessel length in some state-waters Pacific cod fisheries is limited to 58 feet, the maximum length of vessels permitted to seine for salmon in Alaska.[908]

Southeast Alaska. Southeast Alaska is a bit player in Alaska's Pacific cod fishery, accounting for about one-third of 1 percent of the annual Alaska catch. The Pacific cod fishery in Southeast Alaska's internal waters is an entirely state-waters fishery that is managed by the Alaska Department of Fish and Game. In state waters (zero to three nautical miles from the shore) along Southeast Alaska's Gulf of Alaska coast (east of 144° west longitude, approximately Cape Suckling, near

Cordova), the Alaska Department of Fish and Game's management of the Pacific cod fishery parallels that of the National Marine Fisheries Service in the adjacent eastern Gulf of Alaska waters. Fishing with trawl gear, however, is prohibited in these state waters.

Pacific cod in Southeast Alaska's internal waters are primarily caught on longline gear. Fishing with trawl gear is prohibited, but jig gear, dinglebar troll gear, and hand troll gear are legal. Vessel length is not restricted. In 1993, the Alaska Board of Fisheries established a Pacific cod guideline harvest range of 340 to 570 metric tons (750,000 to 1,257,000 pounds) in the internal waters of Southeast Alaska, based on average historic catch levels. The Alaska Department of Fish and Game generally sets the annual guideline harvest level at the midpoint of the range. That level, however, has never been reached. The fishery is open year-round but is subject to closure by emergency order if the guideline harvest level is reached. Most of the Pacific cod harvest occurs during the winter months in the northern waters of the region, particularly in Icy Strait and Frederick Sound. During the years 1997 through 2018, the Pacific cod catch (including discards at sea) in Southeast Alaska averaged 592,000 pounds. The high year was 2015, when 970,000 pounds were caught; the low year was 2018, when 275,000 pounds were caught.

In the Pacific cod parallel fishery along Southeast Alaska's Gulf of Alaska coast, Pacific cod may be taken by longline, dinglebar troll gear, hand troll gear, jigging machines, and pots. The harvest limit in the parallel fishery is incorporated in the National Marine Fisheries Service's total allowable catch for the Eastern Gulf of Alaska Area. Theoretically, the entire Eastern Gulf of Alaska Area catch, except that allocated to the Prince William Sound fishery, could be made in state waters.[909]

Historically, much of the Pacific cod harvested in Southeast Alaska was utilized as bait in fisheries for crabs and halibut. In recent years, however, the Pacific cod harvest has been largely sold for human consumption. The retailer Costco seasonally sells fresh codfish fillets at its Juneau store.

Gulf of Alaska and associated waters. Historically, up to 21 percent of the total allowable catch of Pacific cod in the central and western Gulf of Alaska had been taken in state-waters parallel fisheries.[910] Small-boat fishermen in communities such as Kodiak, Homer, Seldovia, and Chignik, however, had little access to the fishery because the Pacific cod parallel season began in January, and the typically harsh weather during this time of year generally precluded small vessels from participating in the fishery. Usually by mid-March, when the weather was starting to improve, the entire quota had been caught or bycatch limits had been reached.

The fishermen lobbied the North Pacific Fishery Management Council for a separate allocation or to delay the start of the season, but to no avail. Not so with the Alaska Board of Fisheries. Essentially asserting state's rights, in October 1996 the board adopted Pacific cod management plans for five state-waters areas: Prince William Sound, Cook Inlet, Kodiak, Chignik, and South Alaska Peninsula. The board's goal was to provide new entrants—particularly small-scale fishermen from the local coastal communities—access to the groundfish fishery, so participation was not restricted to vessels qualified under the federal moratorium program (see chapter 22).

The board's fishery management plans established fishing seasons and guideline harvest levels (GHLs), and they limited the gear that could be used to pots, jigging machines, and hand troll—gear types that experienced low levels of bycatch. Vessels were limited to sixty pots or five jigging machines. In the Chignik and South Alaska Peninsula areas, vessel length was limited to 58 feet overall. In the Kodiak, Cook Inlet, and Prince William Sound areas, there was no vessel size restriction, but in the Kodiak and Cook Inlet areas vessels within the 58-foot limit received an exclusive allocation of the harvest quota. Moreover, beginning in 1999, vessels more than 58 feet long fishing pot gear in these areas were limited to landing 25 percent of the state-waters guideline harvest level prior to September 1.

To ensure separation of the fisheries and facilitate accurate seasonal catch accounting, the state-waters fishing seasons were scheduled to begin within a week or two after the adjacent federal season closed. (Season opening dates have subsequently been modified numerous times and can now be as little as twenty-four hours after the closure of the adjacent federal season.) Annual guideline harvest levels were a percentage of the allowable biological catch (ABC) in adjacent federal waters. The initial guideline harvest levels reserved 25 percent of the federal eastern Gulf of Alaska ABC for the Prince William Sound Area and 15 percent of the federal western Gulf of Alaska ABC for the South Alaska Peninsula Area. Fifteen percent of the federal central Gulf of Alaska ABC was apportioned among the Kodiak, Chignik, and Cook Inlet areas. To accommodate potential growth in the fishery, the guideline harvest levels were structured to increase incrementally to 20 percent and then to 25 percent of federal ABCs if the Pacific cod harvest in any year was within 10 percent of the established state-waters GHL. The fishery management plans also removed restrictions

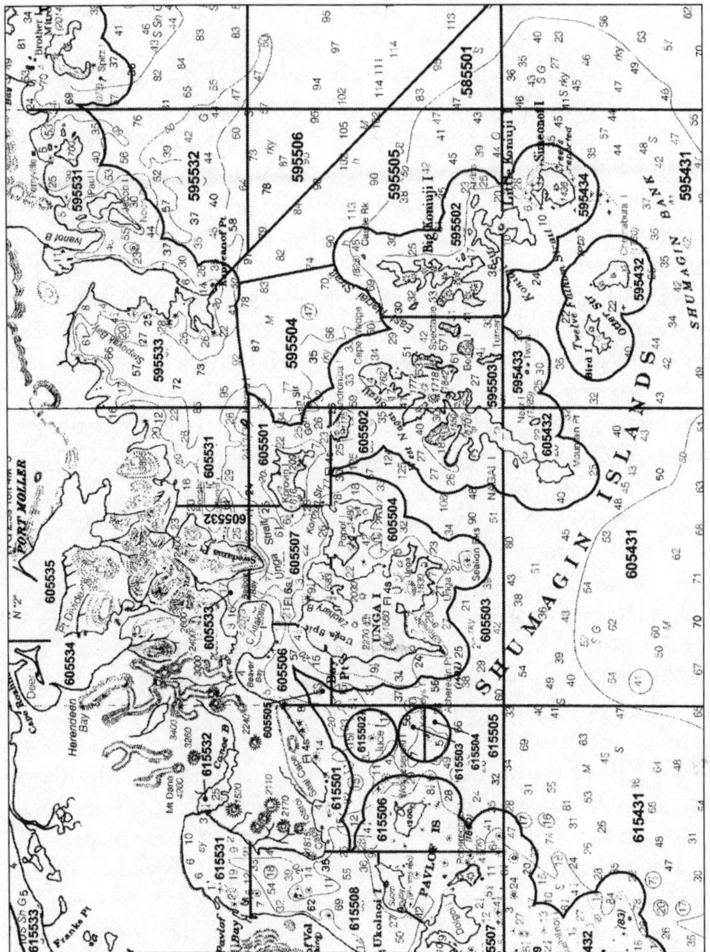

Figure 86. Alaska Department of Fish and Game state-waters groundfish/crab statistical areas, Shumagin Islands, 2001. (Section of: Alaska Department of Fish and Game, *Groundfish/Shellfish Statistical Areas, Chart 9, Alaska Peninsula and Chignik* [2001].)

on vessel size and gear limits after October 31 to increase late-season production. Because these plans reallocated Pacific cod from the EEZ to the state's inshore fishery, they were opposed—to no avail—by other gear groups and even by some federal managers.[911]

The state-waters fisheries were implemented in the spring of 1997, and a considerable number of fishermen took advantage of their new cod-fishing option. Many were salmon fishermen who wanted to augment their income by fishing for cod before and after the salmon fishing season. "It's an option the state has given us, and it's a great one," said one Kodiak fisherman who had installed jigging machines on his salmon seiner. By June 25, 1997, the thirty-four-vessel "pot cod" fleet fishing in the Kodiak Area had landed 4.6 million pounds, a little more than half the guideline harvest level for the area. The ex-vessel price at Kodiak ranged from $0.21 to $0.23 per pound, with one company paying $0.30 per pound for cod delivered the same day they were caught. In the South Alaska Peninsula Area, thirty-two pot boats and nineteen jig boats had delivered more than two-thirds of their 9.4 million pound quota.[912]

In 1999, because the state-waters fleet had shown its ability to harvest Pacific cod, the Board of Fisheries increased the quotas for the Kodiak, Chignik, and Cook Inlet areas to 20 percent of the allowable biological catch in adjacent federal waters. The state-waters cod-fishing effort that year was described as having "exploded." The fishing effort was so great that there was a concern in the industry that limiting entry might become necessary, and vessel owners and processors scrambled to establish records of participation in the fishery.[913] In 1999, to provide a measure of order to the fisheries, the Alaska Board of Fisheries designated the state-waters Pacific cod fisheries at Chignik as a *superexclusive registration area*—a vessel registered to fish for Pacific cod at Chignik is prohibited from harvesting Pacific cod anywhere else in Alaska state waters during the same year. At the same time, the Board of Fisheries designated the Kodiak, South Alaska Peninsula, Cook Inlet, and Prince William Sound areas as *exclusive registration areas*. A vessel registered to fish for Pacific cod in any of these areas may not be used to harvest Pacific cod in any other exclusive or superexclusive state-waters Pacific cod management area during the same year.[914] Although certain licenses are required, there are currently no limits to entry into any of Alaska's state-waters groundfish fisheries.

By 2003, each of the central and western Gulf of Alaska state-waters management areas had reached the maximum GHL increment, and the state-waters South Alaska Peninsula Area GHL was increased to

25 percent of the federal western Gulf of Alaska ABC. Likewise, the combined GHL for the Kodiak, Chignik, and Cook Inlet areas was increased to 25 percent of the federal central Gulf of Alaska ABC. Reflecting the distribution of the catch history and the anticipated fishing effort, the state-waters GHL there was apportioned among the areas as follows: Kodiak, 50 percent; Chignik, 35 percent; Cook Inlet, 15 percent.

In 2014, Alaska's Board of Fisheries increased the South Alaska Peninsula Area GHL to 30 percent of the federal western Gulf of Alaska ABC. During the state-waters season that year, 185 vessels harvested approximately 44 million pounds of Pacific cod—with an ex-vessel value of approximately $13.6 million—in the Kodiak, Chignik, and South Alaska Peninsula areas. The combined state-waters and parallel fisheries in 2014 accounted for approximately 35 percent of the federal central Gulf of Alaska total harvest and 51 percent of the federal western Gulf of Alaska total harvest.[915]

During the years 1997 through 2000, landings in the state-waters Pacific cod fishery in Prince William Sound ranged from 200,520 pounds to 418,994 pounds, with from nine to twelve vessels annually participating in the fishery. Pot gear accounted for 89 percent of the catch. The fishing effort then fell precipitously: the only vessels that fished for Pacific cod in Prince William Sound in 2001 were three jig vessels, which had combined landings of just 228 pounds. There were no Pacific cod landings in the sound in 2002, and the following year the Board of Fisheries reduced the GHL to 10 percent of the federal eastern Gulf of Alaska ABC, with the provision for the allocation to be increased to 15 percent and then 25 percent following years the GHL was reached. The fishing effort continued to languish until 2009, when the Board of Fisheries legalized longline gear in the sound. Subsequently, Pacific cod fishermen there reached the GHL and their allocation was incrementally increased to the 25 percent maximum. In Prince William Sound in 2013, twenty-five vessels—every one of them longliners—landed 1.3 million pounds of Pacific cod.[916]

Super 8 boats

> *Extra wide 58' pot cod boats are able to land large volumes, rivaling the 120' crabbers.*
>
> —*Seafood News*, May 2017[917]

The requirement that vessels fishing for Pacific cod in the Dutch Harbor Subdistrict and the Chignik and South Peninsula areas must be no longer than 58 feet (the maximum length for salmon seiners in Alaska), and that the catch of Pacific cod by vessels longer than 58 feet in the Kodiak and Cook Inlet areas is capped, spurred owners of aging seiners in the fisheries to replace their vessels with larger, more capable vessels that still complied with the 58-foot length limitation. A typical modern 58-foot salmon seiner—commonly referred to as a *limit seiner*—has a beam (width) of about 20 feet and a draft (hull depth below the waterline) of 10 feet. Predecessors to *Super 8s*, as they are called, were three 58-foot "hefty" longliners built by the Hansen Boat Company, in Marysville, Washington, in the late 1980s. The *Celtic*, *Ryan D. Kapp*, and *Tricia Marie* had 22-foot beams, drafts of 10.5 feet, and had hold capacities of 2,350 cubic feet. The *Alaska Fisherman's Journal* characterized the design as being "about as much boat as you can build within the seine-boat limit and still remain good looking."[918]

Good looking or not, Super 8s are considerably larger than these vessels. They have beams and drafts that are "super-sized" relative to the vessels' lengths. An average Super 8's beam is about 26 feet and its draft is about 12.5 feet. Considerably more fishing gear can be carried on the wide aft deck of a Super 8 than can be carried on a traditional limit seiner's aft deck. Additionally, Super 8s' holds average about 3,500 cubic feet—enough room for about 175,000 pounds of Pacific cod in refrigerated seawater, considerably more than the approximately 120,000 pounds that can be carried aboard a typical modern limit seiner. Taken together, these attributes enable Super 8s to catch, hold, and deliver more fish than their counterparts. They are also very stable working platforms and tend to have roomy crew accommodations.

A downside of Super 8s is that they consume more fuel than traditionally configured vessels of similar capacity. A typical Super 8 has a main engine of approximately 625 horsepower, but Super 8s configured to enable trawling may have engines of twice that size.

Most Super 8s were built after 2006, and several others are under construction.[919]

Central and Western Aleutian Islands. In March 2000, the Alaska Board of Fisheries established a state-waters Pacific cod and rockfish fishery in the vicinity of Adak, in the Aleutian Islands. The purpose of the board's action was to foster the establishment of fish-processing facilities at the former U.S. Navy air station at Adak. The Navy had abandoned the station in 1997, and it had come under the ownership of the Aleut Corporation, a for-profit Native corporation created in 1971 under the Alaska Native Claims Settlement Act. Basically, the state-waters area—officially the Bering Sea/Aleutian Islands area—

Figure 87. Unalaska-based Super 8 vessel *Commitment*, with cod pots. (Courtesy of Tammy Rowland.)

stretched about seventy miles along the Aleutian Chain, from the west shore of Adak Island east to Tagalak Island.[920] In its Adak venture, the Aleut Corporation was represented by Clem Tillion, who envisioned Adak becoming, in the words of *Pacific Fishing*, "a colony of resident small-boat fishermen."[921]

Within the Bering Sea/Aleutian Islands area, only vessels less than 60 feet in overall length could be used to take Pacific cod from May 1 to September 15. Legal gear types within this area were pot, longline, mechanical jig, and hand troll. There were no restrictions on the amount of pot, longline, or hand troll gear, but a vessel could operate no more than five mechanical jigging machines. Larger vessels, including trawlers, could be used prior to May 1 and after September 15.[922] The Bering Sea/Aleutian Islands area is managed as a *nonexclusive* registration area. Vessels registered to fish in this area may be used to harvest Pacific cod in one other exclusive or superexclusive state-waters Pacific cod management area during the same year.[923]

As had happened with the 58-foot vessel length limit in other areas, the 60-foot vessel length limit spawned the construction of vessels that are, by traditional standards, disproportionately large for their length. An example is the *Shemya*, which was launched as a freezer-longliner in January 2001. The vessel was built for Alaskan Leader Fisheries, which at that time, in partnership with the CDQ group Bristol Bay Economic Development Corporation, operated two full-size freezer-longliners.

The reason for building the *Shemya*, according to Nick Delaney, one of the partners in Alaskan Leader Fisheries, was that "we liked the thought of the challenge of the smaller vessel." Indeed, the *Shemya* was probably the smallest freezer-longliner ever constructed.[924]

The *Shemya*'s steel hull measures 58 feet, 5 inches long, has a 28½-foot beam (almost half its length), and is 14 feet deep. It was powered by a 615-horsepower Caterpillar diesel engine, carried a crew of ten, and was capable of fishing about 40,000 hooks and freezing 36,000 pounds of H&G Pacific cod daily. The hold's capacity was 172,000 pounds of frozen Pacific cod. As the *Shemya*'s builder, Fred Wahl (Fred Wahl Marine Construction), of Reedsport, Oregon, said, "Every square inch of this boat has been utilized." The *Shemya* fished in both the Bering Sea and the Gulf of Alaska and, in addition to fishing for Pacific cod, fished for halibut and sablefish.[925] In 2006, however, the vessel's processing and freezing equipment was removed. The *Shemya* continued for several years to longline in Alaska as a catcher vessel but subsequently left the state.

Despite the state's efforts, fish processing at Adak has had a short and checkered history. In 1998, the year after the U.S. Navy air station there was closed, Kjetil Solberg, an entrepreneur who had strong ties to the Norwegian fishing industry, founded Adak Fisheries and installed seafood processing equipment in the Blue Shed, a former Navy cold storage building. The company began processing cod in January 1999. The fish, however, wasn't frozen; it was salted for the Portuguese market. By September of that year, the Adak Fisheries had purchased and salted some 6.5 million pounds of cod and had frozen nearly a million pounds of halibut and sablefish. The company's success and Adak's potential attracted the interest of some of the major fish processors that operated in Alaska.[926]

NorQuest Seafoods, an established seafood processing company that operated several shore plants and floating processors in Alaska, was at that time interested in expanding into the groundfish industry and recognized the groundfish potential and the other opportunities that Adak presented. In June 2000, NorQuest became the majority partner in Adak Fisheries, which was renamed NorQuest-Adak, Inc. Solberg managed the new firm.[927] Over the course of the following year, however, the venture lost about $2.5 million, and NorQuest pulled out of the operation in the early summer of 2001.

Solberg soon after formed the Adak Fisheries Development Company and, while courting potential new partners, began operating on his own. In December 2001, Icicle Seafoods, one of the first processors of Pacific cod after the passage of the Magnuson-Stevens Act in 1976 (see chapter 13), became Solberg's partner at Adak. Icicle's primary interests at Adak were cod, halibut, and sablefish, but the company was repeatedly advised of the income potential in processing brown king crab and decided to engage in that fishery.[928]

All was not aboveboard at Adak. Between February 2002 and February 2004, Adak Fisheries processed a little more than 4 million pounds of western Aleutian Islands brown king crab—almost all of which was in excess of Icicle Seafoods' processing cap established under the American Fisheries Act.[929] Added to this, several cod fishermen at Adak complained to the Alaska Department of Environmental Conservation, the state agency responsible for ensuring the wholesome processing of seafood, of disturbing conditions at the Adak Fisheries plant. A state seafood inspector visited the plant in March 2004 and noted a number of sanitation violations and other problems. After testing showed spoilage, the state detained 30,000 to 40,000 pounds of brined Pacific cod for further testing.

Icicle Seafoods pulled out of Adak in June 2004. The company was later fined $3.44 million for its violation of the American Fisheries Act, but the fine was subsequently negotiated down to $615,000.[930]

Pacific cod production in the state-waters fishery at Adak during the years 2001–2004 totaled 27.2 million pounds. Each year, the majority was caught between January 1 and May 15, when the vessel length/gear restrictions were not in effect. Trawl gear was responsible for fully 99 percent of the catch.[931]

In February 2006, to further increase economic opportunities at Adak, the Alaska Board of Fisheries expanded the state-waters Pacific cod fishery in the Aleutian Islands. The new area, the Aleutian Islands District, extended approximately six hundred miles along the arc of the Aleutian Islands, from 170° west longitude (just east of Atka Island) to Attu Island, the westernmost of the Aleutian Islands. The fishery's management plan was adopted by emergency regulation, and fishing commenced on March 15 of that year.[932]

For the 2006 season, regulations permitted bottom trawl, longline, jig, and pot gear. Except for under-60-foot vessel length limitation zones near Adak Island, there were no vessel length limits, and the fishery was open to anyone who chose to participate. To facilitate efficient processing of the catch, each vessel was limited to 150,000 pounds per day and 300,000 pounds per trip. The guideline harvest level (GHL) for Pacific cod within these waters was set at 12.8 million pounds.[933] A unique characteristic of the Aleutian Islands district is that the GHL is not allocated among gear groups.

In 2007, the Board of Fisheries modified the Aleutian Islands

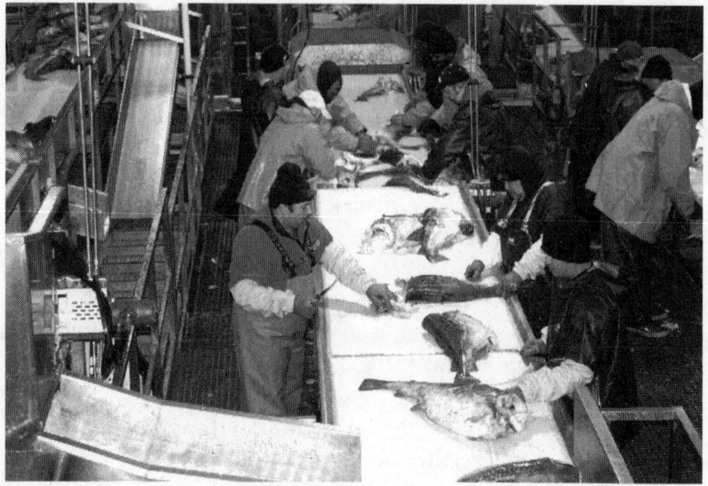

Figure 88. Processing Pacific cod at Adak, circa 2007. (Courtesy of David Fraser.)

District's regulations to limit the overall length of vessels. Pot vessels were limited to less than 125 feet, trawl vessels to less than 100 feet, and longline and jig vessels to less than 58 feet. In 2009, the board established a less-than-60-foot length limit on all gear types during the B season, but the following year allowed pot vessels less than 125 feet to participate in the B season after August 1. In certain circumstances, other vessel length limitations apply.[934]

With various partners, Solberg had continued running the Adak plant. Although there were years in which Adak Fisheries processed more than 20 million pounds of cod, Solberg had financial problems, and in 2009 he shut down the plant and declared bankruptcy. The plant lay idle until it was taken over in 2011 by its former occupant, Icicle Seafoods, which began operating it that year. Icicle Seafoods, however, ceased operations at Adak in April 2013, citing concerns about the short- and long-term health of the region's cod resource and "regulatory uncertainty."[935]

Two months later, the fish-processing equipment at the Adak plant was sold at auction to the City of Adak for $2 million. In October, the city leased the plant and its equipment to the Adak Cod Cooperative, a firm Bristol Bay salmon canneryman John Lowrance had established.[936]

In February 2014, the cooperative began processing Pacific cod in the Blue Shed processing plant. Among the cooperative's products were individual cod fillets, the first produced at the plant. By early May, however, the cooperative shuttered the plant, due in part to competition from American Seafoods, which had sent a processing ship to the area to take deliveries from a handful of catcher vessels. Lowrance left the plant and its equipment in good repair.[937]

In January 2015, the Adak plant lease was assigned to Premier Harvest Seafood. The company, however, did not process any fish. The Adak plant's prospects took a turn for the better in October of that year, when the North Pacific Fishery Management Council provisionally allocated 5,000 metric tons of Aleutian Islands area Pacific cod to trawl catcher vessels for delivery to shore plants west of 170° west longitude. One of the provisions in the allocation was that the fish had to be delivered prior to March 15. The council's intention was to mitigate the impact of the Pacific cod rationalization programs, particularly the Amendment 80 program (see chapter 22), which allowed an influx of processing capacity into the Aleutian Islands Pacific cod fishery, and "to provide stability to Aleutian Islands (AI) shoreplant operations and the communities dependent on shoreside processing activity." Only two shore plants, the Adak plant and the Atka Pride Seafoods plant, at Atka, met the allocation's geographic criterion, and only the Adak plant had a history of processing Aleutian Islands Pacific cod.[938]

The allocation, Amendment 113 to the Bering Sea/Aleutian Islands Groundfish Fishery Management Plan, was implemented in November 2016.[939] Despite this allocation, no cod were processed at the Adak plant in either 2016 or 2017, and the lease held by Premier Harvest Seafood was transferred to Golden Harvest Alaska Seafood, which began processing Pacific cod at the plant in early 2018.[940]

But the Pacific cod allocation upon which Adak depended was not on a sound legal footing. In December 2016, just a month after Amendment 113 was implemented, several fishing organizations that were reliant on Pacific cod from the Bering Sea had challenged it in the U.S. District Court for the District of Columbia. It took more than two years, but in March 2019, Judge Timothy Kelly determined that the National Marine Fisheries Service's approval of the amendment had violated the federal Administrative Procedures Act. Kelly stated that the court would vacate the rule that implemented Amendment 113 and remand the amendment itself to the National Marine Fisheries Service for reconsideration.[941]

Dutch Harbor Subdistrict. In October 2013, the Alaska Board of Fisheries created an open-access state-waters Pacific cod fishery in the eastern Aleutian Islands. The Dutch Harbor Subdistrict comprises state waters in the Bering Sea from just east of the midsection of Unimak Island to about the midsection of Unalaska Island.[942] The only gear permitted in this fishery is pot gear, with a limit of sixty pots per vessel.

Only vessels 58 feet in overall length or shorter can participate. The season opens seven days after the closure of the initial Bering Sea/ Aleutian Islands parallel Pacific cod season for catcher vessels less than 60 feet in length using hook-and-line and pot gear.

The state-waters season closes when the guideline harvest level is reached. However, if a substantial portion of the guideline harvest level remains uncaught on October 1, the sixty-pot limit and the vessel size restriction can be rescinded. The regulatory season closure date—whether the guideline harvest level was reached—is December 31.[943] (In 2018, the Dutch Harbor Subdistrict season opened on January 30, and it closed on March 1.) While fishing, vessel operators are required to report daily to the Alaska Department of Fish and Game regarding the area fished, number of pots lifted, and estimated pounds of cod retained during the previous twenty-four hours.[944] The Dutch Harbor Subdistrict is managed as an exclusive registration area. Once a vessel is registered to fish in this area, it may not be used to harvest Pacific cod in any other exclusive or superexclusive state-waters Pacific cod management area during the same year.[945]

In 2015, fourteen pot boats harvested the Dutch Harbor Subdistrict's 18 million pound Pacific cod quota. For the 2016 season, the Alaska Board of Fisheries nearly doubled the subdistrict's quota, to 35 million pounds. Attracted by the additional quota, twenty-three vessels registered to participate in the fishery. One advocate for limiting entry into the fishery charged that Dutch Harbor was being "invaded by other areas of the state," and called the state-waters pot fishery "a major reallocation" of the Pacific cod resource from the trawlers that had traditionally fished the adjacent federal waters and delivered their catches to shore plants at Dutch Harbor.[946]

Codfish report, April 2017: *Fishermen were pleased with the performance of the Dutch Harbor cod fishery, now in its fourth year.*

"We had good weather, and there was abundant fish" said Unalaska's Dustan Dickerson, owner of the vessel Raven Bay. Not only were the fish plentiful, they were also very big, he said, weighing as much as 45 to 50 pounds.

A typical cod weights about 8.5 pounds, and Dickerson said this year he's seen the most very big cod in a long time. Fishermen were paid a base price of 34 cents a pound, with bonuses based on the size of the delivery. For every 250 pounds, fishers received an extra half-cent, up to a maximum of 37 cents, he said.

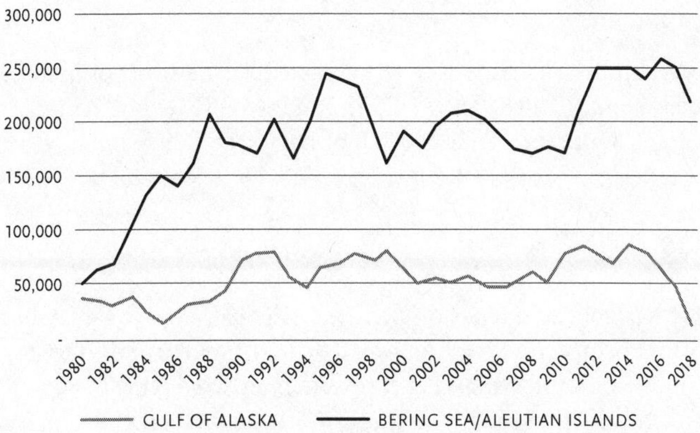

Chart 4. Alaska Pacific cod landings, Gulf of Alaska and Bering Sea/Aleutian Islands area, 1980-2018, round weight, in metric tons.

Data sources: David Witherell and Jim Armstrong (staff, North Pacific Fishery Management Council), Groundfish Species Profiles, 2015, 4, 32, accessed December 2, 2015, www.npfmc.org/wp-content/PDFdocuments/resources/SpeciesProfiles2015.pdf; NOAA Fisheries, Pacific cod landings database, provided by Mary Furuness (NOAA Fisheries).

—Jim Paulin, *Bristol Bay Times–Dutch Harbor Fisherman*, April 2017[947]

The pot cod fleet that fished in the Dutch Harbor Subdistrict nevertheless continued to grow, from twenty-four in 2017, to thirty-two in 2018.[948] And it began to organize. By early 2018, about half the pot cod fleet that fished in state waters in the Bering Sea and Aleutian Islands had organized as the Under Sixty Cod Harvesters. The new organization promptly began lobbying the Alaska Board of Fisheries for an increase in the pot cod fleet's Dutch Harbor Subdistrict allocation. Successfully, as it turned out. At its October 2018 meeting, the board voted to increase the subdistrict's Pacific cod allocation from 6.4 percent of the federal total Bering Sea acceptable biological catch to 8 percent. Moreover, the allocation would then increase 1 percent annually until it topped out at 15 percent. The Board of Fisheries could not, of course, create more cod. The exercise was a zero-sum game because the allocation increase for the state-waters pot cod fleet came almost exclusively at the expense of traditional participants in the federal waters Pacific cod fishery. Jessica Hathaway, editor of *National Fisherman*, called the board's action an "act of cod."[949]

It should be noted that this reallocation occurred during a period of stock decline, and that fishing will happen when cod are spawning.

Catch summary

In 2015, the Alaska state-waters Pacific cod fisheries (not including parallel fisheries) catch of 69 million pounds (round weight) represented 11 percent of the total Alaska catch of Pacific cod.[950]

TAXES

Alaska has collected taxes on fish since the territorial years, when a per-case tax on canned salmon was a primary source of the territory's revenue. Taxes on fish and fish products evolved, and in 1976, the year Congress passed the Magnuson-Stevens Act, the *raw fish tax*, which had been established in 1949, was the state's basic tax on the commercial fishing industry. The tax was paid by the businesses that purchased the fish.

The tax rate for shore-based processors, which were mostly salmon canneries, ranged from 3 percent of the ex-vessel value for salmon, to 1 percent of the ex-vessel value for lower-valued species, such as herring. Shore-based freezer plants paid a *cold storage tax* of 1 percent of the ex-vessel value of all species processed. Floating processors paid the *freezer ship tax* amounting to 4 percent of the ex-vessel value of all species processed.[951]

In 1979, Alaska's legislature adopted the state's current fisheries business tax structure, which applies to fish processed in Alaska, including in state waters. The tax incorporates rates based on whether the utilization of a species is established or developing, whether the fish is processed at a shore-based plant or aboard a floating processor, and the fisheries management area in which the fish is processed. In Southeast Alaska and in the Gulf of Alaska near Yakutat, Pacific cod are considered to be a developing species, and are taxed at a rate of 1 percent when processed at a shore-based plant. (No floating processors currently operate in Southeast Alaska or at Yakutat.) In the rest of Alaska, Pacific cod are considered to be an established species, and are taxed at the rate of 3 percent when processed in a shore-based plant and 5 percent when processed aboard a floating processor.[952]

Beginning in 1981, half of this tax collected in a community has been returned by the state to that community, with the requirement that the money is used to fund fisheries-related infrastructure.[953]

Particularly in the Bering Sea/Aleutian Islands area, much of the groundfish harvest beginning in the early 1980s was by factory trawlers, which were later joined by freezer-longliners. These vessels operated

Figure 89. Kendra Stier, F/V *Julie Anna*, with fifty-five-pound Pacific cod caught on jig gear near Adak, 2008. (Courtesy of Ridgley Stier.)

predominately in the EEZ, but during the fishing season offloaded their production in ports such as Dutch Harbor, where it was loaded aboard ships or barges for shipment to Seattle or foreign ports. While in port, the vessels often also embarked and disembarked crew, took on fuel and supplies, obtained repairs, and discharged waste.

Citing these and other benefits the EEZ catcher-processors received from the affected communities, the burdens placed on the state and on the affected communities by the EEZ catcher-processors, and the need for equal treatment of the EEZ catcher-processor sector and the shore-based sector, Alaska's legislature in 1993 imposed a Fishery Resource Landing Tax, effective January 1, 1994. The rate was 3 percent of the unprocessed value of the "fishery resource." The value was based on the statewide price of fish, in combination with accepted or experienced product recovery rates. For example, if the statewide average price paid to fishermen for Pacific cod was $0.20 per pound, and the accepted recovery rate for skinless/boneless fillets from round fish was 25 percent, an EEZ catcher-processor would have been taxed $0.024 for each pound of Pacific cod fillets landed.[954] [955]

The state returned half of the landing taxes collected to the municipality in which the product was landed. If product is landed outside of a municipality, the Alaska Department of Commerce, Community, and Economic Development distributes half the landing taxes through an allocation program.[956]

During the first year after its implementation (1994), the landing tax generated some $5.8 million, about 87 percent of which was attributed to landings at Unalaska/Dutch Harbor.[957] In 1996, Alaska's legislature restructured the Fishery Resource Landing Tax to mirror the fishery business tax program. It reduced the landing tax to 3 percent for established species and to 1 percent for developing species, where it remains as of this writing.[958]

In 1982, the State of Alaska began levying a 0.3 percent Seafood Marketing Assessment on the value of seafood products processed in Alaska. In 1996, the state amended the seafood market assessment statutes to include fishery resources landed in Alaska. This assessment was increased to 0.5 percent in 2005. The money derived from the assessment helps fund the Alaska Seafood Marketing Institute (see chapter 21).[959]

ENDNOTES

[875] Constitution of the State of Alaska, 1959, Article 8, § 2.

[876] "Hammond unveils Alaska's 20 year bottomfish plan," *Alaska Fisherman's Journal* (August 1979): 16–20.

[877] Institute of Social and Economic Research, University of Alaska, "Prospects for a Bottomfish Industry in Alaska," *Alaska Review of Social and Economic Conditions* (April 1980), 28.

[878] Hal Bernton, "Americans Rush to Bottom Fish. Fishermen Hope Cod, Pollock Will Provide Yearround Activity," *Anchorage Daily News*, December 28, 1986.

[879] Nat Herz, "As the Bering Sea warms, this skipper is chasing pollock to new places," KTOO Radio (Juneau, Alaska), February 6, 2019, accessed February 8, 2019, https://www.ktoo.org/2019/02/06/as-the-bering-sea-warms-this-skipper-is-chasing-pollock-to-new-places/.

[880] Paul Drummond, "Shore-based Preference: Can They Do That," *Alaska Fisherman's Journal* (November 1989): 14–15.

[881] Robert Mann, "Golden Alaska: Another Approach to the Bottomfish Dream," *Alaska Fisherman's Journal* (November 1982): 10–12.

[882] Joel Gay, "Inshore Offshore," *Pacific Fishing* (March 1990): 60–64, 66–69; Staff, NPFMC, and NMFS, *Draft Supplementary Analysis of the Proposed Amendment 18 Inshore/Offshore Allocation of Pollock in the Bering Sea/Aleutian Islands* (September 3, 1992), 1-1.

[883] Joel Gay, "Inshore Offshore," *Pacific Fishing* (March 1990): 60–64, 66–69; "Inshore-Offshore Allocations, Problem Statement," NPFMC newsletter, October 1989, 6–7.

[884] Joel Gay, "Inshore Offshore," *Pacific Fishing* (March 1990): 60–64, 66–69; James Strong and Keith R. Criddle, *Fishing for Pollock in a Sea of Change: A Historical Analysis of the Bering Sea Pollock Fishery* (Fairbanks, Alaska: Alaska Sea Grant, 2013), 57.

[885] Joel Gay, "Inshore Offshore," *Pacific Fishing* (March 1990): 60–64, 66–69; Fishery Conservation and Management Act of 1976, April 13, 1976 (90 Stat. 331), § 2(b)(5).

[886] Joel Gay, "Inshore Offshore," *Pacific Fishing* (March 1990): 60–64, 66–69.

[887] Donna Parker, "Alaska Onshore Wins in Washington," *Pacific Fishing* (May 1992): 28–30.

[888] North Pacific Fishery Management Council, Gulf of Alaska fisheries management plan amendments: GOA amendments 23 and 62.

[889] McDowell Group, *Economic Impact of Inshore Seafood Processing in the Bering Sea/Aleutian Islands Region* (2018), 5, 31.

[890] Clem Tillion, in Alaska Department of Fish and Game, *Sustaining Alaska's Fisheries: Fifty Years of Statehood* (January 2009), 36, accessed January 17, 2017, http://www.adfg.alaska.gov/static/fishing/PDFs/50years_cf/50years_full.pdf.

[891] Wesley Loy, "Congress overhauls Alaska CDQ program," *Pacific Fishing* (September 2006): 9.

[892] Brad Matsen, "CDQs are About More Than Money," *National Fisherman*, March 1993, 6.

[893] Kevin M. Bailey, *Billion-Dollar Fish: The Untold Story of Alaska Pollock* (Chicago: University of Chicago Press, 2013): 182.

[894] Dan Rodricks, "Baltimore Native Son Makes Mark in Alaska,". *Baltimore Sun*, October 18, 1996; Norman Cohen, colleague of Harold Sparck, personal communication with author, January 15, 2017; Don Mitchell, "Congress Recognizes Sparck's Efforts on CDQ's," *Bering Sea Fisherman*, March 1997.

[895] Coastal Villages Region Fund, "Milestones in the Western Alaska CDQ Program," accessed March 13, 2015, http://www.coastalvillages.org/about-us/history. NOAA Office of General Counsel, *A Guide to the Sustainable Fisheries Act, Public Law 104-297* (February 1997), § 111, accessed March 13, 2015, http://www.nmfs.noaa.gov/sfa/sfaguide/111.htm.

[896] Walter J. Hickel, *Crisis in the Commons: The Alaska Solution* (Oakland, Calif.: CS Press, 2002), 210.

[897] Clem Tillion, "Alaska CDQ fish fight: Alaskans vs. Alaskans this time," *Alaska Dispatch News*, October 9, 2012.

[898] Joel Gay, "What's This New CDQ?" *Pacific Fishing* (February 1993): 45.

[899] Laine Welch, "CDQs Spell Success," *Pacific Fishing* (November 1993): 36, 139; National Academy of Sciences, *The Community Development Quota Program in Alaska* (Washington: National Academy Press, 1999), 1.

[900] Sustainable Fisheries Act, October 11, 1996 (110 Stat. 3559), § 111(a); American Fisheries Act, October 21, 1998 (112 Stat. 2681-616), § 206(a).

[901] National Marine Fisheries Service, Alaska Region, *Alaska Groundfish Fisheries, Final Programmatic Supplemental Environmental Impact Statement* (June 2004), C-56; 72 *Fed. Reg.* 50788, September 4, 2007.

[902] Magnuson-Stevens Fishery Management and Conservation Act, § 303A(e).

[903] Coastal Villages Corporation, "Vessels," accessed May 14, 2015, http://www.coastalvillages.org/vessels.

[904] Wesley Loy, "CDQ Groups Gain Clout, Controversy," *Pacific Fishing* (January 2003): 10, 13; Caroline Schultz, "Alaska's Community Development Quota Groups," *Alaska Economic Trends* (November 2014): 12–13, 15; Andrew Olson (Alaska Department of Fish and Game), personal communication with author, March 22, 2019..

[905] Division of Commercial Fisheries, Alaska Department of Fish and Game, *State of Alaska Groundfish Fisheries, Associated Investigations in 1998* (April 1999), accessed April 30, 2016, http://www.psmfc.org/tsc2/99_TSC_rpt/AKTSC981.htm#Pacific.

[906] Mark A. Stichert (Alaska Department of Fish and Game), *Fishery Management Plan for the Chignik Area State-waters Pacific Cod Season, 2011*, ADF&G Fishery Management Rept. No. 11-01 (January 2011), 2.

[907] Trever Shaishnikoff, "Alaskan businesses thrive with state-water cod fisheries," *Juneau Empire*, April 29, 2018, accessed April 29, 2018, http://juneauempire.com/opinion/2018-04-29/alaskan-businesses-thrive-state-water-cod-fisheries.

[908] The 58-foot limit on salmon seiners dates from 1960, the year the State of Alaska took over management of its fisheries from the federal government. The purpose of the length limit was to keep Puget Sound seiners—which tended to be longer—from seasonally coming to Alaska to compete with resident Alaska fishermen.

[909] Eric E. Coonradt (Alaska Department of Fish and Game), *The Southeast Alaska Pacific Cod Fishery*, ADF&G Regional Information Rept. No. 1J02-10 (May 2002), 5–7, 10–11; Alaska Dept. of Fish and Game, *State of Alaska Groundfish Fisheries, Associated Investigations in 2014* (April 2015), 15.

[910] Barbi Failor-Rounds (Alaska Department of Fish and Game), *Bering Sea-Aleutian Islands Area State-Waters Groundfish Fisheries and Groundfish Harvest from Parallel Seasons in 2004*, ADF&G Fishery Management Rept. No. 05-71 (December 2005), 1.

[911] Alaska Department of Fish and Game, *Groundfish Fisheries, Associated Investigations in 1999*, ADF&G Regional Information Rept. No. 1J00-23 (May 2000), 9–10; Nathaniel Nichols, Paul Converse, and Kim Phillips (Alaska Department of Fish and Game), *Annual Management Report for Groundfish Fisheries in the Kodiak, Chignik, and South Alaska Peninsula Management Areas, 2014*, ADF&G Fishery Management Rept. No. 15-41 (October 2015), 2; Joel Gay, "Gulf of Alaska Pot, Jig Boats Get 15%," *Pacific Fishing* (January 1997): 22–24; Charlie Ess, "Jigging Machines: has their moment arrived?" *Pacific Fishing* (March 1997): 38–42.

[912] Charlie Ess, "Skippers Switch to Cod," *Pacific Fishing* (August 1997): 15.

[913] Joe Childers and Brad Warren, "Alaska's New Rush for Access: State-Waters Cod," *Pacific Fishing* (April 1999), 31, 65.

[914] Alaska Department of Fish and Game, *2013–2014 Statewide Commercial Groundfish Fishing Regulations*, 1; Nathaniel Nichols, Paul Converse, and Kim Phillips (Alaska Department of Fish and Game), *Annual Management Report for Groundfish Fisheries in the Kodiak, Chignik, and South Alaska Peninsula Management Areas, 2014*, ADF&G Fishery Management Rept. No. 15-41 (October 2015), 19.

[915] Nathaniel Nichols, Paul Converse, and Kim Phillips (Alaska Department of Fish

and Game), *Annual Management Report for Groundfish Fisheries in the Kodiak, Chignik, and South Alaska Peninsula Management Areas, 2014*, ADF&G Fishery Management Rept. No. 15-41 (October 2015), 2–3.

916 Maria Wessel et al. (Alaska Department of Fish and Game), *Prince William Sound Registration Area E Groundfish Fisheries Management Report, 2009–2013*, ADF&G Fishery Management Rept. No. 14-42 (November 2014), 8, 36.

917 *Seafood News*, News Summary (email), May 1, 2017.

918 "Hefty Longliner Launched at Hansens," *Alaska Fisherman's Journal* (March 1989): 70–71.

919 Craig Farrington, *Super 8 Vessels for CFEC In-house Use*, Alaska Commercial Fisheries Entry Commission Rept. No. 15-5N (December 2015), accessed April 17, 2018, https://www.cfec.state.ak.us/RESEARCH/15-5N/CFEC_Super8s_121015.pdf; Joseph Greinier, former fisherman, personal communication with author, April 17, 2018.

920 The area encompassed all state waters between 175° 30' and 177° W longitude.

921 "Tillion Honored for Achievement," *Homer Tribune*, February 13, 2013, accessed October 17, 2015, http://homertribune.com/2013/02/tillion-honored-for-achievement/; Wesley Loy, "Fish Board steers cod allocation to remote Adak," *Pacific Fishing* (April 2006): 7.

922 Barbi Failor-Rounds (Alaska Department of Fish and Game), *Bering Sea-Aleutian Islands Area State-Waters Groundfish Fisheries and Groundfish Harvest from Parallel Seasons in 2004*, ADF&G Fishery Management Rept. No. 05-71 (December 2005), 1, 7.

923 Alaska Department of Fish and Game, *2013–2014 Statewide Commercial Groundfish Fishing Regulations*, 1.

924 Jenniver Karuza, "Staying Out of Trouble," *National Fisherman* (September 2002): 26–27.

925 Jennifer Karuza, "A Big Little Boat," *National Fisherman* (April 2001): 48–49.

926 Joel Gay, "Adak: Oasis or Mirage of the Aleutian Chain," *Pacific Fishing* (November 1999): 69–71.

927 Wesley Loy, "NorQuest Buys Adak Plant," *Pacific Fishing* (September 2000): 16.

928 National Oceanic and Atmospheric Administration, Docket No. AK035039,

Adak Fisheries, LLC; Adak Fisheries Development, LLC; Icicle Seafood, Inc. (March 12, 2007), 110, 112, 116, 121.

929 National Oceanic and Atmospheric Administration, Docket No. AK035039, Adak Fisheries, LLC; Adak Fisheries Development, LLC; Icicle Seafood, Inc. (March 12, 2007), 1, 12, 142.

930 Wesley Loy, "Adak: Exit Icicle, Enter Swasand," *Pacific Fishing* (July 2004): 9; Andrew Jensen, "NOAA Cuts Deal With Icicle, Reduces $3.44M Fine to $615K," *Alaska Journal of Commerce* (March 2012), accessed April 14, 2015, http://www.alaskajournal.com/Alaska-Journal-of-Commerce/AJOC-March-4-2012/NOAA-cuts-deal-with-Icicle-reduces-344M-fine-to-615K/.

931 Barbi Failor-Rounds (Alaska Department of Fish and Game), *Bering Sea-Aleutian Islands Area State-Waters Groundfish Fisheries and Groundfish Harvest from Parallel Seasons in 2004*, ADF&G Fishery Management Rept. No. 05-71 (December 2005), 7.

932 Charles L. Trebesch (Alaska Department of Fish and Game), *Annual Management Report for the Bering Sea-Aleutian Islands Area State-Waters Groundfish Fisheries and Groundfish Harvest From Parallel Seasons in 2012*, ADF&G Fishery Management Rept. No. 13-33 (September 2013), 6–8.

933 Ibid.

934 Asia Beder and Janis Shaishnikoff (Alaska Department of Fish and Game), *Annual Management Report for Groundfish Fisheries in the Bering Sea–Aleutian Islands Management Area, 2017*, ADF&G Fishery Management Rept. No. 18-18 (September 2018), 40–41.

935 Clem Tillion, personal communication with author, April 13, 2015; Jim Paulin, "Adak Pays $2 Million to Keep Fish Equipment on Island," *Bristol Bay Times* (June 28, 2013); Lauren Rosenthal, "Adak Dealt Setback as Another Processor Folds," *Pacific Fishing* (November 2014): 14–15.

936 Jim Paulin, "Adak Pays $2 Million to Keep Fish Equipment on Island," *Bristol Bay Times* (June 28, 2013); Lauren Rosenthal, "Adak Dealt Setback as Another Processor Folds," *Pacific Fishing* (November 2014): 14–15.

937 Lauren Rosenthal, "Adak Dealt Setback as Another Processor Folds," Pacific Fishing (November 2014): 14–15.

938 Atka Pride Seafoods primarily processes halibut and sablefish.

939 81 *Fed. Reg.* 84434–84455, November 23, 2016.

[940] Jim Paulin, "Cod Allocation Could Revive Adak Processing Plant," *Pacific Fishing, Dutch Harbor Report* (December 2015), accessed October 4, 2017, http://alaskaseafoodcooperative.org/cod-allocation-could-revive-adak-processing-plant/; Rick Koso, Adak Community Development Corp., personal communication with author, October 3, 2017.

[941] *Groundfish Forum et al v. Wilbur Ross et al*, Civ. No. 16-2495 (TJK) (D.D.C., March 21, 2019); Peggy Parker, "Judge Rules Adak Cod Set-Aside Unlawful, Orders NMFS to Rewrite Amendment 113, *Seafood News*, March 22, 2019, accessed March 24, 2019, https://www.seafoodnews.com/Story/1136001/Judge-Rules-Adak-Cod-Set-Aside-Unlawful-Orders-NMFS-to-Rewrite-Amendment-113.

[942] Specifically, the Dutch Harbor Subdistrict consists of state waters between 167° W longitude and 164° W longitude that are south of 55° 30' N latitude.

[943] Charles L. Trebesch (Alaska Department of Fish and Game), *Fishery Management Plan for the Dutch Harbor Subdistrict-State Waters and Parallel Pacific Cod Seasons, 2014*, ADF&G Fishery Management Rept. No. 14-04 (February 2014), 2–3, 5.

[944] Division of Commercial Fisheries, Alaska Department of Fish and Game, "Dutch Harbor Subdistrict State-Waters Pacific Cod Season Opening and Closure of the Parallel Season," Emergency Order No. 4-GF-03-18, news release, January 19, 2018; Division of Commercial Fisheries, Alaska Department of Fish and Game, "Dutch Harbor Subdistrict State-waters Pacific Cod Season Closure," Emergency Order No. 4-GF-04-18, news release, February 28, 2018.

[945] Division of Commercial Fisheries, Alaska Department of Fish and Game, "Dutch Harbor Subdistrict State-Waters Pacific Cod Season Opening and Closure of the Parallel Season," Emergency Order No. 4-GF-03-16, news release, February 4, 2016.

[946] Jim Paulin, "Pacific cod quota increase attracting boats to bay," *Bristol Bay Times–Dutch Harbor Fisherman*, March 25, 2016, accessed March 29, 2016, http://www.thebristolbaytimes.com/article/1612pacific_cod_quota_increase_attracting_boats.

[947] Jim Paulin, "Pot cod fishery numbers all in," *Bristol Bay Times–Dutch Harbor Fisherman*, April 28, 2017, accessed May 1, 2017, http://www.thebristolbaytimes.com/article/1717pot_cod_fishery_numbers_all_in.

[948] Jim Paulin, "Cod caught quickly," *Bristol Bay Times–Dutch Harbor Fisherman*, February 16, 2018, accessed October 24, 2018, http://www.thebristolbaytimes.com/article/1807cod_caught_quickly.

[949] Under Sixty Cod Harvesters to Dan Hull, Chairman, NPFMC, May 30, 2018, accessed October 23, 2018, http://comments.npfmc.org/CommentReview/DownloadFile?p=77a3f321-e45c-4da1-9ec8-8537aaf8cc05.pdf&fileName=USCH%20Comment%20Letter%20to%20NPFMC_Agenda%20Item%20C6_June%202018.pdf; Jessica Hathaway, "An act of cod: Alaska-based fleet gets a gift from Board of Fish," *National Fisherman*, October 23, 2018, accessed October 23, 2018, https://www.nationalfisherman.com/alaska/an-act-of-cod-alaska-based-fleet-gets-a-gift-from-board-of-fish/?utm_source=marketo&utm_medium=email&utm_campaign=newsletter&utm_content=newsletter&mkt_tok=eyJpIjoiWW1SbU4yWTFaVokzWkRBNSIsInQiOiI5dVlKUyswMjVxYWRndkxVMFNGVzdyc2FsTk5SYUhDUVlnbEttNoY3ZVRvRXowYStqcnh3MTBYVGZabXJ4QlJoVGZ6NlNoRTREVWd3eFhpQjhhZ2tRdVhcL1wvZmxzRU9WenhkR1ZVWUNPZjg2VSt2YmVVNDFrRlpwdoJjTzRTeklhIno%3D.

[950] Jennifer Shriver (Alaska Department of Fish and Game), personal communication with author, September 15, 2016.

[951] George W. Rogers, *Development of an Alaskan Bottomfish Industry and State Taxes* (Juneau, Alaska: Institute of Social and Economic Research, undated, but probably 1977), 31–32.

[952] Alaska Department of Revenue, Tax Division, *Fisheries Business Tax, Historical Overview*, accessed July 22, 2017, http://www.tax.alaska.gov/programs/programs/reports/Historical.aspx?60633; Alaska Department of Revenue, Tax Division, "Fisheries Related Taxes, Frequently Asked Questions," accessed July 22, 2017, http://www.tax.alaska.gov/programs/programs/help/faq/faq.aspx?60620.

[953] Alaska Department of Revenue, Tax Division, "Fisheries Business Tax, Historical Overview," accessed July 22, 2017, http://www.tax.alaska.gov/programs/programs/reports/Historical.aspx?60633.

[954] Computation: (1/.25 x 20¢) x 0.03 = 2.4¢

[955] Fishery Resource Landing Tax, chapter 67, *Session Laws of Alaska, 1993*; Alaska Administrative Code, § 43.77.200, Definitions; Alaska Administrative Code, § 43.77.005, Findings, Purpose, and Intent; chapter 81, *Session Laws of Alaska, 1996*, § 21; Alaska Department of Revenue, Tax Division, "Fishery Resource Landing Tax, Historical Overview," accessed June 14, 2018, http://www.tax.alaska.gov/programs/programs/reports/Historical.aspx?60631.

[956] Alaska Administrative Code, § 43.77.060, Fishery Resource Landing Tax.

[957] Alaska Department of Revenue, *Shared Taxes and Fees Annual Report for the Fiscal Year Ended June 30, 1995*, 1, 7.

[958] Fishery Resource Landing Tax, chapter 67, *Session Laws of Alaska, 1993*; Alaska Administrative Code, § 43.77.200, Definitions; Alaska Administrative Code, § 43.77.005, Findings, Purpose, and Intent; chapter 81, *Session Laws of Alaska, 1996*, § 21; Alaska Department of Revenue, Tax Division, "Fishery Resource Landing Tax, Historical Overview," accessed June 14, 2018, http://www.tax.alaska.gov/programs/programs/reports/Historical.aspx?60631.

[959] Alaska Department of Revenue, Tax Division, "Seafood Marketing Assessment, Historical Overview," accessed July 22, 2017, http://www.tax.alaska.gov/programs/programs/reports/Historical.aspx?60636; Alaska Seafood Marketing Institute, *2014 Annual Report*, 25, 30; Alaska Seafood Marketing Institute, *2013 Annual Report*, 1.

Chapter 24
BYCATCH

Again, the kingdom of heaven is like to a net cast into the sea, and gathering together all kind of fishes. Which, when it was filled, they drew out, and sitting by the shore, they chose out the good into vessels, but the bad they cast forth.

—Matthew 13:47–48

Most of our hauls that trip on the Storm Petrel were pollock, though there were enough cod mixed in to send the crew on deck to cull them out, bleed them by slitting their throats just behind the gills, and pitching them into chilled brine in the tanks. The pollock and everything else was squeezed through the scuppers.

—Brad Matsen, *National Fisherman*, 1981[960]

When was the last time the Bering Sea groundfish fleet was shut down for reaching its cod quota? The answer is it isn't anymore. Bycatch triggers cod closures, and the season that used to last the better part of a year now lasts a few months.

—*Alaska Fisherman's Journal*, October 1991[961]

The long-kept, dirty secret of the North Pacific industrial fishing fleet, its 750 million pounds of discards per year, is hardly a secret anymore.

—Joel Gay, *Pacific Fishing*, June 1996[962]

The 800-pound gorilla in the room (though at times it seems like an 8,000-pound gorilla) remains bycatch in our groundfish fisheries.

—Chris Oliver, Director, NPFMC, 2014[963]

The Magnuson-Stevens Act defines *bycatch* as "fish which are harvested in a fishery, but which are not sold or kept for personal use, and includes economic discards and regulatory discards."[964] Basically, bycatch is fish of the wrong species, the wrong size (usually too small), or, in the case of roe fisheries, the wrong sex.[965] Some bycatch survives relatively unharmed if returned to the water in a careful and timely manner. The term *bycatch mortality* denotes bycatch that is killed in the fish-catching process or is statistically expected to die after being returned to the water. Halibut bycatch, because of the amount caught and the commercial value of the species, was early on the most problematic for the groundfish fleet, and it remains so today. For trawlers, bycatch of salmon also became a major problem.

HALIBUT BYCATCH

> *Halibut bycatch issue: a poster child for complex fisheries policy in Alaska.*
>
> —Shirley Marquardt, Mayor, Unalaska, Alaska, 2015[966]

During the decade or so preceding the Magnuson-Stevens Act, the Pacific halibut fishery, which was inaugurated in 1888, ranked third, in terms of value, among Alaska's fisheries, behind salmon and crabs. For a host of reasons—among them the fish itself and the handsome, seaworthy schooners that characterize the distant-waters halibut fleet—the fishery long ago attained iconic status.

Halibut bycatch mortality was relatively small until the dramatic expansion of foreign trawl fisheries in the North Pacific during the 1960s. Bycatch mortality peaked in 1965 at about 21 million pounds. It then declined for several years, but in the early 1970s it increased to about 20 million pounds, after which it again declined.[967] In 1976, the year the Magnuson-Stevens Act became law, halibut bycatch mortality was about 13 million pounds. Halibut that year were at a low point in abundance, and the catch by commercial fishermen of about 20 million pounds in the directed fishery represented approximately 40 percent of the average annual catch during the previous four decades.

In the years following the implementation of the Magnuson-Stevens Act in 1977, the allocation of groundfish to the foreign fleet was slowly phased out while the U.S. fleet developed its ability to harvest groundfish. The foreign fleet's bycatch of halibut was primarily regulated through time/area closures and halibut bycatch limits. To ensure that foreign fishermen had no incentive to target halibut, the retention of halibut was prohibited. Not all these constraints, however, were placed on the domestic fleet, as they were considered overly restrictive of the developing domestic groundfish industry.

The prohibition of the retention of incidentally caught halibut, however, did apply to domestic fishermen. To minimize waste and simultaneously improve their public image, domestic trawler operators had early on begun lobbying for regulations that would allow them to retain incidentally caught halibut for donation to charitable organizations. It wasn't, however, until 1998 that the National Marine

Fisheries Service created the Prohibited Species Donation Program, which allows trawl catcher vessels to retain halibut if the fish will be delivered to an authorized shore-based processor or stationary floating processor for donation to hunger relief agencies, food bank networks, or food bank distributors.[968] No other gear sectors have access to this program.[969]

In 1984, the NPFMC developed long-term management goals for fisheries off Alaska. Among them was to "minimize the catch, mortality, and waste of non-target species, and reduce the adverse impacts of one fishery on another."[970] This statement marked the first time that managers explicitly recognized bycatch as a problem and that one fishery's bycatch could adversely impact the health and catch quotas of other fisheries. Case in point: the International Pacific Halibut Commission (IPHC), the organization responsible for management of the halibut resource in the United States and Canada, deducts bycatch from the available harvest level in establishing the quota for the directed halibut fishery.

Mostly because of constraints on the foreign fleet, halibut bycatch fell from 13 million pounds in 1976 to a record low of about 6.1 million pounds in 1985. But it then began to increase as the growing domestic fleet—unencumbered by adequate halibut bycatch regulations—was allocated increasingly greater percentages of the groundfish harvest. Alaska's groundfish fishery became a fully domestic effort in 1990 (see chapter 15).

During the years 1985–1989, the National Marine Fisheries Service's primary method of regulating halibut bycatch in the domestic industry was based on a statistical model of halibut bycatch mortality. Once an acceptable biological catch for each fishery was determined, the council would use the statistical model to adjust the targeted fish catch downward to keep the estimated halibut mortality below the level the IPHC determined was acceptable. This was likely the first time in the U.S. that bycatch estimates were used in the management of directed fisheries.

In 1985, the National Marine Fisheries Service implemented halibut bycatch limits for groundfish trawl fisheries in the Gulf of Alaska. In 1987, the agency implemented halibut bycatch limits in the Bering Sea/Aleutian Islands area, but these limits applied only to joint-venture fisheries for yellowfin sole and other flatfish. In 1989, the agency established bycatch limits for Pacific halibut (as well as crabs) for all trawl fisheries. The limits were apportioned among the fisheries.

Also in 1989, the NMFS introduced a domestic observer program that put trained observers aboard U.S. fishing vessels to monitor catches.[971]

In the summer of 1989, with less than half its allocations of Pacific cod and flatfish caught, the Gulf of Alaska bottom trawl fishery was shut down because its 2,000-metric-ton halibut bycatch limit had been reached. This was the first time the gulf had been closed to bottom trawling.[972] (Mid-water trawling, because it caught relatively few halibut, was exempt from the closure).

The Alaska Factory Trawler Association attempted to legitimize bycatch as an acceptable cost of doing business. To no one's surprise, this characterization lacked credibility among halibut fishermen, fisheries managers, Congress, environmental groups, and the general public.

In 1991, halibut bycatch mortality ballooned to about 20 million pounds, more than half of which occurred in the Bering Sea. In fact, halibut bycatch mortality in the Bering Sea that year was more than twice the catch by the directed halibut fishery there. As one Bering Sea halibut fisherman pointed out a number of years later, "There has been a de facto reallocation [of halibut] from the directed fisheries to the bycatch fisheries."[973]

Halibut bycatch limits in the Bering Sea/Aleutian Islands area were extended to non-trawl fisheries in 1992. That year, a bycatch mortality limit of 3,775 metric tons (approximately 8.3 million pounds) was imposed on trawlers, and a bycatch mortality limit of 900 metric tons (approximately 2 million pounds) was imposed on non-trawl fisheries. Quarterly bycatch limits were allocated among the subsectors of the groundfish fleet. Once a subsector reached its quarterly limit, directed fishing by that sector was ended for the quarter and potential revenues foregone. To prevent the annual bycatch limit from being exceeded, overage of a limit in one quarter was deducted from the next quarter.

With these regulations in place, management officials hoped each fleet would figure out how to police itself. The reality, however, was that in an open-access race for fish, an effort by an individual fisherman to reduce bycatch usually slowed his catch of the target species and resulted in a monetary loss. On the other hand, a fisherman who chose to catch the target species at the highest rate—regardless of bycatch—was in all likelihood rewarded by having caught a greater share of the target-species quota by the time the fishery closed.[974] As Robert Trumble, of the IPHC, observed in 1992, "Increased competition to maximize harvest before a closure caused individual fishermen to race for fish, at the cost of higher bycatch rates and earlier closures. The vicious circle began with earlier closures causing faster fishing, with higher bycatch rates that caused earlier closures."[975] The following year, the Bering Sea Pacific cod trawl, longline, and pot fleets exceeded their collective annual allowable halibut bycatch on May 11, closing the cod fishery for the year. About 12,000 metric tons of Pacific cod—7 percent of the total allowable catch for all gear types that year—went uncaught.[976]

Halibut bycatch mortality, the *discard mortality rate*, varies among gear groups. Joint studies by the IPHC and National Marine Fisheries Service (ca. early 1990s) estimated that about 60 percent of the halibut caught in bottom trawls died. The halibut bycatch mortality rate for pelagic trawls was about 80 percent, the highest of the gear groups, but because of the very large mesh size (designed to "herd" pollock) in the forward part of the trawls and because the trawls are usually fished in the water column rather than on the seafloor, pelagic trawls catch relatively few halibut.[977]

The mortality rate for longliners was considerably less than for trawl gear. It ranged from about 15 to 25 percent because most halibut brought to the surface on longlines were alive and could be carefully released, although their fate after release was uncertain, especially if the fish had suffered damage by their encounters with hooks (see below).[978]

> *There are several areas in the lower Bering Sea close to Unimak Pass that we can't tow with any success. There is some [Pacific cod] there, but we cannot stay under the bycatch cap rates for halibut and still make any fish out of it. To keep fishing, to fish the springtime and winter fish in the horseshoe area, we are going to have to come up with something.*
>
> —Shawn O'Brien, captain, trawler *Northwest Explorer*, 2001.[979]

In the jig fishery, every halibut caught is brought to the surface promptly, and careful release helps ensure survival. Pot fishing is perhaps the "cleanest" of the methods of catching Pacific cod. Federal regulations prohibit the size of the entry tunnel eyes on cod pots from being greater than nine inches in any one dimension, which prevents all but the smallest halibut from entering (see chapter 18). Because pots catch so few halibut and those small enough to be able to enter a pot can generally be released unharmed, the fishery is exempt from bycatch limits.

In addition to varying by gear group, bycatch varies within the groups. Some vessels fish "dirtier" than others. For example, Clem Tillion recalled that one year in the mid-1990s, a dozen trawlers fishing the Gulf of Alaska caught the entire quarterly quota of Pacific cod in twenty-one days. In doing so, the trawlers also exceeded the cumulative bycatch limits, which forced the National Marine Fisheries Service to close groundfish fisheries all along the coast. Of the dozen trawlers, two were responsible for 80 percent of the bycatch. As Chris Blackburn, director of the Alaska Groundfish Data Bank, a trawl industry organization, said in 1989, "We need to control those who cannot fish in a reasonable range of halibut bycatch." That summer, bottom trawling in the Gulf of Alaska was closed because Pacific cod and flatfish trawlers exceeded their halibut bycatch limit. Some trawl interests attributed this first-of-a-kind closure to "the inexperience of a few vessels new to the fishery."[980]

Halibut bycatch mortality reached 9,199 metric tons (20,293,000 pounds) in 1992, which was followed by a generally decreasing trend that continues to this day. In 2014, the IPHC estimated bycatch mortality to have been 4,223 metric tons (9,315,000 pounds)—still a lot, but less than half that of 1992.[981]

The designation in 1997 of the western Alaska population of the Steller sea lion as endangered (see chapter 25) intensified the halibut bycatch issue in the Gulf of Alaska. Typically, Pacific cod longliners, trawlers, and pot vessels in the gulf began fishing in January of each year and caught their entire Pacific cod annual quota in about three months. In 2001, regulations designed to protect Steller sea lions were implemented. The new regulations divided the fishing season: fishermen could land 60 percent of their cod quota during the winter (A season), but had to wait until after June 10 (B season) to land the remaining 40 percent.[982]

Cod fishermen in January 2001 found few concentrations of cod in the central Gulf of Alaska. Normally, the fleet would have scaled back its effort to give the fish time to school up. Some trawlers did leave the central gulf, but longliners kept exploring and in the process caught 263 metric tons of halibut, substantially exceeding their sector's first trimester Gulf of Alaska halibut bycatch allowance of 175 metric tons and coming close to the annual limit of 290 metric tons. This forced the National Marine Fisheries Service to close the fishery on February 26, a week before the A-season cod quota was anticipated to have been reached, leaving longliners with just 27 metric tons of halibut bycatch allowance for the B season. During the A season, halibut bycatch ranged from 18 to 53 metric tons per week.[983]

The longliner's B season in the central and western gulf, which was scheduled to run from September 1 to December 31, was closed only four days after it opened, when longliners reached their halibut bycatch allowance. Overall, cod fishermen in the central and western gulf in 2001 left some 7,000 metric tons—almost 15 percent of the year's allowable Pacific cod catch—in the water.[984] The longliner's first trimester halibut bycatch allowance was subsequently increased to 250 metric tons.[985]

Halibut bycatch by Pacific cod longliners in Alaska waters has, however, declined. In 2015, the sector's halibut bycatch mortality was estimated at 272 metric tons, a 16 percent decrease from the 326 metric tons estimated for 2014.[986] Part of this decline can be attributed to longliners avoiding areas where halibut concentrate and to improved halibut release practices. Historically low halibut abundance also figures into the equation.

For efficiency purposes as well as for public relations, it has been in the best interest of trawl fishermen, too, to reduce the amount of halibut they catch. In early 2000, at the request of the Groundfish Forum and the At-Sea Processors Association, trade groups that represented Pacific cod trawlers (among others), the National Marine Fisheries Service's Alaska Fisheries Science Center began working with fishermen to design and test a halibut exclusion device for the Pacific cod trawl fishery. Their basic design was based on an excluder that had been successfully utilized in sole fisheries. But keeping halibut out of a Pacific cod trawler's cod end is more difficult than doing so on a trawler targeting sole and other small flatfish because the size difference between Pacific cod and halibut is less, making the excluders more accessible to both species. Nevertheless, as Craig Rose, a National Marine Fisheries Service biologist who helped develop halibut excluders noted, "If they're having significant halibut bycatch, it would be well worth their time to lose 20 percent of their cod catch to get rid of almost all of the halibut." Several variations were produced and tested at sea in 2000 and 2001. These designs were promising—one vessel experienced a 25 percent reduction in its cod catch and a 75 percent reduction in the amount of halibut caught—and they would be refined over the next several years.[987] [988]

In its current design, a halibut excluder consists of escape panels—basically long rectangular grids of semi-rigid elongated slots—that are sewn into each side of the intermediate section of a trawl net just ahead of the net's cod end. The slots are sized so small- and medium-sized halibut can swim through them, as can small cod. Large cod, however, are prevented from doing so because of the size of their heads. Although the excluders are imperfect devices, they have helped considerably to reduce halibut bycatch and to prevent the early closure of Pacific cod (and other) fishing seasons. Excluders are not mandated in regulations, but beginning in about 2005 the trawl catcher-vessel fleet in the Gulf of Alaska broadly embraced their use. The situation is similar in the

Bering Sea/Aleutian Islands area, where most trawlers today carry a variety of excluders, each one designed for specific fishing conditions. A company in Kodiak manufactures the devices.[989]

Deck sorting is another method trawlers use to reduce halibut mortality. The goal is to return halibut to the sea as quickly as possible, while they are alive and, hopefully, vigorous. Prior to the practice of deck sorting, all fish caught by factory trawlers were routed into below-deck tanks before sorting occurred. This took considerable time and substantially reduced a released halibut's chance of survival. In 1993, the IPHC conducted a sorting experiment aboard a factory trawler to devise techniques to reduce halibut discard mortality rates. The commission found deck sorting to be the most effective at returning halibut to the sea in a condition that increased their likelihood of survival. The concept had widespread support among trawlers because improved halibut survival would allow an increased groundfish harvest within a given bycatch mortality limit.[990]

In 1995, the commission proposed to the NPFMC that factory trawlers and catcher vessels that routed their catch to tanks below deck be required to sort halibut bycatch on deck. Regulations, however, required all halibut to be retained until recorded by a federal fishery observer. Concerned that observers wouldn't be able to safely and properly sample a deck-sorted catch, the council rejected the IPHC's proposal.[991]

The halibut bycatch issue did not lessen, but it was 2009 before deck sorting was again revisited. In May and June of that year, the NPFMC investigated deck sorting of Pacific halibut as a means of reducing halibut bycatch mortalities on Amendment 80 vessels (a fleet of factory trawlers that fishes mostly in the Bering Sea/Aleutian Islands area as a cooperative). The investigation determined that by sorting halibut out of the catch on deck and returning them to the sea as quickly as possible, the average halibut mortality rate was 45 percent, compared with the average 75 percent mortality rate that was at the time (perhaps optimistically) assigned to flatfish fisheries in the Bering Sea/Aleutian Islands area. Moreover, most of the handling procedures used in the investigation appeared to be feasible for real-world use.[992]

Nevertheless, authorization of deck sorting as a method for reducing halibut bycatch languished until 2014, when the council requested a study of observer sampling protocols and regulatory changes that would be necessary to allow deck sorting of halibut on factory trawlers in the Bering Sea/Aleutian Islands area.[993]

In 2015, NMFS approved an experimental program in the Bering Sea trawl fisheries to test the feasibility of deck sorting halibut in real-world conditions and granted nine Amendment 80 vessels deck-sorting permits. Preliminary data suggested that deck sorting and other measures to reduce bycatch—such as avoiding halibut "hotspots"—resulted in about 150 metric tons less halibut bycatch that year. Deck sorting, however, came with a cost: each vessel made one less tow per day because of the extra time required to sort each tow.[994] Deck sorting could potentially be extended to trawlers in the Gulf of Alaska, where halibut bycatch has been a chronic problem.

And there is more to the issue than bycatch mortality. Despite the strict requirements for the careful release of incidentally caught (and undersized) halibut imposed on longline fishermen, a high percentage of these fish are injured by their encounters with hooks. An IPHC survey in the Bering Sea/Aleutian Islands area (IPHC Area 4A) in 2014 found that nearly 12 percent of halibut less than thirty-two inches long suffered from prior hook injuries. The IPHC determined that the injuries were "most likely the result of interception of Pacific halibut by Pacific cod groundfish fisheries in those areas" and added that the severity of the problem was probably worse than reflected in the data collected because its observations tallied only the number of injured halibut that had survived their ordeal. According to the IPHC, "Many halibut die from moderate to severe hooking injuries, and those that do survive often stop growing or grow slowly."[995]

In some areas today, the stocks of Pacific halibut are at some of their lowest levels in the ninety-year history of the fishery's management. The growth of individual fish is stunted, possibly attributable to increased competition with other halibut and with a growing population of arrowtooth flounder, as well as the fact that many halibut do not live long enough to reach the directed longline fishery's legal exploitable size (thirty-two inches long). Halibut stocks are not, however, at their lowest level ever, and the recent trend is level.[996]

Rationalization of the trawl fishery in the Bering Sea/Aleutian Islands area and in the Gulf of Alaska—to the extent it has occurred—allows more selective fishing practices that avoid concentrations of halibut. Overall, halibut bycatch in Alaska declined by almost 25 percent during the decade ending in 2014, but because of the broad decline in the halibut stock, the percentage of total removals resulting from bycatch has increased from 12 percent in 2004 to 21 percent in 2014.[997] Between 2014 and 2016, halibut bycatch declined some 2 million pounds, driven primarily by a very substantial reduction of halibut bycatch by trawlers in the central Bering Sea. Preliminary figures indicate halibut bycatch in 2016 will be just over 7 million pounds, the lowest it has been since 1960.[998]

Nevertheless, bycatch of halibut remains an intractable issue. Halibut fishermen—both longliners and sport charter operators—have watched their allowable catches decline and want bycatch eliminated, while bottom trawlers and codfish longliners want liberal bycatch limits that don't threaten to close their fisheries before target species catch limits are reached.

As noted in chapter 18, in late 2015, the NPFMC, in an effort to reduce bycatch of halibut as well as Chinook salmon by trawlers in the Gulf of Alaska, began consideration of a plan to give trawlers the option to voluntarily convert to pot gear to catch Pacific cod.[999]

SALMON BYCATCH

The bane of midwater trawlers that target the region's vast pollock resource was and continues to be salmon bycatch, especially Chinook (king) salmon (*Oncorhynchus tshawytscha*), which are important to commercial and recreational fishermen and are a major food item in Native communities in Western and Interior Alaska. Adding to the problem, some Chinook salmon caught as bycatch are from Pacific Northwest stocks listed under the Endangered Species Act.[1000] Akin to the provision for bycatch halibut under the federal Prohibited Species Donation Program (originally instituted in 1996 as the Salmon Donation Program), salmon caught incidentally by trawl catcher vessels may be retained if the fish will be delivered to an authorized shore-based processor or stationary floating processor for donation to hunger relief agencies, food bank networks, or food bank distributors.[1001]

In the Bering Sea, the only non-pollock trawl fishery that catches a significant number of Chinook salmon is the Pacific cod trawl fishery, which employs bottom trawls. In the Bering Sea during the 1990s, cod trawlers typically caught from 5,000 to 7,000 Chinook salmon per year, while the pollock fishery, which employs pelagic trawls, annually caught approximately 40,000 to 60,000 Chinook salmon. Because of the consistently relatively low bycatch of Chinook salmon by cod trawlers, the North Pacific Fishery Management Council exempted this sector from Chinook salmon bycatch limits.[1002]

Not so in the Gulf of Alaska. From 1997 through 2013, the non-pollock trawl fisheries accounted for approximately 27 percent of the total groundfish trawl fishery Chinook salmon bycatch in the western and central Gulf of Alaska. (A large part of the eastern Gulf of Alaska

Table 2. Catch of Chinook salmon by Pacific cod trawlers in the Gulf of Alaska, 2003–2017.

YEAR	Chinook bycatch by Pacific cod trawlers (number of fish)
2003	3,167
2004	908
2005	41
2006	888
2007	624
2008	436
2009	111
2010	435
2011	1,351
2012	552
2013	386
2014	277
2015	1,171
2016	39
2017	2,136
Average, 2003–2017	835

Data sources: North Pacific Fishery Management Council, *Chinook Salmon Prohibited Species Catch in the Gulf of Alaska Non-Pollock Trawl Fisheries, Secretarial Review Draft* (May 2014), 164; Sam Cunningham, NOAA Fisheries, personal communication with author, April 24, 2018.

has been closed to trawling since 1998.) The average recorded bycatch of Chinook salmon by the non-pollock trawl fisheries during the years 2003–2011 was 5,991 fish, of which 884 (15 percent) were caught by trawlers targeting Pacific cod (see table 2).[1003]

In 2012, the NMFS implemented regulations limiting Chinook salmon bycatch by the pollock fisheries in the western and central Gulf of Alaska, and in January 2015, the agency implemented regulations limiting Chinook salmon bycatch by the non-pollock trawl fisheries to 7,500 fish annually. Of this limit, 2,700 Chinook salmon were allocated to non-rockfish trawl catcher vessels, which mostly targeted Pacific cod.[1004] In early May 2015, the National Marine Fisheries Service closed trawling for Pacific cod in the Gulf of Alaska, ostensibly for the remainder of the year, after cod trawl catcher vessels, with only about half of their cod allocation harvested, exceeded their annual 2,700-fish Chinook salmon bycatch limit.[1005]

The pollock fishery and other sectors subject to Chinook bycatch limits were at that time, however, well below their limits, and in June 2015, the North Pacific Fishery Management Council voted almost unanimously to request the NMFS to, by emergency order, add 1,600 Chinook salmon to the bycatch limit for the cod and flatfish trawlers for the remainder of 2015. Citing its desire to ease the economic impact of the early closure on Kodiak's economy, the NMFS agreed to do so, and in early August the trawl fleet was back on the grounds.[1006] As noted above, in late 2015, the North Pacific Fishery Management Council, in an effort to reduce bycatch of both halibut and Chinook salmon by trawlers in the Gulf of Alaska, began consideration of a plan to give trawlers the opportunity to voluntarily convert to pot gear to catch Pacific cod (see chapter 18).[1007]

ENDNOTES

960 Brad Matsen, "Billion-dollar Bottomfish Dream," *National Fisherman*, Yearbook (1985): 22–25.

961 "Cod Pots on the Rise," *Alaska Fisherman's Journal* (October 1991): 94.

962 Joel Gay, "Fleet Heads Toward Full Retention," *Pacific Fishing* (June 1996): 14.

963 Chris Oliver, "Solutions for Issues New and Old," *North Pacific Focus* (Fall 2014): 22–23.

964 Magnuson-Stevens Fishery Management and Conservation Act, as amended, § 3(2). amended by the Sustainable Fisheries Act, October 11, 1996 (110 Stat. 3561).

965 Seabirds, too, can be bycatch. The short-tailed albatross, a species listed under the Endangered Species Act, is of particular concern. As noted in the introduction to part 3, these birds are attracted to sinking baited hooks and can become hooked and drown. "Takes" of four short-tailed albatross in longline groundfish fisheries within a two-year period can interrupt or even close Alaska's demersal longline fisheries. To deter short-tailed albatross and other birds, longline vessels often trail streamers behind the vessel while setting gear (see chapter 25).

[966] Shirley Marquardt, "Halibut bycatch issue: A poster child for complex fisheries policy in Alaska," *Alaska Dispatch News* (May 25, 2015), accessed December 2, 2016, https://www.adn.com/commentary/article/bsai-halibut-bycatch-issue-poster-child-complex-fisheries-policy-alaska/2015/05/26/.

[967] Gregg H. Williams, "Incidental catch and mortality of Pacific halibut, 1962–2001," in *Int. Pac. Halibut Comm. Report of Assessment and Research Activities 2001* (Seattle: International Pacific Halibut Commission, 2002), 200.

[968] The Prohibited Species Donation program was an expansion of the Salmon Donation Program, which was instituted in 1996.

[969] Al Burch, owner/captain of trawler *Dusk*, personal communication with author, September 20, 2016; 63 *Fed. Reg.* 32144–32145, June 12, 1998; Federal Fishing Regulations for Fisheries of the Exclusive Economic Zone off Alaska, Prohibited Species Donation Program, 50 CFR § 679.26; 61 *Fed. Reg.* 38358, July 24, 1996.

[970] National Marine Fisheries Service, Alaska Region, Alaska Groundfish Fisheries: Final Programmatic Supplemental Environmental Impact Statement, Vol. 6 (June 2004), C-100.

[971] S. Salveson et al., International Pacific Halibut Commission, *Report of the Halibut Bycatch Work Group*, Tech. Rept. No. 25 (1992), 4, 12–13.

[972] Donna Parker, "Gulf Groundfish Shut Down—Kodiak Shut Out," *Alaska Fisherman's Journal* (September 1989): 16.

[973] Jeff Kauffman, in: Laine Welch, "Seattle-based Trawlers Facing Prospect of 50% Halibut Bycatch Cut," *Alaska Dispatch News*, May 15, 2016, accessed March 17, 2019, http://www.adn.com/article/20150515/seattle-based-trawlers-facing-prospect-50-halibut-bycatch-cut.

[974] North Pacific Fishery Management Council, *Current Issues* (March 2010), 21; International Pacific Halibut Commission, *The Pacific Halibut: Biology, Fishery, and Management*, Tech. Rept. No. 59 (2014), 26, 29, 38; National Marine Fisheries Service, Alaska Region, *Alaska Groundfish Fisheries, Final Programmatic Supplemental Environmental Impact Statement* (June 2004), B-32; Christine J. Blackburn and Steven K. Davis, "Bycatch in the Alaska Region: Problems and Management Measures—Historic and Current," in *Proceedings of the National Industry Bycatch Workshop, February 4–6, 1992, Newport, Oregon* (Seattle: National Resource Consultants, 1992), 88–100; Steven K. Davis, former deputy director, NPFMC, email message to Dave Witherell, March 30, 2015; David Witherell and Clarence Pautzke, "A Brief History of Bycatch Management Measures for Eastern Bering Sea Groundfish Fisheries," *Marine Fisheries Review* 1997, 59(4): 26–31; Donna Parker, "Bycatch: A Dirty Word If You're Fishing Dirty,"

Alaska Fisherman's Journal (December 1989): 26–31.

975 Robert J. Trumble, "Looking Beyond Time-Area Management of Bycatch—an Example from Pacific Halibut," in *Proceedings of the National Industry Bycatch Workshop, February 4–6, 1992, Newport, Oregon* (Seattle: National Resource Consultants, 1992), 142–149.

976 Brad Matsen, "Bravo for Bycatch Reduction Gear," *National Fisherman*, August 1993, 6.

977 Joel Gay, "Halibut By-catch Stalls, Mortality Efforts Increase," *Pacific Fishing* (April 1993): 35–36.

978 Ibid.

979 Brian Gauvin, "Save the Halibut!" *Pacific Fishing* (July 2001): 55–58, 60–61.

980 Clem Tillion, "Painful but Necessary, AFA Should Increase Fishery Yields," *Pacific Fishing* (February 1999): 6; Donna Parker, "Gulf Groundfish Shut Down—Kodiak Shut Out," *Alaska Fisherman's Journal* (September 1989): 16.

981 International Pacific Halibut Commission, *Annual Report, 2014* (Seattle: International Pacific Halibut Commission, 2015), 28.

982 66 *Fed. Reg.* 7288, January 22, 2001.

983 Joel Gay, "Halibut Bycatch Hooks Cod Longliners," *Pacific Fishing* (May 2001): 19.

984 North Pacific Fishery Management Council (staff), *Environmental Assessment/Regulatory Impact Review/Initial Regulatory Flexibility Analysis for Proposed Amendment 83 to the Fishery Management Plan for Groundfish of the Gulf of Alaska, Allocation of Pacific Cod Among Sectors in the Western and Central GOA* (May 6, 2011), 35; NOAA Fisheries, Alaska Regional Office, "NMFS Closing Hook-and-Line Gear in the GOA," February 23, 2001; National Marine Fisheries Service, *2001 Gulf of Alaska Groundfish Quotas and Preliminary Catch in Round Metric Tons* (April 3, 2002), accessed December 16, 2016, https://alaskafisheries.noaa.gov/sites/default/files/reports/goa01b.txt.

985 Diana Stram (North Pacific Fishery Management Council), *Halibut Prohibited Species Catch Limits*, appendix C, *Gulf of Alaska SAFE* (November 2003), 791, accessed December 16, 2016, http://www.afsc.noaa.gov/refm/docs/2003/GOAappendixC.pdf.

986 International Pacific Halibut Commission, *Annual Report, 2015* (Seattle: International Pacific Halibut Commission, 2016), 30.

987 To enable measurement of the excluder's effectiveness, the trawl net was fitted with a recapture bag that caught the fish that exited through the excluder.

988 Brian Gauvin, "Save the Halibut!" *Pacific Fishing* (July 2001): 55–58, 60–61; Craig S. Rose, "Development and testing of halibut excluder for Alaska cod trawling," in *Int. Pac. Halibut Comm. Report of Assessment and Research Activities 2001* (Seattle: International Pacific Halibut Commission, 2002), 236–237.

989 Jason Anderson, Beth Concepcion, and Mark Fina, *Alaska Seafood Cooperative Report to the North Pacific Fishery Management Council for the 2014 Fishery* (March 27, 2015), 9, accessed October 1, 2016, https://alaskafisheries.noaa.gov/sites/default/files/reports/asc14.pdf.

990 International Pacific Halibut Commission, *Cruise Report: Halibut bycatch survival/ sorting experiment* (1993).

991 Robert J. Trumble, *New Developments and Management Implications for Noncommercial Removals of Pacific Halibut*, International Pacific Halibut Commission, 72nd Annual Meeting, January 22–26, 1996, accessed December 23, 2016, http://www.iphc.washington.edu/documents/annmeet/1996/BLUEBOOK96.htm.

992 North Pacific Fishery Management Council, "Halibut deck sorting EFP," *News and Notes* (December 2009): 9.

993 North Pacific Fishery Management Council, "BSAI halibut bycatch," *News and Notes* (February 2014): 3.

994 John Sackton, "Halibut Deck Sorting on Track to Deliver Substantial Mortality Improvements This Season," *Seafood News*, December 3, 2015, accessed December 24, 2016, http://groundfishforum.org/halibut-deck-sorting-on-track-to-deliver-substantial-mortality-improvements-this-season.

995 International Pacific Halibut Commission, Annual Report, 2014 (Seattle: International Pacific Halibut Commission, 2015), 46–47.

996 International Pacific Halibut Commission, "Assessment of the Pacific halibut (Hippoglossus stenolepis) stock at the end of 2017," IPHC-2018-AM094-10, December 21, 2017, accessed March 14, 2019, https://iphc.int/uploads/pdf/am/2018am/iphc-2018-am094-10.pdf; International Pacific Halibut Commission, "Possible causes of low growth rates and the effects on future exploitable biomass and spawning biomass," March 2011, accessed March 14, 2019, https://www.npfmc.org/wp-content/PDFdocuments/halibut/IPHC_PSCdiscpaper311.pdf.

997 International Pacific Halibut Commission, *Annual Report, 2014* (Seattle: International Pacific Halibut Commission, 2015), 5.

[998] International Pacific Halibut Commission, *Summary of the 2016 Assessment and Harvest Policy Results* (November 2016), slide 9, accessed December 13, 2016, http://www.iphc.int/meetings/2016im/IPHC-2016-IM092-06-07-AssessmentIMP.pdf; D. J. Summers, "Halibut stock stable, flat harvest likely," *Alaska Journal of Commerce* (December 1, 2016), accessed December 13, 2016, http://www.alaskajournal.com/2016-12-01/halibut-stock-stable-flat-harvest-likely#.WFDLlvkrKM8.

[999] Laine Welch, "Trawlers may convert to pot gear for cod catches," *Alaska Fish Radio*, October 28, 2015, accessed November 1, 2015, http://www.alaskafishradio.com/trawlers-may-convert-to-pot-gear-for-cod-catches/.

[1000] North Pacific Fishery Management Council, *Environmental Assessment/Regulatory Impact Review/Regulatory Flexibility Analysis for a Proposed Amendment to the Fishery Management Plan for Groundfish of the Gulf of Alaska, Chinook Salmon Prohibited Species Catch in the Gulf of Alaska, Secretarial Review Draft* (May 2014), 62.

[1001] 61 *Fed. Reg.* 38358, July 24, 1996; Federal Fishing Regulations for Fisheries of the Exclusive Economic Zone off Alaska, Prohibited Species Donation Program 50 CFR § 679.26.

[1002] 64 *Fed. Reg.* 71390–71395, December 21, 1999.

[1003] 79 *Fed. Reg.* 32526, June 5, 2014; North Pacific Fishery Management Council, *Environmental Assessment/Regulatory Impact Review/Regulatory Flexibility Analysis for a Proposed Amendment to the Fishery Management Plan for Groundfish of the Gulf of Alaska, Chinook Salmon Prohibited Species Catch in the Gulf of Alaska, Secretarial Review Draft* (May 2014), 164.

[1004] GOA Amendment 93, 77 *Fed. Reg.* 42629, July 20, 2012; 79 *Fed. Reg.* 71350–71351, December 2, 2014.

[1005] D. J. Summers, "Council to review Gulf bycatch program, 100% trawl coverage," *Alaska Journal of Commerce*, September 30, 2015, accessed March 14, 2019, http://www.alaskajournal.com/2015-09-30/council-review-gulf-bycatch-program-100-trawl-coverage#.XItAoChKiM8; Laine Welch, "Gulf trawlers get a break from Chinook salmon bycatch closure," *Alaska Fish Radio*, August 12, 2015, accessed August 15, 2015, http://www.alaskafishradio.com/gulf-trawlers-get-a-break-from-chinook-salmon-bycatch-closure/.

[1006] Ibid.

[1007] Laine Welch, "Trawlers may convert to pot gear for cod catches," *Alaska Fish Radio*, October 28, 2015, accessed November 1, 2015, http://www.alaskafishradio.com/trawlers-may-convert-to-pot-gear-for-cod-catches/.

Chapter 25

EFFECTS OF FEDERAL ENVIRONMENTAL LEGISLATION ON ALASKA'S PACIFIC COD FISHERIES

The fear and dread of you will fall on all the beasts of the earth, and on all the birds in the sky, on every creature that moves along the ground, and on all the fish in the sea.

—Genesis 9:1–3

Three pieces of federal environmental legislation—the National Environmental Policy Act, the Marine Mammal Protection Act, and the Endangered Species Act—have directly influenced the management of Alaska's groundfish fisheries, the Pacific cod fishery included. None of these laws conflict with the Magnuson-Stevens Act, which directs fishery managers to consider ecological factors when establishing fishery harvest levels.

NATIONAL ENVIRONMENTAL POLICY ACT (1969)

The National Environmental Policy Act (NEPA), considered the cornerstone of U.S. environmental legislation, requires federal agencies to formally assess the environmental effects of proposed agency actions and federally funded actions that significantly affect the quality of the human environment. Prior to making a decision and before an action is taken, the agencies are required to, using a systematic, interdisciplinary methodology that includes public input, prepare a detailed statement—an environmental impact statement (EIS) or an environmental assessment (EA)—that assesses the potential environmental effects of a proposed action and provides alternatives to the action. The implementation of a regional fishery management plan is considered an action that significantly affects the quality of the human environment and is therefore subject to the requirements of the National Environmental Policy Act.[1007]

MARINE MAMMAL PROTECTION ACT (1972)

In the Marine Mammal Protection Act (MMPA), Congress expressed its concern that "certain species and population stocks of marine mammals are, or may be, in danger of extinction or depletion as a result of man's activities." It was Congress's desire that these species and population stocks "should not be permitted to diminish beyond the point at which they cease to be a significant functioning element in the ecosystem of which they are a part," that measures should be taken to replenish diminished stocks, and that "efforts should be made to protect essential habitats, including the rookeries, mating grounds, and areas of similar significance for each species of marine mammal from the adverse effect of man's actions."[1008]

The National Marine Fisheries Service, an agency of the Department of Commerce, is charged with protecting whales, dolphins, porpoises, seals, and sea lions. Walrus, manatees, otters, and polar bears are protected by the U.S. Fish and Wildlife Service, an agency of the Department of the Interior.

Under the Marine Mammal Protection Act, Steller sea lions are classified as *depleted* because the population is below its optimum sustainable size and because the species is listed as endangered under the Endangered Species Act (see below).[1009] Under the same legislation, Steller sea lions are also classified as a *strategic stock*—a stock for which the level of direct human-caused mortality exceeds the maximum number of animals that can be removed from a population while allowing the population to remain at or to recover to its optimum sustainable size.[1010]

ENDANGERED SPECIES ACT (1973)

The Endangered Species Act defines an *endangered species* as "any species [except certain pest insects] which is in danger of extinction throughout all or a significant portion of its range."[1011] A *threatened species* is "any species which is likely to become an endangered species within the foreseeable future throughout all or a significant portion of its range."[1012] The goals of the Endangered Species Act are to prevent the extinction of species, to foster the recovery of species listed as endangered or threatened, and to preserve the ecosystems of these species.[1013]

The legislation requires federal agencies to ensure that any action they authorize, fund, or carry out "is not likely to jeopardize the continued existence of any endangered or threatened species or result in the destruction or adverse modification of habitat" that is determined to be critical to the species.[1014]

The implementation of a fishery management plan such as those designed for the Bering Sea and Gulf of Alaska groundfish fisheries is considered a major federal action. By law, a fishery management plan contains "'all of the rules, regulations, conditions, methods, and other measures required to maintain the fisheries and prevent long-term adverse effects on fishery resources and the marine environment.'"[1015]

The U.S. Fish and Wildlife Service and the National Marine Fisheries Service are the principal federal agencies responsible for administering the Endangered Species Act. These agencies, directed by the act to use the best scientific and commercial data available, determine whether a species is endangered or threatened because of any of the following factors:

- the present or threatened destruction, modification, or curtailment of its habitat or range
- overutilization for commercial, recreational, scientific, or educational purposes
- disease or predation
- the inadequacy of existing regulatory mechanisms
- other natural or manmade factors affecting its continued existence[1016]

Individuals or organizations may petition the U.S. Fish and Wildlife Service or the National Marine Fisheries Service to have a species listed (or delisted) as threatened or endangered.

On the eastern Bering Sea and western Gulf of Alaska fishing grounds, efforts to ensure the survival of two endangered species, the Steller sea lion and the short-tailed albatross, affect the domestic Pacific cod fishery. On these grounds, fishing is restricted or prohibited in habitat designated as important for the endangered population of Steller sea lions, and the killing during a two-year period of just four short-tailed albatross, which sometimes become hooked and drown while attempting to feed on baited hooks as they are being set by longliners, can potentially shut down the Pacific cod longline fishery. The commercial fishing industry takes the Endangered Species Act very seriously.

Steller sea lion

[Killing of sea lions is permissible] in the necessary protection of property or while the animals are destroying salmon or other food fish.

—Ward T. Bower, U.S. Bureau of Fisheries, 1937[1017]

The only effective remedy for relief from harassment by sea lions was to kill each one.

—Theodore Merrell, U.S. Bureau of Commercial Fisheries, 1963[1018]

Commercial fishermen and Steller sea lions in Alaska often target the same species of fish in the same waters. Fishing activities can disrupt sea lions' routines and can result in injuries to and deaths of the large mammals. The designation of the Steller sea lion in the western Gulf of Alaska and the Aleutian Islands area as an endangered species in 1997 restricted fishermen's access to Alaska's groundfish resources (especially pollock, Atka mackerel, and Pacific cod), and it had the potential to impede the further development of Alaska's groundfish fisheries. The Steller sea lion issue—some called it a crisis—persisted for two decades, during which fishery managers, guided by the federal court and, at one point, Congress, worked to simultaneously accommodate the needs of groundfish fishermen and aid the recovery of Steller sea lions.

The Steller sea lion (*Eumetopias jubatus*) is the largest of the sea lions. Males, which are much larger than females, can be eleven feet long and weigh 2,500 pounds. Steller sea lions inhabit the coastal regions of the North Pacific Ocean from Japan to California and feed mostly on a variety of fishes, among them Pacific cod (especially during the winter months). During late spring and early summer,

Figure 90. Steller sea lions, Prince William Sound, Alaska. (James Mackovjak)

Steller sea lions gather for pupping and breeding on favored beaches, called rookeries, where individual males (bulls) compete to establish a territory and a harem of from three to twenty females. The strongest bulls dominate the rookery and have the largest harems. The bulls aggressively defend their territory and harem; they do not leave their territory and do not feed for the duration of the breeding season, which extends approximately into August.

Females give birth—almost always to single pups—shortly after arriving at a rookery and within a few weeks are ready for mating. Over approximately the next several months, the females periodically leave their pups ashore and go to sea to feed. They may be gone for from several hours to, as the pups grow older, a day or more, before returning to the rookery to nurse their pups. By late summer, the females with their pups, as well as the breeding males, have left the rookery for the open sea, where individuals disperse widely. Steller sea lions, however, are social, and they also gather outside the breeding season. About 70 percent of the world's Steller sea lion population resides off Alaska's coast.[1019]

The predecessor agencies of the National Marine Fisheries Service (established in 1970) began collecting information on the abundance of Steller sea lions during the 1950s, and in 1961 the world population of the species was estimated to be between 240,000 and 300,000 animals.[1020] But during the 1980s, just as Alaska's groundfish fisheries were becoming Americanized and expanding, the western Alaska Steller sea lion population began a precipitous decline. At some rookeries in Alaska, the number of Steller sea lions declined 63 percent between 1985 and 1990. Moreover, the declines were spreading to previously stable areas and accelerating.[1021]

Scientists are uncertain about the exact cause(s) of the Steller sea lion population decline. Climate change, disease, pollutants, and predation by killer whales and sharks are among the factors considered, but the proliferation of the industrial groundfish fisheries along Alaska's coast approximately simultaneous with the decline was considered a likely cause.[1022]

Commercial fishermen interact with sea lions directly and indirectly. Directly, sea lions are injured or killed as a result of being caught by fishing gear—almost exclusively trawls. Based on information collected by on-board observers, the National Marine Fisheries Service estimated that foreign and joint venture commercial trawl fisheries in the Gulf of Alaska and Bering Sea during the years 1973–1988 had incidentally killed a total of 14,000 Steller sea lions. In 1984, the National Marine Fisheries Service issued six general five-year permits that authorized the annual incidental take by commercial fishermen of a total of 2,880 Steller sea lions in the North Pacific Ocean.

Though it is unknown how often this happened, some fishermen intentionally shot Steller sea lions. These shootings occurred at rookeries, haul-out sites, and in the water near boats. (A crew member on a Pacific cod trawler that fished near Kodiak during the late 1970s recalled that the vessel's captain kept a shotgun handy and shot any sea lion that came within range. Few, if any, of the shot animals died immediately, but most seemed to have been blinded and likely died a slow death.)

Indirectly, the commercial fishing industry interacts with sea lions by competing with the species for prey resources and because fishing activities disturb sea lions' routines. These indirect interactions are difficult to document because they are complex and measuring them is a challenge.[1023]

In 1997, based on demographic and genetic dissimilarities, the National Marine Fisheries Service designated two distinct population segments of Steller sea lions. The segments are defined by geography: the western population inhabits the Bering Sea, Aleutian Islands, and the Gulf of Alaska west of the 144° west longitude meridian (approximately Cape Suckling, near Cordova). The eastern population's range begins at this meridian and extends east across the Gulf of Alaska and then south along the coast to California.[1024]

THE STELLER ERA

> *We have to protect the [Steller sea lion] at all costs. The law is very emphatic about this. . . . I was sent up here to scare the fish industry. And there's good reason to be scared.*
>
> —Karl Gleaves, NOAA General Counsel, 1990[1025]

> *[F]rom filing of the suit, Greenpeace v. National Marine Fisheries Service, in April 1998 through March 2003, the court (and Judge Thomas S. Zilly) was the effective manager of the North Pacific commons.*
>
> —Jerry McBeath, University of Alaska, 2004[1026]

What has been referred to as the Steller Era began in November 1989, when the Environmental Defense Fund and seventeen other environmental organizations petitioned the National Marine Fisheries Service for an emergency rule listing all populations of Steller sea lions in

Alaska as endangered. The following April, the agency, which in 1988 had considered listing the Steller sea lion population in Alaska as *depleted* under the Marine Mammal Protection Act, listed the species as *threatened* under the Endangered Species Act. This was the first-ever listing of a coastal fish-eating marine mammal.[1027] The listing obligated the National Marine Fisheries Service to ensure that its actions were not likely to jeopardize the continued existence of this species or to destroy or adversely modify habitat that was critical to its health and survival.[1028] The agency not long after appointed a broad-based Steller Sea Lion Recovery Team to develop a recovery plan for the species. The agency also instituted restrictions to help protect sea lions. Among the restrictions were a reduction of the allowable annual incidental take of Steller sea lions by commercial fishermen from 1,350 to 675 in Alaska waters west of the 141° west longitude meridian (just east of Icy Bay, on the Gulf of Alaska, near Yakutat) and a requirement that vessels stay at least three nautical miles from thirty-two listed Steller sea lion rookery sites.[1029]

In 1993, the National Marine Fisheries Service designated an aquatic zone extending twenty nautical miles seaward of each major rookery and each major haulout in Alaska west of 144° west longitude as *critical habitat*—an "area considered essential for the health, continued survival, and recovery of the Steller sea lion population, [and that] might require special management consideration and protection."[1030] Additional critical habitat was later identified, and by late 2014 Steller sea lion critical habitat in the Aleutian Islands comprised 29,239 square nautical miles.[1031]

There was strong concern within the commercial fishing industry and among its supporters that the listing of the Steller sea lion population as threatened would seriously curtail or possibly end commercial fishing, especially trawl fishing, in the Steller sea lion's range. A 1994 amendment to the Marine Mammal Protection Act, however, allows commercial fishermen, under a permit system administered by the National Marine Fisheries Service, to incidentally *take*—harass (intentionally or unintentionally), capture, or kill—marine mammals while engaged in commercial fishing operations.[1032]

Pursuant to the Endangered Species Act, the National Marine Fisheries Service, in consultation with experts from outside the agency, had beginning in 1990 prepared biological opinion papers (BiOps) regarding the potential negative effects of groundfish fisheries on the Steller sea lion population. The BiOps also recommended actions "to avoid jeopardy for the western population of Steller sea lions and to avoid adverse modification of its habitat."[1033]

The agency's 1996 Steller sea lion BiOp concluded that the current fishery management plans for the Bering Sea and Gulf of Alaska did not jeopardize the Steller sea lion, and the agency continued to manage the fisheries on a fundamentally status quo basis.[1034]

In June 1997, however, the agency reclassified the western Steller sea lion population, which during the 1990s was declining 5 percent annually, as endangered.[1035] Despite this important listing, the National Marine Fisheries Service's focus at that time was on fisheries enhancement, not species preservation.[1036] The agency failed to take immediate measures to protect the endangered Steller sea lion population while it allowed the massive groundfish fisheries to proceed essentially unchanged. This spurred environmental organizations to act. In April 1998, Greenpeace, the Sierra Club, and the American Oceans Campaign jointly sued the National Marine Fisheries Service in the U.S. District Court (Washington), accusing it of implementing groundfish management plans in the absence of a comprehensive environmental impact statement or biological opinions that addressed the full, overall impact of the groundfish fisheries on the endangered Steller sea lion.[1037] [1038]

The organizations asked that all groundfish trawl fishing in designated critical habitat of the endangered Steller sea lions be prohibited until the National Marine Fisheries Service issued a legally adequate biological opinion that addressed the effects of the North Pacific groundfish fisheries on the Steller sea lion and its critical habitat.[1039]

The National Marine Fisheries Service successfully argued that the court's proceedings should be stayed until the agency completed a supplemental environmental impact statement and two new BiOps. The agency released the supplemental material in December 1998.[1040]

The first opinion (BiOp1) specifically addressed the pollock and Atka mackerel fisheries planned for the years 1999–2001 and concluded that the pollock fishery alone was a threat to Steller sea lions.[1041] The second opinion (BiOp2) purported to broadly address the Bering Sea and Gulf of Alaska groundfish fisheries but focused on the total allowable catch specifications for only a single year, 1999. It concluded that, provided measures were taken to address the pollock fishery and other, as yet unspecified, conservation measures were implemented, the Bering Sea and Gulf of Alaska groundfish fisheries were "not likely to jeopardize the continued existence of the Steller sea lion or adversely modify its critical habitat." BiOp2 specifically stated that the Pacific cod fishery was not likely to "appreciably reduce the likelihood of the survival and recovery of Steller sea lions."[1042]

The court, which held several hearings regarding this case,

focused on BiOp2, which, it said, was supposed to address the effects of groundfish fishery management plans in their entirety upon the endangered Steller sea lion population. In January 2000, the court ruled in favor of the environmental organizations, faulting the National Marine Fisheries Service for failing "to critically analyze how core management measures such as the processes for deriving acceptable biological catch, overfishing, and total allowable catch, impact endangered species."[1043] Likewise, the court found that BiOp2 contained "no explanation of how the various fishery management measures interrelate and how the overall management regime may or may not affect Steller sea lions."[1044]

The year 2000 marked the western Steller sea lion population reaching its smallest size, having declined by almost 90 percent throughout its range.[1045] In August of that year, the court enjoined all groundfish trawl fishing within Steller sea lion critical habitat until further order from the court.[1046] The pollock industry was most affected. In November, the National Marine Fisheries Service submitted to the court a plan to mitigate harm to the sea lions, and in early December, the court ruled this plan sufficient, and immediately lifted the injunction.[1047]

The agency's new plan was severe, incorporating, among other provisions, the closure of two-thirds of the Steller sea lion critical habitat to all fishing for pollock, Atka mackerel, and Pacific cod. The Alaska groundfish industry strongly objected to the plan, as did the North Pacific Fishery Management Council, which, as noted earlier (see chapter 12), consists mostly of commercial fishing industry interests. Trawlers at Kodiak called the plan "draconian," and the Groundfish Forum, a Kodiak-based trade group that represents five companies that operate factory trawlers that primarily produced frozen headed-and-gutted (H&G) groundfish, estimated that its sector alone would lose approximately $70 million in revenue annually as a result of the new regulations. At Homer, a fleet of about fifty longliners (many of them operated by Russian-American Old Believers from the nearby community of Nikolaevsk) and another two dozen pot boats anticipated being idled. Alaska senator Ted Stevens characterized the plan as "a crippling blow to the small-boat fishermen and Alaska's coast communities."[1048]

The groundfish industry, with the North Pacific Fishery Management Council, quickly enlisted Senator Stevens, who was at that time chairman of the Senate Appropriations Committee. A must-pass appropriations bill was in the final stages of making its way through Congress, and Senator Stevens attached to the bill a rider that was primarily written by Alaska groundfish industry interests. The language was unacceptable to then-president Bill Clinton, who threatened a veto. After compromise language was incorporated into the rider, the spending bill passed on December 21, 2000.[1049] Among its provisions, the rider ordered that the Bering Sea/Aleutian Islands and Gulf of Alaska groundfish fisheries in 2001 be managed under the plans in effect prior to July 15, 2000—the same plan the court found unacceptable.[1050] To groundfish fishermen, according to *Pacific Fishing*, the rider was like "early Christmas."[1051] The rider also included a $30 million appropriation to compensate "fishing communities, businesses, community development quota groups, individuals, and other entities to mitigate the economic losses caused by Steller sea lion protection measures heretofore incurred," and a $20 million appropriation for the study of Steller sea lions.[1052] [1053]

The rider also directed the National Marine Fisheries Service to work with the North Pacific Fishery Management Council to craft a legally acceptable plan to protect the endangered Steller sea lions.[1054] Considerable back and forth continued, including the issuance of a BiOp by the National Marine Fisheries Service in 2001 that was, like its predecessor, found by the U.S. District Court (Washington) to be deficient.[1055] In April 2003, following a remand order by the court, the National Marine Fisheries Service issued a supplemental BiOp that—with mutual agreement of the plaintiffs, defendants, and intervenors—the court approved as compliant with the National Environmental Policy Act and the Endangered Species Act.

In early 2001, the National Marine Fisheries Service implemented a suite of emergency Steller sea lion protection measures. Among them was the division of the Pacific cod fisheries in the Gulf of Alaska and Bering Sea/Aleutian Islands area into two seasons. Sixty percent of the total allowable Pacific cod catch (allocated among the various sectors) was apportioned to the A season, and 40 percent was apportioned to the B season. The A season opened January 1 and closed June 10 or when that season's total allowable catch was reached, whichever came first, and the B season opened June 10 and closed December 31 or when that season's total allowable catch was reached, whichever came first.[1056]

In past years, the Pacific cod fishery (all gear types) in the Gulf of Alaska and the Bering Sea/Aleutian Islands area Pacific cod trawl fishery typically concluded by mid-March. The intent of the split season and the 60/40 apportionment was to provide temporal dispersion of the harvest to lessen the intensity of competition with Steller sea lions. But there was a price to pay. Fishermen objected to the split season because Pacific cod tend to aggregate (school) during the early part of the calendar year and are relatively easy to locate and catch. Later in the year, the fish become disaggregated and tend to move to deeper water, and catching them requires more fishing time and effort. This may increase bycatch and inhibit full utilization of the total allowable catch.[1057] In several instances, the seasons were later adjusted for various gear types,[1058] but the split-season regime persists to this day.[1059]

The Steller sea protection measures seemed to help, but the issue was far from resolved. During the years 2002–2008, the overall population of western Steller sea lions was estimated to be stable or increasing slightly.[1060] Nevertheless, in a 428-page BiOp released in late November 2010, the National Marine Fisheries Service determined that the ongoing authorization of the North Pacific groundfish fisheries was jeopardizing Steller sea lion recovery, particularly in the westernmost area of the Aleutian Islands (Area 543), where the Steller sea lion population had declined approximately 45 percent during the years 2000–2008. The agency determined that fisheries for Steller sea lion prey might be appreciably reducing the species' reproduction, but it acknowledged that competition with fisheries for prey was likely "one component of an intricate suite of natural and anthropogenic factors affecting Steller sea lion numbers and reproduction."[1061]

The BiOp, which had taken four years to prepare, concluded that Steller sea lions in the Aleutian Islands relied primarily on two key prey species: Atka mackerel and Pacific cod—the targets of the largest fisheries in the central and western Aleutian Islands.[1062] The BiOp contained a number of what the National Marine Fisheries Service considered "Reasonable and Prudent Alternatives" (regulatory changes) for management of the Atka mackerel and Pacific cod fisheries in the central and western Aleutian Islands. Fundamentally, these regulations consisted of a combination of area closures, seasonal restrictions, gear restrictions, and catch limits.

The regulations severely restricted fishing. In the western Aleutians (Area 543), an area almost half the size of Texas, fishing for Atka mackerel and Pacific cod was prohibited entirely. In the central Aleutians (areas 541 and 542), the National Marine Fisheries Service imposed catch limits and an array of temporal and gear restrictions on Atka mackerel and Pacific cod fishing. Among their provisions, the new regulations outlawed commercial fishing year-round for Pacific cod within ten nautical miles of Steller sea lion rookeries and haulouts. In the zone between ten and twenty nautical miles from these sites, seasonal restrictions were established for both trawl and fixed gear.[1063]

The National Marine Fisheries Service believed it was necessary for these regulations to be "implemented quickly in order to halt the immediate effects of the fisheries on the acute population decline in the western portion of the range of the western . . . Steller sea lion." The agency published the regulations on December 13, 2010, and, waiving the standard thirty-day delay in the implementation of a final rule, implemented the regulations on January 1, 2011, the day the Pacific cod hook-and-line, pot, and jig fisheries were scheduled to open.[1064]

Perhaps ironically, the Bering Sea/Aleutian Islands area Pacific cod stock was growing, and the National Marine Fisheries Service had increased the total allowable catch for 2011 fully 35 percent over the previous year's—from 168,780 metric tons in 2010, to 227,950 metric tons in 2011. The following year, the agency increased the total allowable catch to 261,000 metric tons, the highest it had been since 1997. The pollock total allowable catch, too, was growing, but that for Atka mackerel was declining substantially.[1065]

The new sea lion regulations were, to no one's surprise, strongly opposed by the groundfish industry, which stood to potentially lose tens of millions of dollars annually, and its allies. (The industry's most powerful ally, Senator Ted Stevens, lost his senate seat in the 2008 election.)

Among the fishing industry's remaining allies was the State of Alaska, which, hoping to block the new rules, filed suit in the U.S. District Court (Alaska) against the National Marine Fisheries Service the day after the new regulations were published. The state claimed the agency had failed to make a "rational connection" between the information it had considered and its conclusion that the population of Steller sea lions in the central and western Aleutian Islands were declining due to nutritional stress. The state charged that the National Marine Fisheries Service's regulatory action, among other purported legal transgressions, had violated the National Environmental Policy Act because the agency had not supported the new regulations with a full environmental impact statement, and the state asked the court for injunctive relief. In early January 2011, the Alaska Seafood Cooperative, which represented about 90 percent of the Atka mackerel quota in Area 543, and several other seafood organizations filed a similar suit, as did the Freezer Longline Coalition, whose vessels primarily targeted Pacific cod. The environmental groups Oceana and Greenpeace intervened in support of the National Marine Fisheries Service.[1066] At stake in the court proceedings that followed were fish harvests worth many millions of dollars, jobs, the credibility of the National Marine Fisheries Service, and, of course, the health of the Steller sea lion population.

The legal challenges were consolidated in February 2011, and the court in March 2012 determined that the National Marine Fisheries Service's Steller sea lion protection regulations complied with the requirements of the Endangered Species Act, but that in crafting the regulations, the agency had failed to provide sufficient environmental information for informed public comment on the issue, and, moreover, it had not provided for adequate public participation. The court, therefore, left the protective regulations in place but ordered the National Marine Fisheries Service to complete an environmental impact statement for the Steller sea lion protection measures by August 15, 2014.[1067] In July 2013, the U.S. Court of Appeals for the Ninth Circuit affirmed the court's decision upholding the Steller sea lion protection regulations.[1068]

In May 2014, the National Marine Fisheries Service released the court-mandated environmental impact statement and an associated biological opinion. Neither of these documents focused on the Steller sea lion protection measures that had been reviewed and upheld by the court. Instead, the National Marine Fisheries Service charted a completely new course, proposing that Steller sea lion protection measures be substantially relaxed in the Aleutian Islands, particularly for the Atka mackerel and pollock fisheries but also for the Pacific cod fishery. In contrast to earlier BiOps, the 2014 BiOp was premised on the assumption that competition between the fisheries and Steller sea lions was only likely if the sea lions' foraging activity and commercial fishing activity overlapped to a high degree in time, space, depth, and size of prey. Based on the limited data available, the 2014 BiOp concluded that even with a rollback of the protection measures adopted for Pacific cod and Atka mackerel in 2010 and the revival of a pollock fishery that had been banned from critical habitat since 1999, there was not sufficient overlap in the Aleutian Islands fisheries to jeopardize Steller sea lions or to threaten their critical habitat.[1069] According to the National Marine Fisheries Service, it was possible that the fisheries could affect Steller sea lion prey availability, but there was no evidence that nutritional stress was "sufficiently prevalent to reduce birth rate and/or increase the death rate" in the endangered population.[1070]

Fundamental to Pacific cod management, the National Marine Fisheries Service had recently determined that the Aleutian Islands Pacific cod stock was distinct from the Eastern Bering Sea stock, and beginning in 2014 the agency began managing the Aleutian Islands Pacific cod fishery separately from the Bering Sea fishery. In prior years, there had been no limit on the amount of the Bering Sea/

Aleutian Islands area total allowable catch that could be harvested in the Aleutian Islands, which allowed an inordinate amount of fishing pressure on the islands' cod stock. This split resulted in a dramatic reduction in the amount of Pacific cod available to directed fishing in the Aleutian Islands. The 2014 directed Pacific cod-fishing allowance for the entire Aleutian Islands was 4,248 metric tons, a 72 percent reduction in the average Pacific cod catch there during prior years.[1071]

Under the new regulations evaluated in the 2014 BiOp and environmental impact statement, which went into effect in late December 2014 and applied as well to parallel fisheries in state waters (see chapter 23) in the central and western Aleutian Islands, the previous restrictions on fishing for Atka mackerel, pollock, and Pacific cod were relaxed. Regarding Pacific cod, 23 percent more critical habitat area was open to fishing for this species with non-trawl gear than had been open under the regulations implemented in 2011. The new regulations also, among several area-specific critical-habitat provisions, prohibited directed fishing for Pacific cod with trawl gear in waters within three nautical miles of haulouts and within ten nautical miles of rookeries. Additionally, directed fishing for Pacific cod with hook-and-line and pot gear was prohibited in waters within three nautical miles of rookeries.[1072] The National Marine Fisheries Service estimated that the new regulations would relieve roughly two-thirds of the economic burden imposed on Aleutian Islands fishermen by the Steller sea lion protection measures that were implemented in 2011.[1073]

Almost simultaneous with the implementation of the relaxed fishing regulations, the environmental groups Oceana and Greenpeace filed charges against the National Marine Fisheries Service in the U.S. District Court (Alaska) over the agency's authorization of increased "industrial" fishing in the western and central Aleutian Islands. Specifically, the groups charged the National Marine Fisheries Service had used a "novel scientific approach" that was not consistent with previous biological opinions in its decision to substantially scale back Steller sea lion protections.[1074] In August 2015, the court ruled in favor of the agency, stating that the agency, in its preparation of the 2014 BiOp and the environmental impact statement associated with it, had fulfilled its legal obligations under the Endangered Species Act, the National Environmental Policy Act, and the Administrative Procedure Act. In August 2017, the U.S. Court of Appeals for the Ninth Circuit issued a memorandum affirming this ruling.[1075]

Steller sea lions, 2015

The population of Steller sea lion pups is considered a good indicator of the overall Steller sea lion population. In its 2014 assessment of Alaska marine mammals, the National Marine Fisheries Service reported that there was strong evidence that pup counts had increased in the overall western Steller sea lion stock, though there was considerable regional variation. Pup counts were stable in the central Aleutian Islands but were decreasing rapidly in the western Aleutian Islands.[1076]

The accounting is somewhat complicated, but some government officials long involved in the Steller sea lion issue estimate that since 1992 federal funding for Steller sea lion research has totaled about $175 million. Among the agencies and institutions that have been involved in federally funded Steller sea lion research are the National Marine Fisheries Service, Alaska Department of Fish and Game, Alaska SeaLife Center (Seward, Alaska), and the University of Alaska.[1077]

Given this expenditure and the amount of time that has passed, one would expect scientists today to have a better understanding of the cause(s) of the decline of Steller sea lion populations in the western Gulf of Alaska and the Aleutian Islands area, but this is not the case. In fact, their research seems to have generated more questions than answers.[1078] Meanwhile, the Endangered Species Act is among the laws of the land (and sea), and those who earn their livelihoods catching pollock, Atka mackerel, and Pacific cod have their fingers crossed, hoping Steller sea lions thrive.

SHORT-TAILED ALBATROSS: "FEAR OF FEATHERS"[1079]

> *It's unlucky to kill an albatross or a gull at sea, as they host the souls of dead sailors.*
>
> —Maritime superstition[1080]

> *Although seabirds are not included within the Magnuson-Stevens Act's "bycatch" definition, efforts to reduce the incidental take of seabirds in fisheries are consistent with the Magnuson-Stevens Act's objective to conserve and manage the marine environment.*
>
> —National Marine Fisheries Service, 2016[1081]

Worldwide, there are twenty-two species of albatross, three of which are found in the North Pacific. As a group, albatross are the most endangered birds on Earth, with all but one species vulnerable to extinction due to factors that include predation by introduced species (e.g., rats and feral cats), pollution, climate change, and longline fishing.

The short-tailed albatross (*Phoebastria albatrus*) is the largest seabird in the North Pacific. Its wings are adapted to soaring low over the ocean surface and can span over seven feet. Short-tailed albatross feed at the water surface for prey that includes squid, crustaceans, and fish. These birds are slow to mature and long-lived. They first breed when five or six years old and can live to be forty-five years old.[1082]

Once numbered in the millions, short-tailed albatross breed during the fall primarily on several islands—the primary one of which is actively volcanic—south of the main islands of Japan. Females lay only a single egg that, if destroyed, is not replaced. During the late 1800s and early 1900s, feather hunters clubbed short-tailed albatross nearly to extinction. The Japanese government designated the short-tailed albatross as a protected species in 1958.[1083] The State of Alaska, too, recognized the short-tailed albatross's need for protection. In 1972, Alaska listed the bird as endangered under the state's endangered species act, which became law in 1971. The well-intentioned yet largely toothless legislation does little more than direct the commissioners of the Department of Fish and Game and the Department of Natural Resources to "take measures to preserve the natural habitat" of species the commissioner of the Department of Fish and Game recognizes as threatened with extinction.[1084] Technically, the natural habitat of the short-tailed albatross in Alaska, since there is no breeding population, is the open ocean.

The Endangered Species Conservation Act of 1969, predecessor to the Endangered Species Act, maintained two lists of endangered wildlife: one for foreign species and one for species native to the United States. In 1970, the short-tailed albatross appeared only on the foreign species list.[1085] The Endangered Species Conservation Act was superseded by the Endangered Species Act in 1973. The following year, the native and foreign lists were combined into one list of endangered and threatened species. Because of an administrative error, however, the short-tailed albatross was omitted from this list.[1086]

In an attempt to correct this administrative error, in 1980 the U.S. Fish and Wildlife Service proposed listing the short-tailed albatross as endangered throughout its range. No action was taken, however, until two decades later (see below).[1087]

Concern over the short-tailed albatross was elevated in 1987, when a vessel longlining halibut in the central Gulf of Alaska found and reported an unexpected catch on one of its hooks: a short-tailed albatross. This incident, the first reported take of a short-tailed albatross in an Alaska longline fishery, established the potential for short-tailed albatross to be killed in the fisheries.[1088] The worldwide population of short-tailed albatross in 1988 was estimated to be four hundred.[1089]

In 1989, the U.S. Fish and Wildlife Service issued a formal opinion that commercial fishing operations in Alaska could adversely affect the short-tailed albatross. The threat was the hooking and drowning of individual birds on commercial longline gear such as is used in the commercial halibut, sablefish, and Pacific cod fisheries. The birds are attracted to the bait while the gear is being set and, when attempting to seize it, sometimes become hooked and are dragged underwater and drowned. To help reduce potential fisheries-caused deaths of short-tailed albatross, the U.S. Fish and Wildlife Service, in consultation with the National Marine Fisheries Service, that year established an incidental take of two short-tailed albatross per year in the groundfish fisheries. Exceeding this take could trigger additional measures—including the closure of fisheries—to avoid killing short-tailed albatross.[1090] The population of short-tailed albatross, however, was growing; in 1994 it was estimated to be seven hundred.[1091]

The seabird bycatch problem in the groundfish fisheries was highlighted in 1995, when two short-tailed albatross were taken in Alaska's sablefish fishery. Another short-tailed albatross was taken in 1996, the same year U.S. Fish and Wildlife Service designated the species as a candidate for listing under the Endangered Species Act.[1092] At the same time, the incidental take was liberalized somewhat, to four birds in two years.[1093]

Longline fishermen understood clearly the potentially dire impacts an endangered species listing for the short-tailed albatross could have on their industry. Led by Thorn Smith, executive director of the North Pacific Longline Association (now defunct), which represented most of the freezer-longliner fleet, Pacific cod fishermen took a proactive stance to minimize their take of short-tailed albatross and other seabird species. The fishermen worked with government agencies and the Washington Sea Grant program (at the University of Washington) to develop methods to avoid seabird mortality, and they urged the National Marine Fisheries Service to develop regulations to require vessels to use seabird avoidance measures while fishing in Alaska waters.[1094] The North Pacific Longline Association's booth at Seattle's Pacific Marine Expo ("Fish Expo"), an annual exhibition and conference about the marine business, in November 1996 was devoted to avoiding seabird bycatch.[1095] Shannon Fitzgerald, a National Marine Fisheries Service seabird biologist, praised the freezer-longliner fleet as "one of the most proactive fleets anywhere in the world in trying to reduce their bycatch of seabirds."[1096]

The effective and relatively economical deterrent to seabird bycatch was streamer lines, also known as *tori lines* or *bird lines*, as had been required of longline fishermen in Australia and New Zealand for a number of years.[1097] These lines, which are most effective when used in pairs, are bird-scaring devices in which one end of a fifty-fathom line festooned with brightly colored plastic streamers is attached to a high point on a vessel's stern and towed while the vessel is setting gear. The line, a buoy fastened to its end, trails above the gear being set, its streamers whipping and moving unpredictably with the wind and the vessel's motion, scaring birds away. A study testing seabird bycatch deterrents in the Alaska sablefish and Pacific cod fisheries showed that streamer lines deployed in pairs reduced seabird bycatch by 88 to 100 percent.[1098]

In the spring of 1997, the National Marine Fisheries Services issued regulations requiring operators of hook-and-line vessels fishing for groundfish in the Bering Sea/Aleutian Islands area and the Gulf of Alaska to employ specified bird-avoidance techniques to reduce seabird bycatch and incidental seabird mortality.[1099]

In July 2000, with the worldwide short-tailed albatross population numbering approximately 1,200 birds, the U.S. Fish and Wildlife Service listed the species as endangered throughout its range under the federal Endangered Species Act.[1100] Currently, there is no designated short-tailed albatross critical habitat in the United States.

This listing, however, had little effect on the longline fleet, which had already been taking steps to protect the bird. "We've been treating it as endangered anyway," said Thorn Smith.[1101] In 2006, Smith reported that longliners had not taken a short-tailed albatross since 1998.[1102] Although incidental take levels have never been exceeded, the incidental catch of three short-tailed albatross in Alaska's longline fisheries during four months in 2014 came uncomfortably close. The two-year period during which these birds were taken ended in mid-September 2015.[1103]

Credit for success in preventing the killing of short-tailed albatross is also due to the U.S. Fish and Wildlife Service, which beginning in 2000 supplied free paired streamer lines to any commercial longline vessel owner/operator who requested them.[1104] The agency also reimbursed owners of longline vessels that were 100 feet or more in length for half the cost associated with the installation of streamer line deployment booms.[1105]

In 2014, Mustad Autoline, the Norwegian company known for its longline systems, and SeaWave, a Dutch firm, introduced their SeaBird Saver, which uses a laser beam to deter birds. The developers market the device as being most effective at dawn and dusk, during the night, and during rainy and foggy conditions. As of this writing, one freezer-longliner in the Alaska fleet employs a SeaBird Saver, but its effectiveness and whether it may be injuring birds have yet to be definitively determined.[1106]

The 2014 worldwide total population of short-tailed albatross was approximately 4,354 individuals, including approximately 1,928 that were of breeding age. A more recent population model estimated the 2016 to 2017 world population of short-tailed albatross to be approximately 5,144 individuals.[1107]

ENDNOTES

1007 National Environmental Policy Act of 1969, January 1, 1970 (83 Stat. 852), § 102(C).

1008 Marine Mammal Protection Act, October 21, 1972 (86 Stat. 1027), § 2(1–2).

1009 Marine Mammal Protection Act, October 21, 1972 (86 Stat. 1027), § 3(1)(A), § 3(1)(C).

1010 Marine Mammal Protection Act, October 21, 1972 (86 Stat. 1027), § 3(19)(A), § 3(20).

1011 Endangered Species Act of 1973, December 28, 1973 (87 Stat. 884), § 3(6).

1012 Endangered Species Act of 1973, December 28, 1973 (87 Stat. 884), § 3(20).

1013 Lowell W. Fritz, Richard C. Ferrero, and Ronald J. Berg, "The Threatened Status of Steller Sea Lions, *Eumetopias jubatus*, under the Endangered Species Act: Effects on Alaska Groundfish Fisheries Management," *Marine Fisheries Review* (March 1995): 14–27.

1014 Endangered Species Act of 1973, December 28, 1973 (87 Stat. 884), § 7(a)(2).

1015 *Greenpeace v. National Marine Fisheries Service*, 80 F. Supp. 2d 1137 (W.D. Wash. 2000); 16 U.S.C. § 1802(5).

1016 Endangered Species Act of 1973, December 28, 1973 (87 Stat. 884), § 4(b)(1)(A), § 4(a)(1)(A–E).

1017 Ward T. Bower, *Alaska Fishery and Fur-Seal Industries in 1937* (Washington, DC: GPO, 1938), 75.

[1018] Theodore R. Merrell Jr., "Sea Lions Create Havoc," *Pacific Fisherman* (November 1963): 13, 15.

[1019] 58 *Fed. Reg.* 17182, April 1, 1993; NOAA Fisheries, Office of Protected Resources, "Steller Sea Lion (*Eumetopias jubatus*)," fact sheet, accessed March 17, 2019, https://www.fisheries.noaa.gov/species/steller-sea-lion; David Witherell and Jim Armstrong (staff, North Pacific Fishery Management Council), *Groundfish Species Profiles, 2015*, 4, accessed December 2, 2015, www.npfmc.org/wp-content/PDFdocuments/resources/SpeciesProfiles2015.pdf.

[1020] 67 *Fed. Reg.* 956, January 8, 2002; 58 *Fed. Reg.* 17182, April 1, 1993.

[1021] 55 *Fed. Reg.* 12645, April 5, 1990.

[1022] Alaska Fisheries Science Center, "Steller Sea Lions in Alaska: Complexity of the Problem," accessed September 27, 2015, http://www.afsc.noaa.gov/archives/stellers/ssl_entrance.htm.

[1023] 55 *Fed. Reg.* 12647, April 5, 1990; 53 *Fed. Reg.* 16300, May 6, 1988; Lowell W. Fritz, Richard C. Ferrero, and Ronald J. Berg, "The Threatened Status of Steller Sea Lions, *Eumetopias jubatus*, under the Endangered Species Act: Effects on Alaska Groundfish Fisheries Management," *Marine Fisheries Review* (March 1995): 14–27; anonymous by request, personal communication with author, August 28, 2015.

[1024] 62 *Fed. Reg.* 24345, May 5, 1997.

[1025] Krys Holmes, "Sea Lion Cure Could Be Painful . . . And Slow," *Alaska Fisherman's Journal* (April 1990): 8, 11.

[1026] Jerry McBeath, "Management of the commons for biodiversity: lessons from the North Pacific," *Marine Policy* (November 2004): 523–539.

[1027] Mark Buckley, "Steller Era Starts," *Pacific Fishing* (February 1999): 32–34; 53 *Fed. Reg.* 16299, May 6, 1988; 55 *Fed. Reg.* 12645, April 5, 1990.

[1028] 79 *Fed. Reg.* 70286, November 25, 2014.

[1029] 55 *Fed. Reg.* 12645, April 5, 1990.

[1030] 58 *Fed. Reg.* 45269, August 27, 1993.

[1031] 79 *Fed. Reg.* 70286, November 25, 2014.

[1032] Marine Mammal Protection Act, October 21, 1972 (86 Stat. 1027), as amended, § 101(a)(2), § 101(a)(5)(E).

1033 67 *Fed. Reg.* 956, January 8, 2002.

1034 Jerry McBeath, "Management of the commons for biodiversity: lessons from the North Pacific," *Marine Policy* (November 2004): 523–539.

1035 National Marine Fisheries Service, *November, 2010 North Pacific Groundfish Fishery Biological Opinion*, 81; 62 *Fed. Reg.* 30772, June 5, 1997.

1036 Jerry McBeath, "Management of the commons for biodiversity: lessons from the North Pacific," *Marine Policy* (November 2004): 523–539.

1037 The administrative record for this case, which lasted five years, includes more than 50,000 pages of documents.

1038 *Greenpeace v. National Marine Fisheries Service*, 237 F. Supp. 2d 1181 (W.D. Wash. 2002).

1039 *Greenpeace v. National Marine Fisheries Service*, 55 F. Supp. 2d 1248 (W.D. Wash. 1999); *Greenpeace v. National Marine Fisheries Service*, 106 F. Supp. 2d 1066 (W.D. Wash. 2000).

1040 67 *Fed. Reg.* 956, January 8, 2002; *Greenpeace v. National Marine Fisheries Service*, 80 F. Supp. 2d 1137 (W.D. Wash. 2000).

1041 National Marine Fisheries Service, *Endangered Species Act, Section 7 Consultation, Biological Opinion* (walleye pollock and Atka mackerel fisheries, 1999–2002), December 3, 1998.

1042 National Marine Fisheries Service, *Biological Opinion on 1999 TAC Specifications for Groundfish Fisheries in the BSAI and GOA* (December 22, 1998), 116, 119.

1043 *Greenpeace v. National Marine Fisheries Service*, 80 F. Supp. 2d 1137 (W.D. Wash. 2000), at 1148.

1044 Ibid., at 1149

1045 National Marine Fisheries Service, *Marine Mammal Protection Act Section 101(a)(5)(E)—Negligible Impact Determination* (December 13, 2010), 31.

1046 *Greenpeace v. National Marine Fisheries Service*, order, August 7, 2000, C98-482Z (W.D. Wash. 2000).

1047 Ibid.

[1048] Jerry McBeath, "Science and politics in the conservation of biodiversity: The Steller sea lion case," *Environmental Development* 1 (2012): 107–121; "Federal Sea Lion Proposal Criticized by Kodiak Fishermen," *Peninsula Clarion* (Kenai Peninsula, Alaska), December 4, 2000, accessed August 25, 2015, http://peninsulaclarion.com/stories/120400/ala_120400alapm0060001.shtml#.VdylPflViko; Groundfish Forum's press release on NMFS's December 1, 2000, Biological Opinion, December 6, 2000, accessed August 25, 2015, http://groundfishforum.org/groundfish-forums-press-release-on-nmfs-december-1-2000-biological-opinion; Joel Gay, "Cod Fishermen Stung by BiOp, Buoyed by Rider," *Pacific Fishing* (February 2001): 16–18; Lawrence M. O'Rourke, "Fish dispute stalls budget," *Anchorage Daily News*, December 7, 2000.

[1049] Jerry McBeath, "Science and politics in the conservation of biodiversity: The Steller sea lion case," *Environmental Development*, 2012, 1: 107–121.

[1050] An Act Making Consolidated Appropriations for the Fiscal Year Ending September 30, 2001, and for Other Purposes, December 21, 2000 (114 Stat. 2463), § 209(b)(3).

[1051] Joel Gay, "Cod Fishermen Stung by BiOp, Buoyed by Rider," *Pacific Fishing* (February 2001): 16–18.

[1052] Overall, Congress during 2000 appropriated $43.15 million for Steller sea lion protection and research.

[1053] An Act Making Consolidated Appropriations for the Fiscal Year Ending September 30, 2001, and for Other Purposes, December 21, 2000 (114 Stat. 2463), § 209(d), § 209(e); R. C. Ferrero and L. W. Fritz, *Steller Sea Lion Research and Coordination: A Brief History and Summary of Recent Progress*, NOAA Technical Memorandum NMFS-AFSC-129, June 2002; Wesley Loy, "Sea Lion Research Bottleneck," *Pacific Fishing* (March 2001): 19–20.

[1054] An Act Making Consolidated Appropriations for the Fiscal Year Ending September 30, 2001, and for Other Purposes, December 21, 2000 (114 Stat. 2463), § 209(c).

[1055] *Greenpeace, American Oceans Campaign v. National Marine Fisheries Service*, 237 F. Supp. 2d 1181 (W.D. Wash., 2002).

[1056] 66 *Fed. Reg.* 7288, January 22, 2001.

[1057] 66 *Fed. Reg.* 7278, January 22, 2001.

[1058] Most notably, in the Bering Sea/Aleutian Islands area and in the central and western Gulf of Alaska, the opening of the Pacific cod A season for trawl gear was moved to January 20, and the opening of the B season was moved to September 1. In the central and western Gulf of Alaska, the opening of the Pacific cod B season for hook-and-line and pot gear was moved to September 1, while in the Bering Sea/Aleutian Islands area, only the B season for pot gear (for vessels with an overall length of 60 feet or more) opening was moved to September 1.

1059 NOAA Fisheries, Alaska Regional Office, 50 CFR Part 679, *Federal Fishing Regulations for Fisheries of the Exclusive Economic Zone Off Alaska*, December 11, 2017.

1060 National Marine Fisheries Service, Marine Mammal Protection Act Section 101(a)(5)(E)—Negligible Impact Determination (December 13, 2010), table 5.

1061 National Marine Fisheries Service, *November, 2010 North Pacific Groundfish Fishery Biological Opinion*, xxvi, xxxi, 360.

1062 Ibid.

1063 75 *Fed. Reg.* 77535, December 13, 2010; NOAA Fisheries, "NOAA Restricts Commercial Mackerel, Cod Fishing in Western Aleutians to Protect Western Steller Sea Lions," news release, December 8, 2010.

1064 National Marine Fisheries Service, *November, 2010 North Pacific Groundfish Fishery Biological Opinion*, xxxiv, 1; 75 *Fed. Reg.* 77535, December 13, 2010.

1065 74 *Fed. Reg.* 68717, December 29, 2009; 76 *Fed. Reg.* 467, January 5, 2011; *Bering Sea/Aleutian Islands Harvest Specifications, 1986–2015*, accessed September 6, 2015. https://alaskafisheries.noaa.gov/sustainablefisheries/historic/specs/bsai_hs1986-2014.pdf.

1066 *Alaska v. Lubchenco, No. 3:10-cv-00271-TMB (D. Alaska January 19, 2012), aff'd 723 F.3d 1043 (9th Cir. 2013)*, at 24; State of Alaska, "State sues to overturn NMFS decision on western Steller sea lions," news release, December 14, 2010; *Alaska Seafood Cooperative et al. v. NMFS et al.* (No. 3:11-cv-00001- TMB, D.Alaska); *Freezer Longline Coalition v. Jane Lubchenco et al.* (No. 3:11-cv-00004-TMB, D.Alaska); "Fishing co-op sues NMFS over Steller sea lion regulation," KUCB News (Unalaska), January 7 2011, accessed August 29, 2015, http://kucb.org/news/article/fishing-co-op-sues-nmfs-over-steller-sea-lion-regulation/.

1067 *Alaska v. Lubchenco*, No. 3:10-cv-00271-TMB (D. Alaska January 19, 2012), *aff'd* 723 F.3d 1043 (9th Cir. 2013), 77 *Fed. Reg.* 22750, April 17, 2012; National Marine Fisheries Service, *Final Environmental Impact Statement. Steller Sea Lion Protection Measures for Groundfish Fisheries in the Bering Sea and Aleutian Island Areas, Executive Summary* (May 2014), ES-2.

[1068] *Alaska v. Lubchenco*, 723 F. 3d 1043 (9th Cir. 2013).

[1069] National Marine Fisheries Service, *Aleutian Islands Groundfish Fishery Biological Opinion* (April 2, 2014), 249.

[1070] Ibid., 246.

[1071] Ibid., 110, 147, 199–200.

[1072] 79 *Fed. Reg.* 70286, November 25, 2014.

[1073] National Marine Fisheries Service, "NOAA Fisheries Concludes 'No Jeopardy' to Steller Sea Lions from Proposed Fishery Management Changes in the Aleutian Islands," news release, April 2, 2014.

[1074] *Oceana, Inc., and Greenpeace, Inc., v. National Marine Fisheries Service*, No. 3:14-cv-00253 (TMB) (D.Alaska, December 23, 2015), complaint, at 2.

[1075] *Oceana, Inc., and Greenpeace, Inc., v. National Marine Fisheries Service*, No. 3:14-cv-00253 (TMB) (D.Alaska, August 25, 2015); North Pacific Fishery Management Council, Action Memo, Protected Species Rept., File No. REP 17-304, Agenda Date October 2, 2017, Agenda No. B7.

[1076] M. Allen and R. P. Angliss, *Alaska Marine Mammal Stock Assessments, 2014*, NOAA Technical Memorandum NMFS-AFSC-301 (June 2015): 4.

[1077] Robert King, former fisheries aide to Sen. Mark Begich, personal communication with author, October 22, 2015.

[1078] James Strong and Keith R. Criddle, *Fishing for Pollock in a Sea of Change: A Historical Analysis of the Bering Sea Pollock Fishery* (Fairbanks, Alaska: Alaska Sea Grant, 2013), 145.

[1079] International Pacific Halibut Commission, *1998 Annual Report*, 22.

[1080] Laine Welch, "Friday the 13th! Lots of Superstions at Sea,"Alaska Fish Radio, May 12, 2016, accessed May 12, 2016, http://www.alaskafishradio.com/friday-the-13th-superstitions-at-sea-2/.

[1081] A. M. Eich, K. R. Mabry, S. K. Wright, and S. M. Fitzgerald, *Bycatch and Mitigation Efforts in Alaska Fisheries, Summary Report: 2007 through 2015*, NOAA Technical Memorandum NMFS-F/AKR-12 (November 2016), 2.

[1082] National Marine Fisheries Service, *Programmatic Biological Assessment on the Effects of the Fishery Management Plans for the Gulf of Alaska and Bering Sea/*

Aleutian Islands Groundfish Fisheries and the State of Alaska Parallel Groundfish Fisheries on the Endangered Short-tailed Albatross (Phoebastria albatrus) and the Threatened Alaska breeding Population of the Steller's Eider (Polysticta stelleri) (August 2015), 34; U.S. Fish and Wildlife Service, "Species Profile for Short-Tailed albatross (*Phoebastria [=diomedea] albatrus*)," accessed October 18, 2015, http://ecos.fws.gov/tess_public/profile/speciesProfile.action?spcode=-BooY.

1083 U.S. Fish and Wildlife Service, *Biological Opinion on the Effects of the Total Allowable Catch (TAC)-Setting Process for the Gulf of Alaska (GOA) and Bering Sea/Aleutian Islands (BSAI) Groundfish Fisheries to the Endangered Short-tailed Albatross (Phoebastria albatrus) and Threatened Steller's Eider (Polysticta stelleri)* (2003), 5; 63 *Fed. Reg.* 58692, November 2, 1998.

1084 An Act Relating to Endangered Species of Fish and Wildlife, May 27, 1971, chapter 115, *Session Laws of Alaska, 1971*; 5 AAC 93.020.

1085 35 *Fed. Reg.* 8495, June 2, 1970.

1086 63 *Fed. Reg.* 211, November 2, 1998.

1087 63 *Fed. Reg.*, 58692, November 2, 1998.

1088 Ann G. Rappoport, U.S. Fish and Wildlife Service, to Steven Pennoyer, National Marine Fisheries Service, March 13, 1998, attached "Consultation History," accessed September 12, 2015, https://alaskafisheries.noaa.gov/protectedresources/seabirds/section7/pachalibut.pdf.

1089 National Marine Fisheries Service, *Programmatic Biological Assessment on the Effects of the Fishery Management Plans for the Gulf of Alaska and Bering Sea/Aleutian Islands Groundfish Fisheries and the State of Alaska Parallel Groundfish Fisheries on the Endangered Short-tailed Albatross (Phoebastria albatrus) and the Threatened Alaska Breeding Population of the Steller's Eider (Polysticta stelleri)* (August 2015), 33.

1090 U.S. Fish and Wildlife Service, *Endangered Species Act Formal Section 7 Consultation for Pacific Halibut Fisheries in Waters off Alaska* (1998), 1.

1091 National Marine Fisheries Service, *Programmatic Biological Assessment on the Effects of the Fishery Management Plans for the Gulf of Alaska and Bering Sea/Aleutian Islands Groundfish Fisheries and the State of Alaska Parallel Groundfish Fisheries on the Endangered Short-tailed Albatross (Phoebastria albatrus) and the Threatened Alaska breeding Population of the Steller's Eider (Polysticta stelleri)* (August 2015), 33.

1092 U.S. Fish and Wildlife Service, *Endangered Species Act Formal Section 7 Consultation for Pacific Halibut Fisheries in Waters off Alaska* (1998), 17; 63 *Fed. Reg.* 58692, November 2, 1998; 62 *Fed. Reg.* 10016, March 5, 1997.

1093 U.S. Fish and Wildlife Service, *Endangered Species Act Formal Section 7 Consultation for Pacific Halibut Fisheries in Waters off Alaska* (1998), 1.

1094 University of Alaska Fairbanks, "Fishing for Albatross," *Arctic Science Journeys* (radio script), accessed September 12, 2015, https://seagrant.uaf.edu/news/97ASJ/05.19.97_FishAlbatross.html; U.S. Fish and Wildlife Service, *Threatened and Endangered Species, Cooperative Efforts to Conserve Albatrosses and Other Seabirds in Alaska*, accessed September 13, 2015, http://www.fws.gov/alaska/fisheries/endangered/pdf/consultation_guide/55_STAL_Cooperative_Conservation.pdf.

1095 Eric Swenson, "Longliners Tackle Albatross Bycatch," *Pacific Fishing* (November 1996): 20.

1096 Kenny Down, "Even after albatross deaths, longliners have great record" editorial, *Pacific Fishing* (November 2010): 50.

1097 62 *Fed. Reg.* 10016, March 5, 1997.

1098 Edward F. Melvin et al., *Solutions to Seabird Bycatch in Alaska's Demersal Longline Fisheries, Executive Summary* (Seattle: Washington Sea Grant, 2001).

1099 62 *Fed. Reg.* 23176, April 29, 1997.

1100 65 *Fed. Reg.* 46643, July 31, 2000.

1101 Maureen Clark, "Endangered Species Act extended to protect short-tailed albatross," *(Kenai, Alaska) Peninsula Clarion*, August 1, 2000, accessed September 12, 2015, http://peninsulaclarion.com/stories/080100/ala_080100ala0040001.shtml#.VfRhGBFVio.

1102 Laine Welch, "New series of Alaska bird guides," *Alaska Report*, June 5, 2006, accessed September 12, 2015, http://alaskareport.com/guides.htm.

1103 National Marine Fisheries Service, "NMFS Reports the Incidental Take of a Short-Tailed Albatross in the BSAI Hook-and-Line Groundfish Fishery," Information Bulletin No. 49, September 16, 2014, accessed September 14, 2015, http://alaskafisheries.noaa.gov/cm/info_bulletins/bulletin.aspx?bulletin_id=9677.

1104 Edward F. Melvin (Washington Sea Grant), personal communication with author, November 2, 2015.

1105 U.S. Fish and Wildlife Service, "Short-tailed Albatross (*Phoebastria albatrus*)," (fact sheet), Threatened and Endangered Species, 2001; U.S. Fish and Wildlife Service, "Cooperative Efforts to Conserve Albatrosses and Other Seabirds in Alaska," Threatened and Endangered Species, accessed September 13, 2015, http://www.fws.gov/alaska/fisheries/endangered/pdf/consultation_guide/55_STAL_Cooperative_Conservation.pdf.

1106 Mustad Autoline, *SeaBird Saver*, brochure; Edward F. Melvin (Washington Sea Grant), personal communication with author, March 23, 2017.

1107 National Marine Fisheries Service, *Programmatic Biological Assessment on the Effects of the Fishery Management Plans for the Gulf of Alaska and Bering Sea/Aleutian Islands Groundfish Fisheries and the State of Alaska Parallel Groundfish Fisheries on the Endangered Short-tailed Albatross (Phoebastria albatrus) and the Threatened Alaska Breeding Population of the Steller's Eider (Polysticta stelleri)* (August 2015), 33; A. M. Eich, K. R. Mabry, S. K. Wright, and S. M. Fitzgerald, *Bycatch and Mitigation Efforts in Alaska Fisheries, Summary Report: 2007 through 2015*, NOAA Technical Memorandum NMFS-F/AKR-12 (November 2016), 9.

Chapter 26
CASUALTIES

> *The capsizing of the F/V ARCTIC ROSE in 2001 with fifteen fatalities and the fire, explosion, and sinking of the FPV GALAXY in 2002 with three fatalities rank among the most catastrophic accidents in the history of the North Pacific groundfish fisheries.*
>
> —U.S. Coast Guard, 2006[1108]

Fortunately, fishing for cod in Alaska in modern times is far less dangerous than during the schooner era. But it can still be very dangerous. Since the passage of the Magnuson-Stevens Act in 1976, four large vessels have been lost in the fishery: the factory trawler *Aleutian Enterprise*, which sank in 1990; the factory trawler *Arctic Rose*, which sank in 2001; the freezer-longliner *Galaxy*, which burned and sank in 2002; and the pot cod vessel *Katmai*, which sank in 2008. In total, the four accidents cost thirty-four lives.

ALEUTIAN ENTERPRISE

The 162-foot factory trawler *Aleutian Enterprise* was built in Moss Point, Mississippi, in 1980 for the Alaska crab fisheries. In 1983 it was converted into a factory trawler.

Early in the afternoon of March 22, 1990, the *Aleutian Enterprise*, with thirty-one crew aboard, was about twenty miles southwest of St. George Island, in the central Bering Sea. The sky was clear, the temperature was just above freezing, the wind was blowing at about 15 to 20 knots, and the seas were running five to six feet. Under these comparatively moderate conditions, the vessel, with a slight list to port and nearing the end of its trip, was hauling in one of the largest catches of Pacific cod it had ever landed. Mark Siemons, the *Aleutian Enterprise*'s twenty-eight-year-old captain, estimated the catch to be more than forty tons. As the net's cod end was being hauled aboard, however, an intermediate section of the net parted, spilling what Siemons estimated to be about five tons of fish that quickly slid to the vessel's port side. The resulting severe list submerged a through-hull opening (that should have been sealed shut) on the vessel's port side processing deck, allowing seawater to rush in and quickly flood the processing area. Progressive flooding caused the vessel to sink in about ten minutes.

Nine lives were lost. Survivors recalled faulty survival suits and life rafts that failed to inflate. The U.S. Coast Guard later determined that safety equipment, safety training, operating procedures, and structural components on the *Aleutian Enterprise* had been deficient.[1109]

ARCTIC ROSE

The 92-foot *Arctic Rose* was constructed in Biloxi, Mississippi, in 1988 to fish for shrimp in the Gulf of Mexico. In the 1990s, the vessel was converted into a stern trawler and began fishing for flatfish (rex sole, flathead sole, and yellowfin sole) and Pacific cod in the Bering Sea and Gulf of Alaska as a head-and-gut (H&G) factory trawler.

In the early morning darkness on April 2, 2001, the *Arctic Rose*, with fifteen men aboard, was holding an approximate position about two hundred miles northwest of the Pribilof Islands, in the Bering Sea, waiting for daylight to commence fishing. The conditions at the time were determined in retrospect to likely have been severe: sustained gale-force winds and seas up to twenty-four feet high. At 3:35 a.m., the U.S. Coast Guard's command center at Juneau, Alaska, received a distress signal from the *Arctic Rose*'s EPIRB—"emergency position-indicating radio beacon"—a device that can be activated manually or is activated automatically if submerged in water.

After several unsuccessful attempts to contact the *Arctic Rose* via marine radio and email, the Coast Guard dispatched a C-130 search and rescue aircraft from Kodiak. The aircraft arrived at the site of the EPIRB at about 8:40 a.m. and observed a large oil sheen and a debris field of items that had come off the boat when it sank. Subsequently, only one body, that of David Rundall, the *Arctic Rose*'s captain, was recovered. The fourteen missing crew members were presumed dead.

The exact cause of the *Arctic Rose*'s sinking is unknown, but the U.S. Coast Guard's investigation of the incident determined that the vessel was likely jogging (motoring slowly) downwind in following seas or turning when a sudden influx of water from the aft deck rushed through an open aft weathertight door and flooded the fish-processing area and then flowed into the galley and engine room. This caused the *Arctic Rose* to lose stability and then capsize and sink, all possibly in less than five minutes.

The sinking of the *Arctic Rose* was the worst U.S. fishing industry casualty since January 1951, when the trawler *Gudrun* disappeared in heavy weather while returning from the Grand Banks to Gloucester, Massachusetts, with fifteen crewmen aboard.[1110]

GALAXY

> *After the backdraft explosion that had precipitated the crisis, [Captain] Shoemaker made three desperate attempts to enter the smoke-filled wheelhouse, find a radio, and issue a mayday call. Each foray left him puking and gasping, smoke searing his lungs and contact with the steel bulkheads charring his flesh and igniting his clothing.*
>
> —John Sabella, *After the Galaxy: The Legacy Lives On*[111]

The *Galaxy*, a former U.S. Coast Guard buoy tender that was converted to process crab and salmon in 1976, was converted to a freezer-longliner in 1997. Owned by Galaxy Fisheries, a subsidiary of Seattle-based Aleutian Spray Fisheries, the house-aft vessel made its first cod trip in January 1998. In addition to producing H&G product, as did most freezer-longliners, the *Galaxy* was equipped to produce fillets. Doing so required a larger crew than a solely H&G operation, and the *Galaxy* typically carried a crew of about twenty-five. Dave Shoemaker, who oversaw the design and fabrication of the vessel's fish-processing equipment, was the *Galaxy*'s captain during its entire cod-fishing career.

Fire at sea is a mariner's worst nightmare. In the late afternoon on Sunday, October 20, 2002, the *Galaxy* was approximately thirty-five miles southwest of St. Paul Island, in the central Bering Sea. The vessel had just finished hauling a longline set and was running toward a second set. Twenty-six people, including a National Marine Fisheries Service observer, were aboard. The winds were north-northeast at 20 to 30 knots, and the seas were fifteen to twenty feet. The air temperature was 35 degrees (Fahrenheit); the sea temperature was 43. Smoke suddenly filled several decks of the vessel's superstructure. Black smoke poured from the engine room hatch, and moments later a backdraft explosion blew three crew members into the frigid water. Two were rescued immediately; one perished.

Soon after, an explosion in the engine room emitted a huge fireball from the engine room vents and from the accommodation hatches on the forward bulkhead of the wheelhouse. The fireball set the wheelhouse afire, but Shoemaker—severely burning his arm in the process—was able to transmit a Mayday message to the U.S. Coast Guard station at St. Paul. The fireball also separated twenty-one crew members who had gathered on the aft top deck from some forty five survival suits that were stored in a bin on the forward main deck. Four crew members, however, were on the forward main deck.

Shoemaker and several other crew members managed to launch a life raft from the aft top deck, and fourteen crew members attempted to get aboard, primarily by jumping some thirty feet down into the water next to the raft and then making their way onto the raft. Two men drowned in the process, but the dozen who managed to get aboard the raft and an individual who later jumped aboard from the forward main deck were rescued about two hours later by the freezer-longliner *Glacier Bay*.

After the raft was launched, Shoemaker attempted to retrieve survival suits from the crew on the forward main deck, but in the process he fell twenty feet from the top of the wheelhouse onto the deck and broke three ribs. Raul Vielma, the *Galaxy*'s chief engineer, took over directing the abandonment of the vessel.

Five crew members, including Shoemaker, were subsequently lifted off the vessel by a Kodiak-based U.S. Coast Guard helicopter that, by good fortune, had been deployed to Cold Bay—some four and a half hours closer to the *Galaxy*'s position than if it had been in Kodiak.

Two others, Ann Weckback, the National Marine Fisheries Service observer aboard the *Galaxy*, and Ryan Newhall, the vessel's deck boss (and one of the individuals who had been blown overboard in the initial explosion), had jumped overboard when the fire raging on the *Galaxy* expanded and made their position untenable. Weckback had given her survival suit to an injured crew member and, clad only in pajamas and a T-shirt, had tied a buoy around her waist and had jumped overboard first, hoping to make it to the life raft drifting near the *Galaxy*. But she was unable to swim when she hit the water, and began drifting away from the raft. Newhall was wearing a survival suit, but, having just been rescued, he feared getting back into the water. Nevertheless, upon seeing Weckback's peril, he immediately grabbed a life ring and jumped in to try to help. Once next to Weckback, Newhall slid the life ring around her and then, to provide her some insulation from the cold water, wrapped himself around her. The pair survived some two hours before they were rescued by the freezer-longliner *Clipper Express*. Another *Galaxy* crew member was rescued by the freezer-longliner *Blue Pacific* within a few minutes of jumping from the *Galaxy* into the water.

The *Galaxy* eventually sank, and the cause of the fire was never determined. Despite their harrowing experience, twenty-three of the twenty-six people aboard the *Galaxy* survived. An ancillary casualty of the *Galaxy*'s demise, however, was a crewman on the *Clipper Express*. He was washed overboard and lost while securing the *Galaxy*'s life raft.[1112]

As of this writing, Aleutian Spray Fisheries, the parent company of the *Galaxy*'s owner, operates two freezer-longliners, the *Liberator* and the *Siberian Sea*, as well as several trawlers.[1113]

KATMAI

The 92-foot *Katmai* was built in Pensacola, Florida, in 1987. Originally intended to trawl for shrimp in the Gulf of Mexico, the vessel subsequently participated in a number of fisheries worldwide. In 2007, the *Katmai* was modified to fish for Pacific cod using pot gear and to process and freeze its catch onboard. In January 2008, the *Katmai* joined Alaska's H&G fleet.

On the night of October 21, 2008, the *Katmai* was making its way through Amchitka Pass, about one hundred miles west of Adak Island, in the Aleutian Islands, bound for Dutch Harbor. In its hold was about 120,000 pounds of frozen H&G Pacific cod—the first full load since the *Katmai* began fishing for Pacific cod. The load was also about twice the amount of cargo addressed in the vessel's stability report.

There were eleven crew members aboard, and most were asleep. But it wasn't a quiet night. A severe storm was raging, with 60- to 70-knot winds and twenty- to thirty-foot seas. Henry Blake, the *Katmai*'s thirty-nine-year-old captain, was at the helm and at one point sent a message to another vessel in the same area, saying the *Katmai* "was getting beat up."

At about midnight, the *Katmai* lost steering. The vessel's engineer found the lazarette, the aftmost compartment in the ship's hull, which housed the vessel's steering gear, to be flooded. A watertight door had been left open. Blake quickly awoke the crew and had them muster on the bridge and don survival suits. Meanwhile, the engineer started a bilge pump and partially dewatered the lazarette. Somewhat relieved, Blake told the crew they could take off their survival suits but to remain on the bridge until the situation was under control.

The situation, however, quickly became dire. The *Katmai* had developed a starboard list, and a crewman sent to inspect the vessel's engine room reported it was partially flooded. Blake immediately called Mayday on two marine radios and instructed the crew to don their survival suits and prepare to abandon ship. There was no response to either Mayday. (At the time, the *Katmai* was over one thousand nautical miles west of the Coast Guard's communications station at Kodiak, beyond the effective range of one of the radios. The Coast

Guard received the other radio's Mayday, but it lacked the vessel's name, a location, or the nature of the emergency.)

The engine room continued to flood, and the *Katmai*'s aft deck quickly became submerged and water began entering the processing area. Blake manually activated the vessel's EPIRB and approximately ten minutes later ordered the crew to abandon ship.

Seven crew members, including Blake, abandoned the *Katmai* into a fifteen-person life raft. Three crew members intended to abandon the vessel into a ten-person life raft, but none survived and it is not known whether they actually boarded the raft. The engineer apparently went down with vessel, which sank at 12:45 a.m., October 22, 2008.

Shortly after 3:00 a.m., the Coast Guard at Kodiak deployed two search-and-rescue aircraft—an HC-130 airplane and a Jayhawk helicopter—to the accident site. The HC-130 arrived on the scene at about 7:30 a.m. and dropped two life rafts near two strobe lights the crew spotted on the water. The helicopter, after a harrowing six-hour flight that included hurricane-force winds and whiteout conditions, arrived on scene at about 9:30 a.m.

Three of the seven crew members who abandoned the *Katmai* into the fifteen-person life raft were lost at sea when, according to Blake, the heavy seas "threw everyone from the raft . . . it just kept flipping on us and flipping on us and flipping on us." Blake had carried the EPIRB into the life raft, but the device was lost when the life raft overturned the first time.

Finally, at about 4:00 p.m., some fifteen hours after the *Katmai* sank, the four survivors were located and a rescue swimmer was lowered from the Coast Guard helicopter. By turn, the survivors went into the water and were then helped by the swimmer into a basket that was hauled up to the helicopter. All the survivors were cold, hungry, and shivering. But they were alive.[1114]

ENDNOTES

[1108] U.S. Coast Guard, *Alternative Compliance and Safety Agreement (ACSA) for the Bering Sea/Aleutian Islands and Gulf of Alaska Freezer Longliner and Freezer Trawler Fishing Fleets*, June 15, 2006.

[1109] U.S. Coast Guard, *Marine Casualty Report—Uninspected Fishing Vessel Aleutian Enterprise—Sinking in the Bering Sea on 22 March 1990, with 9 Persons Missing and Presumed Dead*, Rept. No. USCG 16732/01HQS91 (April 7, 1991), 113–114; Brad Matsen, "162' Aleutian Enterprise Capsizes, Nine Crewmen Lost," *National Fisherman*, June 1990, 11–13; Charles E. Brown and Jack Broom,

"Aleutian Enterprise Captain Describes Ship's Safety Gear, Sinking," *Seattle Times*, June 2, 1990; Leif Terdal and John Keiter, *Northwest Sea Disasters: Beyond Acceptable Risk* (Victoria, British Columbia: Trafford Publishing, 2005), 95–104.

1110 U.S. Coast Guard, Marine Board of Investigation, *Investigation Into the Circumstances Surrounding the Sinking of the Uninspected Fishing Vessel Arctic Rose, Official Number 931446, in the Bering Sea on April 2, 2001, with One Person Deceased and Fourteen Persons Missing and Presumed Dead* (December 19, 2003), 1, 16–17, 26, 48, accessed May 14, 2016, http://www.uscg.mil/hq/cg5/cg545/docs/boards/Arctic%20Rose%20.pdf; U.S. Coast Guard, Merchant Vessel Inspection Division, *Marine Board of Investigation into disappearance of fishing vessel GUDRUN with all persons on board off Atlantic Coast, January, 1951* (June 5, 1951), 2, accessed May 14, 2016, www.uscg.mil/hq/cg5/cg545/docs/boards/gudrun.pdf; U.S. Coast Guard, Office of Investigations and Analysis, *Analysis of Fishing Vessel Casualties: A Review of Lost Fishing Vessels and Crew Fatalities, 1992–2007* (October 2008), 16, accessed May 14, 2016, http://www.ntsb.gov/news/events/Documents/2010_Fishing_Vessel_Safety_FRM-Panel1c-Christensen1.pdf.

1111 John Sabella, *After the Galaxy: The Legacy Lives On*, excerpt in "Are we going to die today," *Pacific Fishing* (September 2013): 10–14, 16–17.

1112 Report of the Investigating Officer, U.S. Coast Guard Marine Safety Office, Anchorage, Alaska, re 16732/FPV Galaxy (March 2004); "Galaxy Skipper Honored," *Pacific Fishing* (December 2002): 4; Jeb Wyman, "Cod's Next Step?" *Pacific Fishing* (April/May 2003): 32–33, 36; Brad Warren, "Longliner Highlights Efficiency, Comfort," *Pacific Fishing* (November 2002): 37, 100; Christine Clarridge, "Heroism Was Onboard in Fishing-Boat Disaster," *Seattle Times*, February 10, 2003.

1113 Aleutian Spray Fisheries, "Fishing Operations," accessed December 10, 2015, http://www.starboats.com/fishing_operations.php.

1114 Marine Board of Investigation, U.S. Coast Guard, *Sinking of the F/V Katmai in the Amchitka Pass, North Pacific Ocean, on October 22, 2008, with Multiple Loss of Life*, MISLE Activity No. 3351236, April 26, 2010; National Transportation Safety Board, Marine Accident Brief, DCA-09-CM-001, September 13, 2011; Hal Bernton and Mike Carter, "Katmai survivors: Harrowing hours after fishing vessel rolled," *Seattle Times*, October 24, 2008.

AFTERWORD

The BSAI and GOA groundfish fisheries are widely considered to be among the best managed fisheries in the world. These fisheries produce high levels of catch, ex-vessel revenue, processed product revenue, exports, employment, and other measures of economic activity while maintaining ecological sustainability of the fish stocks.

—Ben Fissel et al., Alaska Fisheries Science Center, 2015[1115]

The importance of Pacific cod to the State of Alaska's fisheries and fishery economy cannot be overstated. It is harvested by all gear types, including trawl, longline, pot, and jig, and in both state and federal waters, across all seasons. Pacific cod is an iconic and critical species in the Gulf of Alaska groundfish fishery portfolio.

—City of Kodiak, Kodiak Island Borough, 2017[1116]

The NORTHERN LEADER project [a $35-million freezer-longliner launched in 2013] would not be possible without the many years of diligent efforts by the North Pacific Fishery Management Council to ensure that the fisheries stocks of Alaska remain abundant and fully sustainable. By promoting regulatory stability that encourages long-term investments, companies can now move forward with needed vessel replacements. We believe the NPFMC has done a great job as the steward of Alaska's offshore fisheries while helping maintain the strong economic vitality of our fishing fleets.

—Kenny Down, Executive Director, Freezer-Longline Coalition, 2012[1117]

Likely due to a combination of prudent management and relatively stable environmental conditions, catches of Pacific cod—the second-most-abundant harvested species in Alaska—have remained fairly consistent, averaging 264,000 metric tons (live weight) annually since the fishery became fully Americanized in 1991. The bulk of the fish, an annual average of about 200,000 metric tons, has been harvested in the Bering Sea/Aleutian Islands area. Meanwhile, the annual average harvest in the Gulf of Alaska has been about 64,000 metric tons.[1118]

The 2014 statewide Pacific cod catch of 334,164 metric tons represented 12 percent of the first wholesale volume and 11 percent of the first wholesale value of all Alaska seafood species.[1119] This catch represented about 72 percent of the worldwide Pacific cod catch and about 18 percent of the worldwide catch of Atlantic and Pacific cod combined.[1120]

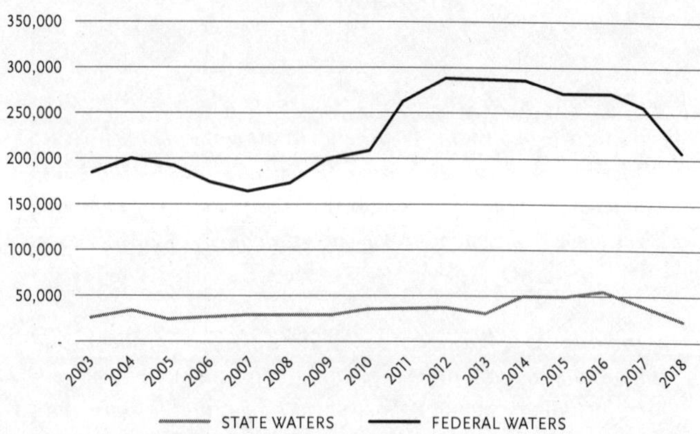

Chart 5. U.S. Pacific cod catch in state and federal waters of the Gulf of Alaska and the Bering Sea/Aleutian Islands area, 2003–2018, round weight, in metric tons. Data sources: David Witherell and Jim Armstrong (staff, North Pacific Fishery Management Council), Groundfish Species Profiles, 2015, 4, 32, accessed December 2, 2015, www.npfmc.org/wp-content/PDFdocuments/resources/SpeciesProfiles2015.pdf; NOAA Fisheries, Pacific cod landings database, provided by Mary Furuness (NOAA Fisheries).

The statewide Pacific cod catch in 2015 declined slightly, to 321,100 metric tons, which represented about 16 percent of the worldwide catch of Atlantic and Pacific cod combined. (Between the years 2010 and 2015, global cod harvests increased 31 percent.)[1121]

The current situation, however, is not without challenges. In August 2017, Chris Oliver, who two months before had resigned his position as director of the North Pacific Fishery Management Council to take the helm of NOAA Fisheries, advised caution:

> We have made great progress, but our achievements have not come easily, nor will they be sustained without continued attention. This is a critical time in federal fisheries management, and we must move forward in a strategic way to ensure our Nation's fisheries are able to meet the needs of both current and future generations.[1122]

In April 2014, the price of frozen headed and gutted (H&G) western-cut, trawl-caught, medium-sized Pacific cod—much of which is shipped to China for reprocessing into fillets and portions that are then sold in European Union countries and the United States—was relatively stable at $2,450 per metric ton, delivered in China.[1123] In July of that year, frozen H&G, eastern-cut, longline-caught, medium-sized Pacific cod were selling in Japan for $3,600 per metric ton, about $700 more per metric ton than had been received the previous year.[1124] During 2015–2016 the Pacific cod industry provided an estimated 5,700 direct jobs (full-time equivalent[1125]) in Alaska.[1126]

CLIMATE CHANGE

> *Call it what you will—global warming, climate change, alarmism, bunk—parts of our environment don't look like they did only 20 years ago. Those changes will accelerate.*
>
> —*Pacific Fishing*, October 2006[1127]

> *Climate change is no longer a distant threat. Climate change is here, it is happening now, and unless there is decisive global action soon, it will likely become far worse.*
>
> —Avery Siciliano et al., Center for American Progress, September 2018[1128]

> *A major impact of climate change in the oceans has been the redistribution of marine organisms, which have generally been shifting poleward or into deeper waters as temperatures warm.*
>
> —James W. Morley et al., 2018[1129]

> *[F]isheries face a serious new challenge as climate change drives the ocean to conditions not experienced historically.*
>
> —Malin L. Pinsky et al., *Science*, June 2018[1130]

> *The blob that cooked the Pacific. When a deadly patch of warm water shocked the West Coast, some feared it was a preview of our future oceans.*
>
> —Craig Welch, *National Geographic*, September 2016[1131]

> *[T]he blob is a dress rehearsal for a future with climate change.*
>
> —Annie Feidt, Alaska Public Radio, July 2018[1132]

New England's cod stocks are on the verge of collapse. During the decade ending in 2014, there was a 75 percent decline in the Atlantic cod stock in the Gulf of Maine. Scientists attribute this decline to a combination of rapidly warming waters and overfishing, the latter of which occurred partly because fisheries managers failed to recognize the effect of warming waters on cod stocks.[1133]

Cod are cold-water fish, and over the past decade, sea surface temperatures in the Gulf of Maine increased faster than 99 percent of the global ocean, in part because the position of the Gulf Stream changed. This warming put Gulf of Maine cod on the southern edge of the species' livable habitat. Because models used by fisheries managers to set fishing quotas for cod did not always account for the effect of rising water temperatures on cod survival, cod stocks were often overestimated, which led to overfishing.[1134]

In Alaska, scientists during the years 2013–2016 got to see firsthand the effects of warm water on codfish. Surveys in the Gulf of Alaska in 2013 showed the Pacific cod 2012 year-class abundance (number of fish) to be in record territory. This was, of course, very encouraging news for the fishing industry because once these fish matured, they were expected to boost the allowable catch. By 2015, however, the population had declined substantially, though it was still above average. Unfortunately, this decline continued and seems to have accelerated. By 2017, the 2012 year class appeared to have been greatly diminished—"wiped out," in the words of one seafood industry reporter. Larval surveys showed the 2013 year class to have also been very large, but it has not been observed as a large component in recent surveys. Ominously, there is no sign of the 2014 and 2015 year classes, and definitive data is not yet available for the 2016 and 2017 year classes.[1135]

The severe decline of young Pacific cod stocks in the Gulf of Alaska corresponded with the string of exceptionally warm years that began in 2013 and continued through 2015. Globally, 2014 was the warmest year on record.

Beginning in the winter of 2013, a mass of very warm water covered much of the northeast Pacific Ocean, including the Gulf of Alaska. Dubbed *The Blob*, this water, which at its peak covered some 3.5 million square miles, was in places more than five degrees Fahrenheit above normal. And the phenomenon wasn't just at the surface; in some places, warm water extended to a depth of 1,300 feet. The Blob—the most intense event of its kind in the northeast Pacific during the observational record—persisted through 2015. Additionally, there was some noticeable warming in the central Gulf of Alaska during the late spring and early summer of 2016.[1136]

The warm water raised the metabolism of cod, increasing their nutritional requirements. In 2015, at three years of age, the 2012 Pacific cod year class was in the eighteen- to twenty-two-inch range

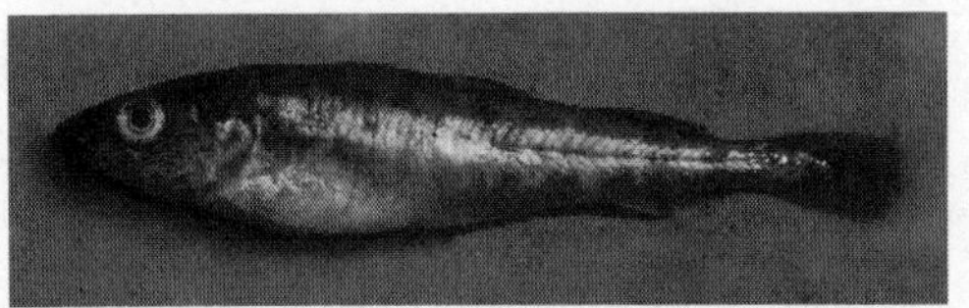

Figure 91. Juvenile Pacific cod. (Courtesy of Wayne Palsson.)

and switching its diet from zooplankton to capelin and small crabs. But there were few capelin or small crabs to be had in the Gulf of Alaska—perhaps due to the exceptionally warm water—and much of the 2012 year class starved. A similar fate seems have befallen the 2013 year class, while the 2014 and 2015 year classes seem to have perished while in the larval stage.[1137]

Steven Barbeaux, a researcher at the Alaska Fisheries Science Center, analyzed Pacific cod measuring less than 80 centimeters (about 31.5 inches) in length that were caught in the 2015–2016 Gulf of Alaska longline and pot fisheries and in the center's 2015 Gulf of Alaska bottom trawl survey. He said the length/weight relationship of the fish was "the worst we've seen. . . . It means they were starving."[1138]

The die-off was not limited to cod; seabirds and large whales perished in unprecedented numbers. In the western Gulf of Alaska, eight whales are found dead in a typical year, but in 2015 the total was 45.[1139] No definitive numbers are available, but the seabird die-off in the Gulf of Alaska during late 2016 and early 2017 was unprecedented in size, as well as in scope and duration.[1140] Millions of seabirds likely perished.

Preliminary estimates of the effect of the die-off of the 2012 and 2013 year classes of Pacific cod is that cod stocks in the Gulf of Alaska in 2017 are less than half of what had been expected, and the estimated total Pacific cod biomass is the lowest ever, having declined fully 78 percent since 2013. The total allowable Pacific cod catch in the gulf in 2017 was 64,442 metric tons. In December 2017, the North Pacific Fishery Management Council—focused on maintaining the spawning stock to increase the likelihood that the Pacific cod fishery will remain viable—determined the allowable catch for 2018 should be 13,096 metric tons, fully 80 percent less than the previous year. Moreover, the failure of the 2014 and 2015 year classes portends an extended scarcity of Pacific cod in the gulf.[1141]

Because of the expected economic hardships that will result from reduced Pacific cod harvests, in early March 2018 Alaska governor Bill Walker and Alaska lieutenant governor Byron Mallott asked Secretary of Commerce Wilbur Ross to declare the 2018 Pacific cod fishery in the Gulf of Alaska a federal fisheries disaster, thus beginning the process of potentially securing federal aid to assist the region's fishermen, processors, fishery support industries, and fishery-dependent communities.[1142]

The Pacific cod situation in the Bering Sea is not as severe but is still serious. Trawl surveys of Pacific cod in the eastern Bering Sea revealed a 46 percent decline in the abundance and a 37 percent decline in the biomass between the years 2016 and 2017—the largest year-to-year declines ever recorded. Consequently, the North Pacific Fishery Management Council determined the 2018 allowable catch of Pacific cod in the Bering Sea should be 188,136 metric tons, 16 percent less than in 2017. The allowable catch in the Aleutian Islands, however, will be the same as in 2017—15,695 metric tons.[1143] Though the exact cause(s) of the decline in abundance of Pacific cod in the Bering Sea is not yet known, it may be because the sea is warming. During the winter of 2017–2018, for the first time on record, there was no sea ice in the northern Bering Sea.

(For research purposes, the boundary between the northern Bering Sea and the southern Bering Sea is considered to be around 60 degrees north latitude, the latitude of Nunivak Island.)

There are indications, too, that Pacific cod may be moving northward, out of the eastern Bering Sea and into the waters around and north of St. Matthew Island. Scientists predict seasonal sea ice in the Bering Sea will decrease by 40 percent by 2050, which will have implications for the timing and location of fishing, as well as the potential catch.[1144]

Of broader concern is the projection contained in the U.S. government's most recent national climate assessment report, which was released in October 2017. According to the report, ocean surface temperatures along Alaska's coast recorded during The Blob's peak could be close to the average by the end of the century, and future blobs could push temperatures even higher. [1145] In the words of Nicholas Bond, a University of Washington climate scientist who assisted in the Gulf of Alaska cod research, "This is really going to be uncharted territory."[1146]

FREEZER-LONGLINER *BLUE NORTH*

> *Known for resuscitating older, dilapidated and bankrupt boats, Blue North Fisheries has now leapfrogged everyone with the brightest, shiniest flagship of all.*
>
> —Bruce Buls, *National Fisherman*, December 2016[1147]

Figure 92. Freezer-longliner *Blue North* entering Dutch Harbor. (Courtesy of Blue North, Inc.)

> *The Blue North will be the most modern, fuel efficient, and technically advanced hook-and-line vessel ever built. It is a game changer in US fishing vessel construction and design.*
>
> —Kenny Down, Blue North Fisheries, October 2015[1148]

In the spring of 2013, Blue North Fisheries, a Seattle-based company that has been operating freezer-longliners since 1994, signed a contract with Dakota Creek Industries, which operates a shipyard in Anacortes, Washington, to construct one of the most environmentally friendly and technologically advanced fishing vessels in the world. The keel of the freezer-longliner *Blue North* was laid in August of that year, and the vessel departed on its maiden voyage to the Bering Sea in September 2016. At 191 feet long and with a frozen cargo capacity of 1.3 to 1.4 million pounds, the *Blue North* is the largest freezer-longliner ever built. With an estimated cost of about $40 million, it is also the most expensive.

The vessel incorporates a *moon pool* in the center line of its hull through which fish are landed at an internal hauling station, enabling the crew to accomplish its work without being exposed to rough seas or freezing temperatures. Norwegian vessels have been fitted with moon pools, but the *Blue North* will be the first U.S. vessel so equipped. Among the other innovations incorporated in the *Blue*

North is a very fuel-efficient diesel-electric propulsion system. (The vessel is among the first constructed in the United States with engines that meet federal Tier III emissions standards.) A heavily weighted (150 tons) box keel, which keeps the vessel's center of gravity low, combined with an anti-roll tank, make the *Blue North* an extremely stable working platform.

The *Blue North* is also the most humane fishing vessel ever built. The owners and managers of Blue North Fisheries are among those who believe, in the words of the company's then president and CEO, Kenny Down, that "wild fish are sentient beings that deserve to be harvested humanely." To help minimize the suffering fish may experience, the *Blue North* is fitted with a stunning table, an innovative device that uses 48-volt DC electric current to quickly stun fish as they are brought aboard. Studies have shown that, in addition to being more humane, stunning, by reducing the stress on fish, results in a higher-quality and healthier end product.[1149] As is the tradition with freezer-longliners, the *Blue North*'s primary product is headed-and-gutted cod, but 20 to 25 percent of the cod caught are filleted and frozen in standard 15-pound shatterpacks. Freezing capacity is 120,000 pounds of finished product per day.[1150]

Forty million dollars is a huge investment that will likely take decades to recoup. The fact that the *Blue North* was constructed is a tribute to the prudence of those who have managed Alaska's Pacific cod fisheries and to the industry that has supported and continues to support truly sustainable management of the Pacific cod resource. The construction of the *Blue North* was also a major vote of confidence in the future. That future, however, has been clouded by the aforementioned Blob.

Figure 93. Coin commemorating christening of *Blue North*.

ENDNOTES

1115 Ben Fissel et al., *NPFMC Draft Stock Assessment and Fishery Evaluation Report for the Groundfish Fisheries of the Gulf of Alaska and Bering Sea/Aleutian Islands Area: Economic Status of the Groundfish Fisheries off Alaska, 2014* (Seattle: National Marine Fisheries Service, 2015), 10, accessed December 9, 2015, http://www.afsc.noaa.gov/REFM/stocks/plan_team/economic.pdf.

1116 City of Kodiak and Kodiak Island Borough to Alaska Governor Bill Walker re 2018 Gulf of Alaska Pacific Cod Fishery Disaster Declaration, December 2017.

1117 Alaskan Leader Fisheries, "Alaskan Fishing Company Contracts Large Eco-Friendly Commercial Fishing Vessel with Tacoma Shipyard," media release, February 14, 2012. accessed August 26, 2016, http://www.freezerlonglinecoalition.com/documents/MediaReleaseNORTHERNLEADER.pdf.

1118 David Witherell and Jim Armstrong (staff, North Pacific Fishery Management Council), *Groundfish Species Profiles, 2015*, 4, 32, accessed December 2, 2015, www.npfmc.org/wp-content/PDFdocuments/resources/SpeciesProfiles2015.pdf.

1119 Alaska Fisheries Science Center, *Wholesale Market Profiles for Alaska Groundfish and Crab Fisheries* (May 2016), ix, 43. This report lists the 2014 Alaska Pacific cod harvest as 333,851 metric tons, which is 313 metric tons less than is listed for 2014 in the National Marine Fisheries Service's 2015 groundfish stock assessment and fishery evaluation (SAFE) reports.

1120 Alaska Fisheries Science Center, *Wholesale Market Profiles for Alaska Groundfish and Crab Fisheries* (May 2016), 42; Alaska Seafood Marketing Institute, *Seafood Market Bulletin*, Spring 2014, accessed October 13, 2015, http://www.alaskaseafood.org/industry/market/seafood_spring14/2014-alaska-seafood-market-outlook.html.

1121 NOAA Fisheries (Alaska Fisheries Science Center), *Pacific Cod Research*, accessed December 20, 2017, https://www.afsc.noaa.gov/species/pacific_cod.php; McDowell Group, *The Economic Value of Alaska's Seafood Industry* (September 2017), 31.

1122 Chris W. Oliver, written testimony, *Hearing on Magnuson-Stevens Fishery Conservation and Management Act*, Sen. Subcommittee on Oceans, Atmosphere, Fisheries and Coast Guard, Committee on Commerce, Science, and Transportation, August 1, 2017, accessed October 2, 2017, http://www.legislative.noaa.gov/Testimony/Oliver080117.pdf.

1123 Masahiko Takeuchi (Minato Tsukiji), "US Pacific cod suppliers selling more to home market," *Undercurrent News*, April 7, 2014, accessed December 11, 2015, https://www.undercurrentnews.com/2014/04/07/alaskan-pacific-cod-suppliers-selling-more-to-north-america/.

1124 Jeanine Stewart, "Pacific cod market strengthening trend looks set to continue," *Undercurrent News*, July 15, 2014, accessed December 10, 2015, http://www.undercurrentnews.com/2014/07/15/pacific-cod-market-strengthening-trend-looks-set-to-continue/.

1125 Full-time equivalent: Many Alaska seafood industry workers are employed seasonally or earn a year's worth of income in less than a year. Full-time equivalent employment figures for the seafood industry facilitate comparisons to other industries.

1126 McDowell Group, *The Economic Value of Alaska's Seafood Industry* (September 2017), 11.

1127 Editor's introduction to: Jennifer Hawks, "Global climate change: Doom or opportunity," *Pacific Fishing* (October 2006): 4–6.

1128 Avery Siciliano et al., *Warming Seas, Falling Fortunes: Stories of Fishermen on the Front Lines of Climate Change*, Center for American Progress, September 2018, accessed September 13, 2018, https://www.americanprogress.org/issues/green/reports/2018/09/10/457649/warming-seas-falling-fortunes/.

1129 James W. Morley et al., "Projecting shifts in thermal habitat for 686 species on the North American continental shelf," *PLoS ONE* 13, no. 5 (May 2018): e0196127, accessed May 25, 2018, http://journals.plos.org/plosone/article?id=10.1371/journal.pone.0196127.

1130 Malin L. Pinsky et al., "Preparing ocean governance for species on the move," *Science* (June 2018): 1189–1191.

1131 Craig Welch, "The blob that cooked the Pacific," *National Geographic* (September 2016), accessed October 7, 2017, http://www.nationalgeographic.com/magazine/2016/09/warm-water-pacific-coast-algae-nino/.

1132 Annie Feidt (Alaska Public Radio), "The Big Thaw: Fishermen in Kodiak cope with record low cod numbers," July 24, 2018, accessed July 29, 2018, https://www.alaskapublic.org/2018/07/24/the-big-thaw-fishermen-in-kodiak-cope-with-record-low-cod-numbers/.

1133 Andrew J. Pershing et al., "Slow adaptation in the face of rapid warming leads to collapse of the Gulf of Maine cod fishery," *Science* (November 2015): 809–812.

1134 Ibid.

1135 "NASA, NOAA Find 2014 Warmest Year in Modern Record," Release 15-010, accessed October 9, 2017, https://www.nasa.gov/press/2015/january/nasa-determines-2014-warmest-year-in-modern-record; Peggy Parker, "Stunning Preliminary Report Shows Gulf of Alaska Cod Stocks Down By a Third From Expected Levels," *Seafood News* (October 4, 2017), accessed October 16, 2017, http://www.alaskafishradio.com/pacific-cod-stocks-gulf-alaska-time-lows-scientists-say/.

1136 Nicholas Bond, National Oceanic and Atmospheric Administration, personal communication with author, October 17, 2017; Hillary A. Scannell, Andrew J. Pershing, Michael A. Alexander, Andrew C. Thomas, and Katherine E. Mills, "Frequency of marine heatwaves in the North Atlantic and North Pacific since 1950," *Geophysical Research Letters*, vol. 43, issue 5, (March 3, 2016): 2069–2076, accessed November 9, 2017, http://onlinelibrary.wiley.com/doi/10.1002/2015GL067308/full.

1137 Steven Barbeaux, Alaska Fisheries Science Center, personal communication with author, October 9, 2017.

1138 Brian Hagenbuch, "Gulf of Alaska cod stocks at all-time lows," *SeafoodSource News*, October 17, 2017, accessed October 19, 2017, https://www.seafoodsource.com/news/supply-trade/gulf-of-alaska-cod-stocks-at-all-time-lows.

1139 Craig Welch, "The blob that cooked the Pacific," *National Geographic* (September 2016), accessed October 7, 2017, http://www.nationalgeographic.com/magazine/2016/09/warm-water-pacific-coast-algae-nino/; Craig Welch, "Mass Death of Seabirds in Western U.S. Is 'Unprecedented'," *National Geographic* (January 2015), accessed October 8, 2017, http://news.nationalgeographic.com/news/2015/01/150123-seabirds-mass-die-off-auklet-california-animals-environment/.

1140 Yareth Rosen, "Scientists think Gulf of Alaska seabird die-off is biggest ever recorded," *Alaska Dispatch News*, January 29, 2016, accessed December 8, 2017, https://www.adn.com/science/article/murre-die-believed-be-biggest-record-and-related-warm-waters/2016/01/29/.

1141 Peggy Parker, "Stunning Preliminary Report Shows Gulf of Alaska Cod Stocks Down By a Third From Expected Levels," *Seafood News* (October 4, 2017), accessed October 16, 2017), http://www.alaskafishradio.com/pacific-cod-stocks-gulf-alaska-time-lows-scientists-say/; Peggy Parker, "Latest Stock Assessment Shows 80% Drop in GOA Pacific Cod ABC for 2018," *Seafood News* (November 15, 2017), accessed November 18, 2017, http://www.seafoodnews.com/Story/1082755/Latest-Stock-Assessment-Shows-80-percent-drop-in-GOA-Pacific-Cod-ABC-for-2018; 82 *Fed. Reg.* 60327, December 20, 2017; North Pacific Fishery Management Council, *Attention Cod Fishermen!* (flyer), December 2017.

1142 Alaska Governor Bill Walker and Alaska Lieutenant Governor Byron Mallott to Secretary of Commerce Wilbur Ross, March 8, 2018, accessed March 14, 2018, http://s3-us-west-2.amazonaws.com/ktoo/2018/03/State-cod-disaster-request-030818.pdf?_ga=2.81859860.365992459.1521088561-674714349.1518369241.

1143 Peggy Parker, "Bering Sea Pacific Cod Quota May Drop by a Third for 2018," *Seafood News* (October 13, 2017), accessed October 16, 2017, http://www.seafoodnews.com/Story/1079074/Bering-Sea-Pacific-Cod-Quota-May-Drop-by-a-Third-for-2018; Aaron Bolton, "Cod numbers in the Gulf of Alaska fall dramatically," KBBI Radio, Homer, Alaska, November 1, 2017, accessed November 5, 2017, http://kbbi.org/post/cod-numbers-gulf-alaska-fall-dramatically; Peggy Parker, "Bering Sea Pollock and Cod in Good Shape But Could Be Moving North," *Seafood News* (November 16, 2017), accessed November 18, 2017, http://www.seafoodnews.com/Story/1082921/Bering-Sea-Pollock-and-Cod-in-Good-Shape-But-Could-Be-Moving-North.

1144 Brian Hagenbuch, "Lack of ice in Bering Sea casts uncertainty over future of Alaska fish stocks," *SeafoodSource News*, November 13, 2018, accessed November 17, 2018, https://www.seafoodsource.com/news/environment-sustainability/lack-of-ice-in-bering-sea-casts-uncertainty-over-future-of-alaska-fish-stocks?utm_source=marketo&utm_medium=email&utm_campaign=newsletter&utm_content=newsletter&mkt_tok=eyJpIjoiWmpRMll6VmlZalZpWXpjdyIsInQiOiJxczFXNzkoVW5QdlNlNnc4ODVoKzN2T25vSkpmdXc4Q3FnRldzZFBGRXdTRUtSWUI4eVZLVXZGZmRqcjJETo1kXC9RVlwvTDduaGEwTjJXOEZET1ltNHIxQlZyM3dXRWREWlhQSDlSUW9NRWpnQkxtMDlBR1J3RzUoR2c3MUo2TmRoIno%3D; Alan Haynie and Lisa Pfeifer, "Climate Change and Location Choice in the Pacific Cod Longline Fishery," in Ben Fissel et al., *NPFMC Draft Stock Assessment and Fishery Evaluation Report for the Groundfish Fisheries of the Gulf of Alaska and Bering Sea/Aleutian Islands Area: Economic Status of the Groundfish Fisheries off Alaska, 2014* (Seattle: National Marine Fisheries Service, 2015), 369, accessed December 9, 2015, http://www.afsc.noaa.gov/REFM/stocks/plan_team/economic.pdf; Peggy Parker, "Are Bering Sea Cod and Pollock Shifting Northward?" *Seafood News*, December 13, 2017, accessed December 13, 2017, http://www.seafoodnews.com/Story/1085662/Are-Bering-Sea-Cod-and-Pollock-Shifting-Northward.

1145 D. J. Wuebbles, D. W. Fahey, K. A. Hibbard, D. J. Dokken, B. C. Stewart, and T. K. Maycock (eds.) (U.S. Global Change Research Program), *Climate Science Special Report: Fourth National Climate Assessment*, Volume I (2017), 367.

1146 Hal Bernton, "Climate change preview? Pacific Ocean 'blob' appears to take toll on Alaska cod," *Seattle Times*, November 4, 2017.

1147 Bruce Buls, "For the Cod," *National Fisherman*, December 2016, 38–39, 42–44.

1148 Ross Davies, "Blue North to receive flagship vessel in April 2016," *Undercurrent News*, October 22, 2015, accessed December 27, 2015, https://www.undercurrentnews.com/2015/10/22/blue-north-to-receive-flagship-vessel-in-april-2016/.

1149 "Blue North introduces initiative to ensure 'humane harvesting'" *Undercurrent News*, March 11, 2015, accessed December 5, 2017, https://www.undercurrentnews.com/2015/03/11/blue-north-introduces-initiative-to-ensure-humane-harvesting/; Jeanine Stewart, "Blue North targets 'next generation' in protein trends with new vessel," *Undercurrent News*, June 5, 2015, accessed December 27, 2015, www.undercurrentnews.com/2015/06/05/blue-north-targets-next-generation-in-food-protein-trends-with-new-vessel/; Ross Davies, "Blue North to receive flagship vessel in April 2016," *Undercurrent News*, October 22, 2015, accessed December 27, 2015, https://www.undercurrentnews.com/2015/10/22/blue-north-to-receive-flagship-vessel-in-april-2016/.

1150 Dakota Creek Industries, "FV Blue North Freezer Longliner Vessel," accessed December 27, 2015, http://dakotacreek.com/dci/projects/new-construction/fv-blue-north-freezer-longliner-vessel/; *Blue North*, christening brochure, September 2016.

BIBLIOGRAPHY
(Chronological)

1887

Tarleton Bean. *The Cod Fishery of Alaska.* In George Brown Goode, *The Fisheries and Fishery Industries of the United States, Section V: History and Methods of the Fisheries.* Vol. 1. Washington, DC: GPO, 1887 [1880], 26 pp.

1890

Z. L. Tanner et al. *Explorations of the Fishing Grounds of Alaska, Washington Territory, and Oregon, During 1888, by the U.S. Fish Commission Steamer* Albatross, *Lieut. Comdr. Z. L. Tanner, U.S. Navy, Commanding.* In *Bulletin of the United States Fish Commission.* Vol. 8, 1888. Washington, DC: GPO, 1890, 92 pp.

1892

J. W. Collins. "The Cod Fishery," *Report on the Fisheries of the Pacific Coast of the United States, Appendix 1, Report of the United States Commissioner of Fish and Fisheries for the Fiscal Year Ending June 30, 1889.* Part 16. Washington, DC: GPO, 1892, 18 pp.

Z. L. Tanner. *Report Upon the Investigations of the U.S. Fish Commission Steamer* Albatross *for the Year Ending June 30, 1889, Appendix 4, Report of the United States Commissioner of Fish and Fisheries for the Fiscal Year Ending June 30, 1889.* Part 16. Washington, DC: GPO, 1892, 117 pp.

1894

Richard Rathbun. *Summary of the Fishery Investigations Conducted in the North Pacific Ocean and Bering Sea from July 1, 1888, to July 1, 1892, by the U.S. Fish Commission Steamer* Albatross. In *Bulletin of the United States Fish Commission.* Vol. 12, 1892. Washington, DC: GPO, 1894, 74 pp.

1915

E. Lester Jones. *Report of Alaska Investigations, 1914.* Washington, DC: GPO, 1915, 58 pp.

1916

John N. Cobb. *Pacific Cod Fisheries.* Bureau of Fisheries Doc. No. 830, Appendix 4 to *Report of the U.S. Commissioner of Fisheries for 1915.* Washington, DC: GPO, 1916, 111 pp.

1926

Jimmy Crooks. "Schooner *Fanny Dutard*." Typed recollection of 1926 cod-fishing voyage to the Bering Sea aboard the schooner *Fanny Dutard*. San Francisco National Marine Historical Park, HDC379, 19 pp.

1927

John N. Cobb. *Pacific Cod Fisheries*. Rev. ed., 1926. Bureau of Fisheries Doc. No. 1014, Appendix 7 to *Report of the U.S. Commissioner of Fisheries for 1926*. Washington, DC: GPO, 1927, 114 pp.

1958

Harlan Trott. *The Schooner That Came Home: The Final Voyage of the C. A. Thayer.* Cambridge, Md: Cornell University Press, 1958, 117 pp.

1989

Russ Hofvendahl. *Hard on the Wind*. Dobbs Ferry, N.Y.: Sheridan House, 1989, 251 pp.

2001

Ed Shields. *Salt of the Sea: The Pacific Coast Cod Fishery and the Last Days of Sail.* Lopez Island, Wash: Pacific Heritage Press, 2001, 238 pp.

2006

Joe Follansbee. *Shipbuilders, Sea Captains, and Fishermen: The Story of the Wawona.* New York: iUniverse, 2006, 205 pp.

2013

James Strong and Keith R. Criddle. *Fishing for Pollock in a Sea of Change: A Historical Analysis of the Bering Sea Pollock Fishery.* Fairbanks, Alaska: Alaska Sea Grant, 2013, 177 pp.

Kevin M. Bailey. *Billion-Dollar Fish: The Untold Story of Alaska Pollock*. Chicago: University of Chicago Press, 2013, 263 pp.

2015

Ben Fissel et al., North Pacific Fishery Management Council draft *Stock assessment and fishery evaluation report for the groundfish fisheries of the Gulf of Alaska and Bering Sea/Aleutian Islands area: economic status of the groundfish fisheries off Alaska, 2014*. Seattle: National Marine Fisheries Service, November 2015, 406 pp.

Note: page numbers in italics refer to figures; those followed by t refer to tables; those followed by n refer to notes, with note number.